William Humble

**Dictionary of Geology & Mineralogy**

Comprising Such Terms in Natural History as are Connected With....

William Humble

**Dictionary of Geology & Mineralogy**
*Comprising Such Terms in Natural History as are Connected With....*

ISBN/EAN: 9783337026295

Printed in Europe, USA, Canada, Australia, Japan

Cover: Foto ©berggeist007 / pixelio.de

More available books at **www.hansebooks.com**

# DICTIONARY

OF

# GEOLOGY AND MINERALOGY

COMPRISING SUCH TERMS IN

# NATURAL HISTORY

AS ARE CONNECTED WITH

## THE STUDY OF GEOLOGY

BY

# WILLIAM HUMBLE, M.D., F.G.S.

THIRD EDITION, REVISED

LONDON
CHARLES GRIFFIN AND COMPANY
10, STATIONERS' HALL COURT

# PREFACE.

In submitting the following pages to public approbation, or public censure, I avail myself of the accustomed privilege to offer a few prefatory observations; explanatory, on the one hand, of the motives which led to their preparation; and deprecatory, **on the other,** of severity of criticism.

The labours of the lexicographer greatly differ from those of authors generally. Dr. Johnson has observed, " every other author may aspire to praise; the lexicographer can only hope to escape reproach, and even this negative recompence has been yet granted to very few. It is the fate of those, who toil at the lower employments of life, **to be rather driven by** the fear of evil, than attracted by **the prospect of good** ; **to be exposed to** censure, without hope of praise; to **be disgraced by** miscarriage, **or** punished for neglect, where success would have been without applause, and diligence without reward. Among these unhappy mortals is the writer of dictionaries; whom mankind have considered, not as the pupil, but the slave of science, the pioneer of literature, doomed only to remove rubbish and clear obstructions from the paths through which learning and genius press forward to conquest and glory, without bestowing a smile on the humble drudge that facilitates their escape."

When I commenced collecting materials for the present work, I was induced to undertake the **labour** from a conviction that something of the kind was greatly needed. **At entering on the study of** geology, scarcely had I read through a single page, ere I found my difficulties much enhanced by the non-existence of a dictionary, containing such technological terms as are peculiar to this branch of science, and, for a time, I was frequently obliged to pass over words, without any distinct comprehension of their force or application. Assuredly, some writers on geology have appended a glossary to their productions; but, I need scarcely say, these are, for the most part, necessarily meagre and ineffectual. The very necessity, also, for their insertion, I may, perhaps, claim as one of the strongest arguments in justification of my present attempt.

It can hardly, however, be adduced as a charge of inattention to the wants of the student, against the **writers on** geology, that no dictionary relating to its nomenclature has already appeared. Geology may still be regarded as in its infancy; it is, as it were, almost a creation of the present century; it may, not inaptly, be termed a new science; for, although Pythagoras, and Aristotle, and Strabo, were, to a certain extent, geologists; although Ovid puts into the mouth of the Samian philosopher—

> "Vidi factas ex æquore terras:
> Et procul a pelago conchæ jacuere marinæ;
> Et vetus inventa est in montibus anchora summis;
> Quodque fuit campus, vallem decursus aquarum
> Fecit; et eluvie mons est deductus in æquor:
> Eque paludosa siccis humus aret arenis;
> Quæque sitim tulerant, stagnata paludibis hument."

although, from time to time, theories of the earth have been published, and hypotheses the most crude, and fanciful, and illusory have been propounded; although men have been found so blind as to argue in favour of a plastic force; although, almost even in our own days, Vulcanist would have submerged Neptunist in his own aqueous deposits, and Neptunist would have torrefied Vulcanist in the igneous causes which he advocated; although, for upwards of two thousand years, geology may be said to have had its students and its advocates; yet, till within the last half century, it has never deserved the name of a science. Mixed up and confounded with cosmogony, it continued in a state of flux and reflux, at one time making advances, at another retrograding, till Hutton, in 1795, declared that "geology was in no ways concerned with questions as to the origin of things." Nor was it till, throwing aside all preconceived notions, geologists determined to found, and gradually advance, step by step, their theories on sound induction, that geology, in the magnitude and sublimity of the objects of which it treats, second only to astronomy, assumed its proper position in the order of scientific pursuits.

With the great increase of knowledge in geology, there necessarily sprang up a new nomenclature, and although this particular branch of technological lexicography may, and does, admit of much modification, it appears to me that it has at this time become sufficiently established to warrant, and call for, the issuing of a dictionary of geological terms.

Nomenclature being in itself an important part of science, I trust I may be excused for offering in this place a few observations on the subject. It is perhaps a very natural weakness that men should desire to distinguish things by names of their own appointing; but, inasmuch as a redundancy of names is prejudicial to the interests of science; perplexing, and often disgusting, to the student; and, in fact, raises an unnecessary obstruction in the path of knowledge; it becomes a subject of grave consideration whether the imposition of a new name, in lieu of one already become conventional, though that which has become conventional may, probably, not be the best or most appropriate that could have been chosen, be not a hazardous and injudicious course. It has been remarked by one of the most scientific men and greatest philosophers of the present day, Sir John Herschel, "it appears doubtful, whether it is desirable, for the essential purposes of science, that extreme refinement in systematic nomenclature should be insisted on. In all subjects where comprehensive heads of classification do not prominently offer themselves, all nomenclature must be a balance of difficulties, and a good, short, *unmeaning* name, which has once obtained a footing in usage, is preferable almost to any other."

These remarks are the more readily offered, in consequence of a fear, which I trust is groundless, arising out of, and caused by, the occasional observations of some of our most able geologists. Thus, I find one author objecting to the term *tertiary*, as applied to the supra-cretaceous deposits, stating it to be exceedingly objectionable: I turn over the pages of another great luminary, and I find that "the name of tertiary has been given with much propriety; that the name of super-cretaceous is peculiarly inappropriate, and that if a new name were necessary, post-cretaceous should have been chosen." Every neophyte in geology now knows that the tertiary deposits have been divided into eocene, miocene, and pliocene, the last being subdivided into older and newer pliocene: this also is objected to, and it is said, "if it be considered convenient to divide the supra-cretaceous rocks of Europe into three or more sub-groups, names which imply their actual geological position in the series, such as 'superior,' 'medial,' and 'inferior,' 'upper,' 'medial,' and 'lower,' or others of the like kind, would seem preferable to those derived only from a per-centage of certain organic contents." To multiply instances of this kind, would, however, be useless, and the sole motive for adducing the above, springs from a desire of restraining, as far as may be, a too natural fondness for innovating on established nomenclature.

It is most desirable that geologists should endeavour to avoid a very great evil which has gradually obtained in, and now sadly clogs, the pursuit of mineralogy. The redundancy of terms there introduced is most painfully bewildering, as the following instance will illustrate:—"The nomemclature of most minerals is at present so encumbered with synonyma, that it has become exceedingly perplexing to the student. The mineral which is called *epidote* by Haüy, is named *pristazit* by Werner, *thallite* by Lemetherie, *akanticone* by Dandrada, *delphinite* by Saussure, *glassy actinolite* by Kirwan, *arendalit* by Karsten, *glassiger strahlstein* by Emmerling; *la rayonnante vitreuse* by Brochant, *prismatoidischer augit-spath* by Mohs, &c., &c."*

To enter, here, on any defence of geology, against the groundless objections of weak, but amiable opponents, would be to travel out of the record. Happily, the mists of delusions, and the prejudices consequent on long-cherished and preconceived notions, are rapidly clearing away before the lucid and delightful, and unanswerable statements and views of the galaxy of learned, and scientific, and pious geologists of the present day. I trust I may be permitted to quote from one of these a most happily conceived and beautifully expressed passage:—"How then can they, by whom the magnificent truths of elapsed time and successive generations have been put in clear and strong evidence—how can they be expected to yield to false notions of philosophy, and narrow views of religion, the secure conviction that, in the formation of the crust of the earth, Almighty wisdom was glorified, the permitted laws of nature were in beneficent operation, and thousands of beautiful and active things enjoyed their appointed life, long before man was formed of the dust of the ancient earth, and endowed with a divine power of comprehending the wonders of its construction? It is something worse than philosophical prejudice, to close the eyes of reason on the evidence which the earth offers to the eyes of sense; it is a dangerous theological error, to put in unequal conflict a few ill-understood words of the Pentateuch, and the thousands of facts which the finger of God has plainly written on the book of nature; folly, past all excuse, to suppose that the moral evidence of an eternity of the future should be weakened by admitting the physical evidence for an immensity of the past."†

It remains for me to add a few words only, as deprecatory of severe criticism. No one can be more aware than myself of the numerous errors and deficiencies everywhere pervading this small volume: for these I urge nothing, even in extenuation. For myself, it is sufficient that I have derived from its preparation much information, great gratification. I entertain no morbid sensitiveness respecting the fate that awaits it. With

---

\* Professor Cleaveland. † Professor Phillips.

our prince of lexicographers I may say, "I dismiss it with frigid tranquillity, having little to fear or hope from censure or from praise."

The necessity created, by preparing such a book for the press, for carefully looking into and examining the opinions of various authors on the same subject, has made me acquainted with many works which, otherwise, I never might have perused; and I have, from this circumstance alone, reaped a rich reward of the purest pleasure; and, though critics may in unmeasured terms condemn, it is more than probable that

> "Seu me tranquilla senectus
> Exspectat, seu mors atris circumvolat alis;
> Dives, inops, Romæ, seu fors ita, jusserit, exsul;
> Quisquis erit vitæ, scribam, color."

# PREFACE TO THE THIRD EDITION.

A FEW years since, I was informed by my publisher that every copy of the two editions of my Dictionary had been disposed of, that the work was in demand, and urging me to prepare a third edition. Although I had by me a large mass of additional matter, I was, at that time, so unceasingly occupied with professional duties, that I had not leisure even to arrange the collected materials. Subsequently, from an affection of the eyes, I abandoned all idea of again taking up my pen as an author. A desire to aid in promoting the spiritual welfare of my fellow creatures, in the erection of a new church, has now induced me to offer a third edition; an edition by no means such as I could have wished, inasmuch as I have been compelled to abbreviate many portions, and to omit others, lest I should swell the volume to a magnitude beyond the reach of purchasers generally.

With these few prefatory observations I close my lexicographical labours; not from disinclination, but from increasing inability consequent on impaired vision.

Stonelands, February 16th, 1860.

# A DICTIONARY.

## A

**A.** In words of Greek derivation, A is used privatively, or in a negative sense;—as acephalous, without a head; acaulous, having no stem; apetalous, having no petals; acotyledonous, having no cotyledons, &c.

**ABARTICULA'TION.** (from *ab* and *articulatio*, Lat.) That kind of articulation which admits of manifest motion; it is also called diarthrosis, from the Greek word διάρθρωσις, and dearticulation.

**ABBRE'VIATED.** (*abbreviatus*, Lat.)
1. In botany, an epithet for the perianth. An abbreviated perianth is shorter than the tube of the corolla, as in Pulmonaria maritima.
2. Shorter than a correspondent part.

**ABDO'MEN.** (*abdomen*, Lat. *abdomen*, Fr. *abdomine*, It.) The large cavity commonly known as the belly, containing the organs more immediately concerned in the process of digestion, as the stomach, liver, spleen, pancreas, bowels, &c.

**ABDO'MINAL.** (from *abdomen*.)
1. Pertaining to the abdomen or belly.
2. Fishes belonging to the order abdominales, or the fourth order of Linnæus.

**ABDOMINA'LES.** The fourth order of fishes in the arrangement of Linnæus; they have ventral fins behind the thoracic, or fins placed on the belly, and the branchia ossiculated; they chiefly inhabit fresh water. The salmon, trout, smelt, &c., are examples.

**ABDU'CENT.** (from *abduco*, Lat.) The name given to those muscles which serve to open or draw back parts of the body; their opposites, or antagonists, are called adducent.

**ABDU'CTOR.** (*abducteur*, Fr. *abduttore*, It.) The same as abducent.

**ABERRA'TION.** (*abberatio*, Lat.)
1. A certain deviation in the rays of light, from the true or geometrical focus of reflection or refraction, in curved specula or lenses.
2. A deviation from the ordinary course of nature.

**ABNO'RMAL.** } *abnormis*, Lat.) Irregular; unwonted; unnatural.
**ABNO'RMOUS.** }

**ABNO'RMITY.** Departure from natural formation; irregularity.

**ABRA'NCHIA.** (from *a*, priv. and βράγχια, Gr.) Animals destitute of gills, and having no apparent external organs of respiration.

**ABRANCHIA'TA.** The third order of articulata, having no apparent external organs of respiration, but seeming to respire, some, by the entire surface of the skin, others, by internal cavities. The abranchiata are divided into two families; the first, Abranchiata setigera, comprising the Lumbrici and Naides of Linnæus; the second, Abranchiata asetigera, comprising the Hirudo and Gordius of Linnæus.

ABRU'PT. (*abruptus*, Lat.) Broken; craggy; steep; precipitous. In botany, applied to leaves, when the extremity of the leaf is, as it were, cut off by a transverse line.

ABRU'PTLY PIN'NATE. Applied to pinnate leaves, terminated neither with a leaflet nor a tendril.

ABSO'RBENT. (from *absorbeo*, Lat.)
1. Any substance possessing the property of absorbing, or sucking up, fluids, or neutralizing acids; as the earths, alumina, magnesia, &c.
2. In anatomy, the absorbents are small pellucid tubes, which have been discovered in most parts of the body, and are supposed to exist in all. The absorbents begin by numberless open mouths, too minute to be visible to the naked eye; by the assistance, however, of glasses, the orifices of the lacteals have been seen in the human body, and those of the lymphatics in certain fishes.

ABSO'RBENT SYSTEM. This consists of the absorbent vessels and conglobate glands; the former are divided into lymphatics and lacteals, and the thoracic duct, or common trunk, in which they terminate.

ACA'LEPHA. (ἀκαλήφη, a nettle, Gr.) A class of zoophytes found swimming in the waters of the ocean. The acalephæ are divided into orders, namely, the A. simplicia, and A. hydrostatica. The latter, or hydrostatica, are recognized by one or more bladders commonly filled with air, by the aid of which they suspend themselves in their watery element. "In all parts of the ocean," says Professor Grant, "are numerous gelatinous animals, for the most part of a simple and transparent texture, such as the Medusa and Portuguese man-of-war, and numerous other genera. For the most part, those soft gelatinous animals (which are entirely aquatic, and all of them marine) possess a property by which they excite inflammation, when they touch the surface of the skin, like nettles. The Acalephæ scarcely possess a trace of a nervous system, and are all aquatic. They float, for the most part, by the action of their own muscular power, or by air-sacs, or by cilia. They feed upon animals of extreme minuteness—for the most part, on the microscopic crustaceous animals, which abound in the ocean as in other waters." Cuvier places the Acalephæ in the third class of the fourth great division of the animal kingdom, or Radiata. Linnæus placed them in the class Zoophyta, order Gelatinous zoophytes. See Malactinia.

ACALY'CINOUS. A term applied to plants which want a calyx.

ACA'MAS. A genus of fossil multilocular, straight and conical shells; mouth round and horizontal; siphuncle central; summit pierced by eight small tuberculated apertures, disposed round a stelliform figure; the septa conical, plaited at the bottom, and plain at the edge. The substance spathose, similar to that of belemnites. *Parkinson.*

ACANA'CEOUS. (from ἄκανος, Gr.) Armed with spines or prickles.

ACA'NTH. The *acanths* constitute a family of ichthyolites, of the Old Red Sandstone, rich in genera and species, forming an intermediate link between the placoids and the ganoids.

ACA'NTHA. (ἄκανθα, Gr.)
1. In botany, the spine or prickle of a plant.
2. In zoology, a term for the prickly fins of fishes.
3. In anatomy, the spinous process of a vertebra.

ACANTHA'CEOUS. (*acanthace*, Fr.) Possessing spines or prickles.

ACAN'THODES. The name assigned to a genus of ichthyolites of the old red sandstone, characterised by having a spine or thorn on each fin.

ACANTHOPTER'YGII. (from ἄκανθα, a thorn, and πτέρυξ, a wing, Gr.) Thorny-finned fish. Cuvier divided all recent bony fishes into two great orders, the Malacopterygii, and the Acanthopterygii. In the acanthopterygii the rays are stiff continuous spikes of bone, each standing detached as a spear, without joint or branch. The perch will serve as a familiar illustration of this order.

ACANTHOPTERY'GIOUS. (from ἄκανθα and πτέρυξ, Gr.) A term applied to those fish whose back fins are osseous and prickly. The acanthopterygii are the first, and by far the most numerous division of ordinary fishes.

ACANTHOCE'PHALA. (from ἄκανθα and κεφαλή, Gr,) An intestinal worm belonging to the order Parenchymata, class Entozoa.

ACAN'TICONE. A sub-species of prismatoidal augite, occurring principally in primitive rocks, such as mica-slate, gneiss, &c. Known also as pistacite and epidote.

A'CARUS. (ἄκαρι, Gr. animal minutissimum.) The tick or mite. A genus of insects belonging to the order Aptera. The Acarus has eight legs, two eyes, and two jointed tentacula. The female is oviparous, and excessively prolific. Authors estimate the number of species variously; Linnæus enumerates 35, and Gmelin 82 species. Most of these are very small and almost microscopical. Some are parasitical, living in the skin of animals: some of the coleoptera are found covered with them.

ACAU'LOUS. (from a, priv, and καυλός, Gr.) A term applied to plants, the flowers of which have no pedicle or stalk.

ACCI'PITRES. (Lat. from *accipiendo*.) Rapacious birds; the first order of birds in the Linnæan system of ornithology. The Accipitres are known by their hooked beak and talons; they feed upon other birds, as well as upon the weaker quadrupeds, and **reptiles**. They have been divided into two families, the diurnal and the nocturnal.

ACCI'PITRINE. Belonging to the order Accipitres; rapacious.

ACCLI'MATED. (*acclimaté*), Fr.) Accustomed to **a** climate not natural to it.

ACCLIMA'TION. Neutralization to foreign climate.

ACCLI'MATIZE. (*acclimater*, Fr.) To accustom to foreign climate; **to** accustom to the temperature of **a new climate**. A term applied both to persons and things; to animals and plants.

ACCRE'TION. (*accretio*, Lat.) Increase, or growth, by the accession of new parts. Bacon says, plants do nourish; inanimate bodies do not; they have an *accretion*, but no alimentation.

ACCRE'TIVE. Increasing, or growing, by the accession of **new parts**.

ACE'PHALA. (*a*, priv. and κεφαλή, the head, Gr.)
1. A class of animals not having any head, but merely a **mouth**, concealed between the folds **of their** mantle; the mouth is always edentated: the oyster furnishes an example. All the acephalæ are aquatic.
2. In entomology, an order of insects.

ACE'PHALOUS. (*acephale*. Fr. ἀκέφαλος, Gr.) Headless; this term was given by Cuvier to animals not having any head.

ACE'RB. (*acerbus*, Lat. *acerbe*, Fr. *acerbo*, It.) Acid with an addition of roughness: the taste of an **unripe** sloe is a familiar and good illustration.

ACE'RBITY. (*acerbitas*, Lat. *acerbité*, Fr.) Sourness combined with roughness of taste.

ACERO'SE. } (*acerosus*, Lat.)
A'CEROUS. }
1. Chaffy, branny.
2. In botany, leaves linear, needle-shaped, everywhere of an equal breadth, mostly acute, and rigid.

ACEROTHE′RIUM. An extinct genus of animals, constituting a link connecting Palœotherium with Rhinoceros.

ACE′SCENT. (*adcescens*, Lat. *acescent*, Fr.) That has a tendency to become sour spontaneously, or by spontaneous decomposition.

ACETA′BULUM. (Lat.)
1. A cavity in a bone formed for receiving the head of another bone, and thus named from its cup-like shape; it is more particularly used for expressing the cavity in the os innominatum, which receives the head of the femur, or thigh-bone.
2. In botany it is used for the cotyledon.

ACE′TARY. (from *acetum*, Lat.) An acid pulpy substance found in some fruits, especially the pear, surrounding the core; it is enclosed in a congeries of small calculous bodies towards the base of the fruit.

A′CETATE. Any salt formed by the union of acetic acid with a salifiable base, as acetate of iron, acetate of potash, &c. The acetates are all soluble in water; many of them are deliquescent, and crystallizable with difficulty; they are decomposed by the sulphuric acid.

ACHA′TES. (ἀχάτης, Gr.) The agate.

ACEOUS. Terminations in aceous and icius, express a resemblance to a material; those in eous indicate the material itself; as membranaceous, resembling membrane; tufaceous, resembling tufa; membraneous, skin itself.

ACHA′NIA. (from ἀχανής, Gr. ab α, priv. et χαίνω.) In botany, plants whose corolla does not open. Order Polyandria, class Monodelphia; natural order Columniferæ.

ACHELOIS. A genus of fossil shells, described by De Montfort as being of a conical form, with conical septa, but Parkinson considers its characters not sufficiently ascertained at present to warrant its being considered a distinct genus.

A′CHIRITE. Emerald malachite.

ACI′CULA. (Lat.) A prickle or fine spine.

ACI′CULAR. In the shape of a needle: rocks of granite, having sharp, needle-like summits, are thus named.

ACI′CULARLY. Needle-like.

ACID. (*acidum*, Lat. *acide*, Fr.) The word *acid*, originally synonymous with *sour*, and applied only to bodies distinguished by that taste, has been gradually extended in its signification, and now comprehends all substances possessed of the following properties:—
1. When applied to the tongue, they excite that sensation which is called *sourness*.
2. They change the blue colours of vegetables to a red.
3. They unite with water in almost any proportion, with a condensation of volume, and evolution of heat.
4. They combine with all the alkalies, producing effervescence during the combination, and with most of the metallic oxides and earths, and form with them those compounds which are called salts.

The acids terminating in *ous* produce compounds to which the termination *ite* is given, as, e. g. the combination of sulphurous acid and potassa is a sulphite of potassa; the acids ending in *ic* form compounds to which the termination *ate* is applied; the combination of sulphuric acid and potassa is a sulphate of potassa.

ACIDASPIS. (ἀκίς, cuspis, and ἀσπίς, clypeus vel scutum, Gr.) The name assigned by Sir R. Murchison to a genus of trilobites, the generic characters of which he thus describes; "capitis scutum marginatum, antice subtruncatum, trituberculatum: tuberculo media postice in mucronem desinente." Although, says the talented author of the Silurian system, most unwilling to multiply names, the very remarkable form of the head or shield of this trilobite, the posterior end of its central lobe projecting over the body in the form of a stomacher, and rendering it totally distinct from any published figure, induces me to propose it as a new genus." Two species have been described, namely Acidaspis Brightii and Abispinosus. They occur in the Wenlock limestone of the Malvern Hills.

ACI'DIFIABLE. Any substance capable of being converted into an acid, by the union of an acidifying principle without decomposition.

ACI'DULOUS. (*acidulus*, Lat.) Slightly acid; sub-acid; sourish.

ACINA'CIFORM. (from *acinaces* and *forma*, Lat.) Cimiter shaped; a term applied to leaves, one edge of which is straight and thick, the other curved and thin.

A'CINOSE. Iron ore. A variety of iron ore found in masses, and commonly lenticular. Colour, generally, brownish red: lustre metallic: texture granular: hardness 5 to 9: brittle.

A'CINUS. (Lat.) Acini Pl. Each separate part of a compound berry containing a seed: compound berries consist of many simple acini united together, as the raspberry, blackberry, &c.

ACOTYLE'DON. (from *a*, priv. and κοτυληδών, Gr.) A plant whose seeds have no cotyledon, or side-lobes.

ACOTYLE'DONOUS. Plants, whose embryos have no lobes, or seminal leaves; not having cotyledons, or seed lobes.

ACROCERAU'NIAN. (from ἄκρος and κεραυνός, Gr.) A term given to some mountains, supposed to be especially subject to the effects of lightning.

A'CROGEN. (from ἄκρος and γεννάω, Gr.) An acrogen is a cylindrical plant growing at its point only, and never augmenting in thickness after once formed.

ACRO'MION. (from ἄκρος and ὦμος, Gr.) The humeral extremity of the spinous process of the scapula; situated over the upper end of the humerus, and contributing to the protection of the shoulder joint.

A'CROSPIRE. (from ἄκρος and σπείρα Gr.) The shoot or sprout of a seed, also called the plume or plumule.

ACTEONELLE. A genus of cylindrical smooth univalves, with elevated spire, aperture entire, lengthened, moderately wide anteriorly, narrow posteriorly; columella smooth, rounded; lip thin. It occurs fossil in the oolite.

ACTI'NIA. The sea-anemony, a genus of the order Vermes mollusca. The fleshy body of the actinia is frequently ornamented with bright colours, and exhibits numerous tentacula placed round the mouth in several ranges, like the petals of a double flower, from which it has obtained its name of sea-anemony.

ACTINO'CERAS. A genus of chambered fossil shells belonging to the family of the Orthocerata, and established by Brenn; it is a straight conical camerated shell with septa, as in Orthoceras, the siphuncle is external but divided into chambers, and within it is a continuous tube which appears to have been capable of contraction and expansion, and is furnished with verticillate radii which connect the tube with the walls of the syphon. *Lycett.*

ACTINO COMAX. A genus of fossil shells, having the form of belemnites.

ACTINOLEPIS. An ichthyolite of the old Red Sandstone.

ACTINOCRINITES. (from ἀκτίνωτος, radiated, and κρίνον, a lily Gr.) Radiated, lily-shaped animals. A genus of fossil encrinites established by Miller. There are many species, some of which are found in the Silurian rocks and in the carboniferous limestone. Mr. Phillips says that Mr. Miller himself is in error when he speaks of the species A. Moniliformis as belonging to the Mountain Limestone. The generic characters of the genus are thus described by Mr. Miller. A crinoidal animal, with a round column composed of numerous joints, perforated by a round alimentary canal. At the summit of the column is placed an operculum formed of three plates on which five first costals and one irregular costal adhere; which are succeeded by the second costals and intercostals and the scapula, from whence five arms proceed, forming two hands, with several tentaculated fingers. Round side arms proceed at irregular distances from the column which terminates at the base in a fascicular bundle or root of fibres.

ACTI'NOLITE. (from ἀκτὶν and λίθος, Gr.) A variety of hornblende. Its constituent parts are silica 46·26, magnesia 19·03, lime 13·96, alumina 14·48, protoxide of iron 3·43, protoxide of manganese 0·36, fluoric acid 1·60, water, &c. 1·04. This variety of hornblende rarely occurs in the secondary rocks, being principally confined to those of the primary class. It is of a green colour. Cleaveland says " this mineral possesses all the essential characters of hornblende. In fact common hornblende and actynolite insensibly pass into each other. The actynolite has usually a greater translucency, a more lively green colour, arising from the chrome which it contains, and differs in the result of fusion by the blowpipe. Jameson places actynolite in the hornblende family but, as a distinct species, divided it into four subspecies, 1. asbestous actynolite, 2. common actynolite, 3. glassy actynolite, and 4. granular actynolite. He remarks " Haüy is of opinion that actynolite and hornblende belong to the same species, because his observations shew an identity in their primitive forms. Count de Bournon, on the contrary, proves that their primitive forms are not the same; hence he infers that they are distinct species. These two minerals are further distinguished by their colour, fracture, and crystallizations; those of actynolite being few in number and simple; whereas those of hornblende are more numerous and complex; and lastly, they differ remarkably in geognostic situation." Phillips places actynolite amongst the varieties of hornblende, and divides it into crystallized asbestiform, and glassy actynolite.

ACTI'NOLITE-SCHIST. A metamorphic rock, consisting principally of actinolite, with an admixture of mica, quartz, or felspar; its texture is slaty and foliated.

ACTINOLI'TIC. Containing actinolite; of the nature of actinolite.

ACU'LEATE. } (*aculeatus*, Lat.)
ACU'LEATED. } Prickly; having spines or prickles. Applied to leaves armed with prickles. Used to denote prickles, fixed in the bark, in distinction from thorns which grow from the wood.

ACU'LEUS. A prickle or spine, arising from the bark only, and not growing from the wood.

ACU'MINATED. (*acuminatus*, Lat.) Ending in a point; sharp-pointed, the decrease being very gradual.

A'DAMANT. (ἀδάμας, Gr.) A name given to different stones of excessive hardness, as to the diamond.

ADAMA'NTINE SPAR. Imperfect corundum; a variety of rhombohedral corundum, nearly analogous to perfect corundum, containing from 3 to 5 per cent. of silica, and 1 to 2 of oxide of iron. It occurs massive and in crystals. The crystals brought from India are the most pure.

ADA'MIC EARTH. A name given to red clay.

A′DAPIS. One of the extinct pachydermata, found in the gypsum quarries of Montmartre. The form of this creature most nearly resembled that of a hedge-hog, but it was three times the size of that animal: it seems to have formed a link connecting the pachydermata with the insectivorous carnivora. — *Buckland*.

ADDU′CENT. (from *adduco*, Lat.) A name given to those muscles which bring forward, close, or draw together, the parts of the body to which they are attached; their antagonists are termed abducent.

ADE′NOID. (from ἀδήν, a gland, and εἶδος, form, Gr.) Glandiform; having the shape of a kernel, almond, or gland; glandulous.

ADIPOCERA′TION. The process of being converted into adipocire.

A′DIPOCERE. } (*adeps*, fat, and *cera*,
A′DIPOCIRE. } wax, Lat.) A substance resembling spermaceti, produced by the conversion of animal matter exposed to running water; in this way animal matter may be converted into a soft, unctuous, or waxy substance in the space of a little more than a month; but adipocire has also been produced, though not so rapidly, by the heaping together large masses of putrefying animal matter, as was discovered on the removal of a very great number of bodies from the burial ground of the Church des Innocens at Paris, 1787. Adipocire possesses many of the properties of fat combined with a portion of ammonia. It was first discovered by Fourcroy.

ADIPOCE′RE MINERAL. A fatty matter found in the argillaceous iron ore of Merthyr: it is fusible at about 60°, and is inodorous when cold, but when heated it emits a slightly bituminous odour.

A′DIT. (*aditus*, Lat.) The shaft or entrance into a mine, usually made in the side of a hill, for the conveyance of ore, and the carrying off of the water. It is the first object of a miner, in the working of a mine, to drive a passage or *adit* from the nearest low ground or valley to meet the shaft, for the purpose of conveying off the water, which is raised to the *adit* level by the means of the steam engine. It will therefore be obvious that the depth of the adit from the surface of the mine, must depend on the height of the ground in which the mine is, and the depth of the neighbouring valley. The depths of different parts of mines are usually dated from the adit level.

ADMI′XTION. (*admisceo*, Lat.) The union of various bodies, or substances, by mingling them together. In admixtion each body retains its own character, and does not undergo any chemical change, as in composition.

ADNA′TA. (*adnatus*, Lat.)
1. Those parts of animal, or vegetable bodies which are natural, as the nails, hair, &c.
2. Accidental parts, as fungi, misletoe, &c.
3. The external coat of the eye.

A′DNATE. Growing to; adhering. In botany, it is used when a leaf adheres to the branch or stem by the surface or disk itself; applied to stipules when they are fixed to the petioles.

ADULA′RIA. (from *adula*, the summit of a Swiss mountain.) Moonstone; a transparent white-coloured variety of feldspar, with a silvery, or pearly opalescence.

Æ′DELITE. A stone found in Sweden, and thus named by Mr. Kirwan. Its form is tuberose and knotty. Texture striated; sometimes resembles quartz. Lustre from 0 to 1. Specific gravity 2.515 after it had absorbed water. Colour light grey. Before the blow-pipe it intumesces, and forms a frothy mass. Acids convert into a jelly. A specimen, analysed by Bergman, contained 69 silica, 8 lime, 20 alumina, 3 water.—*Thomson*.

Aë'lodon. A fossil saurian of the oolite and lias.

A'erate. To combine with carbonic acid.

A'erated. Combined with carbonic acid, or fixed air.

A'eration. The combining with fixed air, or carbonic acid; the saturation of a liquid with air.

Aë'rial acid. A name given by Bergman to carbonic acid or fixed air; aerial acid is of greater specific gravity than atmospheric air, and extinguishes flame.

Aë'rolite. (from ἀὴρ and λίθος, Gr.) Called also meteorite. A name given to meteoric stones, which, occasionally, fall to the earth. Nothing is positively known as to the origin of aërolites: by some authors they have been supposed to come from the moon, being projected by volcanic force beyond the sphere of the moon's attraction; by others they have been thought to be children of the air, created by the union of simpler forms of matter. They do not resemble any other substance found on the earth, and it has been indisputably proved that they are not of terrestrial formation. The fall of these bodies has been well ascertained, and has occurred at different times, and in various parts through many ages. Some of these aërolites are immensely large, from 300 lbs. downwards. From an analysis of them, they are found to agree in their component parts. They are covered with a thin crust of a deep black colour, their exterior is roughened with small projections, and they are destitute of gloss. Internally their texture is granulated, and of a greyish colour. When carefully examined, they appear to be composed of a number of small spherical bodies and metallic grains embedded in a softer matter, composed, according to the Hon. Mr. Howard, who diligently and carefully studied them, of silica, magnesia, iron, and nickel. In addition to these substances, Vauquelin found chrome, and Stromeyer discovered cobalt, in aërolites: lime, alumine, and manganese, have also been detected in them. Meteoric iron has been imitated by fusing iron with nickel. When it is considered how many of these bodies have been seen, or heard, to fall through the air, we must conclude that they are very numerous especially when we reflect on the small proportion which must be observed, and the small comparative portion of the globe which is inhabited, or habitable, by man. The fall of meteoric stones, in the opinion of some writers, is much more frequent than is generally supposed, hardly a year passing without some instances occurring.

Aero'meter. (from ἀὴρ and μέτρον, Gr *acrometre*, Fr.) An instrument for ascertaining the weight, or density, of the atmosphere.

Aero'metry. (*aérométrie*, Fr.) The science which treats of the properties of the air; it comprehends not only the doctrine of the air itself, considered as a body, but also its pressure, elasticity, rarefaction, and condensation.

Aero'scopy. (from ἀὴρ and σκοπέω, Gr.) The observation of the air.

Aerostation. (from ἀὴρ, the air, and ἵστημι to weigh, Gr.) Primarily, the science of weights suspended in the air, but in the modern acceptation of the term, the art of navigating in the air.

Æru'ginous. (from *æruginosus*, Lat.) Partaking of the nature of the rust of copper.

Æru'go. Verdigrease, or verdigris; rust of copper, formed by the combination of an acid with copper. Impure subacetate of copper. Verdigris is inodorous, and when first applied to the tongue is nearly insipid, though strongly styptic; it leaves a metallic taste in the mouth. It is poisonous; sugar acts as a specific against its poisonous effects.

ÆSTIVA'TION. (*æstivatio*, Lat.)
1. The effect produced by summer heat.
2. The mode in which the parts of a flower, taken separately, are ar-arranged in the bud.

Æ'TITES. (ἀετὸς, Gr. *astite*, Fr.) Eagle-stone; a variety of oxide of iron mixed with clay. It is found in masses, generally under the form of a rounded knob, something resembling a kidney. It prevails in the coal formations of England, Wales, and Scotland, has a rough surface, and is of a brown colour. Specific gravity 4 to 7. Lustre of the exterior metallic. It frequently contains a sort of kernel, which rattles on being shaken. It was formerly in repute for several extraordinary magical as well as medical properties, such as preventing abortion, discovering thieves, &c. It derives its name from a popular notion that it was found in eagles' nests, where it was supposed to prevent the eggs from becoming rotten. See *Nodular Iron Ore*.

AFFI'NITY. (*affinis*, Lat. *affinité*, Fr. *affinità*, It.) The tendency which bodies, dissimilar in their composition, have to unite and form new compounds. Different bodies are possessed of different attractive powers, and if several be brought together, those which have the strongest mutual affinities enter first into union. Affinity, like sensible attraction, varies with the mass and the distance of the attracting bodies. That the course of affinity increases as the distance diminishes, and the contrary, is obvious; for it becomes insensible when the distance is sensible, and exceeding great when the distance is exceedingly diminished. Affinity agrees with sensible attraction in every point which it has been possible to determine.

AGALMA'TOLITE (from ἄγαλμα and λίθος, Gr.) Figure-stone. A sub-species of talc-mica, of different colours, as white, red, brown, green, and grey. It occurs massive. It feels greasy, is translucent, and has a conchoidal fracture. The finest specimens are brought from China. It does not contain any magnesia, but in other respects it has the character of talc.

A'GARIC. (*agaricus*, Lat.) The generic name for the mushroom, a genus of the order Fungi, class Cryptogamia. Gmelin enumerates nearly 400 species.

A'GARIC MINERAL. A variety of soft carbonate of lime. It is found in the clefts of rocks, or the bottom of lakes, in pieces loosely cohering, and it is so light as nearly to swim upon water. It obtains its name from its resemblance to a fungus in colour and texture.

AGAR'ICIA. A genus of lamellated stony polypifers, fixed with flat, subfoliaceous expansions, the upper surfaces only having stelliferous grooves: the stars sessile, lamellous, and in rows. This genus, which much resembles *Pavonia*, differs in having one surface only of its foliaceous expansions furnished with stelliferous grooves.

A'GATE. (*achates*, Lat. ἀχάτης, Gr. *agate*, Fr. *agata*, It.) A siliceous, semi-pellucid gem, of which there are many varieties, not of great value. Agates are principally composed of quartz with various colouring matters. Agates may be artificially coloured by immersion in metallic solutions. Mr. Allan says "Agate is an impure variety of calcedony, of frequent occurrence in the vesicular cavities of amygdaloidal rocks. It presents the most brilliant and the most varied colours, and, from its hardness and compact structure, being capable of receiving a high polish, occupies a distinguished position in most collections." Professor **Jameson** places Agate amongst the members **of the** Quartz family, constituting it a distinct species. He says "it is not, as some mineralogists **maintain**, a simple mineral, but is composed **of** various species of **the** quartz family, intimately joined **together.** It is principally composed of calcedony, with flint, horn**stone**, carnelian, jasper, cacholong, amethyst, and quartz." Professor Cleaveland says " **the calcedony is, however,** the most common **and** abundant ingredient, and may frequently be considered the *base* of the agate; in fact, some agates are composed entirely of calcedony differently coloured." Werner divided agates into Ribbon Agate; Fortification Agate; Brecciated Agate; Moss Agate; Landscape Agate; Tubular Agate; and Jasper Agate. As precious stones, agates are now less esteemed than formerly; the most valuable **are** the oriental. When cut and polished, agates present an appearance of waving lines, sometimes accurately parallel, sometimes varying in breadth, and some**times** containing a resemblance to vegetable forms, as mosses, ferns, &c. Small agates are frequently found in common gravel.

A'GATY. Of the nature of agate.

A'GGREGATE. (*aggregat*, Fr.)

1. The complex, **or collective** re-result of the conjunction, **or** acervation, of many particulars: it differs from a compound body, inasmuch **as the** union in the last is more intimate than between the parts of an aggregate.

2. In botany, a term **used to** express flowers **composed of many** small florets, **having a common** undivided **receptacle;** the anthers separate and **distant, the florets** commonly standing on stalks, each having a single or double partial calyx. They are opposed to simple flowers, and are usually divided into seven kinds.

A'GGREGATED. Collected; accumulated; heaped together. A rock is said to be aggregated when the **several** parts of which it is composed merely adhere and **may be** separated from each other by *mechanical* means: thus in **granite**, the several parts of which, quartz, mica, and felspar, may be mechanically separated, we have an instance of an *aggregated* rock.

AGGREGA'TION. (*aggregation*, Fr. *aggregazione*, It.) The collection into one mass of bodies having no natural connexion, but, by a species of union, made to constitute one body.

AGNO'STUS. (ἄγνωστος, Gr.) A fossil genus of trilobites, established by Brongniart, the *Battus* of Dalman. Agnosti are found in the lower Silurian rocks, especially in the Llandeilo flags. In Norway, says Sir R. Murchison, the Agnosti **occur** in millions, but in **our rocks** they are much less frequent. Macleay does not consider Agnostus to belong to the Trilobita. The discovery of **an** entire one in Bohemia, **by Dr.** Beyrich, proves the agnostus to be a true trilobite.

AGNOTHE'RIUM. An extinct animal of the miocene period, order Mammalia, allied to the dog, but of very large size. One species only **has** been found, at Epplesheim, in Germany.

AIGUE MARI'NE. A variety of topaz, of a blueish **or** pale green colour.

AIGUI'LLES. (*aiguilles*, Fr.) The needle-like points, or tops, of granitic rocks.

AIGUI'LLE DE DRU. A pyramidal granitic mountain, according to Bakewell, the most remarkable at present known; the upper part, or spire, rises above its base nearly to a point, in one solid shaft, **more** than 4,000 feet; the summit being 11,000 feet above the level of the sea.

A'LA. (*ala*, Lat.)
1. In botany, a term used for the hollow, which either the leaf, or the pedicle of the leaf, makes with the stalk; the hollow turning, or **sinus**, placed between the stalk or branch of a plant and the leaf, and whence a new offspring generally rises. Sometimes it is used for those parts of leaves otherwise called lobes **or** wings. Those petals **of** papilionaceous flowers placed between those other petals, distinguished as the vexillum and carina, and which constitute the top and bottom of the flower, are also called alæ.
2. In anatomy, the **lobes of the** liver, the cartilages **of the nostrils**, and the cartilaginous **parts of the** ears, are called alæ.

A'LABASTER. (*Alabaster*, Lat. ἀλάβαστρον, Gr.) Granular or massive sulphate of lime. Alabaster **is** found in this country accompanying the salt deposits in Cheshire. It is also most abundant at Montmartre, in the neighbourhood of Paris. At Montaiont, in Italy, it is found in blocks of such magnitude, that statues of the size of life are occasionally cut from them. Being semi-transparent, it has sometimes been employed for windows instead of glass, and a church at Florence is still illuminated by alabaster windows. Instead of panes of glass, there are slabs of alabaster 15 feet high, each of which forms a single window, through which the light is conveyed. Alabaster may be turned by the lathe, and is thus formed into a great variety of ornamental articles.

ALAB'ASTRITES. (*alabastrites*, Lat. ἀλαβαστρίτης, Gr.) Alabaster stone; a kind **of** marble, whereof the ancients **made** vessels for ointment; by Horace called **onyx**.

A'LALITE. Called also Diopside, a variety of augite. It occurs massive, disseminated, and crystallized, with a vitreous external, and pearly internal lustre; it is translucent, and either white or **of** a pale green colour. It was named by Bonvoison, from his finding **a** variety of it near the village **of** Ala, in Piedmont.

ALA'SMODON. A species of shells of the genus Unio, having cardinal, but no lateral teeth.

A'LATE.  } (*alatus*, Lat.) Winged.
ALA'TED. } In conchology, applied to shells having **an** expanded lip, or when any portion of them is much expanded.

A'LBITE. Tetarto-prismatic felspar; soda felspar. A name given **to** felspar, whose alkali is soda instead of potash. Colour generally white, sometimes grey, green, or red. Lustre upon faces of cleavage pearly, in other directions vitreous. Albite forms a constituent part of the greenstone rocks in the neighbourhood of Edinburgh. It is composed of silica, alumina, **and** soda, with a trace of lime.

ALBI'TIC. **Of** the nature of albite; containing albite.

ALBUGI'NEA. [from *albus*, Lat.]
1. The fibrous membrane in the eye, situate immediately under the tunica conjunctiva.
2. One of **the** tunics of the **testis**.

ALBU'GINEOUS. [*albugineus*, Lat.]
1. The aqueous humour of the eye.
2. Resembling the white of an egg.

A'LBUM GRÆ'CUM. The excrement of dogs, wolves, hyenas, &c., feeding or living on bones. It principally consists of the earth of bones or lime, in combination with phosphoric acid.

ALBU'RNUM. [*alburnum*, Lat.] Called also sap-wood; the interior white bark of trees: it is this which yearly becomes new wood; the last formed wood of the trunk of trees and woody plants. It appears probable, that the new layers of alburnum and liber, which are produced each year on the outside of all that preceded, are formed by the descending fibres, or roots, of the leaf-buds.

ALCYO'NIUM. A genus of zoophytes, the characters of which are, that the animal grows in the form of a plant; the stem, or root, is fixed, fleshy, gelatinous, spongy, or coriaceous, with a cellular epidermis, penetrated with stellated pores, and shooting out tentaculated oviparous hydræ.—*Encyclop.*

From the experiments of Hatchett, it appears that these animals are composed principally of carbonate of lime and a little gelatinous matter. The alcyonium belongs to the class Vermes, order Zoophyta. Cuvier places the alcyonium in the order Coralliferi, class Polypi. Much obscurity prevails in the distribution even of the recent species of the families of *alcyonium* and *spongia*. Ellis makes their distinction to consist in the presence of polypi, as inhabitants of the cells of alcyonia, and believes the sponges to possess none of these animalcules, but to be simply invested with a living gelatinous flesh. Lamark, however, supposes the sponges to have polypi, like the alcyonia, differing only in the greater solidity of the fleshy parts of the latter, which permit them to be observed when removed from the water; while those of the former dry up instantly on being taken out of their natural element.

ALCY'ONITE. Alcyonites are fossil alcyonia, or zoophytes nearly allied to sponges, the production or habitation of polypi.—*Bakewell*.

ALE'MBIC. (*alambic*, Fr. *lambicco*, It. *alembicum*, Lat.) A vessel used in the process of distillation, usually made of glass or copper. Of alembics there are two different forms, the beaked and the blind, the former having communication with the receiving vessel, the latter being without such. The use of the alembic has yielded to that of the retort.

ALE'PIDOTE. (from a, priv. and λεπις, squama, a scale.) Any fish destitute of scales, as the eel, cod-fish, &c.

A'LGA. (Lat.) Sea-weed.

A'LGÆ. An order, or division, of the Cryptogamas class of plants. It is one of the seven families, or natural tribes, into which Linnæus distributed the vegetable kingdom. The whole of the sea-weeds are comprehended under this division. The plants belonging to this order are described as having their leaf, stem, and root all one. The depths at which, according to Syell, some of the algæ live, is extremely great, being no less than one thousand feet; "and although in such situations there must reign darkness more profound than night, at least to our organs, many of these vegetables are highly coloured."—*Principles of Geology*.

M. Lamouroux states that the groups of algæ, or marine plants, affect particular temperatures or zones of latitude, though some few genera prevail throughout the ocean. Some of the algæ grow to the enormous length of several hundred feet, and are all highly coloured, though many of them must grow in the deep caverns of the ocean, in total, or almost total, darkness. From the observations of Humboldt, who discovered green plants growing in complete darkness at the bottom of one of the mines of Freyberg, it may be concluded that light is not the only principle on which the colour of vegetables depends.

ALGALMA'TOLITE. Figure-stone. A mineral, the finest varieties of which we receive from China. A sub-species of talc-mica. See *Agalmatolite*.

A′LGOUS. (*algosus*, Lat.) Having the nature, or characters, of sea-weed.

A′LKALI. (from the Arabic word *kali*, with the usual prefix *al*; the name given by the Egyptians to the plant called by us glasswort.) Any substance which, by uniting with an acid, neutralizes or impairs its activity, and forms a salt. Alkalies possess the property of converting vegetable blues to green, and yellows to red. There are three kinds of alkalies: 1. The vegetable alkali, or potash; 2. The mineral alkali, or soda; 3. The animal, or volatile alkali.

ALKALI′NITY. The property of changing vegetable blues into green.

ALKALI′METER. An instrument for ascertaining the proportion of alkali contained in any substance.

A′LKALOID. A body possessing some of the properties of an alkali.

A′LKANET. The name of a plant the root of which yields a fine red, and is much used by dyers.

A′LLAGITE. A mineral; colour brown or green; massive; semi-opaque; fracture conchoidal: it is a carbo-silicate of manganese.

A′LLANITE. An orthitic melane-ore. The cerium oxydé siliceux of Haüy. A mineral brought from Greenland, and thus named after Mr. Thomas Allan, of Edinburgh, who first distinguished it as a peculiar species. According to the analysis of Dr. Thomson, allanite was found to contain silica 35·4, oxide of cerium 33·9, oxide of iron 25·4, lime 9·2, alumina 4·1, moisture 4.

It is of a black colour, inclining to grey or brown. It is found massive, or in acicular crystals. External lustre imperfect, metallic; internal, shining. Fracture conchoidal. Opaque. Streak greenish or brownish-grey. It is a siliceous oxide of cerium.

ALLIACEOUS. (from *allium, garlick*, Lat.) Resembling garlick; a term usually applied to substances which, on being heated, emit the odour of garlick.

A′LLOCHROITE. A mineral variety of the dodecahedral garnet. It is found massive, of a green, brown, grey, or yellowish colour; lustre glimmering. It consists of silica, lime, carbonate of lime, oxide of manganese, oxide of iron, alumina, and moisture. Before the blow-pipe it melts into an opaque black enamel. It was first discovered and described by Dandrada, and hitherto it has only been met with in an iron mine, near Drammen, in Norway. Sp. gravity 3·58.

A′LLOPHANE. A mineral of a blue, green, or brown colour, occurring massive, or in imitative shapes. It is rather hard and brittle. It gelatinizes in acids. According to the analysis of Stromeyer, it consists of alumina, silica, carbonate of copper, lime, sulphuric acid, and water.

ALL′OTROPISM. (from $ἀλλότροπος$, Gr. that can be turned from one thing into another.) A term in mineralogy, as opposed to isomorphism. A modification in the properties of a body, not resulting from chemical combination.

ALLO′Y.

1. A mixture of different metals: it must however be kept in mind, that when mercury forms one of the metals, the mixture is called amalgam.

2. The metal of inferior value, which is used to deteriorate, or give new properties to, another metal.

ALLU′VIAL. That is carried by water to another place, and lodged upon something else.

ALLUVIAL DEPOSITS. These consist in the accumulation of sand, shingle, and debris, along the sea-coast; in the formation of new lands on the banks of rivers and lakes by the alluvial depositions they carry down; in the growth and increase of tracts of marsh land; and in the accretion of calcareous tufa. "These formations appear to have proceeded uninterruptedly, as at present, from the period when our continents assumed their present form."—*Rev. J. Conybeare.*

ALLUVIAL EPOCH. Some writers have attempted, says Sir C. Lyell, to introduce into their classification of geological periods an alluvial epoch, as if the transformation of loose matter from one part of the surface of the land to another had been the work of one particular period. With equal propriety might they have endeavoured to institute a volcanic period, or a period of **marine or** fresh-water deposition, for alluvial formations must have originated in every age since the surface of the earth **was first** divided into land and sea.

ALLU'VION. } (*alluvio*, Lat. *alluvion*, Fr.
ALLU'VIUM. } *alluvione*, It.) Earth, sand, gravel, stones, or other transported matter which has been washed away and deposited by water upon land not permanently submerged beneath the waters of lakes or seas.—*Lyell.*

Alluvium has been **divided into** modern and ancient. The modern, characterized by the remains of man, and **contemporaneous** animals and plants; the ancient, by an immense proportion of large mammalia and carnivora, both of extinct and recent genera and species. The term alluvium has been assigned to the partial debris occasioned by causes still in operation; such as the wear produced by the present rivers, the more violent action of torrents, &c. The Hon. W. Fox Strangways dates the commencement of the alluvium from the period of the retreat of the last waters that have covered the earth, and includes under it

1. Drift sand, marine, or inland.
2. Marsh land, composed of mud deposited by rivers.
3. Peat.
4. Calcareous tufa.

To these may be added **beds of** gravel, produced locally by **torrents** and rapid rivers. All these formations are referable to causes that are still in daily action, and they may, by careful investigation, be always distinguished from the gravel which is strictly diluvian, as the latter may be distinguished from those more completely rolled pebble beds of antediluvian origin that occur among the regular strata which compose the crust of our globe.

A'LMANDINE. A precious **stone,** having some of the characters of the garnet. Almandine is much **valued** as a precious stone, its principal colour is red of various shades, having sometimes a tinge of yellow or blue, or a smoky aspect. Sp. gr. 4·3, and it is fusible into a black enamel. The most beautiful are brought from Sirian, the capital of Pegu.

A'LPINE. (*alpinus*, Lat.) This term is not confined merely to the Alps, and the things therewith connected, but is applied to any lofty or mountainous country, and to the productions of elevated situations.

ALTE'RNATE. (*alternus*, Lat.) Being by turns; one after another; reciprocal. In botany, applied **to** leaves when they stand singly **on** the stem or branches, alternately first on one side, then on the other; to branches when placed round the stem alternately, one above the other; to flowers placed in regular succession, one above another.

ALTI'METER. (from *altus*, Lat. and μέτρον, Gr.) An instrument by which the heights of bodies **may be** ascertained.

ALTI'METRY. (*altimétrie*, Fr. *altimetria*, It.) The art of measuring altitudes or heights, whether accessible or otherwise.

A′LUM. (*alumen*, Lat. *alun*, Fr. *allume*, It.) A triple sulphate of alumina and potassa. Alum is both native and factitious. The common mode of obtaining alum is by roasting and lixiviating certain clays containing pyrites; to the leys a certain quantity of potassa is added, and the triple salt is obtained by crystallization. Alum has a sweetish astringent taste. It dissolves in five parts of water at a temperature of 60°, and the solution reddens blues.

A′LUM-STONE. The shale from which alum is extracted. An enormous bed of alum-shale exists near Whitby, in Yorkshire; from this bed, alum has been manufactured ever since the days of Queen Elizabeth. The depth to which this stratum reaches, says Mr. Winch, has never been ascertained, but, to give an idea of its thickness, I need only mention that the cliffs at Bowdly are 600 feet high, 400 of the lower part being an entire mass of shale; and to what depth it may extend below the level of the sea, remains to be proved.

ALU′MINA. } This substance obtained
A′LUMINE. } the name of *alumina*
A′RGIL. } from its forming the base of common alum, and that of *argil* from the Lat. *argilla*, clay, on account of its being a constituent of clays; clays are termed argillaceous substances. It is found in the greatest purity in corundum and its varieties; it is a sesquioxide of aluminium. Pure argillaceous earth, or alumina, is a substance which in a mixed state is well known, but pure and unmixed, is one of the rarest substances in the mineral kingdom. This earth is soft, smooth, and unctuous, to the touch. Combined with other earths, or rocks, it communicates to them some of these properties; such rocks are termed argillaceous. Alumina constitues some of the hardest gems, such as the ruby and sapphire, the latter being crystallized alumine. According to the analysis of Klaproth, the sapphire contains 95 per cent. of pure clay. Alumina was considered an elementary substance till Sir Humphry Davy's electrochemical researches led to the opinion of its being a metallic oxide. Next to silicium, aluminum would appear to be the most important base of the earths on the face of the globe. Its collective amount is by no means so great as that of silicium, but it is quite as widely spread. There is scarcely one among the mechanical rocks that does not contain alumina. It constitutes the base of the various clays, and must be regarded as a very abundant and important constituent part of rocks. It contains 46·8 per cent. of oxygen.—*De la Beche.*

ALU′MINITE. Sub-sulphate of alumine. A white mineral, dull, opaque, and having an earthy fracture. This mineral occurs massive, in veins, and in tabular and tuberose masses; the former frequently attaining a length of several feet, and the latter a weight equal to three or four pounds. It appears to have been of stalactitical origin, and is supposed to result from the decomposition of iron pyrites, and the reaction of other substances. It is infusible at 166° of Wedgewood, but fuses rapidly when exposed to the stream of the hydro-oxygen blow-pipe. According to the analysis of Stromeyer, it consists of alumine 30, sulphuric acid 25, water 45.

A′LUMINOUS. Having the properties of alum; containing alum; resembling alum.

ALU′MINUM. The metallic base of alumina. The metal itself has not yet been obtained in a separate state, but the analyses to which alumina has been subjected have clearly shewn that it is a metallic oxide.

ALU'MOCALCITE. An earthy mineral consisting of silica 86, alumina 3, lime 7, and water 4. Spec. gravity 2·174. It is of a milk-white color, inclining to blue. Fracture conchoidal. It adheres to the lip when moistened. It is met with in the clefts of iron-stone veins.

ALVE'OLAR. (*alveolus*, Lat.) Containing sockets, pits, hollows, or cavities.

ALVE'OLATE. Pitted or honey-combed.

ALVEOLI'NA. A genus of microscopic foraminiferous shells.

ALVE'OLITES. (The name given by Lamarck to a genus of corals. One species, found in the Upper Ludlow Rock and Aymestry Limestone, has been named A fibrosa, by Mr. Lonsdale.) A lapidaceous polypifer, either incrusting or in a free mass, composed of many concentric tables, involving each other. The tables are formed of tubulous, alveolar, prismatic, short, contiguous, and parallel cellules, connected externally in a net-work.—*Lamarck.*

ALVE'OLUS. (*alveolus*, Lat. *alvéole*, Fr. *alveolo*, It.) A socket for a tooth; a small cavity or cell; the cell of the honey-comb.

AMA'LGAM. (from ἅμα, together, and γαμέω, to marry.) A compound of any metal with mercury. When two or more metals, neither being mercury, are mixed together, the compound is termed an alloy, but when mercury enters into the composition it is called an amalgam, and its derivation has been supposed to be from μάλαγμα, or μαλάσσω, to soften, which derivation appears to be more correct than that of Johnson, and lexicographers generally.

AMA'LGAMATE. (*amalgamer*, Fr. *amalgamare*, It.)
1. To mix mercury with any other metal.
2. To mix any two substances capable of uniting into one body.

AMALGAMA'TION. (*amalgamation*, Fr. *amalgamazione*, It.)
1. The act of mixing mercury with other metals.
2. The act of blending different bodies.

AMA'LTHUS. A species of ammonite, established by Montford.

A'MAZON-STONE. A variety of prismatic felspar, of a blue or green colour.

A'MBER. (*ambar*, Arab.) A fossil resin. For a great length of time, various were the opinions as to the nature and composition of amber, but it is now well ascertained to be a fossilized vegetable resin. It is found in similar localities with coal and jet. It is brittle, easily cut with a knife, of various shades of yellow, sometimes nearly white, and semi-transparent: insects are frequently found enclosed in it, and Jussieu states that these are not European. M. de France mentions a piece of amber, about the thickness of one's thumb, in which twenty-eight insects were distinctly to be seen, such as ants, tipulæ, small coleoptera, and a curculio. Its constituent parts are carbon 70·68, hydrogen 11·62, oxygen 7.77. Amber is found in nodular masses, which are sometimes eighteen inches in circumference; that which is found on the eastern shores of England, and on the coasts of Prussia and Sicily, is derived from beds of lignite in tertiary strata. Fragments of fossil gum were found near London, in digging the tunnel through the London clay at Highgate. In the royal cabinet of Berlin there is a lump of amber, discovered in Lithuania, weighing eighteen pounds. Amber is one of the most electric substances known; when submitted to distillation, it yields an acid sublimate, which has received the name of succinic acid. Ten pounds of amber yield about three ounces of purified succinic acid.

AMBE'RGRIS. (from *amber* and *gris*, or grey.) A concretion from the intestines of the physeter macrocephalus, or spermaceti whale. It was long doubted of what ambergris consisted; **and Todd,** in his last edition of Johnson's Dictionary, retains, without any comment or observation, the absurd opinions of former days, stating that "some imagine it to be the excrement of **a** bird, which, being melted by the **heat of the** sun, and washed off the shore by the waves, is swallowed by whales, who return it back in the condition we find it." Neumann absolutely denies it to be an animal substance, as not yielding in the analysis any one animal principle. He concludes it to be a bitumen issuing out of the earth into the sea; at first of a viscous consistence, but hardening, by its mixture with some liquid naphtha, into the form in which we find it. It is stated by Sir E. Home that this substance is only found in the unhealthy animal, but whether the cause or the effect of disease is not well ascertained. When the pieces of ambergris are large, they are found to contain beaks of the sepia octopedia, **or cuttlefish,** the usual food **of the spermaceti** whale. Ambergris **is a solid, opaque, ash-coloured, inflammable substance,** variegated **like marble, remarkably** light, its **specific gravity ranging** from 780 to 926; rugged, and, when heated, emitting **a fragrant odour.** It is sometimes **found in** masses **of** two hundred pounds weight and upwards. It breaks easily, but cannot be reduced **to** powder; melts like wax, and is soluble in ether and the volatile oils, **and,** assisted by heat, in alcohol, ammonia, and the fixed oils. It has been employed in medicine, but is **now** quite laid aside. In consequence of its fragrance, it enters into the composition of many articles **of perfumery.**

A'MBIT. (*ambitus,* Lat.) The compass or circuit of anything; the line that encompasses anything.

2. In conchology, the circumference or outline of the valves.

AMBLY'GONITE. An earthy mineral, so named from the Greek, in allusion to the obtuse angles of its prism. It occurs in rhombic prisms of 106° **10'** and 73° 50', rough externally, and of a greenish white, or sea-green colour.

AMBLY'PTERUS. A genus of fishes whose duration was limited to the early periods of geological formations; and which are marked **by** characters that cease after the deposition of the magnesian limestone. This genus occurs only in strata of the carboniferous order, and presents four species at Saarbrück, in Lorraine; **it is** found also in Brazil. The character of the teeth in Amblypterus shews the habit of this genus **to** have been to feed on decayed sea-weed, and soft animal substances at the bottom of the water; they are all small and numerous, and set close together like a brush. The form of the body, being not calculated for rapid progression, accords with this habit. **The vertebral column continues** into the upper lobe of the tail, which **is** much longer than the lower lobe, and is thus adapted to sustain the body in an inclined position, with the head and mouth nearest to the bottom. This remarkable elongation of the superior lobe of the tail is found in every bony fish of strata anterior to, and including, the magnesian limestone —*Buckland.*

AMBLYRRHY'NCHUS. (from $ἀμβλύς$, blunt, and $ῥυγχος$, rhynchus, snout.) The amblyrrhynchus constitutes a genus of lizards, established by

Bell, and so named from their obtusely truncated head and short snout. "The amblyrrhynchus cristatus, (says Mr. Darwin) is extremely common on all the islands throughout the archipelago. It lives exclusively on the rocky sea-beaches: its usual length is about a yard, but some attain to four feet long. It is of a dirty black colour, sluggish in its movements on the land; but when in the water, it swims with perfect ease and quickness, by a serpentine movement of its body and flattened tail, the legs at this time being motionless and closely collapsed on its sides." "On a comparison of this animal with the true iguanas, (says Mr. Bell) the most striking and important discrepancy is in the form of the head. Instead of the long, pointed, narrow muzzle of those species, we have a short, obtusely truncated head, not so long as it is broad, the mouth, consequently, only capable of being opened to a very short space. These circumstances, with the shortness and equality of the toes, and the strong curvature of the claws, evidently indicate some striking peculiarity in its food and general habits." Mr. Darwin says, "I opened the stomachs of several, and in each case found it largely distended with minced sea-weed, of that kind which grows in thin foliaceous expansions of a bright green or dull red colour. The stomach contained nothing but the sea-weed. The intestines were large, as in other herbivorous animals."

The only existing marine lizard now known.—*Lyell.*

AMBLYRHY'NCHUS SUBCRISTATUS. A species of lizard, thus named by Gray. This is a terrestrial species, confined, says Mr. Davison, to the central islands of the Galapagos Archipelago. "These lizards," says that naturalist, "like their brothers the sea-kind, are ugly animals; and from their low facial angle, have a singularly stupid appearance. The colour of their belly, front legs, and head, (excepting the crown, which is nearly white), is a dirty yellowish orange; the back is a brownish red, which in the younger specimens is darker. In their movements they are lazy and half torpid. When not frightened, they slowly crawl along, with their tails and bellies dragging on the ground.

They inhabit burrows. The individuals which inhabit the lower country, can scarcely taste a drop of water throughout the year; but they consume much of the succulent cactus, the branches of which are occasionally broken off by the wind. They eat very deliberately, but do not chew their food. I opened the stomachs of several, and found them full of vegetable fibres, and leaves of different trees, especially of a species of acacia.

The meat of these animals when cooked is white, and by those whose stomachs rise above all prejudices, it is relished as very good food."

A'MENT. } (*amentum*, Lat.) A catkin, one kind of inflorescence. When the bracteæ on the principal stalk are close, and overlap one another, or are imbricated with the flowers sessile in their axillæ, the spike is termed an *amentum, or catkin*, and the peduncle is always articulated with the main stem of the plant. Aments, or catkins, are generally pendent, while spikes are for the most part erect.
A'METHYST. (ἀμέθυστος, Gr. contrary to wine, or drunkenness, so called, from a supposed virtue it possessed of preventing inebriation.) Called also, Violet Quartz. The Gemeiner Amethyst of Werner; Quartz hyalin violet of Haüy; Quartz hyalin Améthyste of Brouginart. Quartz, coloured by a minute portion of iron and manganese. The finest

specimens come from India, Spain, and Siberia, but the amethyst is commonly found in most countries. The amethyst is a transparent gem of a purple or violet-blue colour; it is sometimes found naturally colourless, and may at any time be made so by putting it into the fire. When deprived of its colour, it greatly resembles the diamond. Some derive the name amethyst from its colour, which resembles wine mixed with water; whilst others, with more probability, think it obtained its name from its supposed virtue of preventing drunkenness; an opinion which, however imaginary, prevailed to that degree among the ancients, that it was usual for great drinkers to wear it about their necks. It occurs massive, in rolled pieces, in angular pieces, and crystallized. In the massive specimens several colours occur together. Its characters generally are those of common quartz. According to Rose, it contains silex 97·50, alumine 0·25, oxide of iron 0·50, oxide of manganese 0·25. The oriental amethyst is a sapphire; the green variety is the chrysolite of some authors.

AME'THYSTINE. Possessing the properties of an amethyst; of the colour of an amethyst.

A'MIANTH. } (*amiante*, Fr. *amianto*,
AMIA'NTHUS. } It.) A variety of asbestos, or flexible asbestus; an incombustible mineral composed of very delicate and minute fibres, which were sometimes, according to Dioscorides, worked into a cloth capable of resisting the action of fire. It is unctuous to the touch; has a shining or silky lustre; and is slightly translucent. Although in mass it fuses with difficulty, when in single fibres it melts in the flame of a lamp.

AMIA'NTHIFORM. Having the form or likeness of Amianthus.

AMIANTHOID. A variety of asbestiform actinolite, so named by Haüy. See Asbestiform actinolite.

AMMONA'CEA. According to the arrangement of De Blainville, a family of the order Polythalamacea. It embraces the genera Discorbis, Scaphites, Ammonites, and Simplegas. In the Lamarckian system the ammonacea is a family of the order Polythalamous cephalopoda, embracing the genera Ammonites, Ammonoceras, Baculites, and Turrilites.

AMMO'NIA, or Volatile alkali, when pure, is in a gaseous form. It consists of hydrogen and nitrogen, in the proportions of 1·76 of hydrogen, and 98·24 of nitrogen.

AMMONELLIPSI'TES. A genus of fossil, multilocular, flatly discoidal, and elliptically spiral shells; the turns contiguous and apparent on both sides; the chambers separated by winding septa; the siphuncle marginal. *Parkinson.*

A'MMONITE. (from *Jupiter Ammon.*) An extinct and very numerous genus of the order of molluscous animals called Cephalopoda, allied to the modern genus Nautilus, which inhabited a chambered shell, curved like a coiled snake. Species of it are found in all geological periods of the secondary strata; but they have not been seen in the tertiary beds. They are so named from their resemblance to the horns on the statues of Jupiter Ammon.—*Lyell.*

One hundred and seventy-three species are mentioned as having been discovered in the oolitic group, and upwards of three hundred have been described. The ammonites are a genus of shells of the class of univalves, the characters of which are a discoid spiral, with contiguous turbinations all apparent, and the internal parieties articulated by sinuous sutures; they have also transverse partitions, lobated through their centre, and pierced by a marginal tube.

The ammonite differs greatly from the chambered nautilus, the whorls, or turns, being all distinct, and in the same plane, and the cells very small. The family of ammonites extends through the entire series of the fossiliferous formations, from the transition strata to the chalk inclusive. M. Brochant, in his translation of De la Beche's Manual of Geology, enumerates 270 species; these species differ according to the age of the strata in which they are found, and vary in size from a line to more than four feet in diameter. The geographical distribution of ammonites in the ancient world, seems to have partaken of that universality we find so common in the animals and vegetables of a former condition of our globe, and which differs so remarkably from the varied distribution that prevails among existing forms of organic life. We find the same genera, and, in a few cases, the same species of ammonites, in strata apparently of the same age, not only throughout Europe, but also in distant regions of Asia, and of North and South America. Dr. Gerard has found at the elevation of 16,000 feet in the Himalaya Mountains, species of ammonites, identical with those of the lias at Whitby and Lyme Regis. The ammonite, like the nautilus, is composed of three essential parts: —1st. An external shell, usually of a flat discoidal form, and having its surface strengthened and ornamented with ribs. 2nd. A series of internal air chambers, formed by transverse plates, intersecting the inner portion of the shell. 3rd. A siphuncle, or pipe, commencing at the bottom of the outer chamber, and thence passing through the entire series of air chambers to the innermost extremity of the shell. The most decided distinction between ammonites and nautili is founded on the situation of the siphon. In the ammonite, this organ is always on the back of the shell, but never so in the nautilus. —*Buckland*.

The opinions of geologists and conchologists have greatly varied as to the situation and use of the shell of the ammonite; Cuvier, Lamarck, Bakewell, and others, have supposed that the shell was an internal one; but the reasoning of Buckland on this subject seems conclusively and indisputably to prove that the shell was external.

AMMONITI′FEROUS. Containing the remains of ammonites.

AMMONO′CERAS. } (from *ammon*, AMMONOCE′RATITES. } and κέρας, Gr.) The shells of this genus resemble ammonites in their internal structure, but that they are only curved instead of being spirally convolute.

A′MPELITE. (from ἄμπελος, Gr. a vine.) A kind of aluminous slate, belonging to both the fossiliferous and metamorphic series of rocks.

AMPHI′BIA. (from ἀμφί and βίος, Gr.) A class of animals possessing the property of living either in the water or on dry land; undergoing a metamorphosis whereby the gills become obliterated and the lungs developed, while the heart, from being bilocular, or possessing two cavities only, obtains three cavities, namely, two auricles and one ventricle.

The fourth class of the subkingdom Vertebrata, kingdom Animalia. In this class are comprised four orders, namely, Labyrinthodonta, Batrachia, Saurobatrachia, and Ophiomorpha. The lungs of the amphibia differ greatly from those of animals of the classes aves and mammalia. Their body is covered with a shell, or with scales, or is quite naked. They have neither hair, mammæ, feathers, nor radiated fins: they are oviparous or viviparous, and are divided into reptiles and serpents; or reptilia pedata, and serpentes apodes, the former being furnished with teeth, and the latter being destitute of them. The amphibia possess the extraordinary property of reproducing parts, such as their legs, tails, &c., if destroyed.

AMPHIBIOLI'THUS. (from ἀμφίβιος and λίθος, Gr.) Fossil amphibiæ. The amphibiolithi form a very large and important class of fossils.

AMPHI'BIOUS. (ἀμφίβιος, Gr. *amphibie*, Fr. *anfibo*, It.) That partakes of two natures, being able to live either in the air or in the water.

A'MPHIBOLE. (ἀμφίβιος, **Gr.**) **The** name given by Haüy, **and the** French, to hornblende: **for particulars** see *Hornblende*.

AMPHI'BOLITE. Any rock whose basis is amphibole or hornblende.

AMPHIDE'SMA. (from *ampho*, both and *desmos*, ligament.) A genus of bivalve shells belonging to the family mactucea; it is equivalve, oval or rounded, nearly equilateral; hinge with one or two cardinal teeth in each valve, and two elongated lateral teeth distinct in one valve, nearly obselete in the other; ligament separated from the cartilage, which is **elongated** and placed **obliquely in an excavation** of the hinge. **The separation of** the cartilage and **ligament have** given origin **to the name, which** signifies *double ligament*, an arrangement of parts which distinguishes it from Tellina. The recent species are few, but upwards of twelve are enumerated **fossil in** the British Isles, viz., four in the crag, one in the greensand, one in the coral rag, one in the cornbrash, one in the inferior oolite, and four in the carboniferous rocks; but we may be allowed **to** express some doubt as to the correctness **of all these** identifications.—*Lycett*.

AMPHIPNE'USTA. The second order of the class Arachnida, comprising the spiders.

A'MPHIGENE. (from ἀμφί and γένος, Gr.) Trapezoidal zeolite, or leucite. This mineral, also called Vesuvian, occurs in embedded grains or crystals, in the more ancient lavas, and is found mixed with garnet, hornblende, quartz, &c. in the ejected masses of old volcanoes.

A'MPHITHERE. } The name assigned by Professor
AMPHITHE'RIUM. } Owen to a fossil genus of insectivora found in the Stonesfield slate. This genus differs from Didelphys, both in the number and size of its teeth, being smaller in size, and in the lower jaw thirty two in number.

A'MPHISBÆNA. (from ἀμφίς and βαίνω, Gr. to walk both ways.) The **name** given to a genus **of serpents, natives** of South America.

A'MPHITRITE. A genus of **Tubicola,** of the division Articulata.

AMPHO'DELITE. An earthy mineral described by Nordenskiold and occurring in the limestone of Finland. Its form is crystalline, resembling that of felspar. Its colour is a light red; fracture uneven and splintery and possessing two cleavages which meet at an angle of 94° 19'. Specific gravity 2·76. Hardness 4·4.

AMMPLEXICAU'LENT. (from *amplexus* and *caulis*, Lat.) Stem-clasping; embracing the stem.

AMPLE'XUS. A singularly-formed fossil, resembling a coral or madrepore, found in the Dublin limestone: it is described as being nearly cylindrical, divided into chambers by numerous transverse septa, which embrace each other with reflected margins.—*Sowerby*.

AMPULLA'RIA. (from *ampulla*, Lat.) A ventricose, subglobose, univalve, with an umbilicated base; the opening oblong and entire, with no thickening on the left lip. The ampullaria is a river shell of warm climates. **Its spire,** which always slightly projects, distinguishes this genus from Planorbis; and there being no thickening on the left lip marks it from Natica.—*Parkinson.* Lamarck places the genus in the family Peristomata, order Trachelipoda. There are many species, as the Ampullaria patula, Ampullaria Sigaretina, &c.

AMY'GDALOID. (αμυγδάλη, an almond, and εἶδος, Gr. *amygdaloïde*, Fr.) A volcanic, or igneous, rock of any composition, containing nodules of minerals, scattered through its base, of a roundish shape; cellular volcanic rock, having its cells, or cavities, occupied with nodules of a dissimilar substance.

AMYGDALO'IDAL. Containing rounded, or kernel-shaped, cavities, filled with mineral matter of a different character from the substance generally.

AMY'GDALYTE. Almond-stone. Of the nature of, or resembling almonds.

A'NAL. Pertaining to the anus; the fin between the vent and the tail.

ANA'LCIME. A simple mineral, a variety of zeolite, with which it was formerly confounded; it is also called cubizite. It occurs regularly crystallized in angulo-granular concretions, and massive. Specific gravity above 2. When rubbed, it acquires only a small degree of electricity, and with difficulty. It is composed of silica 55·07, alumina 20·22, soda 14·71, moisture 8·28. It is found, in secondary greenstone rocks, in various parts of Scotland, more especially near Edinburgh. This mineral, also called Cubizite, has been regarded by mineralogists as having the cube for its primitive form. Analcime has certainly no cleavage planes, and it must be regarded at present as forming in this respect as great an anomaly in crystallography as it does in optics by its extraordinary optical phenomena. The most common form of analcime is the solid, called the icositetrahedron, which is bounded by twenty-four equal and similar trapezia; and we may regard it as derived from the cube, by cutting off each of its angles by three planes equally inclined to the three faces which contain the solid angle. The Abbé Haüy first observed in this mineral its property of yielding no electricity by friction, and derived the name of analcime from its want of this property.

A'NALOGUE. (*analogue*, Fr. *On le fait quelquefois substantif. Ce sont deux analogues.*) Any body which corresponds with, or bears great resemblance to, some other body. A recent shell of the same species with a fossil shell, is an Analogue of the latter.—*Lyell.*

ANA'MESITE. A mineral, a fine-grained dolerite. Its colour is dark grey, or greenish or brownish black. It forms the intermediate step between dolerite and basalt.

ANANCHY'TES. A helmet-shaped echinus, a fossil of the chalk formation.—*Bakewell.* It approaches near to the form of Spantangus globosus. Of the genus Ananchytes, eight species have been determined as occurring in the chalk deposit: one species, ananchytes bicordatus, has been found in the Oxford clay, a member of the oolitic group.

ANASTOMO'SIS. (*anastomose*, Fr. *anastomoso*, It. from ἀνὰ and στόμα, Gr.) The running of vessels one into another, or communication by inosculation, as of the arteries into the veins.

ANASTOMO'SING. Communicating by anastomosis. Applied to vessels, threads, or fibres, which by meeting or touching in separate points only, form a sort of net-work, or reticulation.

A'NATASE. (ἀνάτασις, Gr. extension.) Pyramidal titanium; this mineral is nearly of the same nature as

Titanite. It is found in Dauphiny, Bavaria, Norway, Switzerland, Spain, and Brazil. It is a pure octahedral oxide of titanium. Its colours are brown and blue; structure lamellar; lustre splendent and adamantine; it scratches glass. Specific gravity 3·80. The Abbé Haüy states the primitive crystal of anatase to be an acute octohedron; the common base of the pyramids is square: the reflecting goniometer gives the measurement 136°47'.

ANA'TIFA. A cuneiform multivalve, composed of several unequal valves, five or more, united together at the extremity of a cartilaginous tube, fixed at its base. The opening without an operculum.—*Parkinson*. The genus comprises several species. It belongs to the class Cirrhopoda. The anatifæ are often found adhering to rocks, pieces of wood, the bottoms of ships, &c.

ANA'TIFER. (from *anas*, a duck, and *fero*, to bear, Lat.) A name given to the barnacle, or pentelasmis. The same as Anatifa.

ANATI'NA. A genus of bivalve shells, belonging to the family Myaria; it is thin, transparent, inequilateral, transverse; hinge with a spoon-shaped process containing the cartilage, and a small shelly moveable appendage.—*Sowerby*. A single species is recorded fossil in the Oxford clay and **great oolite**.—*Lycett*.

ANCI'LLA. } An oblong subcylin-
ANCILLA'RIA. } drical univale, **with** a short spire, not channelled: the aperture effused, and its base slightly notched.—*Parkinson*. The ebuana glabrata, or ivory shell, belongs to the genus Ancilla. It is found both fossil and recent.

ANDALU'SITE. A massive mineral of a red or grey colour; it occurs also crystalized. Lustre shining, glistening, **and** vitreous. Fracture uneven; is easily broken. Feebly translucent. Specific gravity 3·160.

Constituent parts, alumina 52, silica 32, potash 8, oxide of iron 2. It was first found in Spain; it occurs in gneiss in England, Ireland, and Scotland.

A'NDESIN. (from *Andes*, the mineral having been found in that mountain range.) A variety of feldspar.

A'NDESITE. When feldspar is of the variety called Andesin, and the rock contains hornblende, some mica, and a little quartz, it is called *Andesite*.—*Jukes*.

ANDRE'OLITE. Thus named from its having been first found at Andreasberg, in the Hartz; called also Harmotome, and, sometimes, from form of its crystals, cross-stones. Its crystals are two four-sided flattened prisms, terminated **by** four-sided pyramids, intersecting each other at right angles; **the** plane of intersection passing **lon**gitudinally through the prisms. Texture foliated. Colour milk-white. Constituent parts, silica 44, alumina 20, barytes 20, water 16. It effervesces with borax and microcosmic salt, and is reduced to a greenish opaque mass. **With** soda it melts into a frothy **white** enamel. When its powder is thrown on a hot coal, it emits a greenish yellow light.—*Thomson*.

ANDRO'GYNOUS. } (from ἀνήρ and γύνη,
ANDRO'GYNAL. } Gr.) Having two sexes; being both male and female; hermaphroditical. Plants bearing male and female flowers on the same root are thus called.

ANEMO'METER. (from ἄνεμος and μέτρον, Gr. *anémomètre*, Fr. *anemometro*, It.) **An instrument for measuring the strength or velocity** of the wind.

ANEN'CHELUM. A genus of fossil fish, found in the locality of Glaris, and described by Blainville.

ANEN'TERA. The name assigned to a group of monads from their having no intestinal canal. In this the simplest form of animalcules there is but one general orifice to the alimentary cavities, the several stomachs opening into the buccal-orifice.

ANEN'TEROUS.
An anenterous monad with a single cavity presents the simplest form of the digestive apparatus known among animals.

ANGIOSPE'RMIA. (from ἀγγεῖον, a receptacle, and σπέρμα, seed, Gr.) In the artificial system of Linnæus, an order of plants of the class Didynamia. In a more recent arrangement, the order Apetalæ is sub-divided into two sub-orders, the Gymnosperms and the Angiosperms; of the Angiosperms, the nettle, spurge, oak, elm, &c., are examples. In the first order, or those having naked seeds, the plants are mostly wholesome and aromatic. In the second, where the seeds are enclosed in a seed-vessel, we find the Digitalis, and other poisonous plants.

ANGIOSPE'RMOUS. (angiosperme, Fr.) Belonging to the order Angiospermia; having the seeds enclosed in a seed-vessel.

ANGIO'STOMA. } A family of univalve
ANGYO'STOMA. } shells, in the order of Siphonobranchiata. It includes many genera, as the Conus, Cypræa, Terebellum, &c.

ANGUI'LLIFORM. (from anguilla, an eel, and forma, Lat.) A term given to fishes having the form of an eel.

ANGUL'ITHES. A genus of multilocular shells, possessing many of the characters of nautilus, the mouth being of a triangular form.

ANGU'STATE. (angustatus, Lat.) Beginning with a narrow base, which base then dilates and thickens.

ANHY'DRITE. (A substance so named in consequence of no water entering into its composition.) The Count de Bournon has proposed to substitute the term Bardiglione for Anhydrite. It is a combination of lime and sulphuric acid, in the proportion, according to Vauqeulin, of 0·40 lime, and 0.60 sulphuric acid. It has obtained various names, as Chaux Sulfateé Anhydre, by Haüy; Chaux Sulfatine by Brongniart; Anhydrite by Werner; Muriacite by Poda and Klaproth; Pierre de Vulpino by Fleurian, and Marno Bardiglio di Berganio by the Italian statuaries. Anhydrous gypsum. A variety of sulphate of lime, called anhydrous gypsum, or anhydrite, in consequence of its being quite free from water. It is harder than selenite, and sometimes contains chloride of sodium, when it is called muriacite. Its colours are white, blue, red, and grey. It occurs both massive and crystallised. Lustre alternates from splendent to glistening, and is pearly. Fracture splintery and conchoidal. Specific gravity 2·85. There are six varieties of this mineral.

ANHY'DROUS. (from a priv. and ὕδωρ, water.) Without water in its composition; containing no water.

ANIMA'LCULE. (animalculum, Lat. animalcule, Fr. animaletto, It.) An exceedingly small animal, scarcely discoverable by unaided vision, but which, by the help of the microscope, is found both in solids and fluids. The simplest gelatinous animalcules, which possess no internal cavity, are reduced to super-absorption, and thus form a transition to the mode of nourishment of the vegetable kingdom.—*Professor Grant.*

A'NKERITE. Paratomous limestone, a species of limestone thus named after Prof. Anker. It is found in the mines of Styria.

A'NIMAL KINGDOM. The animal kingdom comprehends beings the most diversified as to form, structure, and the media in which they live. The great naturalist Cuvier arranged the animal kingdom under four great divisions, or sub-kingdoms, but recent authorities have deemed it necessary to divide it into five, splitting the Radiata into two, and re-arranging some of its constituents. The five sub-kingdoms are:—1. Vertebrata. 2. Annulosa. 3. Mollusca. 4. Cœlenterata. 5. Protozoa. 1. Vertebrata, divided into five classes; namely, Mammalia, Aves, Reptilia, Amphibia, and Pisces. 2. Annulosa, divided into **seven classes, Insecta,** Myriapoda, **Arachnida, Crustacea,** Annulata, **Scolecida, and** Echinodermata. 3. **Mollusca,** consisting **of** three classes, Cephalophora, Conchifera, and Molluscoidea. 4. Cœlenterata, comprising two classes, Actinozoa, and Hydrozoa. 5. Protozoa, also containing two classes, **Stomatoda,** and Astomata.

ANNE'LIDANS. ⎫ (from *annellus*, a small
ANNELI'DES. ⎭ ring, Lat.) Worms with, in general, red blood, whose bodies are composed of rings. Annelida, in the classification of some authors, constitutes the fifth class of the animal kingdom, and comprises three orders, namely, Tubicola, Dorsibranchiata, and Abranchiata. Professor Buckland observes, "We have abundant evidence of the early and continued prevalence of that order of Annelidans, which formed shelly, **calcareous** tubes, in the occurrence **of fossil** serpulæ, **in** nearly **all formations,** from **the** transition **periods to** the present time." **The shores of** the sea, the moist sands of coasts, as well as the soils of all countries, **are** inhabited by myriads of worms, which **are** found **to** contain a red-coloured fluid, circulating in veins **and** arteries. These constitute the red bloody worms of naturalists, **the** "vers à sang rouge" of Cuvier. The term *annelida* is most frequently applied to them, from their being surrounded by rings, extending from the anterior to the posterior part of the body.—*Professor Grant.*

"Annelidans," says Mr. MacLeay, "differ from true annulosa in being hermaphrodite, and in general red-blooded." All the annelidans are not red-blooded. Mr. MacLeay divides annelidans into two groups, the first, or normal group, consisting of marine animals, having their body provided with distinct feet, contains Nereidina and Serpulina; the second or aberrant group, Apoda, the body being without feet or **a** distinct head, comprises Lumbricina, Nemertina, and Hirudina. If the conclusion of Philippi be correct, the shells of annelides can have no palæontological bearings, further than as affording indications of the presence of their order and class.

A'NNOLIS. An American animal, resembling a lizard.

ANNULA'RIA. A species of **phalæna,** of the geometra section.

ANNULA'TA. In the classification by Busk, Annulata forms the fifth class of the sub-kingdom Annulosa, and comprises five orders; namely, Polychœta, Oligochœta, Discophora, Tardigrada, and Sagittida.

ANNULOSA. A sub-kingdom of **animalia,** comprising seven classes, namely, Insecta, Myriapoda, Arachnida, Crustacea, Annulata, Scolecida, and Echinodermata.

A'NNULOSE. Furnished with rings; composed of rings. The annulose animals form two great series; those without jointed feet, viz., vermes, annulosa, cirripeda; and those with jointed feet, namely, insecta, myriapoda, arachnida, crustacea.

ANOCY'STI. The incongruous assemblage of fossil substances, termed *echinites,* have been arranged by Leske into two classes: the first class is that of the *anocysti,* the vent of which is in the vertex. This class is arranged under two divisions, Cidaris and Clypeus.

ANODO'NTA. A form of bivalvular mollusc, with a transverse shell, having three muscular impressions; the hinge plain, having no appearance of a tooth.

A′NOGENS. A class of plants comprising two orders, the Hepaticæ, or liver-worts, and the musci, or mosses.

ANO′MIA. A genus of molluscous bivalve. The anomiæ are inhabitants of every sea, and are found adhering to foreign bodies by means of an operculum, or valve. Recent and fossil.

ANO′MITE. A fossil shell of the genus Anomia.

ANOMORHOMBOI′DA.  
ANOMORHOM′BOID. } (from ἀνόμοιος, irregular, and ῥομβοειδής, of a rhomboidal figure.) A genus of pellucid, crystalline spars, of no determinate, regular, external form, but always fracturing into regular rhomboidal masses. Of this genus there are five known species, all possessing, in some degree, the double refraction of the island crystal.

ANOPLOTHE′RE.  
ANOPLOTHE′RIUM. } (from ἄνοπλος, unarmed, and θηρίον, a wild beast.) A fossil extinct quadruped, belonging to the order Pachydermata, resembling a pig. Five species of Anoplotherium have been found in the gypsum of the neighbourhood of Paris. The largest (Anaplotherium commune) being of the size of a dwarf ass, with a thick tail, equal in length to its body, and resembling that of an otter; its probable use was to assist the animal in swimming. The posterior molar teeth in the genus Anoplotherium resemble those of the rhinoceros; their feet are terminated by two large toes, like the ruminating animals, whilst the composition of their tarsus is like that of the camel. The place of this genus stands, in one respect, between the rhinoceros and the horse; and in another, between the hippopotamus, the hog, and the camel.—*Buckland.*

Cuvier has shown that the structure of the hind foot alone is sufficient to prove, that the Anoplotherium was of a species at present unknown. He divides the genus into three sub-genera, namely, the anoplotheria, properly so called, the xiphodons, and the dichobunes.

The anoplotherium possesses two characters distinguishing it from all other animals; feet with two toes, in which the bones of the metacarpus and metatarsus remain distinct, and are not soldered together as in the ruminantia; and teeth in a continued series, without any intervening gap.—*Griffiths.*

Professor Owen says "the anoplotherium appears to have been one of the earliest forms of hoofed quadrupeds introduced upon the surface of this earth, and that the ancient herbivore presents, in comparison with living species, no indications of an inferior or rudimental character in any known part of its organization, and that, with regard to its dentition, it not only possessed incisors and canines in both jaws, but that those teeth were so equally developed, that they formed one unbroken series with the premolars and true molars, which character is now manifested only in the human species."

ANOPLOTHE′RIAN. Relating to the anoplothere. The Xiphodon is a long and slender *anoplotherian* animal.—*Owen.*

ANOPLOTHE′RIOID. (from *anoplotherium* and εἶδος, Gr.) Resembling the anoplotherium. The animal belonged to that group of the Anoplotherioid family which includes the genera dichobune and xiphodon of Cuvier.—*Owen.*

ANO′RTHITE. The mineral to which this name is given is thus called from the absence of right angles in its fracture, which circumstance serves to distinguish it. It is a variety of felspar, and has been described by Rose. Its specific gravity is 76·3. Its constituent parts are silica 44·49, alumina 34·4, lime 15·68, magnesia 5·3, oxide of iron, under 1.

Anorthite is found at Monte Somma, the ancient crater of Vesuvius. Monticelli described it at the same time with Rose, under the name of Christianite, after Prince Christian of Denmark.

ANORTHI′TIC MELANE-ORE. A species of melane-ore, called also Allanite, which see.

ANTA′GONISM. Opposition of action.

ANT′AGONIST. (*antagonista*, Lat. *antagoniste*, Fr. *antagonista*, It.) A term applied to such muscles as oppose, or counteract, others.

ANTA′RCTIC. (from ἀντί, against, and ἄρκτος, the bear, or northern constellation.
1. The southern pole, so called as opposite to the northern.
2. One of the lesser circles, drawn on the globe, at the distance of twenty-three degrees and a half from the antarctic, or south pole.

ANTE′CIAN. (from ἄντοικος, Gr. living opposite.) Those who live under the same meridian east or west, but under opposite parallels of latitude north and south. The word is also written Antœcian.

ANTEDILU′VIAL. } (from *ante*, before,
ANTEDILU′VIAN. } and *diluvium*, a deluge.)
1. Existing before the deluge.
2. Relating to things existing before the deluge.

ANTEDILU′VIAN. One that lived before the deluge.

ANTEMU′NDANE. (from *ante*, before, and *mundus*, the world.) That existed before the creation of the world.

ANTE′NNÆ. (*antenna*, Lat. *antennes*, Fr. This word appears by all lexicographers to be given in the plural only.) Those delicate moveable horns with which the anterior part of the heads of insects are furnished. These are peculiar to this order of beings, and are easily distinguished from the tentaculæ of vermes, in being crustaceous; and from the palpi of insects, by their situation being nearer the mouth. The antennæ rarely exceed two in number, though in some insects of the apterous kind they amount to four, or even six. Of the uses of the antennæ we are still ignorant.

The antennæ are jointed organs, placed one on either side of the head between the angle of the mouth and the eyes; the variations in their structure are very great. Those which consist of equal joints are called equal; those whose joints are dissimilar are called unequal. The inequality of antennæ proceeds chiefly from the differing form of their second and last joint. Antennæ which consist of but one joint are called exarticulate; those with two joints, biarticulate; with three, triarticulate; while those whose joints are numerous are called multiarticulate. The great majority of antennæ are completely naked; others have a clothing consisting of shorter or longer hair.

ANTE′RIOR. In conchology, the anterior of bivalves is the side opposite to the hinge; of a spiral univalve, that part of the aperture most distant from its apex; of a symmetrical conical univalve, that part where the head of its inhabitant lies.

A′NTHER. (*anthera*, Lat. ἀνθηρά, Gr.) That part of the flower which contains the fertilizing dust, pollen, or farina, which, when mature, it scatters. The anther forms the essential part of the stamen. Anthers differ greatly as regards their figure, number, and situation. The most common form of the anther is that of a grain of corn, only smaller; it has a crease, or line, down it, as the grain has, at which it opens when bursting; this is generally turned inwards towards the axis of the flower; but in some plants, as the cucumber, iris, ranunculus, &c., it is turned outwards. The anther is generally fixed immoveably to the filament in various ways; but in most of the grasses, and many other plants, it is attached by its middle, and the filament being very thin, it is moved by the slightest air.

AN'THOLITE. ⎫ (from ἄνθος, a flower,
AN'THOLITHE. ⎭ and λίθος, a stone.)
The name applied to a fossillized flower. It has been doubted whether fossillized flowers are ever met with, on the ground that the succulent substance of the stamens and pistils must be too delicate to undergo the lapidifying or carbonizing process; but there exist impressions on shale and sandstone in the British Museum, on viewing which it is difficult to resist the conviction that they exhibit some kind of stellate blossoms.—*History of Fossil fuel.*

ANTHOPHY'LLITE. (from ἄνθος, a flower, and φύλλον, a leaf, Gr. Or, according to others, from the resemblance of its colour to that of the flower Anthophyllum. Its name was assigned to it by Schumacher, who first described it.) A mineral occurring both crystallized and massive, of a yellowish grey, or brownish colour. Its constituent parts are, silica 54·0; alumina 3·0; magnesia 23·0; lime 2·0; oxide of iron 13·0; oxide of manganese 4·0.—*Gmelin.* Specific gravity 3·2. Alone, infusible before the blowpipe; but with borax it yields a grass-green transparent head. It is the prismatic schiller-spar of Mohs. It is found in Invernessshire and in Norway.

A'NTHRACITE. (from ἄνθραξ, Gr., *anthrax*, Lat.) A shining substance like black-lead; a species of mineral charcoal; a mineral approaching to the state of plumbago; it consists nearly of pure carbon, is hard to ignite, and has frequently a semimetallic lustre. The coal in the extensive coal-formation in Pennsylvania is called anthracite, because it emits but little smoke in burning; but it is only a variety of common coal, containing but little bitumen, and is not the true anthracite of mineralogists. From the same circumstance, also, it has become a common thing to call the Welsh coal anthracite. In the vicinity of some trap dikes, coal is found converted into anthracite. Some anthracite contains 97 per cent of carbon. Hardness from 2·0 to 2·5. Specific gravity from 1·3 to 1·6.

AN'THRACITIC. Partaking of the nature of anthracite.

ANTHRA'COLITE. The same as anthracite.

ANTHRA'CONITE. A variety of calcareous spar, of a black colour, with a compact fracture, of a glimmering lustre, and which, on rubbing, yields a sulphureo-bituminous odour.

ANTHRACOTHE'RIUM. (from ἀνθράκιος, and θηρίον, wild beast, Gr.) A name given to an extinct mammifer, thus named by Cuvier, supposed to belong to the Pachydermata, the bones of which, changed into a kind of coal, have been found in the lignite and coal of the tertiary strata. This genus was first discovered in the lignite of Cadibona, in Liguria: seven species are known, some approximating to the size and appearance of the hog; others approaching that of the hippopotamus. The jaw teeth of this extinct genus exhibited considerable analogies with those of the chæropotamus and the dichobunes. But besides that these molars presented of themselves specific distinctions, the large and projecting canines with which they were accompanied left no doubt of the existence of a new and distinct genus.

Baron Cuvier considered this genus to have held an intermediate place between the palæotheria, anoplotheria, and swine.

The anthracotherium has hitherto been found fossil only, and in strata more recent than the chalk formation.

ANTHRO'POLITE. (from ἄνθρωπος, a man, and λίθος, a stone, Gr.) A petrifaction of the human body; a fossil human skeleton. Several skeletons of men, more or less mutilated, have been found in the West Indies; these still retain some of their animal matter, and all their phosphate of lime. One of them may be seen in the British Museum, and another in the Royal Cabinet at Paris.

ANTHROPOMO'RPHOUS. (ἀνθρωπόμορφος, from ἄνθρωπος, a man, and μορφή, form, Gr.) Having a form resembling the human: man-like.

A'NTICHRONISM. (from ἀντὶ, against, and χρόνος, time.) Deviation from the right order, or account, of time.

ANTICLI'NAL. If a range of hills, or a valley, be composed of strata, which on the two sides dip in opposite directions, the imaginary line that lies between them, towards which the strata on **each side rise,** is called the anticlinal axis. In a row of houses, with steep roofs facing the south, the slates represent inclined strata dipping north and south, and the ridge is an east and west anticlinal axis.—*Lyell.*

In most cases an anticlinal axis forms a ridge, and a synclinal axis a valley.

ANTIMO'NIAL. Made of antimony; having the properties of antimony.

ANTIMO'NIATE. } A salt formed by the
ANTI'MONITE. } combination of antimonic acid with a salifiable base.

A'NTIMONY. (*antimoine,* Fr. *antimonio,* It.) The derivation of this word is not agreed on, some lexicographers stating it to be from ἀντὶ and μόνος, two Greek words, signifying that it is never found alone: Dr. Johnson, however, on the **authority of** Furetiere, refers it to a ludicrous story related by Basil Valentine, a German, who appears to have been the discoverer of the metal in 1620. It is stated that he was a monk, and practised as a physician, and having thrown some of it to the hogs, he observed that after it had purged them, they immediately fattened; imagining that the effect on bipeds would be similar, he administered a like dose to his fellow monks. The experiment, however, proved rather an unfortunate one; for, in consequence of the dose being too large, they all died of it, and the substance thenceforth obtained the name of Antimoine, *i. e.* Antimonk. A metallic **ore,** consisting of **sulphur combined** with the **metal** which **is properly** called antimony. This **metal is of** a blueish-white colour, **and** considerable brilliancy, with a **specific** gravity of 6·712. It fuses **at a** temperature of 900, but requires **a** greatly increased heat to volatilize it. It is not malleable, being so brittle as to be easily reduced to powder by trituration, and its ductility is inconsiderable. The most abundant ore of antimony **is** that in which it is found combined with sulphur, and called sulphuret of antimony. Antimony combines with chlorine so rapidly as to produce a shower of fire, if it be poured, finely powdered, into a glass jar filled with that gas. It unites with many metals, some of the alloys being useful. That with lead is used for the plates on which music is engraved. With tin it forms a kind of pewter, and with lead and copper it forms printer's type metal. Native, or rhombohedral, antimony occurs in metalliferous veins in primitive rocks in Sweden, and in the mountains of Hanover, Dauphiny, Hungary, Brazil, and Mexico.

ANTI'PATHES. A genus of fixed, subdendroidal polypifers, composed of a central axis, and a corticiform, fugacious, and deciduous crust. The axis is flattened and fixed at its base; it is caulescent, subramose, horny, solid, flexible, rather fragile, and mostly set with small spines. According to Parkinson, seventeen species have been distinguished.

AN'TIPODE. (ἀντίποδες, Gr. *antipode*, Fr. *antipodi*, It.) Although this word is occasionally, and with propriety, used in the singular, yet it is more commonly used in the plural number; antipodes.

Those people who, from their situation on the globe, have their feet opposed directly to each other.

AN'TIQUATED. In conchology, longitudinally furrowed, but interrupted by transverse furrows, as if the shell had acquired new growth at each furrow.

A'NTRUM. (*antrum*, Lat. *antre*, Fr. *antro*, It.)
1. A cavern; a cave; a den.
2. The maxillary sinus, situate above the molar teeth of the upper jaw.

A'NUS. (*anus*, Lat. *anus*, Fr. *ano*, It.)
1. The termination downwards of the intestinal canal.
2. In conchology, a depression on the posterior side near the hinge of bivalves.

AO'RTA. (ἀορτή, Gr. **aorte**, Fr. **aorta**, It.) The principal artery of the body, which arises from the left ventricle of the heart.

A'PATITE. A genus of calcareous and brittle earths, composed of lime 55·75 and phosphoric acid 44·25. Apatites are white, green, blue, red, brown, and yellow; they occur both crystallized and massive. Fracture conchoidal and uneven; lustre resinous. Specific gravity 3·1. The crystals are six-sided prisms, low, and sometimes passing into the six-sided table. One set of varieties, in which the cleavage is very distinct, is named foliated apatite; another, in which the fracture is conchoidal, is called conchoidal apatite; and such varieties as display an uneven fracture have obtained the name of phosphorite. The crystallized variety is found, extremely beautiful, in Devon and Cornwall.

A'PENNINES. (*Apenninus*, Lat.) A chain of mountains extending through Italy. What now constitutes the central calcareous chain of the Apennines must for a long time have been a narrow ridgy peninsula, branching off, at its northern extremity, from the Alps near Savona. This peninsula was afterwards raised from one to two thousand feet, by which movement the ancient shores, and, for a certain extent, the bed of the contiguous sea, were laid dry, both on the side of the Mediterranean and the Adriatic.—*Lyell*.

APE'TALÆ. In botany, an order of plants belonging to the class Exogens. Apetalæ are divided into Gymnosperms and Angiosperms.

APE'TALOUS. (from *a*, priv. and πέταλον, a flower-leaf, or petal, Gr.) Without flower-leaves, or petals; not having petals.

APE'TALOUSNESS. The state of being without flower-leaves, or petals.

A'PEX. (Lat.) The tip or point of any thing; the highest point of a hill or mountain. This word makes *apices* in the plural, and not *apexes*.

A'PHANITE. (from *a*, priv. and φαίνω, Gr., luceo.) A mineral, a variety of amphibole.

A'PHIS. Plural, aphides. The puceron, or plant-louse. Class Insecta, order Hemiptera. The numerous tribes of this family of insects are most annoying to the florist, and most destructive to the plants. The best means of destroying them is either by fumigations of tobacco, or by watering the plants with a week solution of the chloride of lime. They are astonishingly prolific; they live in society on trees and plants, of which they suck the juices with their trunk.

A′PHRITE. (from ἀφρὸς, spuma, Gr.) A species of stone composed of carbonate of lime, and thus named from its frothy, silver-white appearance.

A′PHRIZITE. A variety of black tourmaline.

APHYLLA′NTES. (from α, priv. φύλλον, a leaf, and ἄνθος, a flower.) An apetalous flower; a genus of plants, class Hexandria, order Monogynia.

APHY′LLOUS. (from α, priv. and φύλλον, a leaf.) Without leaves; leafless.

APIOCRINI′TES ELLIPTICUS. The oval-columned pear-like encrinite. A species of Apoicrinites, thus named, from the elliptical shape of its columnar joints. It has hitherto been found fossil only, and in a mutilated state, in beds of the chalk formation. Miller thus describes its specific characters, "a crinoidal animal, having a column composed of oval joints articulating by a transversely-grooved surface; the two upper joints of the column enlarged, sustaining the pelvis, costæ, &c. The columns provided with auxiliary side-arms. The base formed by numerous irregular columnar joints sending off fibres for adhesion."

APIOCRINI′TES ROTUNDUS. The round-columned pear encrinite, so named from the remains of the animal possessing a pear-like form. This is a species of the genus Apiocrinites. Its specific characters are thus stated by Miller: "A crinoidal animal, with a round column composed of joints adhering by radiating surfaces, of which from ten to fourteen gradually enlarge at its apex, sustaining the pelvis, costæ, and scapulæ, from which the arms and tentaculated fingers proceed. Base formed by exuding calcareous matter, which indurates in laminæ, and permanently attaches the animal to extraneous bodies. The Apiocrinites rotundus, or round-columned Pear encrinite, has been plentifully found in the neighbourhood of Bradford, near Bath, and at Pfeffingen, in Germany. In reference to this species of encrinite, Professor Buckland thus writes: "When living, their roots were confluent, and formed a thin pavement at this place, over the bottom of the sea, from which their stems and branches rose into a thick submarine forest, composed of these beautiful zoophytes. The stems and bodies are occasionally found united, as in their living state; the arms and fingers have almost always been separated, but their dislocated fragments still remain, covering the pavement of roots that overspreads the surface of the adjacent oolitic limestone rock."

APLOME. A species of the common garnet, thus named by Haüy, but differing in this respect, that although it commonly occurs in rhombic dodecahedrons, its planes are striated parallel with their lesser diagonal. Weiss considers that there is no reason why this substance should be deemed a distinct species.

A′PODA. An order of animals belonging to the class Echinodermata; division Radiata. They are distinguished from Pedicellata by the absence of the vesicular feet, which peculiarly belong to animals of that order.

APOPHY′LLITE. A mineral whose constituent parts are silica 50·76, lime 22·39, potash 4·18, water 17·36, and a trace of fluoric acid.

This substance is called also Ichthyophthalmite, and Fish-eye stone. It occurs both massive and regularly crystallized. It is found in Sweden, in secondary traprocks in Scotland and the Hebrides, and in Iceland, whence the finest specimens are obtained.

APO′PHYSIS. (ἀπόφυσις, Gr. *apophyse*, Fr.) A process of a bone, and part of the same bone; herein differing from ephiphysis, which is a process attached to a bone, and not a part of the same bone, an excrescence.

APORRHA′IS. A genus of shells found fossil and recent. The Aporrhais first appears fossil in the coralline crag epoch.

APPE′NDAGE. Something added to another thing, without being necessary to its essence. In botany, applied to additional organs of plants, which are not universal or essential; neither is any one plant furnished with them all. Botanists distinguish seven kinds of appendages, namely, stipules, floral leaves, thorns, prickles, tendrils, glands, and hairs.

APPENDI′CULATE. Appendicled, or appended. Applied to flowers furnished with some addition distinct from the tube; to petioles with leafy films at the base; to seeds furnished with hooks, scales, &c.

APPRE′SSED. (*appressus*, Lat.) In botany, applied to leaves pressed to the stem; also to peduncles.

A′PSIDES. (*apsides*, Fr. from ἀψίς, Gr. The plural of *apsis*,) Those two points in the orbit of a planet, one of which is the farthest from, and the other the nearest to, the sun. The motion of the apsides may be represented, by supposing the planet to move in an ellipse, while the ellipse itself is slowly revolving about the sun in the same plane. This motion of the major axis, which is direct in all the orbits except that of the planet Venus, is irregular, and so slow, that it requires more than 109,830 years for the major axis of the earth's orbit to accomplish a sidereal revolution.

A′PSIS. (ἀψίς, Gr. *apsis*, Lat.) A term used indifferently for either of the two points of a planet's orbit, where it is at the greatest or least distance from the sun or earth; and hence the line connecting those points is called the line of the apsides. The *apsis* at the greatest distance from the sun is called the *aphelion*, and at the greatest distance from the earth is called the *apogee*; while that at the least distance from the sun is termed the *perihelion*, and at the least distance from the earth, the *perigee*.

A′PTER. } (from *a*, priv. and πτερὸν,
A′PTERA. } a wing.) Insects which have no wings, forming, according to the Linnæan system, the seventh order of insects.

A′PTEROUS. Destitute of wings, wingless.

APTIEN TERRAIN. The continental name for certain beds contemporaneous with the Lower Greensand of England.

A′PULUM. The name of a metallic substance obtained from alumina.

APY′ROUS. (from *a*, priv. and πῦρ, Gr. *apyre*, Fr.) Capable of resisting the action of fire.

ARA′CHNIDA. } (from ἀράχνη and
ARA′CHNIDAN. } εἶδος, Gr. resembling a spider.) The arachnida are members of that series of annulose animals possessing jointed feet, and belong to the third class of articulated animals. In the animal kingdom, the third class of annulosa, comprising the following orders, namely, Pulmonata, Amphipneusta, Trachearia, and Pycnogonida. The two great families in the higher order of living arachnidans are spiders and scorpions.—*Buckland*. In the arrangement of

Cuvier, the arachnidans compose the second class of articulated animals provided with moveable feet. They have no wings, and do not undergo any metamorphosis, merely casting their skin. The majority of the arachnidans feed on insects; some are parasitical, living on vertebrated animals; others are found in flour, in cheese, and on vegetables. Cuvier has divided the arachnidans into two orders, Pulmonariæ and Trachcariæ; the former he subdivided into two families, Araneides and Pedipalpi; the latter into three families, Pseudo-Scorpiones, Pycnogonides, and Holetra.

Ara'chnoid. ⎫
Arachnoï'des. ⎭
1. A cobweb-like membrane, forming one of the tunics or coats of the brain.
2. One of the tunics, or coats, of the eye.
3. A species of fossil madrepore.

Arb'oreous. (*aboreus*, Lat.)
1. Belonging to trees, resembling trees.
2. A term used to distinguish such mosses, or funguses, as grow upon trees, from those that grow on the ground.

Arb'orescence. (from *arboresco*, Lat.) The likeness of a tree, frequently observed in crystallizations and in mineral productions.

Arc of a circle. An arc of a circle is any part of its circumference; and the chord, or subtense of an arc, is a straight line joining the two extremities of that arc.

A'rca. A transverse inequilateral shell: the beaks distinct; the hinge with many teeth disposed in a straight line. These are marine shells. Lamarck particularizes seven species. Found recent and fossil.

Arca'cea. A genus of shells distinguished by having numerous small penetrating teeth disposed on both valves in a straight or bent line.

Arca'cea. A family of bivalve conchifera, according to Lamarck, consisting of shells provided with a linear series of teeth on the hinge. In this family Lamarck places the genera Arca, Cuculloea, Pectunculus, and Nucula; to which may be added Crenella (Brown), Solenetta (Sowerby), Myopora (Lea), Limopsis (Sassi), Byssoarca (Swainson), and Macroden (Lycett). —*Lycett*.

A'rctic. ⎫ (from ἄρκτος, ursus;
A'rctick. ⎭ *arctique*, Fr. *artico* It.) Northern; lying under the arctos, or bear.

A'rctic circle. One of the lesser circles of the sphere, twenty-three degrees and twenty-eight minutes from the north pole. The circle at which the northern frigid zone begins. This and its opposite, the antarctic, are called the two polar circles.

A'rcuapure. The curvature of an arch.

Arena'ceous. (*arenaceus*, Lat.)
1. Sandy; having the properties, or appearance, of sand.
2. Growing in sand, a term applied to certain plants which are called arenaceous. "The surface of the bank is covered with various *arenaceous* plants."

Arena'ceous Quartz. Sand. There are varieties of sand which are of a pure white, or nearly so, as some of the sands of Alum Bay in the Isle of Wight, and which appear to have resulted from the destruction of quartz.

Aren'dalite. The name given by Karsten to the mineral more commonly known as Epidote. For description, see *Epidote*.

Arenili'tic. Resembling sandstone; having the quality of sandstone; composed of sandstone.

Areo'meter. (from ἀραιός and μετρέω, Gr. *areomètre*, Fr.) An instrument

F

for measuring the density or weight of any liquid.

A′RFWEDSONITE. The name given by Brooke, in honour of Professor Arfwedson, to a mineral now separated from hornblende, of which it was wont to be deemed a ferriferous variety. It occurs in Greenland. Colour black; and opaque. It contains above 35 per cent of iron.

A′RGAL. Crude tartar, as deposited by vinous fermentation.

ARGE′NTAL. (from *argentum*, Lat.) Containing silver; combined with silver.

ARGENTI′FEROUS. (from *argentum* and *fero*, Lat.) Producing silver.

ARGENTI′NA. A genus of fishes of the order abdominales.

ARGE′NTINE. Slate-spar; a mineral of a lamellated, or slaty structure; a nearly pure sub-species of carbonate of lime.

A′RGIL. ⎫ *argilla*, Lat. ἀργιλλος or
A′RGILL. ⎭ ἀργιλος, Gr. *argille*, Fr. *argilla* It. See Alumine.) In 1754, Margraff showed that the basis of alum is an earth of a peculiar nature different from every other; an earth which is an essential ingredient in clays, and gives them their peculiar properties. Hence this earth was called argill; Morveau afterwards gave it the name of alumina, because it is obtained in the state of greatest purity from alum.

ARGILLA′CEOUS. (*argillaceus*, Latin.) Clayey; of the nature of argil; containing argil. The rocks which are termed argillaceous have been all arranged into one suite, and present the following gradation; clay slates, shale of the coal-measures, shale of the lias, clays alternating, in the oolite series, and that of the sand below the chalk; and clays above the chalk.

ARGILLA′CEOUS-SCHIST. Clay slate. An indurated clay, or shale, common to the fossiliferous and metamorphic series.

ARGILLI′FEROUS. (from *argilla* and *fero*, Lat.) Producing or yielding clay.

A′RGILLITE. Argillaceous-schist, or clay-slate. Slate is a very extensive formation, composing entire mountains in many alpine districts. The prevailing colours are bluish, or greenish grey: it has a silky lustre.

ARGI′LLOUS. (*argillosus*, Lat. *argilleux*, Fr. *argilloso*, It.) Containing clay; of the nature or quality of clay.

ARGONA′UTA. The Paper Sailor; a genus of animals; class Vermes, order Testacea. An involuted univalve, both recent and fossil, the spire turning into the opening; very thin, with a tubercular double dorsal keel. There are several species, but the most remarkable one is the Argonauta Argo, or Paper Nautilus. "Doubts still exist whether the Sepia found within this shell be really the constructor of it, or a parasitic intruder into a shell formed by some other animal not yet discovered. Broderip, Gray, and Sowerby, are of opinion, that this shell is constructed by an animal allied to Carinaria.—*Buckland.*

Cuvier placed argonauta among the subgenera of Sepia, and Dr. M'Murtrie, in his translation, says, "These mollusca are always found in a very thin shell, symmetrically fluted and spirally convoluted, the last whorl so large that it bears some resemblance to a galley, of which the spine is the poop. The animal makes a consequent use of it, and in calm weather whole fleets of them may be observed navigating the surface of the ocean, employing six of their tentacula as oars, and elevating the two membranous ones by way of sail. If the sea become rough, or they perceive any danger, the argonaut withdraws all its arms, concentrates itself in its shell, and descends to the bottom."

ARMADI'LLO. *(armadille,* Fr.) The Dasypus of Linnæus, and placed by him in the order Bradypoda, class Mammalia. Cuvier has placed the armadillo in the order Edentata, or quadrupeds having no front teeth, class Mammalia. The armadillo is constructed with unusual adaptations to the habit of burrowing in search of its food, and shelter in the sand; its fore feet forming instruments of peculiar power for the purpose of digging; and presenting an extraordinary enlargement and elongation of the extreme bones of the toes, for the support of long and massive claws. The armadillo and chlamyphorus are the only known animals that have a compact coat of plated armour. There are several sub-genera.

ARI'LLUS. } *(arillus,* Lat.) A substance
A'RIL. } enclosing the seed in some plants: it is either a complete or partial covering of a seed, fixed to its base only, and more or less loosely or closely enveloping its other parts. Mace is the *arillus* of the nutmeg: the red arillus of the seed of the common spindle-tree is well known, and is very ornamental in our hedges in the autumn.

A'RMATURE. *(armatura,* Lat. *armature,* Fr.)
1. That by which the body is protected from injury.
2. Weapons of attack.
3. A piece of soft iron applied to a loadstone, or connecting the poles of a horse-shoe magnet.

ARME'NIAN STONE. A blue mineral, or earth, variously spotted. It much resembles Lapis lazuli.

ARRA'GONITE. A variety of carbonate of lime, found originally in Arragon in Spain, from which circumstance it has obtained its name. Its colours are white, grey, green, and blue; it is found both crystallized and massive. It is frequently combined with a small proportion, about four per cent. of carbonate of strontites. Beautiful specimens of this rare mineral, of a snow white colour and satin-like lustre, have been found in the lead-mines of Cumberland.

ARSE'NIATE. A compound of arsenical acid with a metallic oxide; many arseniates are found native; when heated along with charcoal powder, they are decomposed, and arsenic sublimes.

A'RSENIC. (ἀρσενικὸν, Gr. *arsenic,* Fr. *arsenico,* It.) Native arsenic is a mineral found in Germany, France, and England. It occurs generally in masses of various shapes; its colour is that of blue steel; it is brittle; its surface readily tarnishes on exposure to the atmosphere. When struck, it gives a smell resembling garlic; before the blowpipe it emits a white smoke, burns with a blueish flame, gives a strong garlicky smell, and deposits a white powder. This metal and all its compounds are virulent poisons. Combined with sulphur it forms orpiment or realgar, or the yellow and red sulphurets of arsenic. The term ἀρσενικὸν, from which the word arsenic is derived, was an ancient epithet, applied to those natural substances which possessed strong and acrimonious qualities, and as the poisonous quality of arsenic was found to be remarkably powerful, the term was especially applied to orpiment, the form in which this metal more usually occurred. Dr. Paris, from whose work the above is quoted, states that in the celebrated plague of London, amulets of arsenic were worn, suspended over the region of the heart, as a preservative against infection; on the principle, so prevalent at one period, that all poisonous substances possess a powerful and mutual elective attraction for each other.

A′RSENITE. A name given by Fourcroy to the combinations formed between oxide of arsenic, or arsenious acid, and the earths and alkalies. Arsenite of potassa is the active ingredient in Fowler's Ague Drop, and in the Liquor Potassæ Arsenitis of the Pharmacopœia.

ARTE′MIS. A genus of fossil shells, found in the glacial beds of Scotland, Iceland, the north of England, and the Isle of Man. Recent, according to Adamson, it ranges as far south as Senegal, and is found, according to Phillippi, in the Red Sea.

A′RTERY. (from $ἀὴρ$, and $τηρέω$, Gr. *artère*, Fr. *arterio*, It. Thus called because the ancients thought that only air was contained in the arteries.) The arteries are strong elastic canals, which convey the blood from the heart to the different parts of the body, and, during life, are distinguished from the veins by their pulsation. The original trunks of the arteries are two in number, and from these all the other arteries are derived.

ARTE′SIAN WELLS. Springs of water, or fountains, obtained by boring through strata destitute of water into lower strata loaded with this fluid, to sometimes great depths; thus named from its having been first practised at Artois, the ancient Artesium, in France. In forming an Artesian well, if the boring penetrate a bed containing impure water, it should be continued deeper until it arrive at another stratum containing pure water; the bottom of the pipe being plunged into this pure water, it ascends within it, and is conducted to the surface through whatever impurities may exist in the superior strata. The impure water, through which the boring may pass in its descent, being excluded by the pipe from mixing with the pure water ascending from below. The height to which these springs will rise above the surface must depend on the quantum of hydrostatic pressure from below; this is sometimes very great. The water of an Artesian well in Roussillon rises from thirty to fifty feet above the surface. At Perpignan and Tours, M. Arago states that the water rushes up with such extreme force as violently to eject a cannon ball placed in the pipe. An economical and easy method of sinking Artesian wells has recently been practised. Instead of the tardy and costly process of boring with a number of iron rods screwed to each other, one heavy bar of cast iron, about six feet long, and four inches in diameter, armed at its lower end with a cutting chisel, and surrounded by a hollow chamber, to receive, through valves, and bring up the detritus of the perforated stratum, is suspended from the end of a strong rope, which passes over a wheel or pulley fixed above the spot in which the hole is made. As this rope is moved up and down over the wheel, its tortion gives to the bar of iron a circular motion, sufficient to vary the place of the cutting chisel at each descent. When the chamber is full, the whole apparatus is raised quickly to the surface to be unloaded, and is again let down by the action of the wheel.—*Buckland.*

According to the observations of M. Arago, the greater the depth of these wells, the higher is the temperature of the waters that flow from them. In an Artesian well in Aberdeen, the bore is eight inches in diameter, and 250 feet deep, and the temperature of the water three degrees above the average temperature of the locality.

ARTI′CULAR. (*articularis*, Lat. *articulaire*, Fr. *articolaire*, It.) Belonging to the joints.

ARTICULA'TA. The first division of the sub-kingdom Annulosa, comprising four classes, namely; Insecta, Myriapoda, Arachnida, and Crustacea.

ARTI'CULATED. Jointed; having joints; united by joints. In botany, the term articulated is applied to leaves, when one leaflet, or pair of leaflets, grow out of the summit of another, with a sort of joint; to stems divided by joints or knots, or divided from space to space by contractions: to culm with joints.

ARTI'CULATING. Fitting by means of joints.

ARTICULA'TION. (*articulatio*, Lat. *articulation*, Fr. *articolazione*, It.) The juncture or joint of bones. There are three kinds of articulation, 1. Immoveable, called Synarthrosis; 2. Moveable, or Diarthrosis; 3. Mixed, or Amphi-arthrosis.

ARTIODA'CTYLA. The fourth order of the class-mammalia, and so named from having an even number of toes, two or four. This order comprizes ron-ruminantia, as the hippopotamus and pig; and ruminantia, as the camel, stag, sheep and cow.

ARTICULO'SA. See *Articulata*.

ARVI'COLA AGRESTIS. The field vole. A species of fossil rodentia, discovered in the caves at Torquay and Kirkdale.

ARVI'COLA AMPHIBIA. The water vole.
ARVI'COLA PRATENSIS. The bank vole.
Two species of fossil rodents, found in Kent's hole.

ARUNDINA'CEOUS. (Lat.) Resembling reeds.

ARUNDI'NEOUS. } (*arundineous*, Latin.)
ARU'NDINOSE. } Reedy; abounding in reads.

ARYTÆ'NOID. (from ἀρύταινα, a ewer, and εἶδος, resemblance, Gr.) A name given to some of the cartilages, glands, and muscles of the larynx.

A'SAPHUS. (ἀσαφής, Gr. obscurus.) A genus of Trilobites, thus named by Brongniart. Professor Buckland observes, in writing of Trilobites, "Fossils of this family were long confounded with insects, under the name of Entomolithus paradoxus; after many disputes respecting their true nature, their place has now been fixed in a separate section of the class Crustaceans, and although the entire family appears to have been annihilated at so early a period as the termination of the carboniferous strata, they nevertheless present analogies of structure, which places them in near approximation to the inhabitants of the existing seas.

The generic characters of asaphus are thus described by Brongniart. "Corps large et assez plat; lobe moyen, saillant et très distinct. Flancs ou lobes latéraux ayant chacun le double de la largeur du lobe moyen. Expansions submembraneuses depassant les arcs des lobes latéraux. Bouclier demi-circulaire, portant deux tubercules oculiformes réticulés? Abdomen divisé en huit ou douze articles." Sir R. Murchison's Silurian system. There are many species; fourteen are figured in Sir R. Murchison's work.

In some parts of Wales, that species of Asaphus known as A. Debuchü, is so abundant that the laminæ of the slates are charged with them, so that millions of them have probably lived and died not far distant from those places where we now discover them.

ASBE'STINE. Incombustible; partaking of the properties of asbestos.

ASBE'STINITE. A species of asbestos. This mineral is amorphous. Texture foliated. Lustre silky, 3. Specific gravity, 1·880. Colour white, with shades of red, yellow, blue, and green. At 150° Wedgewood, it melts into a green glass.

ASBE'STOID. A mineral, thus called from its resemblance to Asbestos.

It is amorphous. Texture foliated or striated. Specific gravity from 3· to 3·30. Colour olive or green. It consists of silica 46, oxide of iron 25, lime 11, oxide of manganese 10, magnesia 8.

ASBE'STOS. } ἀσβεστος, Gr. nomen
ASBE'STUS. } lapidis, unde telæ fiunt, quæ non comburuntur in igni; *asbeste*, Fr. *asbeste*, It.) A mineral of which there are several varieties, all marked by their fibrous flexible quality. Asbestos is itself a variety of hornblende. It was well known to the ancients, by whom a kind of **cloth** was made of one of its varieties, which was esteemed to be incombustible. It is found abundantly in most mountainous countries, and in the isle of Anglesea it lies in considerable quantities between the beds of serpentine. Veins of asbestus may be seen in almost all the serpentine formation of the Lizard; some of them are as wide as half-an-inch; but their width, in general, is that of a line, running in all directions. Veins of asbestus occur also in the greenstone. Although fire acts slowly on its fibres, yet it will, in the course of time, consume them. It is commonly amorphous. Texture fibrous. Lustre **from 0** to 2. Hardness from 3 to 7. It absorbs water. Colours white, green, blue, yellow, and brown. Its constituent parts are, silica 60, magnesia 30, lime 6, alumina 4. It feels soapy or greasy. For one of its varieties, flexible asbestos, see *Amianthus*. Another variety has obtained the name of mountain cork, from its swimming when thrown into water. This variety has a strong resemblance to common cork. Its fibres are interwoven. Specific gravity from 0·6806 to 0·9933. It feels meagre; yields to the fingers like cork, and is somewhat elastic. Colour white or grey. Its constituent parts are, silica 62, carbonate **of** magnesia 23, carbonate of lime 12, alumina 2·7, oxide **of** iron 2·3. One variety **is** called rigid or common asbestos. Of this the colours are usually green, and disposed in straight, pearly, rigid fibrous concretions. Soapy or unctuous to the feel. Another variety is known by the names of rockwood, mountainwood, or ligneous asbestos. The colour of this variety is brown, and its general appearance greatly resembles fossil wood.

ASBE'STIFORM ACTYNOLITE. } A subASBE'STOUS ACTYNOLITE. } species or variety of the mineral actynolite, of a green, greenish-grey, or brownish-green colour. It occurs in beds in gneiss, mica slate, and granular limestone. Specific gravity 2·5 to 2·8 Those varieties of asbestiform actynolite which occur in very thin scopiformly aggregated acicular elastic flexible crystals, have by some mineralogists been deemed a distinct species, and named, by Saussure, Byssolite, by Haüy, Amianthoid, and by others, Asbestoid.

A'SCARIS. (ἀσκαρίς, Gr.) Cuvier placed the ascaris in **the** order Nematoidea, class Entozoa. The thread-worm.

ASCI'DIA. A genus of animals found in the sea, adhering to the rocks. Class Vermes, order Mollusca.

ASCI'DIOIDA. Having the characters of ascidia; resembling ascidia. Just as we see the ascidioida and helianthoida of our seas fixed to the boulders and rocky skerries.—*Hugh Miller.*

Ascidioida, or Tunicata, constitutes the third order in the class Molluscoidea; these have no fossil representatives, as they have no hard parts likely **to be** preserved.

ASCITI'TIOUS. (*ascititius*, Lat.) Supplemental; additional; not originally forming part of.

ASHBURNHAM BEDS. The **lowest** division of the Wealden strata, consisting of inferior limestones and shale. Dr. Mantell states the Ashburnham beds to consist of a series of highly ferruginous sands, alternating with clay and shale, containing ironstone and lignite; and shelly limestone, alternating with sandstone, shale, and marl, and concretional masses of grit. The organic remains consist of ferns and carbonized vegetables; cypris; shells of cyclas and cyrena; and lignite.

A'shlar. A name given to freestone **as it is taken** from the quarry.

Asiphonibranchia'ta. In De Blainville's system, the second order of the class Paracepholophora Dioica, comprising the genera Goniostomata, Cricostomata, Hemicyclostomata, Ellipsostomata, and Oxystomata.

A'sphalt. } ($ἄσφαλτος$ Gr. bitumen,
Aspha'ltos. } *asphalte*, Fr.) A bituminous substance,
Aspha'ltum. } found abundantly on the shores of the Dead Sea; in the island of Trinidad, in China, America, and various parts of Europe. Its colour is brown or black; it is lighter than water, and easily soluble in naphtha, but quite insoluble in water. Fracture conchoidal. Brittle. Feels smooth, but not unctuous. Does **not** stain the fingers. On the surface of the Dead Sea it is found floating in a state of liquidity, but exposure to the air soon renders it hard. It melts easily when heated, and, if pure, burns without leaving any ashes.

Asperg'illum. A genus of bivalve shells of the order tubicola. The Serpula penis of Linnæus.

Aspidorhy'nchus. The name given to a fossil Sauroid fish from **the** lime stone of Solenhofen. An example of this is given by Professor Buckland in his Bridgewater Treatise, pl. 27 *a*, fig. 5.

Assa'y. **The** operation of determining the proportion of precious metal contained **in any** mineral or metallic compound, by analyzing a portion thereof.

A'stacid. } The craw-fish, or lobster;
A'stacite. } a genus of the family Macroura; it is divided into four sections, each consisting of many subgenera. The lobster, crab, craw-fish, prawn, and shrimp are **in**cluded.

A'stacus. (from $ἀστακὸς$, Gr. *astacus*, Lat.) The lobster or craw-fish.

Asta'colite. (from $ἀστακὸς$ and $λίθος$, Gr.) Fossil or petrified craw-fish, or lobster.

Asta'rte. (from the Sidonian goddess; the Crassina of Lamarck.) A genus of bivalve shells; equivalve, inequilateral, thick, compressed; hinge with two large diverging teeth in the right valve, one in the left; ligament external. The **spe**cies are numerous, more **especially** the fossil ones; forty-six are enumerated in Morris's catalogue of English fossils, and many others have not been described.—*Lycett*.

Aste'ria. (*asterias*, Lat. *asterie*, Fr.) A variety of sapphire, or bastard opal.

Aste'ria. (from $ἀστήρ$, **Gr. a star.**) The star-fish, or sea-star, a genus of animals, class Vermes, order Mollusca. These animals have their mouth in the centre, and placed downwards; from their bodies five or more rays, or arms, **are** given off, furnished with numerous retractile tentacula. They have the power of reproducing their rays if destroyed. They are all inhabitants of the sea, and they are frequently found fossil, in great perfection, in the chalk. Some remarkably fine impressions have been discovered in flint. The whetstone of Devonshire affords similar remains. Linnæus has placed them in the order Pedicellata, class Echinodermata.

Two species of this genus of radiata have been found fossil in the chalk, one of which has been named by Goldfuss A. quinqueloba, the other has not yet been determined. Eight species found in the oolite of Germany have been thus named by Goldfuss, Munster, and Schlotheim, A. arenicula, A. jurensis, A. lancelota, A. lumbricalis, A. prisca, A. scutata, A. stellifera, A. tabulata. Of the Linnæan genus Asteria, M. Lamarck has formed a family to which he has assigned the name Stellerida, comprehending four genera.

ASTE′RIATED. Radiated.

ASTE′RIALITE. (from ἀστήρ, and λίθος, Gr.) Fossilized, petrified, or silicified asterias, or star-fish.

ASTE′RIDÆ. ⎱ The fourth order of the
ASTERID′EA. ⎰ class Echinodermata, comprising the star-fish, &c.

A′STERITE. ⎫ (astroite, Fr. Espèce de
A′STRAITE. ⎪ madrepore ou de corps
A′STRITE. ⎬ marin, sur lequel on
A′STROITE. ⎭ voit représenté la figure d'une étoile.) Star stone.

This name is also given to certain varieties of the perfect corundum.

A′STEROID. (from ἀστήρ and εἶδος, Gr.) The name assigned by Herschel to some newly discovered planets.

ASTERO′IDAL. Resembling a star-fish.

ASTEROLE′PIS. A genus of ichthyolites of the old red sandstone, known also as chelonichthys, which latter name has now been supplanted by the former. Many species have been described by Agassiz.

ASTEROPHY′LLITE. (from ἀστήρ and φύλλον, Gr.) A plant discovered in the coal formation, and thus named from the stellated disposition of the leaves around the branches.

ASTO′MATA. The second class of the sub-kingdom Protozoa, comprising the four orders, Spongiadæ, Foraminifera, Thalissicolidæ, and Gregarinidæ.

ASTRE′A. A genus of saxigenous polyi. The appearance of groups of *astreæ*, and other corals, is described as being most beautiful when viewed with the animals alive and in activity; looking down through the clear sea-water, the surface of the rocks appears one living mass, and the polypi present the most vivid hues.

It is stated, by those who have opportunities of observing, that the species of saxigenous polypi which constantly form the most extensive coral reefs and islands belong to the genera *Meandrina*, *Caryophyllia*, and *Astrea*, but especially to the latter; and that these are not found at depths exceeding a few fathoms. M. M. Quoy and Gamard observe, that neither with the anchor nor the lead, have they ever brought up fragments of astreæ, except where the water was shallow, about twenty-five or thirty feet in depth, though they found that the branched corals, which do not form solid masses, lived at great depths.

Fifteen species of the genus Astrea have been determined by Goldfuss as met with in the members of the chalk group: twenty-four species also are described as found in the oolite: one belongs to the muschelkalk, namely, Astrea pediculata: one species, A. undulata has been found in the carboniferous limestone: three species have been discovered in the grauwacke.

Parkinson thus describes the genue *Astrea*: "A stony polypifer, fixed, conglomerated, incrusting other bodies, or formed in a subglobose but rarely lobated mass. The upper surface set with sessile, lamellated, round or subangular stars, which are circumscribed. The stars in this genus are circumscribed. The substance is never raised in extended expansions, or developed in leaves, as in the *Explanariæ*, or ramified like the *Madrepores*." The same author describes 31 species.

ASTRI′FEROUS. ⎱ Bearing stars; having
ASTRI′GEROUS. ⎰ stars; carrying stars.

A′STRITE. See *Asterite*.

A'STROITE. See *Asterite*. A name given by some writers to corals of the genus *Astrea*.

ATA'CAMITE. Prismatoidal green malachite. Native muriate of copper, of a green colour, occuring both massive and crystallized. It consists of oxide of copper 76·6, muriatic acid 12·4, water 12. It has obtained its name from having been found in alluvial sand in the river Lipas, in the desert of Atacama in Peru. It has also been found in some of the Vesuvian lavas. The primitive form is an octoëdron. It is the Cuivre muriaté of Haüy.

ATMO'METER. (from $\dot{a}\tau\mu\grave{o}s$, vapour, and $\mu\epsilon\tau\rho\acute{\epsilon}\omega$, to measure, Gr.) An instrument contrived by Professor Leslie, for ascertaining the quantity of moisture exhaled from a damp surface in a given period.

ATMOS'PHERE. (from $\dot{a}\tau\mu\grave{o}s$, vapour, and $\sigma\phi a\hat{i}\rho a$, sphere, Gr. *atmosphère*, Fr. *atmosféra*, It.) The gaseous compound which surrounds the earth: it is a thin, transparent, invisible, and elastic fluid, essentially composed of oxygen and nitrogen, yielding by analysis 79 parts of nitrogen to 21 of oxygen, and containing in every 1000 parts three or four of carbonic acid gas. These proportions are found to be the same at all heights hitherto attained by man. The atmosphere, from its powers of refraction, has been calculated to extend upwards about forty-five miles. That the atmosphere is a ponderous body, was first suspected by Galileo, who found that a copper ball, in which the air had been condensed, weighed heavier than when the air was in its ordinary state of tension. The fact was afterwards demonstrated by Torricelli, whose attention was drawn to the subject by the attempt of a well digger to raise water by a sucking pump to a height exceeding 33 feet.

The pressure of the atmosphere is about fifteen pounds on every square inch; so that the surface of the whole globe sustains a weight of 11,449,000,000 hundreds of millions of pounds. Shell-fish which have the power of producing a vacuum, adhere to the rocks by a pressure of fifteen pounds upon every square inch of contact.

When in equilibrio, the atmosphere is an ellipsoid flattened at the poles, in consequence of its rotation with the earth.

The barometer, by its fall and rise, indicates a corresponding change in the density of the atmosphere. At the surface of the earth the mean density or pressure is considered equal to the support of a column of quicksilver 30 inches high; at 1000 feet above the surface of the earth the column of quicksilver is found to fall to 28·91 inches; at one mile to 24·67 inches; at two miles to 20·29 inches; at three miles to 16·68 inches; the density decreasing upwards in geometrical progression. The air even on mountain tops is sufficient to diminish the intensity of sound, to effect the breathing, and to produce a loss of muscular power. The greatest elevation attained by man, is 4·36 miles; that height M. M. Gay-Lussac and Biot ascended in a balloon, but they suffered severely from the rarity of the atmosphere.

A'TOLL. The name given to a ring-formed coral reef.

ATRY'PA. (from *a*, priv., and $\tau\rho\upsilon\pi a$, a foramen, Gr.) A genus of fossil shells found in the Lower Silurian Rocks. A sub-division of the family of Terebratula. "This genus is divided from *Spirifer*, and includes those species which have a short hinge line without a large area, and are either destitute of a foramen or possess only a small triangular one. They are rounded shells, and are not furrowed like the typical species of Spirifer. The species of this sub-division have generally been described as

Terebratulæ by British authors, but they have acute, not perforated beaks."—*Sir R. Murchinson. Silurian System.*

ATTE′NUATED. (*attenuatus*, Lat. *atténué*, Fr.) Thin; slender; tapering. An epithet for a leaf tapering at one or both extremities.

AU′GITE. (ἀυγὴ, Gr. splendour.) A mineral of a dark green, brown, or black colour, found in volcanic rocks. It is the Pyroxéne of Haüy, the Paratoma augit of Mohs, and the augit of Werner. Its fracture is conchoidal and uneven. It generally crystallizes in six or eight-sided prisms, terminated by dihedral summits. It is commonly attracted by the magnet. Scarcely fusible by the blow-pipe. With borax it melts into a yellowish glass, which while hot appears red. There are many varieties of augite, as the Diopside, Musite, Alalite, Sahlite, Pyrogome, Fassaite, Malacolite, Common Augite, Conchoidal Augite, Granular Augite, Coccolite and Amianthus. Augite consists of silica 52, lime 13, protoxide of iron and manganese 16, magnesia 10, alumina 9.

AUGI′TIC PO′RPHYRY. A rock with a dark grey, or greenish, base, containing crystals of augite and Labrador felspar.

AULO′PORA. The name given by Goldfuss to a genus of corals. Four species are described as found in the Wenlock Limestone. Three of them have been named by Goldfuss, namely, A. conglomerata, A. serpens, and A. tubæformis; the fourth A. consimilis has been described and thus named by Mr. Lonsdale. Species of the genus Aulopora are found in other strata.

AU′RATED.
1. Resembling gold.
2. Eared; having ears, as in the scallop-shell.

AURE′LIA. The first change of the eruca, or maggot, of any kind of insect; a chrysalis, having a golden hue, previous to its becoming the perfect insect.

AU′RICLE. (*auricula*, Lat.)
1. That part of the ear which is prominent from the head.
1. A cavity of the heart. The heart is divided into four cavities, or chambers, namely, two auricles and two ventricles.

AU′RICLED. Having ears; having appendages resembling ears; applied to leaves when they are furnished with a pair of leaflets, generally distinct, but sometimes joined with them.

AURI′CULA. An ovate or oblong pyramidal marine univale, with the spire extruded: the opening entire, oblong, and narrowed upwards; the columella plicated, with different plicæ in the opposite lip. Lamarck has placed those shells whose openings are entire, but whose columellæ are plicated, under this genus, *Auricula.—Parkinson.* Fossil auriculæ occur in the chalk marl and in the Shanklin sand. Seventeen species have been recorded by Deshayes: of many species recorded from the neocomian rocks of France, the genus is doubtful.

AURI′CULATE. Ear-shaped.

AU′RIFORM. (from *auris* and *forma*, Lat.) Having the form of an ear; in the shape of an ear; the haliotis is an example.

AU′STRAL. } (*australis*, Lat. *austral*,
AU′STRINE. } Fr. *australe*, It.) Southern; southward.

AUTO′MALITE. (Octahedral corundum of Mohs. Spinelle zincifère of Haüy. The Automolith of Werner; sometimes also called Gahnite, from Gahn, its discoverer. The name automalite was given to it by Eckeberg.) A variety of corundum containing oxide of zinc. It occurs imbedded in talc, and associated with lead-glance. It is crystallized in regular octahedrons, or in tetrahedrons with truncated angles.

Automalite is placed by some mineralogists in the ruby family; by others it is considered to be **a** variety of spinel, and from the large quantity of oxide of zinc which enters into its composition it has been termed zinciferous spinel. Externally it is glistening; lustre pearly, inclining to semi-metallic. Internally it is shining on the principal fracture, but glistening on the cross fracture, and the lustre resinous. Fracture foliated, exhibiting a four-fold cleavage parallel with the planes of the octahedron. It is heavier than spinel, its specific gravity being 4·1 to 4·3, from which it also differs in being nearly opaque, and of a dark bluish-green colour by transmitted light. It is **of** sufficient hardness to scratch quartz. It is brittle and easily frangible. Before the blow-pipe it is infusible. It is a non-conductor of electricity. Its constituent parts are alumina, the oxides of zinc and iron, silica, and sometimes magnesia. It has been found in America and in Sweden.

AUTO′MOLITE. See *Automalite*.

A′VALANCHE. (*avalange, ou avalanche*, Fr.) A mass of snow which, detached from any mountainous height, by rolling onwards accumulates frequently prodigious bulk and acquires **great** momentum. Avalanches are **in** mountainous countries productive of the direst misfortunes, sweeping before them in their irresistible and destructi**ve** progress every impeding object; breaking off large masses of rocks, uprooting, or tearing away, the noblest trees, damming up river courses, and burying beneath their volumes villages, with their whole population.

AV′ANTURINE. } Quartz Hyalin Aven-
AV′ENTURINE. } turiné of Haüy. A variety or sub-species of quartz, exhibiting numerous points or spots that glitter like gold. This appearance is generally owing to the intermixture of small laminæ of mica; in some instances it is caused by reflexion from numerous small rents or fissures, in the stone. The name is said to be derived from the French, par avanture; a workman having first found it by accident.

AVELL′ANA. A genus of shells formerly attributed to Auricula, occurring in the Lower Chalk **of** Lyme Regis, the Gault of Folkstone, and in the Greensand.

A′VES. The second of the sub-king**dom** vertebrata, comprising eight orders of birds.

AVER′TEBRATE. Having no vertebræ; destitute of vertebræ.

AVI′CULA. (from *avis*, Lat.) A genus of inequivalved, fragile, rather smooth bivalves; the base transverse and straight, with produced extremities and caudiform anteriorly; the left valve notched; the hinge linear; with a tooth in each valve beneath the beaks; the ligamental area marginal, narrow and grooved, not traversed by the byssus.—*Parkinson*. Many species of auricula have been recorded from the Neocomian rocks of France, but the genus is doubtful.—*Lycett*.

AXE-STONE. A mineral found in New Zealand and the islands of the Pacific, and by the inhabitants made into axes and other cutting instruments, from which circumstance it has obtained its name. It is a sub-species of jade, and in many respects resembles nephrite, or nephritic stone. See *Jade*.

AXI′LLA. (*axilla*, Lat.)
1. In anatomy, the arm-pit.
2. In botany, the angle formed by the stalk of a leaf with the stem.

AXI′LLARY. (*axillaris*, Lat. *axillaire*, Fr.)
1. In botany, applied to peduncles when proceeding from the angle made by the leaf and stem, or branch and stem; also to flowers,

and to spikes of flowers, proceeding from either of the above situations. 2. In anatomy, pertaining to the axilla, or arm-pit. 3. In entomology, applied to parts which spring from the point of union of two other parts.

A′xinite. The thumerstein or thumerstone of Werner. It has obtained the name of axinite in consequence of the axe-like shape of its crystals. Its colours are brown, grey, black, and blue. The name of thumerstein was given to it by Werner, from its having been found near Thum, in Saxony. It occurs massive, often disseminated, but most generally crystallized. Specific gravity from 3·21 to 3·29. Hardness = 6·5 — 7·0. The crystals are sometimes tabular, and are often so arranged as to form small cells. It is transparent or translucent, sometimes at the edges only, or is quite opaque. Sometimes one part of a crystal is violet and nearly transparent, while the other is green and nearly opaque. Texture foliated. Fracture conchoidal. Before the blow-pipe it froths like zeolite, and melts into a hard black enamel. It has been found sparingly in Cornwall, but in no other part of Great Britain. A specimen analysed by Vauquelin was found to consist of silica 44, alumina, 18, lime 19, oxide of iron 14, oxide of manganese 4.

Some of the porphyritic rocks of Cornwall, says Dr. Boase, are very beautiful, having a violet coloured basis, which appears to be occasioned by the compact felspar passing into massive *axinite*; for the latter mineral, in crystals, frequently occurs in small irregular veins, which traverse this rock, after the manner of calcareous spar in limestone.

A′xinitic. (from *axinite*.) Containing axinite.

A′xinus. A fossil genus of bivalve shells, described by Sowerby as equivalve, transverse; posterior side very short, rounded, with a long ligament placed in a furrow, extending along the whole ridge; anterior side produced, angulated, truncated, with a flattish lunette near the beaks. One species has been recorded in the London clay, four in the new red sandstone, one in the magnesian limestone, and one in the carboniferous shale.— *Lycett.*

A′xis. (*axis* Lat.) 1. The line, real or imaginary, that passes through anything on which it may revolve. 2. In botany, the imaginary central line of different parts of a plant, round which leaves, or modified leaves, are produced. The stem is also so called, for this reason.

Axo′tomous. (from ἄξων, and τέμνω, Gr.) A mineralogical term, signifying cleavable in one particular direction.

A′ymestry Limestone. So named from the village of Aymestry, where it laid open. One of the sub-divisions (the central) of the Ludlow rocks. A sub-crystalline grey or blue argillaceous limestone, a marine formation.

This sub-division of the Ludlow rocks, says Sir R. Murchinson, is, in the neighbourhood of Aymestry, arranged in beds of from one to five feet thick, dipping to the south and south-east at slight angles; the laminæ of deposit being marked by layers of shells, and sometimes of corallines. When quarried into, the rock is of an indigo or bluish grey colour, in parts mottled by the mixture of white carbonate of lime, both crystalline and compact. In nearly all the quarries between Norton Camp and Aymestry, the rock is charged with a profusion of that remarkable shell called *Pentamerus* Knightii. This species is confined almost exclusively to the limestone of this sub-division. The following fossils also characterise the calcareous zone: Lingula Lewisii; Terebratula Wilsoni; Bellerophon Aymestriensis; Avicula reticulata; and the coral Favosites Gothlandica.

The Aymestry Limestone differs in lithological aspect and useful properties, from any under or overlying calcareous rocks, being much less crystalline and pure than the mountain or **carboniferous** limestone, **and inferior in quality to the lower limestone of Wenlock.** Its earthy character renders it of very great value as a cement, particularly in subaqueous operations, the mortar formed of it setting rapidly under water.—*Sir R. Murchinson. Silurian System.*

Azo'ic. (Destitute of animal remains.) Those rocks in which no remains of animal existences have been discovered.

A'zote. (from *a*, priv. and ζωή, life, Gr.) A constituent part of the atmosphere, receiving its name from its fatal effects on animal life. It is now usually called Nitrogen, which see.

Azo'tic. Consisting **of azote**; resembling azote in **its properties**; destructive of life.

A'zurite. Another name for Lazulit or Lazulite; a combination of phosphoric acid, alumina, magnesia and water. Primary form a right rhombic prism of 121° 31'.

A'zure Stone. Lapis Lazuli; the Lazurstein of Werner, and Lazulite of Haüy. A mineral consisting of silica, alumina, lime, oxide of iron, soda, magnesia, and sulfuric acid. Specific gravity 2·95. The finest specimens come from China, Persia, and Siberia.

# B

B'abingtonite. An earthy mineral, so named by Levy, in honour of Dr. Babington. It occurs in distinct crystals at Arendal in Norway. It is of a dark, greenish black colour.

Ba'cca. (*bacca*, Lat.) A fruit; a berry.

Bacci'ferous. (from *bacca*, a berry, and *fero*, to bear, *baccifère*, Fr.) Berry-bearing; that produces berries.

Ba'culite. (from *baculus*, Lat. So named from its resemblance to a straight staff.) A fossil, straight chambered, conical, elongated and symmetrical shell, depressed laterally, and divided into numerous chambers by transverse, sinuous, and imperfect septa; the articulations, or sutures, being indented in the manner of the battlements of a tower. The external chamber is considerably larger than the rest, and capable of containing a considerable portion of the animal. Five species, namely, Baculites Faujasii, B. obliquatus, B. vertebralis, B. anceps, and B. triangularis, are described as occuring in the cretaceous **group**. Sir H. De La Beche says, **it** was once considered that the genus Baculites was confined to the cretaceous group, but, though more abundant in it than in any other series, it is not thus limited. The remains of baculites have been hitherto found in the chalk formation only, and the baculite appears to have become extinct simultaneously with the last of the ammonites, at the termination of the chalk formation. This fossil may be seen beautifully figured in Professor Buckland's Bridgewater Treatise.

BAG'SHOT BEDS. } One of the groups of
BAG'SHOT SAND. } the Eocene deposits in the London basin. These Eocene deposits have generally been divided into three groups, namely, the plastic clay, the London clay, and the Bagshot sand. Of these, the Bagshot sand is the uppermost, in the order of superposition: it rests conformably upon the London clay, and consists of siliceous sand and sandstone, without any cement, with some thin deposits of marl associated. The sandstone of the Bagshot sand is so hard as to require blasting with gunpowder, and is used for paving and building. Windsor Castle is constructed of it. This deposit is destitute of mineral contents.

BAI'KALITE. A variety of augite, of a whitish, or yellowish white, and pale green colour.

B'ALA LIMESTONE. A group of **a few** beds found in North Wales, **rarely** exceeding twenty feet **in** thickness.

BA'LANITE. (*balanites*, Lat.) A fossil belonging to the genus balanus.

BAL'ANUS. (*balanus*, Lat. βάλανος, Gr.) A sessile, affixed, conical univalve, the apex truncated; the base closed by an adherent testaceous plate; the opening subtrigonal or elliptical, with four moveable valves, inserted near the inner base of the shell. The balani are not to be considered among those fossils which are frequently found.—*Parkinson*.

The recent balanus is observed on rocks and shells at a depth ranging to ten fathoms; and affixed to bottoms of ships and other floating bodies.—*De La Beche*.

Balanus is the only genus of sessile cirrhipedes, the shells of which consist of six principal valves, except Coronula.—*Sowerby*.

BA'LASS. } (Called also Spinel and
BA'LLASS. } Zeilanit by Werner.) A sub-species of corundum; it **is** found in crystals of a regular octahedron, composed of two four-sided pyramids applied base to base. Colour red. Balass is chiefly found in Ceylon, and the dark and black varieties have obtained the name Ceylanite. It ranks among **the** precious stones, and when of a certain size is deemed very valuable.

BALI'STES. **The file-fish;** a cartilaginous **fish** belonging to the fourth class. **Professor** Buckland, in his chapter **on** Ichthyodorulites, or fossil spines, states that the spines of balistes have not their base, like that of the spines of sharks, simply imbedded in the flesh, and attached to strong muscles; but articulate with a bone beneath them. The spine of balistes is also kept erect by a second spine behind its base, acting like a bolt or wedge, which is simultaneously inserted or withdrawn by the same muscular motion that raises or depresses the spine.

BA'LKSTONE. A provincial name given to an impure stratified limestone.

BA'NNER. The upper large petal of a papillionaceous flower.

BA'OBAL. A stone which has obtained its name from its resemblance to the fruit of the baobal tree.

BARB.
1. That which grows in place of a beard.
2. A sort of pubescence in plants.

BA'RBATE. } (from *barbatus*, Latin.)
BA'RBATED. } Bearded; awned.

BA'RBED. Bearded; awned.

B'ARDIGLIONE. A name given by the Count de Bournon to Anhydrite. See *Anhydrite*.

BA'RIUM. The metallic basis of baryta, discovered by Sir H. Davy.

BARK. In botany, the covering of plants, composed of woody fibres, situated above the wood and under the cellular integument, consisting of from one to many layers, according to the age of the plant or branch, an additional layer being produced every year.

BA'RNACLE. (The Lepas balanus of Linnæus. *Barnacle*, Fr. *barnacla*, It.) A species of shell-fish, a pedunculated cirrhipede, frequently found adhering to the bottoms of ships in such prodigious numbers, and of so great a length, as to materially impede their progress through the water. Some very fine specimens may be seen in the British Museum The barnacle is known by the names Anatifer, and Pentelasmis In the arrangement of Cuvier the barnacle is placed in the sixth class, namely, Cirrhopoda, of Mollusca. Linnæus comprised them all in one genus, Lepas, which Brugiéres divided into two. The name anatifer, from the two Latin words *anas* and *fero*, signifying duck-bearing, was given to the barnacle from a ridiculous notion, formerly entertained, that they enclosed in an embryo state the young of the barnacle duck. Sowerby states that fossil specimens of this marine genus are found in the calcaire-grossièr, of Paris, and in other similar beds. Parkinson observes that anatifa lævis and anatifa striata are both said by Bose to be found fossil; the latter is also said by Gmelin to be sometimes found fossil, but that he believes neither of these statements to be supported by sufficient authority. He, however, gives a representation of what he believes to be a fossil barnacle in a flint stone which he found in the gravel pits near Hackney-road, and the opinion he first formed, not only of its having derived its figure from animal organization, but of its affinity to the barnacle, received corroboration from different specimens which he subsequently met with. Large bunches of barnacles attached to pieces of wood are frequently thrown up by the waves upon our coasts.

BA'ROLITE. (from $\beta\alpha\rho\dot{\upsilon}\varsigma$, heavy, and $\lambda\iota\theta$ος, a stone, Gr.) Carbonate of barytes. The Baryte carbonatée of Haüy. Barolite is found native, or it may be artificially prepared. It was first discovered native by Dr. Withering, from whom it was named Witherite. According to the analysis of Dr. Withering, carbonate of barytes consists of barytes 80, and carbonic acid 20. It is soluble in dilute nitric acid. It is poisonous. It occurs abundantly in lead veins, that traverse a secondary limestone in Cumberland and Durham, and at Anglesark, in Lancashire.

Barolite or Witherite occurs in various forms; it is found in irregular stalactitical minute crystals, opaque and white; in dodecahedral crystals formed of two hexahedral pyramids, of a pale wine-yellow colour; and in elongated hexahedral pyramids or spiculæ of a chalk white colour. It is also met with incrusting fragments of galena, blende, and limestone; sometimes forming crystallized balls of a dirty white colour, with a striated fracture; radiating from a centre. In irregular six sided prisms without pyramids, and perfectly transparent, occurring occasionally in the centre of the balls above mentioned.

BAROSE'LENITE. (from $\beta\alpha\rho\dot{\upsilon}\varsigma$, heavy, and selenite.) Heavy spar; native sulphate of barytes, or boroselenite. The Baryte sulphatée of Haüy. A mineral, found abundantly in this and other countries. It occurs both massive and crystallised. The varieties of its crystals are very numerous. It is of various colours, white, grey, yellow, brown, red, green, blue, and black. It consists of 66 per cent. of barytes and 34 per cent. of sulphuric acid. Its texture is generally foliated. When heated it decrepitates. It is soluble in dilute sulphuric acid. It is found in veins, in primary, transition, and secondary rocks.

BA'ROTE. A name given to barytes by Morveau. See *Barytes*.

BARR LIMESTONE. An argillaceous, concretionary limestone, nodular and interstratified with beds of shale. A rock of the Upper Silurian series.

BARYSTRO'NTIANITE. (from βαρὺς, heavy, and strontian.) This mineral has also obtained the name of Stromnite, from its being found at Stromness, in the island of Pomona. Its principal constituent is carbonate of strontia, of which it contains nearly 70 per cent., combined with sulphate of baryta and a small proportion of carbonate of lime and oxide of iron. It occurs massive, of a greyish colour externally, and of a yellowish white internally.

BARY'TA. } from βαρὺς, heavy, Gr.)
BA'RYTE. } Barytes has been also
BARY'TES. } called ponderous spar, heavy spar, and barote. The first account of the properties, &c. of barytes, was published by Scheele in his dissertation on Manganese. It has obtained its name from its great specific gravity, which is about 4, being the heaviest of all the known earths. It was called barote by Morveau, and barytes by Kirwan. Barytes converts vegetable blues to green. When exposed to the atmosphere it attracts moisture, and when water is poured upon it the same appearances present themselves as in the slaking of lime, with the evolution of great heat, the process being more rapid, and the evolution of heat greater. Barytes is found in two natural combinations only, with the sulphuric and carbonic acids, forming sulphate and carbonate of barytes. The primitive crystal of sulphate of barytes is, according to Haüy, a right prism with rhombic bases; the measurements by the reflecting gonionometer, according to Prof. Phillips, are 101°42′ on the one angle, and 78°11′ on the other. It is a violent poison. Nearly all the compounds of barytes are poisonous, the best antidotes being dilute sulphuric acid, or sulphate of soda in solution.

BARY'TIC. Containing barytes; resembling barytes; having the properties of barytes.

BA'SALT. (Said to be derived from an Ethiopian word, *basal,* signifying iron.) A variety of trap-rock of a dark green iron grey or brownish black colour, composed of augite and felspar with some iron and olivine, the predominant mineral being felspar. Sir H. de la Beche says that basalt is supposed to be essentially composed of augite, and titaniferous iron. Berthier says " mais c'est surtout dans les roches volcaniques que le titane abonde; il y est toujours combiné avec le fer et le manganèse."

The following are some of the usual characters of the basaltic rocks of this country; considerable tenacity and hardness, a sharp, and sometimes conchoidal fracture; a granular aspect, often reflecting light from a number of brilliant spots or striæ, some of which seem to be felspar, others hornblende or augite; very subject to superficial decomposition, in which case the colour passes to a rusty brown, often mingled with spots of green, arising apparently from grains of hornblende. Basaltic rocks are fusible at a low degree of heat, and attract the needle strongly. Sp. gr. of the basalt of Staffordshire is 2.86.

Basalt occurs, sometimes, in veins or dykes, which traverse rocks of all ages, filling up fissures or crevices, and at others, in layers spread over the surface of the strata, or interposed between them. Many modern lavas differ so little from basalt, that it is unnecessary to adduce proof of the volcanic nature of this rock. It often occurs in the form of regular pillars, or columns, clustered together; or, in scientific language, has a columnar structure, a character also observable in some recent lavas. The columnar arrangement may be more or less observed in all the trappean rocks. Basaltic columns are often curved, they are also frequently articulated. This structure is found, by some highly interesting and philosophical experiments, to have originated from the manner in which refrigeration took place. Mr. Gregory Watt melted seven hundred weight of basalt, and kept it in the furnace several days after the fire was reduced. It fused into a dark-coloured vitreous mass, with less heat than was required to melt pig-iron; as refrigeration proceeded, the mass changed into a stony substance, and globules appeared; these enlarged till they pressed laterally against each other, and became converted into polygonal prisms. The articulated structure and regular forms of basaltic columns have, therefore, resulted from the crystalline arrangements of the particles in cooling; and the concavities, or sockets, have been formed by one set of prisms pressing upon others, and occasioning the upper spheres to sink into those beneath. —*Mantell.*

On examination with a lens, even the more compact varieties of basalt are seen to be composed of minute crystalline grains. Basalt, in enormous masses, often covers the primary mountain in the Andes, and arranged in regular columns, which to the eye of the traveller appear like immense castles lifted into the sky. Basaltic dykes intersect both primary and secondary rocks. Few countries in the world present more magnificent basaltic columnar ranges than the north part of Ireland, and some of the Hebrides. The Giant's Causeway constitutes a small part of a vast basaltic range, along the north coast of Ireland, in the county of Antrim. The promontories of Fairhead and Borgue, in the same range, are situated eight miles from each other: these capes consist of various ranges of pillars and horizontal strata, which rise from the sea to the height of five hundred feet.—*Bakewell.*

The ancient inhabitants of Britain formed the heads of their battle axes, which the people called celts, from this stone. These celts resemble in shape the tomahawks brought from the South Sea Islands.

Mr. Powlett Scrope has distinguished basalt under the following arrangement; 1. Common basalt, composed of felspar, augite, and iron. 2. Leucitic basalt, when leucite replaces the felspar. 3. Olivine basalt, when olivine replaces the felspar. 4. Hauyine basalt, when hauyine replaces the felspar. 5. Ferruginous basalt, when iron is a predominant ingredient. 6. Augite basalt, when augite composes nearly the whole rock.

BASA'LTIC. Composed of basalt; resembling basalt; containing basalt.

BASA'LTIFORM. Resembling basalt in its columnar form, or structure.

BASA'LTIC HORNBLENDE. } Two names
BASA'LTINE. } given to the same mineral. A variety of common hornblende to which these names have been given from its having been found commonly in basaltic rocks. The primitive form of its crystals is a rhomboidal prism. It has by analysis been found to consist of silica 58, alumina 27, iron 9, lime 4, and magnesia 1. Its colour is black, dark-green, or yellowish-green. Texture foliated.

BA'SANITE. (βάσανος, Gr. *lapis quo probatur aurum*, *lapis Lydius*.) Lydian stone, a variety of schistose hornstone. This stone acquired

H

its name from its having been formerly used as a touchstone in trying the purity of metals: it also was called Lydian stone, from its being found abundantly in Lydia. According to an analysis of it, its constituents are, silica 75 per cent., lime, magnesia, carbon, and iron. See *Lydian Stone*.

BASILOSAU'RUS. The name of an enormous fossil reptile, described by Dr. Harlan of Philadelphia. Neither the relation of the basilosaurus to other species, nor its geological position, has been accurately determined.—*Mantell.*

BASE. (from *basis*, Lat. βάσις, Gr. *base*, Fr. *basa*, It.)
1. The bottom, or lowest part, of any thing.
2. In conchology, that part of the shell in univalves by which they are attached to rocks, or other substances; in multivalves, the opposite extremity to the apex.
3. The substance to which an acid is united; as in oxide of copper, the copper is the base.

BA'SIN. (*bassin*, Fr. *bacino*, It.) In geology, large concavities filled with deposits, as the London basin, the Paris basin, &c. are called basins. The surface of the earth is covered with a series of irregular depressions or basins, divided from one another, and sometimes wholly surrounded by projecting portions of subjacent strata, or by unstratified crystalline rocks, which have been raised into hills and mountains of **various** degrees of **height**, direction, and continuity. This disposition in the form of basins, which is common to **all** formations, has been more particularly observed in the carboniferous series, from the beds of coal contained therein having been **wrought** throughout their whole **extent**. In consequence of this **basin-shaped** disposition of the carboniferous strata a most beneficial result

obtains, namely, that these strata, which an uninterrupted inclination in one direction only would soon have plunged into depths inaccessible to man's greatest efforts, are, by their being placed around the circumference of the basin, all brought sufficiently near the surface to be attainable, and are thus made subservient to his benefit and comfort.

BA'SSET. A term, used by miners, to express an upward slanting direction of a vein, from below to the surface.
2. The mergence of strata in succession from each other.
The angle of inclination between these planes and that of the horizon, is called their dip or pitch.

BA'SSETTING. Slanting upwards.

BATH-STONE. A species of limestone, called also Bath-oolite. This member of the oolite formation has been called the great oolite; it is of considerable thickness, and yields an abundant supply of freestone for building. It has obtained the name of oolite from its being composed of small rounded grains, or particles, supposed to resemble the roe of a fish. Bathstone consists of minute globules, cemented together by yellowish earthy calcareous matter, and a considerable portion of broken shells. Each of the little grains, rarely exceeding the size of a pin's head, would, if broken open, be found to be composed of concentric coats or shells, sometimes hollow in the centre, sometimes enveloping a small grain. When these grains are larger, and of the size of peas, the rock is called Pisolite, or Peastone. When first quarried, Bathstone is soft, but it soon becomes hard by exposure to the atmosphere.

BATH-OOLITE. See *Bath-stone*.

BATRA'CHIA. (from βάτραχος, a frog, Gr.) The second order of the class Amphibia, comprising the frog, toad, &c.

BA'TRACHITE. (from βατράχειος, Gr. *batrachites*, Lat.) A fossil of the colour of a frog; a fossil frog; a fossil resembling a frog, either in form or colour.

Under the name of batrachite or bufonite, are found figured, in the works of some oryctologists, a great number of fossil bodies, more or less rounded and shining, which are evidently portions of the teeth, or deretary palates, of fishes. The above names were given because it was imagined that they had been engendered in the heads of toads or frogs. See Bufonite.

BEAK.
1. In conchology, the continuation of the body of univalves in which the canal is situate.
2. In ornithology, the bill, or horny mouth of a bird.
3. In botany, applied to an elongation of the seed-vessel; proceeding also from the permanent style; also to naked seeds.

B'BAR. The name given for a block of sandstone, which, having been exposed to an intense heat in the furnace, has become converted into a substance resembling quartz rock.

BEARD. (from *barba*, Lat. *barbe*, Fr. *barba*, It.)
1. In botany, a bristle-shaped projection, growing out from the glume or chaff, in corn and grasses; called also the awn.
2. In conchology, the process by which some univalves adhere to rocks, &c.

BED. A stratum of considerable thickness. It is desirable that the geological student should draw a distinct line between the words bed and stratum. Whenever a layer, or stratum, is of the thickness of two yards or more, it should be denominated a bed, but otherwise a stratum. There are sometimes found many distinct strata in the thickness of an inch; to denominate these as beds would be absurd. Let it therefore be kept in mind that the words bed and stratum are not synonymous. By a *bed* there should be understood a series of layers, or a succession of deposits of earthy matter so continuously formed that the whole adhere more or less firmly together.

BE'ETLE. A coleopterous insect, the scarabæus of Linnæus. Remains of beetles have been found in the oolite: wing covers of beetles occur in the shale of the Danby coal-pits, in the eastern moorlands of Yorkshire.

BE'ETLE. To jut out; to hang over: thus rocks are said to beetle.

BE'ETLE-STONE. A name given to coprolites, from their falsely imagined insect origin.

BEHE'MOTH. A huge animal spoken of in the Scripture, supposed by some to mean the elephant, by others the ox, and by Bochart the hippopotamus.

BE'LEMNITE. (from βέλεμνον, Gr. a dart.) Belemnite, thunderstone, or arrow-head. An extinct genus of chambered molluscous aminals, having a straight tapering shell. Belemnites are found in the secondary formation only, the lowest stratum containing their remains being the muschel-kalk, and the highest the upper chalk of Maestricht. M. De Blainville has given a list of ninety-one authors, from Theophrastus downwards, who have written on the subject of belemnites. The most intelligent of these agree in supposing these bodies to have been formed by cephalopods allied to the modern sepia. That fossil which is called a belemnite was a compound internal shell, made up of three essential parts, which are rarely found together in perfect preservation. The belemnite is one of the most common fossils of the chalk, it resembles an elongated conical stone, of a crystalline, radiated structure, and is generally of a brown colour: some limestones on the continent of Europe are almost wholly composed of them. Ink-bags, resembling those of the Loligo, have been found in connection with belemnites in the lias at Lyme Regis; these, in some instances, are nearly a foot long, and prove that the animal to which they belonged must have been of great size. The fact of these animals having been provided with a reservoir of ink, affords an *à priori* probability that they had no external shell, but recent discoveries decide the question, two specimens having been found each containing an ink-bag within the anterior portion of the sheath; and, consequently, all the species of belemnites may henceforth with certainty be referred to a family in the class of Cephalopods. Eighty-eight species of belemnites have already been discovered; and the vast numerical amount to which individuals of these species were extended, is proved by the myriads of their fossil remains that fill the oolitic and cretaceous formations.—*Buckland, Bakewell, Mantell.*

BELE′MNO-SE′PIA. The name proposed to be given by Professor Buckland, in concurrence with M. Agassiz, to a new family of cephalopods, to which family may be referred every species of belemnites.

BELLE′ROPHON. An extinct genus of mollusca, belonging to the order Heteropoda, found in the Silurian and Carboniferous rocks, the shell of which was without chambers. De Montfort placed the bellerophon among chambered shells; De Blainville assigned their position next to Bulla.

BELO′PTERA. A genus of fossil fishes established by Deshayes, of which no recent species is known; Cuvier considered that the remains thus placed in a distinct genus were merely portions of some sepia.

BI′VALVE. (*bivalvis*, Lat. *bivalve*, Fr.) An animal having two valves, shells, or shutters, as the oyster, muscle, &c.

BE′MBRIDGE BEDS. These are of the Eocene period, and consist of the upper marls, the lower marls, the oyster-bed, and the limestone bed.

BE′RENICIA. The name given by Lamouroux to a genus of fossil corals. One species, B. irregularis, is found in the Wenlock limestone, at Dudley. It is thus described—"opening of the cells round, distant where the surface is flat, generally near together where it is uneven; more or less regularly disposed from a centre.—*Murchison. Silurian System.*

BERG′MANNITE. A mineral so called in honour of the celebrated chemist Bergmann. It occurs massive and is of a greenish or greyish white: it has been found in Norway.

BE′RYL. (*beryllus*, Lat. *beryl*, Fr.) A crystallised compound of the earth glucina with silica, alumina, lime, and oxide of iron. The beryl is a gem, or precious stone, of the genus emerald, but less valuable than the emerald. It differs from the precious emerald in not possessing any of the oxide of chrome, from the presence of which the emerald obtains its splendid green colour. The aqua-marine is a variety of the beryl, having a more transparent texture. The beryl is of a greyish-green colour, blue, yellow, and sometimes nearly white; occasionally different colours appear in the same stone. Beryl is found in many parts of the world, but the finest specimens are brought from Siberia. Vauquelin first discovered the earth glucina from analyzing the beryl. Some mineralogists consider beryl and emerald to differ merely in their colours, presenting, as they state, an uninterrupted series; others, however, constituting emerald a species, divide it into two sub-species, namely, precious emerald and beryl. Specific gravity from 2·65 to 2·75. Hardness = 7·5 to 8.

BE′RYX LEWESIE′NSIS. A fossil discovered in the Lewes chalk quarries, of the length of twelve inches, greatly resembling the dory, and by the workmen called the Johnny Dory. This is the most abundant of the Sussex ichthyolites; its scales are very frequent in all the pits of the South Downs, as well as in those of Surrey and Kent.—*Mantell.* Cuvier places the beryx in the family Percoides, order Acanthoptergii.

BE′RYX RA′DIANS. A fossil fish from the chalk-marl, of the length of seven inches. This, like the Beryx Lewesiensis, last described, belongs to the family Percoides, order Acanthoptergii.

BICA′PSULAR. Having two capsules, or seed vessels.

BICI′PITAL. } (from *biceps,* Lat.)
BICI′PETOUS. } Having two heads. It is a term applied to muscles, which have two distinct origins.

BICO′RNUS. (*bicornis,* Lat.) Having two horns.

BICU′SPID. (from *bis* and *cuspis,* a spear, Lat.) Two pointed; two-fanged.

BI′FID } (from *bifidus,* Lat.) Cleft,
BI′FIDATED } or cloven, into two; opening with a cleft; two-cleft, but not very deeply divided.

BIGE′MINATE. In botany, applied to a compound leaf, having a forked petiole, with several petioles, or leaflets, at the end of each divsion.

BILA′BIATE. (from *bis* and *labium,* a lip, Lat.) Two-lipped; furnished both with an outer and inner lip.

BI′LDSTEIN. (from *bild,* shape, and *stein,* stone, German.) A massive mineral, with sometimes an imperfect slaty structure. It is also called agalmatolite. By M. Brongniart it has been named steatite pagodite, but it is wanting in magnesia, which is present in all steatites.

BILO′BED. } (from *bis* and *lobus,* Lat.)
BILO′BATE. } Divided into two lobes.

BILO′CULAR. (from *bis* and *loculus,* Lat.) Two-celled; divided into two cells.

BIMA′RGINATE. In conchology, furnished with a double margin as far as the lip.

BIMA′NA. (from *bis* and *manus,* Lat.) The first order of the class mammalia; this order consists of but one species, viz., man. Some naturalists, amongst whom are Ray, Brisson, Pennant, Swainson, Daubenton, &c., &c., would exclude man from the pale of the animal kingdom; others have regarded him as an example of an order per se. From one of the structural peculiarities of the race, the possessing two hands, Cuvier has applied to the order the term Bimana, which is now very commonly adopted.

BI′NARY. (*binareus,* Lat.) Arranged by twos; containing two units.

BI′NATE. (from *binus,* Lat.) Two and two; by couples; growing in pairs; a fingered leaf of two leaflets, inserted at the same point, precisely on the summit of the petiole.

BIND. Called also clunch; a name given to the soil on which the coal strata rest. An argillaceous shale, more or less indurated, sometimes intermixed with sand resembling sandstone, but generally decomposing into a clayey soil on exposure to the atmosphere.—*Bakewell.*

BINO′XIDE.—When oxygen combines with another substance in the proportion of two equivalents of oxygen to one of the other, the result is a binoxide, or dentoxide.

BI′OTITE. A mineral, called also Magnesia Mica, a variety of mica, containing from 10 to 25 per cent. of magnesia, and only 12 to 20 of alumina.

BIPA′RTITE. (from *bis* and *partitus*, Lat.) Having two correspondent parts; an epithet for the corolla, leaf, and other parts of plants, when divided into two correspondent parts at the base.

BIPE′NNATE. } (*bipennis*, Lat. Having
BIPE′NNATED. } two wings.

BIPE′TALOUS. (from *bis*, Lat. and πέταλον, Gr.) Consisting of two flower leaves; having two petals.

BIPI′NNATE. (*bipinnatum*, Lat.) Doubly pinnate: applied to a compound leaf, having a common petiole, which produces two partial ones, upon which the leaflets are inserted.

BIPINNA′TIFID. Having pinnatifid leaves on each side the petiole.

BIRA′DIATE. } (from *bis* and *radiatus*,
BIRA′DIATED. } Lat.) Consisting of two rays.

BIRHOMBOI′DAL. Having a surface of twelve rhombic faces.

BIRO′STRATE. (from *binus* and *rostrum*, Lat.) Having a two-beaked prominence; two beaked.

BIROSTRI′TES. A fossil bicornuted bivalve with conical umbones.

BI′SMUTH. (*bismut*, German, *bismuth* and *bismut*, Fr.) A metal of a reddish-white, or cream colour. It is neither malleable nor ductile, its specific gravity is 9·8, it fuses at a temperature of 476 Fahrenheit. In hardness it is intermediate between gold and silver. Bismuth unites with most metals, rendering them generally more fusible, and in some cases remarkably so. Eight parts of bismuth, five of lead, and three of tin, constitute what has been called Sir I. Newton's fusible metal, which liquefies at the temperature of boiling water, 212°, and may be fused over the flame of a candle in a piece of stiff paper. Bismuth was discovered in the early part of the sixteenth century, and is mentioned by Bermannus. It occurs in veins in primitive rocks, as gneiss, granite, mica-slate, and clay-slate, in Saxony, Bohemia, France, Sweden, and Cornwall. One part of bismuth with five of lead and three of tin form the soft solder used by pewterers; it is also used in the manufacture of printer's types. It is with a compound of two parts of bismuth, one of lead, one of tin, and four of mercury, the whole being fusible at a temperature under that of boiling water, that glass globes are silvered on the inside; a piece of this compound being placed within the globe, the latter is plunged into hot water, the metallic compound readily melts, and the globe being turned round, the fluid metal is spread over the internal surface.

BISU′LCATE. } (*bisulcus*, Lat.) Cloven
BISUL′COUS. } footed, as the ox, or the pig.

BITTER SPAR. See *Lolomite*.

BITU′ME, } (*bitumen*, Lat. *bitume*, It
BI′TUMEN. } *bitume*, Fr, *matière liquide, épaisse, noire et inflammable, qui se trouve dans le seine de la terre, et dont on prétend qu'on se servoit autrefois au lieu de ciment*.) The term bitumen is applied to a number of inflammable substances found in the earth, or issuing from the earth's surface, and these are known under their names of naphtha, petroleum, mineral tar, mineral pitch or maltha, asphalt, elastic bitumen, jet, mineral coal, amber, and mineral tallow. These, however, may perhaps be more correctly called bituminous varieties Bitumen is a substance of a peculiar kind, seeming to partake both of an oily and resinous nature, and is found either buried in, or proceeding from, different parts of the earth, in different states of consis-

tence. Bitumen is composed of carbon and hydrogen. It appears that formerly bitumen was generally used instead of mortar, and authors suppose that the tower of Babel, the walls of Babylon, of Sodom, and other places, were built of bricks cemented together by bitumen, and that the ark of Noah, and the vessel of bull-rushes in which Moses was exposed, were coated with this substance. Bitumen, when fluid, has been called by some Latin writers, oleum vivum. Lyell says that the tar-like substance, which is often seen to ooze out of the Newcastle coal when on fire, and which makes it cake, is a good example of bitumen. Mr. Hatchett says "we may, with the greatest probability, conclude that bitumen is a modification of the resinous and oily parts of vegetables, produced by some process of nature, which has operated by slow and gradual means on immense masses, so that, even if we were acquainted with the process, we should scarcely be able to imitate the effects, from the want of time, and deficiency in the bulk of the materials. But although bitumen cannot at present be artificially formed from the resinous and other vegetable substances by any of the known chemical processes, yet there is every reason to believe that the agent employed by nature in the formation of coal and bitumen has been either muriatic or sulphuric acid." The varieties of bitumen will be separately described under their different names.—*Parkinson. Lyell. Bakewell.*

BITU′MINATED. (*bituminatus*, Lat.) Prepared with bitumen; impregnated with bitumen.

BITUMINI′FEROUS. Yielding bitumen; containing bitumen.

BITU′MINISE. To prepare with, or coat with, bitumen. 2. To convert into bitumen.

BITUMINISA′TION. The preparing, or impregnating, with bitumen.

BITUMINIOUS SHALE. (The Brandschiefer of Werner: le schiste bitumineux of Brochant: schiste argileux bitumineux of Brongniart.) An argillaceous shale, much impregnated with bitumen, very common in the coal measures. A blackish-brown, or greyish, subvariety of shale; fracture slaty; usually soft and unctuous to the touch. Specific gravity about 2. When placed in the fire, it blazes, crackles, and gives out a black smoke and bituminous odour, and is converted into a whitish or reddish ash: it sometimes effervesces slowly with acids.

BITU′MINOUS SPRINGS. We are informed by Sir C. Lyell that springs impregnated with petroleum, and the various minerals allied to it, are very numerous, and are, in many cases, undoubtedly connected with subterranean fires. The most powerful yet known, are those on the Irawadi, in the Burman empire, which, from one locality, are said to yield 400,000 hogsheads of petroleum annually.

BLACK CHALK. (the Zeichen scheifer of Werner; argile schisteuse graphique of Haüy; ampelite graphite of Brongniart; le schiste à dessiner of Brochant.) A clay of a bluish black colour, extremely soft, a quality which it owes to the presence of about twelve per cent. of carbon. That most esteemed is found in Italy and Germany, and takes its name from those countries respectively. It is massive, opaque, soils lightly, and writes; retains its colour in the streak, and becomes glistening; is soft, sectile, does not adhere to the tongue, feels fine, but meagre, and is infusible. According to Wiegleb, a variety from Bayreuth contained silica 64·50, alumina

11·35, oxide of iron, 2·75, carbon 11, water 7·5.

BLACK-JACK. A name given by miners to a sulphuret of zinc. See *Blende*.

BLACK-LEAD. The substance about to be described has been thus named from its leaden appearance, or general resemblance to lead, but it does not in fact contain a single particle of lead in its composition. It is the same as plumbago and graphite. Black-lead is a compound of carbon, with a small portion of iron, and some earthy matters. It is of a dark steel-grey colour, inclining to iron-black; it occurs regularly crystallised; in granular concretions; massive and disseminated; it has a greasy feel, and blackens the fingers, or any other substance to which it is applied; it is infusible; and burns with much difficulty. According to Vauquelin its constituent parts are carbon 92., iron 8.; but according to Allen and others, it contains only five parts per centum of iron. Its nature was first investigated by Scheele, who, by combustion, converted nearly the whole into carbonic acid gas, the residuum being oxide of iron. Black-lead, or carburet of iron, is used for many domestic purposes, but its principal use is in its manufacture into black-lead pencils. It is found in the primary, transition, and secondary rocks. Anthracite resembles and appears to pass into plumbago, or black-lead; common coal, also, according to Bakewell, sometimes graduates into plumbago.

BLACK-WADD. An ore of manganese, used as a drying ingredient in paints.

BLASTOIDE'A. An extinct order of the class Echinodermata. The genera, which are all extinct, are Pentremites, Eleacrinus, Codonaster, Zygocrinus, and Phyllocrinus.

BLA'TTA. (*blatta*, Lat.) The cockroach, placed by Linnæus in the second order, Hemiptera, of the seventh class, Insecta.

BLENDE. (from *blenden*, Germ. to dazzle, or blind.) Sulphuret of zinc; a metallic ore whose constituent parts are zinc, iron, sulphur, and a trace of quartz. Blende is called by the English miners black-jack; it occurs in the lead mines. The primitive form of its crystals is a rhomboidal dodecahedron; there are several varieties known, as brown blende, yellow blende, black blende.

BLOOD-STONE. (See *Heliotrope*.) Hematites; a variety of agate to which the name bloodstone has been applied from some absurd notion of its efficacy in restraining hemorrhage.

BLUE-JOHN. A name given by the miners to fluor spar; called also Derbyshire spar, in which county it occurs in great abundance. It is manufactured into vases and ornamental figures, being capable of being turned by the lathe. Bakewell, in describing the blue-john, or fluor spar mine near Castleton, in Derbyshire, observes, "the crystallizations and mineral incrustations on the roof and sides of the natural caverns which are passed through in this mine, far exceed in beauty those of any other cavern in England; and were the descriptions of the grotto of Antiparos translated into the simple language of truth, I am inclined to believe it would be found inferior in magnificence, and splendour of mineral decoration, to the natural caverns of the fluor mine."

BLUE VITRIOL. Sulphate of copper.

BLUFF. Any high head-land, or bank, presenting a precipitous front.

BOG-IRON-ORE. } Iron ore peculiar to
BOG-ORE. } boggy land. Sir C. Lyell observes, "at the bottom of peat mosses there is sometimes found a cake, or pan, of oxide of iron, and the frequency of bog-iron

ore is familiar to the mineralogist. From what source the iron is derived is by no means obvious, since we cannot in all cases suppose that it has been precipitated from the waters of mineral springs. It has been suggested that iron, being soluble in acids, may be diffused through the whole mass of vegetables, when they decay in a bog, and may, **by its** superior gravity, sink to the **bottom**, and be there precipitated, **so as to form bog-iron ore**. Dr. Mantell **observes**, " the formation of what is termed bog-iron ore, found in marshes and peat bogs, is supposed to have been derived from the decomposition of rocks over which water has flowed; but the observations of Ehrenberg, seem to indicate a different origin." Ehrenberg discovered that bog-iron consists of innumerable articulated threads, of a yellow-ochre colour, composed partly of flint, and partly of oxide of iron; these threads being the cases of a minute animalcule, termed Gaillonella ferruginea.

BOLE. (from βῶλος. Gr.) A friable clay, or earth, often highly coloured with iron. It occurs in solid amorphous masses of a yellow red, or blackish-brown colour, or pitch **black; it is** found in wacke, and basalt, **from the** decomposition **of** which it may **arise**.

BOLE'TUS. A genus of mushrooms, **of** the order Fungi.

BOLO'GNA STONE. } A variety of sulBOLO'GNIAN STONE. } phate of barytes, possessing phosphoric properties. These properties were first discovered accidentally by Vicenzo Casciarolo, an Italian shoemaker.

BORA'CIC ACID. A compound of boron and oxygen, containing about 26 per cent. of boron and 74 of oxygen. It is found native on the edges **of** certain hot mineral springs in Italy. It occurs in scaly crusts, or small pearly scales, and massive. **Homberg was the** discoverer **of boracic acid. Boracic** acid may be obtained by **adding to** a solution of borax half **its weight** of sulphuric acid. The term **Sassolin** has been applied to **boracic** acid, from its presence **in the hot** springs of Sasso.

Bo'RACITE. Borate of magnesia; a combination of boracic acid with magnesian earth. Boracite is found imbedded in gypsum, in Hanover and Holstein; its colours are white and greyish; it is generally of a cubic form, and possesses, when heated, strong electrical properties. Borate of magnesia may be artificially obtained. Before the blowpipe boracite froths, emits a greenish light, and is converted into **a** yellowish enamel.

Bo'RATE. A combination of boracic acid with any saturated base; a salt formed by the **combination of any base with the acid of borax.—** *Parkes*.

Bo'RAX. (*borax*, Lat. *borax*, **Fr.** *borace*, It.) Subborate or borate **of** soda; a salt of soda formed of the alkali with boracic acid. It is prepared artificially by purifying the natural borate of soda, a mineral found in Thibet, where it is held in solution in the waters of a lake, which also contains common salt. Borax in its impure state is called *tincal*—and is purified by calcination, solution, and crystallization, after its importation. Borax is an important article in the arts, as a flux in the reduction of many metals, especially in assaying; it is also used in medicine. According to Berzelius, borax consists of soda 16·31, boracic acid 36·59, water 47·1. Bergmann states the proportions to be boracic acid 34, soda 17, water 49; and Kirwan gives as his analysis, boracic acid 84, soda 17, water 47.

BORE. A violent rush of tidal water.

I

Bo′RON. The undecomposable base of boracic acid; this may be obtained by heating in a copper tube two parts of potassium with one of boracic acid previously powdered and fused. Boron is a dark olive coloured substance, having neither taste nor odour, insoluble in water, and a non-conductor of electricity. It is about twice the weight of water. Its symbol is B.

Boss. (*bosse*, Fr.) A protuberance or swelling; a kind of knob or stud.

BOSSED. Knobbed or studded.

BOTHRIOLE′PIS. The name assigned to a genus of ichthyolites of the Old Red Sandstone, and described by Agassiz in his Poissons Fossiles.

BOTHRODE′NDRON. (from $βόθρον$ and $δένδρον$, Gr.) An extinct genus of coniferæ belonging to the coal formation. The bothrodendron has a stem not furrowed, covered with dots. Scars of cones, obliquely oval. The stems are marked with deep oval or circular concavities, which appear to have been made up by the bases of large cones. These cavities are ranged in two vertical rows, on opposite sides of the rock, and in some species are nearly five inches in diameter.— *Prof. Buckland.*

BO′TRYOID. } (from $βότρυς$, a bunch
BOTRYOI′DAL. } of grapes, and $εἶδος$, form.) Resembling a bunch of grapes; clustered like grapes.

BOTRY′OLITE. (from $βότρυς$, a cluster of grapes, and $λίθος$, Gr.) "A mineral presenting an aggregation of large sections of numerous small globes is termed *botryoidal*; but when the globes are larger, and the portions are less and separate, the appearance is expressed by the term mammillated." *Phillips.* Grape stone; a variety of prismatic datolite, occurring in mammillary concretions.

BO′VEY COAL. A name given to wood-coal, from its having been found in abundance at Bovey Heathfield near Exeter; called also brown-coal. In wood-coal we may almost seize nature in the act of making coal, before the process is completed. These formations of coal are far more recent than those of common coal, and have been referred to the first, or Eocene, period of the tertiary formations. Heat and pressure appear to be required to convert wood coal into mineral coal. Bovey-coal contains carbon 77·19, oxygen 19·34, hydrogen 2·54, earthy parts 1.

"The precise age of the Bovey coal," says Sir H. de la Beche, "cannot at present be well determined. A body of water has passed over it, working hollows in the clay, and leaving a large deposit of transported substances in some situations. It also appears to have been tranquilly deposited in a previously existing depression. The area comprising the surface of this deposit is far more extensive than is usually given, and it has certainly once occupied a greater elevation, as a mass, than it now does, the upper portion having been removed by denudation. The principal deposit of lignite occurs near Bovey Tracy, in Devonshire, at the north-western end of the deposit. Beneath about twenty feet of what the workmen term the *head*, there is an alternation of compressed lignites; shales or clays. The lignite is composed of dicotyledonous trees, many of which are knotted. The most useful product of this deposit is a clay used in the potteries, in some cases so fine as to constitute what is termed pipe-clay. Lignite more or less accompanies the clay throughout, occurring either in beds or in small detached pieces. Animal remains must be rare; for I could not, after diligent search, obtain any traces of them, though I was told some shells had been seen at Teignbridge.

According to Mr. Whiteway and Mr. Kingston, the Bovey deposit consists chiefly of five clay beds, and as many of gravel, the latter varying from 100 to 50 feet in width. The clay beds are described as undulating, like the waves of the sea; and it is stated that beneath the four more western beds the Bovey-coal is found."

BOU′LDER. ⎱ Large fragments, or roun-
BO′WLDER. ⎰ ded masses of any rock found lying on the surface, or, sometimes, imbedded in soil, and differing from the rocks where they are found; these fragments, or outlying boulders, are of no determinate size, they are supposed to have been transported by the force of water, and are occasionally found at very great distances from their parent rocks. The mass of rock on which is placed the statue of Peter the Great at St. Petersburgh, is a detached block of granite, or a boulder, forty-two feet in length, twenty-seven feet broad, and twenty-one feet high, and was removed from the Gulf of Finland.

The Hon. Fox Strangways says "the celebrated block out of which the pedestal which supports the statue of Peter the Great is cut was a rolled boulder of the red Finland granite. It was not brought, as has been asserted, from Siberia, nor, by human means at least, from Finland: but was found, among many others of smaller size, in a bog between Petersburg and Oesterbeck. It was diminished two thirds, before placing the statue on it."

BRACHELY′TRA. A family of coleopterous insects, having but one palpus to the maxillæ. It comprises only one genus, namely Staphylinus.

BRA′CTEA. (*bractea*, Lat.) In botany, a leafy appendage to the flower or stalk, differing from the other leaves of the plant in form or colour; the floral leaf. Bracteæ vary greatly in appearance; most commonly they are green and herbaceous. The leaf, in the axilla of which a flower-bud is produced, is called a bractea. The most remarkable sort of bractea is that called spathe; the spathe or bractea of many flowers is membranous. When two or more bracteæ, instead of appearing singly on the principal flower-stalk, are opposite, or verticillate, they form an involucrum.

BRA′CHIAL. (from *brachium*, an arm, Lat. *brachial*, Fr. *del braccio*, It.) Belonging to the arm.

BRA′CHIATE. (*brachiatus*, Lat.) Four-ranked; applied to stems, when they divide and spread in four directions, crossing each other.

BRANCHI′FERA. In the conchological system of De Blainville, we find branchifera placed in the order Cervicobranchiata, and it comprises three genera, namely, Fissurella, Emarginula, and Parmophorus.

BRACHIO′PODA. (from $\beta \rho \alpha \chi \iota \omega \nu$, an arm, and $\pi o \hat{\upsilon}s$, a foot, Gr.) Animals having arms instead of feet. The brachiopoda are all bivalves. Brachiopoda forms the first order of the class Molluscoidea.

BRACHIO′PODOUS. Having arms in the place of feet and legs; belonging to the class Brachiopoda.

BRACKLESHAM BEDS. These beds constitute a portion of the Bagshot series, lying between the Lower Bagshot and the Barton Clay. They take their name from Bracklesham in Sussex, but are best seen in the Isle of Wight: their thickness in some places is about 110 feet.

BRADFORD CLAY. A blue unctuous clay, occurring at Bradford, a member of the Cornbrash group; it is full of Apiocrinites Parkinsoni.

BRADY′PODA. Slow-moving animals, with their bodies generally covered by a hard crust. Some want the incisor teeth; others want the incisors and cuspidati; in others, the jaws are destitute of teeth. Placed by Linnæus in the class Mammalia, and composing the third order. The order Bradypoda includes the genera Bradypus, or Sloth, Mermecophaga, Ant-eaters, Manis, Scaly Lizard or Pangolin, Dasypus, Armadillo, and Ornithorrhynchus. Duck-billed animal.

BRA'DYPUS. (βραδύπους, Gr.) The sloth, a genus of the order Bradypoda, class Mammalia. These animals have no fore-teeth; they have six grinders in either jaw; and their bodies are covered with hair. There are several species.

BRA'NCHIA. (from βράγχια, Gr.) This word is rarely used substantively; it makes branchiæ in the plural. Branchiæ are filamentous organs for breathing in water; gills. Respiration is effected by the transmission of water through the mouth, over the surface of the fringe-like branchiæ; and the blood is transmitted to the gills from the ventricle, whence, instead of returning immediately to the heart, it is conveyed by the branchial veins to the body; these veins after giving branches to the anterior parts, unite to form the aorta, which sends the arterialized blood through the rest of the system, without the aid of a systemic heart.

BRANCHIO'PODA. (from βράγχια, and πούς, Gr.) The third order of the class Crustacea.

BRANCHIO'PODOUS. Gill-footed; belonging to the order Branchiopoda.

BRANCHIO'STEGI. (from βράγχια, gills, and στέγος, or στέγη, a covering.) A term used to express one of the orders of fishes, the characters of which are, that the rays of the fins are of a bony substance.

BRANCHIO'STEGOUS. Having the characters of the branchiostegi; belonging to the order Branchiostegi; having the gills covered.

BRA'NCHIPUS. The cancer stagnalis of Linnæus; an animal belonging to the crustaceans, having the legs reduced to soft paddles, and combining the functions of respiration with those of locomotion. In the branchipus, we find antennæ, but no crustaceous legs. The soft branchiæ of branchipus perform the double office of lungs and feet.

BRAU'NITE. (The Brachytypes manganerz of Mohs.) A mineral of a brownish black colour, occurring massive and crystallized, consisting of protoxide of manganese 87 per cent.; oxygen 10 per cent.; baryta 2.26 per cent.; and water nearly 1. It has been thus named after Mr. Braun of Gotha.

BRE'CCIA. (Ital.) Any rock composed of angular fragments cemented together.

BRE'CCIATED. Composed of angular fragments united into a mass by cement.

BRE'CCIATED AGATE. A beautiful variety of agate, consisting of fragments of ribbon agate united by a base of amethyst, it is found in Saxony.

BREI'SLAKITE. A Vesuvian mineral, thus named after Breislak.

BRE'WSTERITE. An earthy mineral, so named after Sir D. Brewster. It occurs in small white, or yellowish coloured crystals, and consists of silica 53.65, alumina 17.4, strontia 8.32, baryta 6.75, lime 1.34, water 12.58, oxide of iron 0.29. Sp. gr. 2.1, hardness 5 to 5.5.

BRI'LLIANT. (brillant, Fr. brillante, It.) A diamond cut into angles, by which the rays of light are refracted, and a greater brilliancy is obtained.

BRI'NDED. } In conchology, streaked.
BRI'NDLED. }

BRI'STOL-STONE. } Rock-crystal,
BRI'STOL-DIAMOND. } or crystallized quartz. Very fine specimens are found in the rocks near Bristol, and these have thence obtained the name of Bristol diamonds. They

are pure silica, crystallized in six-sided prisms, terminated by six-sided pyramids.

BROME. } (from βρῶμος, odour, Gr.)
BRO′MINE. } A simple, or elementary substance, being non-metallic. Bromine exists in sea water, either as bromine of sodium or bromine of magnesium; it was discovered in 1826 by M. Balard, of Montpellier. At common temperatures it is liquid, of a dark red colour by reflected, of a hyacinth-red by transmitted light. At 116° Fah. it boils; between 0 Fah. and —4 it congeals. The density of its gas is 5·54; its equivalent 126·3; its symbol Br. It acts powerfully on animal substances, and is extremely poisonous. It unites with all the simple bodies, and with the metals, forming, with the latter, a class of compounds called bromides.—*Phillips.* Bromine has obtained its name from its powerful and unpleasant smell; its odour is suffocating, and its taste pungent.

BRO′NTIA. (from βροντή, Gr. thunder, from its being supposed that these fossils were thrown to the earth by thunder.) A fossil echinite of the family Cidaris.

BRO′NZITE. A mineral called by Werner Blättriger anthophyllite, and by Haüy Diallage metalloide. It has a yellowish brown colour, with a semi-metallic lustre. It is found in serpentine, in Shetland and in Upper Styria, and in greenstone in many other places. According to Klaproth, it consists of, silica 60; magnesia 27·5; oxide of iron 10·5; water 0·5. Hardness = 4—5. Specific gravity = 3—3·3.

BROWN-COAL. A fresh-water formation of the tertiary series, but to which sub-division of the tertiary period it may belong is considered uncertain by Sir C. Lyell, from the extreme rarity of shells found in it. Professor Buckland states, " In some parts of Germany this brown-coal occurs in **strata of more** than thirty feet **in thickness, chiefly** composed **of trees which have been** drifted, apparently by fresh **water,** from their place of growth, **and** spread forth in beds, usually alternating with sand and clay, at the bottom of then existing lakes or estuaries." The varieties of wood found in the brown coal strata are said to belong entirely to dicotyledonous trees; but among the impressions of leaves some have been referred to a palm, by Prof. Lindley, and others resemble the Annamomum dulce, and Podocarpus macrophylla, all indicating a warm climate.

"The place in **the series of the** supracretaceous rocks," says Sir H. de la Beche, "to which **the brown** coal formation of Germany **should** be referred, does not appear **to be** as yet well determined. This deposit is characterized **by an** immense quantity of vegetable remains, and is probably **of** different ages. The brown-coal may be traced from the environs of Aix-la-Chapelle to the Rhine. It there occurs in a narrow plateau between the latter and the Erst, and acquires a thickness of above 100 feet between Bonn and Cologne, without any extraneous bed. The brown-coal deposit rests on the declivity of the grauwacke mountains on the right bank of the Rhine, and is connected with the trachytic conglomerates and basaltic formations of the Siebengebirge. It extends to the country around Leipsic, to the Elbe as far as Forgan, and occurs frequently in the low tracts between Magdeburg and the Hartz. It is found in the level country between the Elbe and the Oder. It generally rests upon a compact tenacious clay, and is covered by large masses of sand."—*Geological Manual.*

In this country the brown-coal is represented by the strata of bituminous wood, called Bovey-coal, which exhibits a series of gradations from the most perfect ligneous texture to a substance nearly approaching the characters of pit-coal.

BRU′CITE. Called also Chondrodite and Hemiprismatic Chrysolite. This mineral was named Brucite, after Bruce, an American mineralogist; it occurs massive and in small grains; colours from a pale yellow to a brownish red; it possesses a vitreous lustre, is translucent, with a fracture imperfectly conchoidal. It contains a small portion of fluoric acid, and scratches glass. It is found in America, Scotland, and Finland. It consists of magnesia, silica, fluoric acid, about four per cent. oxide of iron, potash, and water.

BRYOZO′A. See *Polyzoa*.

BUCA′RDIUM. An acephalous bivalve, having powers of locomotion.

BU′CCAL. (*bucca*, Lat. the cheek; *buccale*, Fr. as *glandes buccales, artère buccale.*). Pertaining to the cheek.

BU′CCINUM. (*buccinum*, Lat.) The whelk. An ovate elongated univalve; opening oblong, notched in the lower part, and with no canal; columella convex, full and naked. —*Parkinson*.

Linnæus places this genus, the Buccinum, or Whelk, in the order Gasteropoda, class Mollusca: it comprises all the shells furnished with an emargination inflected to the left, and in which the columella is destitute of plicæ.. Many fossil species have been discovered, the greater number in the crag; some in the London-clay; six species have been found in the environs of Paris. De Blainville places the buccinum in the family Entomostomata. The greater part of this genus may be considered littoral; they are found however at depths varying to ten fathoms.

BU′CCINITE. The fossil remains of the *buccinum*. The greater part of the genus buccinum is littoral.

BUCCINOI′DA. The third family in the order Pectinibranchiata, division Mollusca.

B′UCKLANDITE. (So named, by Levy, after Prof. Buckland.) This rare mineral, like Babingtonite, occurs at Arendal, in Norway, and was distinguished and described by Levy. It is of a dark brown or nearly black colour, opaque, having a vitreous lustre with an uneven fracture. It greatly resembles augite.

BU′FONITE. (from *bufo*, Lat.) Fossil teeth of fishes belonging to the family of Pycnodonts; they occur in great abundance throughout the oolite formation. These bufonites have been also called Serpent's-eyes, Batrachites, and Crapaudines, from the notion of their having been formed in the heads of serpents, toads, and frogs; and, from presumed virtues which it was thought they possessed, they were worn in rings and as amulets.

BULI′MUS. A fossil ovate or oblong subturrited shell: the opening entire, oblong, and longitudinal, and this is the chief characteristic of this genus. The bulimus is a land shell. It is found both recent and fossil. Many species have been distinguished.—*Parkinson*.

BU′LLA. An ovate, gibbous, and cylindrical univalve: the fossil occuring in tertiary formations: the spire not standing out, but concealed: the opening the length of the shell: the lip acute. The recent bulla is marine, and found in sands and sandy mud, at depths varying to twelve fathoms.—*Parkinson. De la Beche.*

BU′LLITE. The fossil remains of the bulla.

BUMA'STUS. (from βούμαστος, ex βοῦ et μαστός, Gr.) The name given by Sir R. Murchison to a genus of trilobites, who describes one species only, which he has named B. Barriensis, the specific name being derived from the locality where it was found, near Barr, in Staffordshire. The generic characters of Bumastus are thus given by Sir R. Murchinson in his splendid work The Silurian System: "Pars anterior capitis rotundato-convexa, subæqualis : **oculis lunatis**, glabris, remotis. **Pars** costalis s. corpus sulcis longitudinalibus vix apparentibus, costis decem. Pars posterior maxima, rotundato-tumida, æqualis." This remarkable crustacean, says the same author, has been hitherto known in England as the Barr trilobite, having been found near the village and beacon of Barr, in Staffordshire. A very large specimen of it, five inches long by 3½ inches wide, has been lithographed.

BUNT. In conchology, an increasing cavity; a tunnel.

BUNTER SANDSTEIN. The name given by the Germans to an extensive arenaceous formation of red and variegated stones, composing one of the sub-divisions of the Red Sandstone group. In Germany, this formation is distinctly separated from the Keuper by the Muschelkalk **or** shelly limestone. The Bunter Sandstein of the Germans is the same **as** the Grès bigarre of the French, and the New Red Sandstone of English geologists. See *New Red Sandstone*.

BURRH-STONE. This word **is** sometimes written *buhr-stone*. Millstone. The substance of burrhstone, or mill-stone, when unmixed, **is** pure silex; it has generally a reddish or yellowish colour, but that of the best quality is nearly white; it is full of pores and cavities, which give it a corroded and cellular appearance. Burrhstone is a vesicular and corroded variety of common quartz. It has been hitherto found only in France.

BYSSACAN'THUS. A genus of fossil ichthyolites of the Old **Red** Sandstone, described by **Agassiz** in his Poissons Fossiles.

BYSSOA'RCA. A genus of bivalve conchifera, assigned by **Swainson**, belonging to the family **Arcacea**.

BY'SSUS. (from βυσσὸς, Gr.) **A** beard, as in the mytilus and pinna. The *byssus* is peculiar to bivalves.

BY'SSOLITE. (from βυσσὸς, flax, **and** λίθος, a stone.) So named by Saussure. A rare mineral, occurring massive, in short, delicate, and stiffish filaments, of an olivegreen, or brownish colour, with a silky lustre, a variety of asbestiform actinolite.

# C

C'ACHOLONG. (called also mother of pearl opal: the quartz-agathe cacholong of Haüy; the silex cacholong of Brongniart.) A milkwhite variety of quartz, having **a** pearly or glistening lustre, a flat conchoidal **fracture**, and perfect opacity. It is found on the river Cach, in Bucharia, where it occurs in loose masses, and obtains its name from that river, and cholong, the Calmuc word for stone. It is said by Brongniart that the *cacholong* has been found in calcareous breccia in France. Prof. Jameson places cacholong as a sub-species of opal, stating that it is distinguished from calcedony by lustre, fracture, hardness, and specific gravity. Prof. Cleaveland, however, deems it a variety of calcedony, as do Messrs. Phillips and Allan. Its specific gravity is from 2·2 to 2·27. It is opaque or slightly translucent at the edges. Fracture flat conchoidal. Infusible before the blow-pipe.

CA'DMIUM. A metal of a blueish-white colour, with a specific gravity of 8·6. It is the least malleable and ductile of all metals which possess those properties. It was discovered by Prof. Stromeyer, of Gottingen, in 1817, in some oxide, or carbonate, of zinc; it has subsequently been found in the silicates of zinc of Derbyshire, and in ores of that metal found in other situations. The equivalent number of cadmiun is 52·5; its symbol Cd. Cadmium has not as yet been usefully employed in the arts.

CADU'CI-BRAN'CHIATE. (from *caducus*, perishable, and *branchia*, a gill.) A term applied to certain amphibious animals, as the frog, toad, newt, &c., who, at a certain period of their existence, undergo a species of metamorphosis, their gills, or branchiæ, becoming obliterated, and their lungs developed.

CADU'COUS. (*caducus*, Lat.) In botany, applied to leaves falling before the end of summer; to corollas falling off before the dropping of the stamens; to perianths falling before the corolla is well unfolded.

CAIRN. (Gael.) A name given to a heap of stones covering a dead body.

CAIRN-GO'RUM.  
CAIRN-GORM. } Yellow quartz.

CA'LAITE. Called also odontalite and mineral turquoise: the uncleavable azure spar of Mohs. This earthy mineral has no cleavage. Hardness = 6. Specific gravity = 2·8 — 3. Massive disseminated; colour blue or greenish-blue passing into sky-blue and into apple-green. Streak white. Feebly translucent on the edges, or opake. Fracture conchoidal. The Russian chemist John states calaite to consist of alumina 73; oxide of copper 4·5; oxide of iron 4; water 18. This mineral may be distinguished from malachite, with which it has sometimes been confounded, by its yielding a white streak, while that of malachite is green.

CA'LAMAR. A name given to the cuttle-fish.

CA'LAMINE. (*calamine*, Fr. *giallamina*, It.) Carbonate of zinc. Calamine is found either loose, or in masses, or crystallized; colour white, grey, or yellow. Before the blow-pipe it decrepitates, but does not melt. It is used in the manufacture of brass. It consists of oxide of zinc 65, and carbonic acid 34. It contains also some sesquioxide of iron.

CA'LAMITE. (from *calamus*, Lat. καλαμος, a reed, Gr.)

1. A genus of fossil equisetaceæ. Calamites abound universally in the most ancient coal formations, occur but sparingly in the lower strata of the secondary series, and are entirely wanting in the tertiary formations, and also on the actual surface of the earth. Brongniart enumerates twelve species of calamites. Calamites are characterized by large and simple cylindrical stems, articulated at intervals, but either without sheaths, or presenting them under forms unknown among existing equiseta; they however most differ from equiseta in their height and bulk, sometimes exceeding seven inches in diameter, whilst that of a living equiseta rarely exceeds half an inch. A calamite fourteen inches in diameter has lately been placed in the museum at Leeds.

2. A mineral variety of hornblende, called also Actynolite, or Actinolite. See *Actinolite*.

CALC-SINTER. Stalactitical or stalagmitical carbonate of lime. This is so called from the German kalks, lime, and sintern, to drop. It is deposited from thermal springs, holding carbonate of lime in solution.

CALCAI'RE GRO'SSIER. (Fr.) A coarse limestone, often passing into sand, and abounding in marine shells; it contains by far the greater number of the fossil shells which characterize the Paris basin. Not fewer than four hundred distinct species have been found in a single locality near Grignon. Dr. Berger observes that this variety of limestone was thought to be very scarce in England, but that he is disposed to believe, upon a further examination, it will be found to occupy a great extent of country. To M. Brongniart we are indebted for much valuable information concerning the calcaire grossier. The mean specific gravity of specimens of the calcaire grossier from different parts of England, Dr. Berger gives as 2·579, ranging from 2·466 to 2·666. The calcaire grossier corresponds with our Bembridge series.

CALCAI'RE MOELLON. The name given by M. Marcel de Serres to certain marine beds of limestone found in the south of France and comprised in the supracretaceous group. The calcaire moellon is commonly worked as a building stone in the south of France.

CALCAI'RE SILI'CEUX. (Fr.) A compact silicious limestone, occupying, according to the early opinion of Brongniart, the place of the calcaire grossier where that is wanting, who discovered in it the silicate of magnesia. Like the calcaire grossier, the calcaire siliceux belongs to the eocene tertiary period; it has been found in the Paris basin, in the Isle of Wight, and in several parts of France.

CALCA'REOUS ROCK. Limestone.

CALCA'REOUS SPAR. Crystallized carbonate of lime. Calcareous spar occurs crystallized in a vast variety of forms, but its primitive form is invariably a rhomboid with obtuse angles of 105°5′, and 74°55′, the crystals break easily with the stroke of a hammer. It consists of nearly 57 parts of lime and 43 parts and a fraction of carbonic acid. The finest specimens are brought from Derbyshire, but it is found in almost all parts of the globe. Calcareous spar is often as transparent as rock-crystal, but it is usually coloured of various tints by the presence of oxide of iron. All its forms, which amount to nearly 500, are derived from the rhomb. When in irregular forms, it may always be distinguished from quartz by its admitting of being scratched with a knife, and by its effervescing when weak nitrous acid is applied to it. Iceland spar is this mineral in its purest form, and affords the readiest means of observing the optical phenomenon of double refraction. Its almost universal diffusion is probably owing to its partial solubility in water; in this condition it filters through the crevices abounding in all strata, and recrystallizes as the water evaporates. When this filtration continues so uninteruptedly as to prevent by its mechanical action crystallization from taking place, and yet so gradually as to admit of the solid mineral being deposited from the water, those curious and beautiful concretions called stalactites are produced, which ornament, in so singular a manner, most caverns in rocks or mountains formed principally of limestone.

CALCA'REOUS TUFA. Beds of calcareous tufa are sometimes formed in valleys, and at the bottom of lakes, by a process which bears some resemblance to chemical formations. Springs containing carbonic acid, that issue from limestone strata, contain particles of carbonate of lime chemically dissolved in water; but on exposure to air and light, the carbonic acid, which had but a slight affinity for the particles of limestone, separates, and the particles of lime are precipitated, and form calcareous incrustations: these, in the course of time, form beds, and occasionally are of sufficient hardness to be used for architectural purposes. Thermal springs holding in solution calcareous earth, rapidly deposit beds of calcareous tufa. At a temperature of 60° Fah., lime is soluble in 700 times its weight of water; if to this solution a small quantity of carbonic acid be added, a carbonate of lime is formed, and precipitated in an insoluble state. If, however, the carbonic acid be in such quantity as to supersaturate the lime, it is again rendered soluble in water; and it is thus that carbonate of lime, held in solution by an excess of fixed air, not in actual combination with the lime, but contained in the water, and acting as a menstruum, is commonly found in all waters. Hence it is obvious, that a deposition of carbonate of lime from water may be occasioned by either an absorption of carbonic acid, or from the loss of that portion which exists in excess.

CALC-DIABASE. A finely-grained, or entirely compact diabase, with round grains of calc-spar.

CALCE'DONY. } (*calcedonius*, Lat. *calce-*
CHALCE'DONY. } *doine*, Fr. *calcedonia*, It. Sometimes spelt chalcedony.) The Gemeiner Kalzedon of Werner. Quartz agathe calcedoinæ of Haüy. Silex calcedoine of Brongniart. A semi-transparent and translucent variety of quartz, to which this name has been applied, from its having been formerly found at Calcedon. It is a simple, siliceous, uncrystallized mineral. Flint nodules are frequently calcedonic, the calcedony occupying the hollows of such flints, and being either mammillated, botryoidal, or stalactitical. It has been observed that although, in the present compact state of the matter of flint, it is not easy, though possible, to force a fluid through its pores, yet it is probable that before its consolidation was complete, it was permeable to a fluid whose particles were finer than its own; and that the particles of calcedony, while yet in a fluid state, being finer than those of common flint, did thus pass through the outer crust to the inner station they now occupy; and that these particles of calcedony permitted a passage through their interstices to the finer particles of still purer siliceous matter which, in the form of crystallized quartz, are often found crystallized in the centre, enveloped both by calcedony and common flint. When flints contain calcedony, there may generally be perceived some small bubbles, or a mammillated appearance, in some part of the exterior of the flint: between calcedony and flint there is a near resemblance, being only different modes of the same substance. Specific gravity 2·56. There are several sub-species; the beautiful apple-green is called chrysoprase; the grass-green varieties, plasma; those with red, brown, yellow, and green tints, carnelian; others are known as heliotrope, jasper, onyx, agate, &c.; these will all be described under their several names, and in their proper order. Of common calcedony the most usual colour is grey, passing into blue and brown by every intermediate variety. It is harder than flint, and infusible before the blow-pipe without addition. When

calcedony is held between the eye and the light, it is characterized by a cloudy or milky appearance, resembling milk diluted with water. Professor Jameson divides calcedony into four sub-species, namely, common calcedony, chrysoprase, plasma, and carnelian. Calcedony was first accurately described by Werner. Phillips says that onyx, plasma, heliotrope, chrysoprase, cacholong, carnelian, and agate, are considered to be varieties of calcedony.

CALCE′OLA. A fossil genus of bivalve shells, belonging to the family Rudistes of Lamark. It is equilateral, very inequivalve, triangular, the umbones are separated by a large triangular disk in the lower valve; the hinge margin is straight and dentated; the upper valve is flat and semi-orbicular, forming a kind of operculum to the lower, which is much larger and deep. Calceola is now generally regarded to belong to the Brachiopoda. The C. sandalina is found in the Devonian system of rocks.—*Lycett.*

CA′LCINATE. (*calciner,* Fr. *calcinare,* It.) To calcine; to burn by fire to a calx, or friable substance.

CALCINA′TION. (*calcination,* Fr. *calcinazione,* It.) The reduction by the action of fire of any substance to a condition that it may be converted into a state of powder.

CA′LCINE. (*calciner,* Fr. *calcinare,* It.) To burn by fire to a calx, or friable substance.

CALCITE. Another name for calc-spar.

CA′LCIUM. The metallic base of lime; this metal was obtained by Sir H. Davy from lime by means of galvanic agency. Being received during the process into a vessel filled with naphtha, it was excluded from oxygen, and consequently retained its metallic appearance, which resembles that of silver. But no further investigations can be made, in the present state of science, regarding its properties as a metal, for the instant atmospheric air is admitted to it, it absorbs oxygen rapidly, burns with an intense white light, and re-produces lime, which is a protoxide of calcium.

CALC-TUFF. A deposit of carbonate of lime from calcareous springs. See *Calcareous Tufa.*

CA′LIX. } (κύλιξ, Gr. *calix,* Lat. *calice,*
CA′LYX. } Fr. *calice,* It.) The calyx, or flower-cup, is the outer expanded part, or external covering, of a flower, generally resembling the leaves in colour and texture; there are seven kinds of calyxes, or calyces, namely, periantheum, amentum, spatha, gluma, involucrum, volva, perichætium.

The calyx is the outer set of the floral envelopes, when there are more than one verticil of these. It is composed of two at least, but generally more, leaves, called sepals. When the sepals are distinct, or separate from each other, the calyx is said to be *polysepalous.* In many plants the sepals are joined together, more or less, by their edges, so as to form one piece in appearance; in this case the calyx is said to be *monosepalous.* When all the sepals are alike in size and form, the calyx is said to be *regular.* When the sepals vary in size or form, the calyx is said to be *irregular.* When the calyx has one of its sepals hollowed out into a long thin tube, as in the larkspur, geranium, &c., it is said to be *spurred.* When the calyx dies off soon after or immediately on its expanding, it is termed *deciduous;* this is commonly the case with polysepalous calyces. When the calyx survives the rest of the flower, either enclosing or forming part of the fruit, it is said to be *persistent:* most monosepalous calyces are persistent.

CA′LLIARD. A provincial name for a fine grained silicious stone, a member of the coal series: it is used for mending the roads, and often exhibits films and encrustations of coal. It is also called ganister and galliard.

CALO'RIC. (from *calor*, Lat. heat.) An imaginary fluid substance, supposed to be diffused through all bodies, the sensible effect of which is termed heat.

CALORI'METER. An instrument for measuring the degree of caloric.

CALP. A sub-species of carbonate of lime; an argillo-ferruginous limestone. The name given to a member of the Irish carboniferous series; it consists of black limestones.

CALY'MENE. (from κεκαλυμμένη, Gr. concealed.) A genus of trilobites, which appears to have been annihilated at the termination of the carboniferous strata. Fossils of this family were long confounded with insects under the name of Entomolithus paradoxus. The following is M. Brongniart's description of Calymene: "Corps contractile, en sphère presque hemicylindrique. Bouclier portant pulsieurs tubercles ou plis, deux tubercules oculiformes réticulés. Abdomen et post abdomen à bords entiers, l'abdomen divisé en douze ou quatorze articles. Point de queue prolongée." The Calymene has also been called the Dudley fossil: several species have been described; six are figured in Sir R. Murchison's Silurian System, namely, C. Blumenbachii, C. Downingeœ, C. Tuberculata, C. Macropthalma, C. variolaris, and C. Punctata. The last of these is stated to have been found in the lower, the other five in the upper Silurian rocks.

CALY'PTRA. (καλύπτρα, Gr. *calyptra*, Lat.) The calyx of mosses, according to some writers. In the mosses, the organs of reproduction consist of sporules, contained within an urn, or theca, placed at the top of a thin stalk: this is closed with a lid, called an operculum, and that again is covered with a hood termed a *calyptra*.

CALYPTRE'A. A fossil conoidal univalve, with the apex entire, erect, and somewhat pointed, the cavity furnished with a spirally convoluted lip; also recent.—*Parkinson*.

CA'MBIUM. (*cambium*, Latin.) In botany, a juice exuded between the bark and the alburnum, supposed to serve the purpose of nourishing the fibres of the leaf buds.

CA'MBRIAN. (from *Cambria*, a name for the principality of Wales.) A name given by Professor Sedgwick to a group of rocks, placed below the Silurian, from their being largely developed in North Wales; they principally consist of slaty sandstone and conglomerate. The Cambrian rocks belong to the Grawacke group, and geology owes much to Professor Sedgwick for the valuable information he has supplied in relation to them. The Cambrian strata are not only conterminous with the Silurian system, but are in several parts seen to rise from beneath its lowest beds, and to unite with them.

CAMPA'NULATE. (from *campanula*, Lat.) Bell-shaped; in the form of a bell. A term applied to the calyx or corolla.

CAMPANULA'RIA. A zoophyte, found abundantly on our shores, and thus named from its bell-shaped cells placed on footstalks.

CANALI'CULATED. (from *canaliculatus*, Latin.) Channelled; furrowed; made like a pipe or gutter. Applied to any distinct groove or furrow in shells.

CANCELLA'RIA. A genus of shell comprising many species. It is an ovate, or subturrated univalve, with the lip internally sulcated; the base of the opening slightly channelled. The columella having sharp, but compressed, plicæ. Fossils of this genus have been found in the London clay and calc-grossièr of Paris. The recent cancellaria is found in sandy mud at depths varying from 5 to 15 fathoms.

CA'NCELLATED. (from *cancelli*, lattice-work, Lat.) Cross-barred; marked with lines crossing each other. In conchology, surrounded with arched longitudinal ribs.

CA'NDLE-COAL. ⎫ Called also splint or
CA'NNEL-COAL. ⎭ parrot coal. (This substance has probably obtained its name from the bright flame, unmixed with smoke, which it yields during combustion, lighting a room as with candles, candle being provincially pronounced *cannel*.) Candle, or cannel, coal is a bituminous substance, next in purity to jet. It is black, opaque, compact, and brittle; breaking with a conchoidal fracture. Cannel-coal does not soil the fingers when handled, is susceptible of polish, and is capable, like jet, of being worked into trinkets and ornaments. The difference between jet and cannel-coal appears to consist entirely in the presence or absence of foreign earthy matters. When these are absent, or exist in minute proportion only, the bituminous mass is so light as to float on water, and then the term jet is properly applicable; but when the presence of foreign earthy matters is considerable, and the mass is specifically heavier than water, and does not readily manifest electric properties, it is with more propriety termed cannel-coal. According to Dr. Thomson, cannel-coal contains 21·56 of hydrogen, whereas Newcastle coal contains only 4·18 per cent. "Cannel coal never manifests internally any traces of vegetable structure, but sometimes bears on its surface evident marks of impressions formed on it whilst in a soft state."—*Parkinson*.

CAPRI'NA. A genus of fossil bivalve shells belonging to the family Rudistes of Lamark. M. D'Orbigny has separated this genus from Diceras (see that genus). One species, the C. inœquirostratus, is recorded from the chalk of Norfolk.—*Lycett*.

CAPRO'TINA. (See *Diceras*.) A genus established by M. D'Orbigny. The C. Lonsdalei is found in the lower green sand near Calne, in Wiltshire. —*Lycett*.

CAPSA. A genus of transverse, equivalved, close shells, of the order Nymphaceœ; the hinge has two teeth on the right valve, and one on the left; there are no lateral teeth; ligament external.

CA'PSTONE. The name for a fossil echinite, or that genus of echinite known as *conulus*.

The capstone, thus called from its supposed resemblance to a cap, rises from a circular base into a cone, with an acute or obtuse vertex, from which five pairs of punctuated or crenulated lines pass, dividing the shell into five large and five small areœ, that in which the anal aperture is placed being rather the largest.—*Parkinson*.

CA'PSULE. (*capsula*, Lat. *capsule*, Fr.) 1. In botany, a membranous or woody seed vessel, internally consisting of one or more cells, splitting into several valves, and sometimes discharging its contents through pores or orifices, or falling off entire with the seed.

2. A membranous or ligamentous bag.

CAOU'TCHOUC. 1. Vegetable caoutchouc, called also elastic-gum, and India-rubber, is the milky exudation from certain trees, more especially the Hœvea caoutchouc and the Iatropha elastica, but it is obtained from several others.

2. Mineral caoutchouc. A bituminous fossil, elastic when soft, but brittle when hard. It was discovered in 1786 near Castletown in Derbyshire. In its appearance

it much resembles India-rubber, whence it has obtained its name.

CA'RAPACE. The upper shell of reptiles.

CARADOC SANDSTONE. The name given to a formation constituting the upper member of the Lower Silurian Rocks. "The name has been selected, says Sir R. Murchison, because the strata of which it is composed constitute a number of eminences, which abut against the remarkable chain of trap hills called the Caradoc. Unlike the sandstones of the upper Silurian rocks, this formation is composed essentially of sandstones of different colours, with an occasional subordinate course of calcareous matter. The Caradoc sandstone formation is made up of beds of red, green, and purple sandstones, some of which it is difficult, upon first inspection, to distinguish from strata of the old, or even the new red sandstone. Its best and clearest distinctions consist in its order of infra-position to the upper Silurian rocks, and in its organic remains, nearly all of which are dissimilar from the fossils of the formations which immediately overlie it.

Among the fossils characteristic of this formation may be enumeted Productus sericeus; Bellerophon bilobatus and Bellerophon acutus; Littorina striatella; Orthis alternata; Orthis collactis, and Orthis canalis; Pentamerus loevis and Pentamerus oblongus. The trilobites common to the upper Silurian rocks are here wanting, and instead are found other forms, including the Trinucleus, a genus not observed in the upper but abounding in the lower Silurian rocks: also a large species of Asaphus, named by Sir R. Murchison Asaphus Powisii. In other beds of this formation there are found Avicula orbicularis and Avicula obliqua; Orthis actonia, and Orthis grandis; Orthis Anomala; Orthis vecten; Orthis flabellulum; and Orthis vespertilis; Terebratula anomala, and Terebratula anguis; Pentamerus oblongus, and the plumose coral Calamopora fibrosa.

Malachite, or the green carbonate of copper, occurs in films and nests. Thin strings of galena, with some associated crystals of blende, have been also found. For a full description of this formation, together with illustrations of the contained fossils, the reader is referred to Sir R. Murchison's splendid work on the Silurian System, from which the above is principally extracted.

CA'RBON. (*carbo*, Lat. *carbon*, Sp. *carbone*, It. *charbon*, Fr.) Caradoc limestone. The pure inflammable principle of charcoal. If a piece of wood, or any vegetable matter, be placed in a closed vessel, and kept red-hot for some time, it is converted into a shining black brittle substance, possessing neither smell nor taste, commonly known as charcoal. Charcoal is infusible, insoluble in water, is capable of combining both with hydrogen and sulphur, is a conductor of electricity, and has a powerful affinity for oxygen. Carbon is obtained nearly pure in charcoal; but, what is astonishing, the diamond appears to be this elementary substance in its purest known form. Why it is, or how it is, that the same elementary substance can, with little or no addition, form two bodies so dissimilar in every respect as charcoal and diamond, the one a soft, black, brittle mass, the other the clearest and hardest body we know of, is a mystery beyond our weak comprehensions to understand. Carbon enters as a constituent part into many of the slate rocks, to which it generally communicates a dark colour: it forms also regular beds of considerable thickness, being the principal constituent part of coal. Of Newcastle caking coal it constitutes 75·28 per cent., of cannel coal 64·72 per cent. The carbon in the atmosphere is not considerable, but without it vegetation could not exist. Saussure ascertained that 10,000 parts of atmospheric air contained, as a mean, 4.9 of carbonic acid. Combined with oxygen, carbon forms carbonic acid or fixed air.

CA'RBONATE. A combination of carbonic acid with a base. Carbonic acid is capable of combining with earths, oxides, and alkalies, and to these combinations the term carbonate has been applied; thus we have the carbonate of lime, carbonate of magnesia, carbonate of lead, carbonate of iron, carbonate of ammonia, &c., &c., &c.

CA'RBONATE OF LIME. A union of carbonic acid and lime, consisting of 57 parts of lime and 43 of carbonic acid. The form of the integrant molecule of carbonate of lime was decided by the Abbé Haüy to be an obtuse rhomboid. This conclusion was implicitly adopted by mineralogists, till the Count de Bournon announced the discovery of cleavages passing through the long diagonal of its rhomboidal faces, and maintained that the form of its integrant molecule was a trihedral prism with inclined bases. Dr. Brewster has since satisfactorily proved that the cleavages obtained by Count Bournon exist only in those specimens which are crossed by interrupting veins, and therefore that the trihedral prism is not, but that the obtuse rhomboid is the form of the integrant molecule of carbonate of lime. Carbonate of lime, under the several names of chalk, limestone, marble, &c., is found most abundantly throughout nature. All limestones effervesce when a drop of strong acid is thrown on them, and they entirely dissolve in nitric or muriatic acid.

It is a difficult problem, says Professor Buckland, to account for the source of the enormous masses of carbonate of lime that compose nearly one eigth part of the superficial crust of the globe; some have referred it entirely to the secretions of marine animals; an origin to which we must obviously refer those portions of calcareous strata which are composed of comminuted shells and corallines; but until it can be shown that these animals have the power of forming lime from other elements, we must suppose that they derived it from the sea, either directly, or through the medium of its plants. [The presence of carbonate of lime in a rock may always be ascertained by applying to the surface a drop of diluted sulphuric, muriatic, or nitric acids: the lime having a stronger affinity for any one of these than for the carbonic acid, unites itself immediately with them, and the carbonic acid being liberated escapes in a gaseous form, frothing up or effervescing as it makes its way in small bubbles.]

CA'RBONIFEROUS GROUP. } The carboniferous group
CA'RBONIFEROUS SERIES. } or series comprises the coal measures, the mountain or carboniferous limestone, and the old red sandstone. The rocks of this group, says Sir C. Lyell, consist of limestone, shale, sandstone and conglomerate; interstratified with which are large beds of coal. Several hundred species of plants have been found in the shales and limestones associated with the coal, all of which are, with few exceptions, of species differing widely from those which mark the vegetation of other eras. It is in this formation chiefly, that the remains of plants of a former

world have been preserved and converted into beds of mineral coal. The most characteristic type that exists in this country of the general condition and circumstances of the strata composing the great carboniferous order, is found in the north of England. According to Mr. Forster's section of the strata from Newcastle-upon-Tyne to Cross Fall, in Cumberland, it appears that their united thickness along this line exceeds 4,000 feet. This enormous mass is composed of alternating beds of shale, sandstone, limestone, and coal.

The above group constitutes in the Rev. J. Conybeare's arrangement the Medial order; he says "this series of rocks is by some geologists referred to the flœtz, by others to the transition class of the Wernerians; we have preferred instituting a particular order for its reception, a proceeding justified by its proportional importance in the geological scale, its peculiar characters, and the many inconveniences arising from following either of the above conflicting examples."

Sir R. Murchison in his splendid work "The Silurian System," says "Being convinced that the Old Red Sandstone is of greater magnitude than any of the overlying groups, I venture for the first time in the annals of British geology to apply to it the term *system*, in order to convey a just conception of its importance in the natural succession of rocks, and also to show, that as the carboniferous system, in which previous writers have merged it, (but from which it is completely distinguishable, both by zoological contexts and lithological characters) is surmounted by one-red group, so it is underlaid by another, this lower red group being infinitely thicker than the upper."

Mr. Mushet, from a careful examination of the Forest of Dean, has given a list of the various beds of the coal measures, carboniferous limestone, and old red sandstone, constituting a mass of 8,700 feet in thickness: the coal measures being 3060 feet in depth. The whole reposes on the grauwacke limestone of Long Hope and Huntley.

CARBONIFEROUS LIMESTONE. The name proposed by Mr. Conybeare to be given to mountain limestone. See *Mountain Limestone*.

CARBO'NIC ACID. A compound of carbon and oxygen; it has been called aërial acid, fixed air, cretaceous acid, and mephitic gas. Carbonic acid is very plentifully disengaged from springs in almost all countries, but especially near active or extinct volcanoes. This elastic fluid has the property of decomposing many of the hardest rocks with which it comes in contact, particularly that numerous class in whose composition felspar is an ingredient. In volcanic countries these gaseous emanations are not confined to springs, but rise up in the state of pure gas from the soil in various places. The Grotto del Cane, near Naples, affords an excellent example. The acid is invisible, is specifically heavier than atmospheric air, and on this account accumulates in any cavities on the surface of the ground. It may be dipped out of any excavations in which it has accumulated, poured into a bottle, like water, corked, and carried to any distance. It is fatal to human life when breathed undiluted, and by the miners it is called chokedamp. Carbonic acid gas is evolved abundantly in coal pits. M. Bischof calculates that the exhalation of carbonic acid gas, in the vicinity of the lake of Laach, amount to 600,000 pounds daily, or 219,000,000 pounds in a year.

CA'RBURET. A compound formed by the combination of carbon with any metal, alkali, or earth.

CA'RBURETTED HYDROGEN GAS. The fire-damp of miners.

CA'RBUNCLE. (*carbunculus*, Lat.) A precious stone, supposed by some authors to be the ruby, by others the garnet; called by the Greeks anthrax.

CA'RDIAC. } (from καρδία, the heart, CARDI'ACAL. } Gr. *cardiaque*, Fr.) Relating to the heart, as the cardiac nerves, &c.

CARDIA'CEA. In Cuvier's arrangement a family of bivalves of the class Testacea, comprising Venericardia, Cardita, Cardium, Cypricardia, and Trocardia.

CARDINIA. A genus of fossil bivalve shells separated by M. Agassiz from the unios; Cardinia is thick, equivalve, inequilateral, rather compressed, and transverse; the hinge is remarkable for the size, prominence, and lengthened form of the lateral teeth. The species of cardinia are numerous, five are recorded from the coal measures, nine from the lias, and two from the inferior oolite in England.—*Lycett.*

CARDIOMORPHA. A genus of fossil bivalve shells, hitherto found only in the carboniferous limestone. M. Köninck describes it as being equivalve, inequilateral, thin, oblique, and transversely elongated; the hinge without teeth, linear, with a smooth cardinal lamina extending from the umbones to the extremity of the margin; the ligament linear and external. Dublin, Holland, Kildare and Visé are given as the localities. One species only is known, the C. oblonga.—*Lycett.*

CARDI'TA. An inequilateral bivalve, found at various depths to thirteen fathoms, in mud and sands; sometimes attached to stones. The hinge with two unequal teeth; the hinge tooth the shortest, beneath the beak; the other longitudinal, beneath the insertion of the cartilage. Lamarck places Cardium in the family Cardiacea.

CA'RDIUM. The cockle; animal a tethys. A genus of bivales, the shells of which are characterised by the teeth of their hinge, and by the projection of their beaks; the latter giving them a cordiform appearance. They are generally ornamented with longitudinal ridges, and frequently with striæ, scales or spines. The different species are found at depths varying to thirteen fathoms, in mud, sands, and gravel. This genus belongs to the class Vermes, order Testacea. Many of the species, as the C. aculeatum, arcuatum, ciliare, discors, edule, elongatum, lævigatum, nodosum, spinosum, &c. &c. are found on our coasts. The cardium, with the exception of one species, cardium fluviatile, has only been found to inhabit the ocean; generally they live just under the surface of the sand. Fifty-two species have been described.

CARI'NA. (Lat.) The keel; a term applied to two of the petals in papilionaceous flowers. The carina is composed of two petals, separate or united, and encloses the internal organs of fructification.

CARINA'RIA. A very thin univalve, in the form of a cone flattened on its sides, the apex terminating in a very small convoluted spire, and the back having a dentated keel. De Blainville places Carinaria in the family Nectopoda. It derives its name from its dorsal keel. Parkinson states that it has not been found fossil, nor is its inhabitant known. Sowerby mentions that this beautiful shell was once so rare, that a specimen would fetch one hundred guineas.

CA'RINATED. (*carinatus*, Lat). Keel-shaped; in conchology, having a

L

longitudinal prominence like the keel of a vessel.

CARIN'THINE. A variety of hornblende, of a dark green or black colour. It obtains its name from being found in Carinthia.

CARN. } The name assigned to small
KARN. } round hills in some parts of England; in others these are called tors. Carn-Marth and Carn-Brea are two small hills near Redruth, in Cornwall.

CARNA'RIA. (*carnarius*, Lat.) Flesh-eating animals. In Cuvier's arrangement, the third order of Mammalia.

CARNE'LIAN. } (*cornaline*, Fr. *cornalina*,
CARNE'LION. } It. The Karneol of Werner, Quartz-agathe Cornaline of Haüy.) A precious stone of various colours, as red, brown, yellow, and white. It is a variety of rhombohedral quartz. The finest specimens are brought from India. Carnelian is composed of 94 parts silica, 3·50 alumina, and a trace of oxide of iron. Carnelian differs from calcedony only in being more or less transparent. It varies in its constituents from being nearly pure silex, to a mixture of this earth with alumine and iron, in small quantities. Some particulars relative to the carnelians of Cambay are given in a memorandum from the minute book of the Geological Society. "These carnelians are all procured from the neighbourhood of Broach, by sinking pits during the dry season in the channels of torrents. The nodules which are thus found lie intermixed with other rolled pebbles, and weigh from a few ounces to two or three pounds. Their colour when recent is blackish olive passing into grey. The preparation which they undergo is, first, exposure to the sun for several weeks, and then calcination. This latter process is performed by packing the stones in earthen pots, and covering them with a layer five or six inches thick of dried goats' dung; fire is then applied to the mass, and in twelve hours time the pots are sufficiently cool to be removed. The stones which they contain are now examined, and are found to be some of them red, others pink, and others nearly colourless, the difference depending partly on the original quantity of colouring matter and partly on the difference of heat applied."

CARNI'VORA. (from *caro carnis*, and *voro*, Lat.) 1. Animals which subsist solely on flesh. They belong to the third order, class Mammalia. This order is divided into three families, Digitigrada, Plantigrada, and Pinnigrada.

CARNI'VOROUS. (*carnivorus*, Lat. *carnivore*, Fr.) Living on flesh; devouring flesh.

CA'ROTID. (from καρωτίδες, Gr. *carotides*, Lat. *carotides* Fr.) The name given to certain arteries of the neck.

CA'RPAL. (from *carpus*. Lat. καρπὸς, Gr.) Relating to the wrist.

CARPE'LLUM. (from καρπὸς, Gr.) In botany, a leaf in a particular state of modification. Each modified leaf which forms the pistil, is called a *carpellum*, and has its under side turned outwards, and its upper inwards, or towards the centre of the flower. The *carpella* are folded so that the margins of the leaf are next to the axis, or centre: from these a kind of bud is produced, which is the seed. On the form of the carpella, on their number, and on their arrangement around the centre, depends, necessarily, the form of the pistil.

CA'RPOLITE. } (from καρπὸς, *fructus*,
CA'RPOLITH. } and λίθος, *lapis*.)
CARPOLI'THUS. } Any fruit which by silification has been converted into stone.

CARPO'LOGY. (from καρπὸς and λόγος, Gr.) That branch of the science of botany which treats of fruits.

CA'RTILAGE. (*cartilago*, Lat. *cartilage*, Fr. *cartilagine*, It.) Smooth, solid, animal matter, softer than bone, and harder than ligament; gristle.

CARTILA'GINOUS. (*cartilagineux*, Fr. *cartilaginoso*, It.) Consisting of cartilage; resembling cartilage; gristly.
2. A name given to all fish whose muscles are supported by cartilages instead of bones.
3. A term applied to leaves, the borders of which are hard and horny.

CARYOPHY'LLIA. "A stony fixed polypifer, simple or ramified; the stem and branches rather turbinated, and striated longitudinally, each being terminated by a cell, radiated in a stelliform figure.—*Parkinson*. Lamarck separated Caryophyllia from Madrepora. In the caryophilliæ possessing more than one cell, each receptacle contains a polypus. A branched madrepore with a star at the end of each branch; each star having a mouth and tentacula.—*Bakewell*.

CA'SPIAN. The name given to a large body of salt water not communicating with the main ocean. "Masses of salt water are sometimes included in the dry land, which have been termed Caspians, from the Caspian Sea, the largest of them. These bodies of salt water have been variously accounted for; some supposing that they have been left isolated by a change in the relative level of land and water, while others imagine their saltness to arise from their occurrence in countries impregnated with saline matter."—*M. Le Beche, Geological Manual*.

CASSIDA'RIA. A genus of univalve molluscs found both recent and fossil. The recent species are found near the shore, and at small depths from the surface. The fossil specimens occur in the tertiary strata.

CA'SSIS.
1. The helmet-stone. An echinite, a section of the class of Catocysti.
2. A gibbose ventricose univalve; the aperture longitudinal and sub-dentated, and terminating in a short reflected canal. The columella plicated in its lower part; the left lip flattened, and forming a ridge on the body of the shell.—*Parkinson*.
This genus of shells is found both recent and fossil: the recent is an inhabitant of tropical seas; the fossil occurs in the tertiary deposits. Some species are figured in Parkinson's Organic Remains.

CA'SSIDITE. A fossil shell of the genus cassidaria. The hills of Tuscany yield these fossils.

CA'TACLYSM. (from κατακλυσμὸς, Gr. *cataclysme*, Fr.) A great inundation or deluge: used generally to describe the Noachian deluge.
The sect of Stoics taught that catastrophes were of two kinds; the *cataclysm* and the *ecpyrosis*; the former sweeps from the face of the earth the whole of the animal and vegetable productions; the latter destroys the globe itself.

CAT'S-EYE. The katzenange of Werner; quartz agathe chatoyant of Haüy; quartz hyalin chatoyant of Brongniart. A beautiful mineral, a variety of rhombohedral quartz, having an opalescence resembling the light from the eye of the cat; whence its name. The peculiar play of light arising from the structure of this stone, is better known than susceptible of description. The *chatoyement* of this mineral is supposed to arise from fine fibres of asbestus or amianthus included in it. This chatoyement, or opalescence, is increased by the mineral being cut in a spherical form. The colours of cat's-eye are greenish or yellowish-grey, yellowish-brown, reddish-brown, and greyish-white, with intermediate shades. The finest specimens are brought from Ceylon. Cat's-eye is harder than quartz, and consists of silex 95, alumine 1·75, lime 1·25, oxide of iron 0·25.

CATARHINI. A family of the order Quadrumana, comprising the monkeys of the old world.

CATILLUS. A genus of bivalve couchifera, considered by some authors to be identical with Inoceramus, but is separated by M. Deshayes for reasons apparently conclusive; it is oval or oblong, nearly equivalve, inequilateral, the umbones being more or less prominent, the hinge is straight, nearly perpendicular to the longitudinal axis, its border being furnished with a series of small cavities; the shell is very thick, sometimes of enormous dimensions, being mentioned as attaining many feet in length; the structure of the test is fibrous. The white chalk of France and England has produced all the known species.—*Lycett.*

CATO'DON. A name given to the spermaceti whale. Of the genus Catodon, Ray mentions a large one stranded on the coast of Holland.

CATOCY'STI. The second great division, or family, of Echini. The catocysti have the opening for the vent in some part of the base of the shell; they are divided into fibulæ, cassides, scuta, and placentæ.

CAU'DA. (Lat.) In conchology, the elongated base of the ventre, lip, and columella.

CAU'DEX. (Lat.) The stock or trunk, the stem or body of a tree.

CA'VERNOUS (*cavernosus*, Lat. *caverneux*, Fr. *cavernoso*, It.) Full of caverns and hollows.

CAWK. A name for sulphate of barytes.

CA'YMAN. The name assigned to a sub-genus of crocodiles. Authors are not agreed as to the origin of this name; Rochefort says it was original with the native islanders of the Antilles.

CELA'CANTH. Hollow spine. A family of fossils of the old red sandstone, deriving their name of hollow-spine from this peculiarity.—*Hugh Miller.*

CELE'STINE. Sulphate of strontia, or sulphate of strontites. It has obtained the name of *celestine* from being frequently found possessing a blue colour; but as it does not invariably possess that colour, and is often found either colourless or red, the name appears to be inappropriate. It occurs both massive and crystallised. It is composed of 58 parts of strontia, and 42 of sulphuric acid. It is found in Scotland, in Yorkshire, and in Somersetshire, near Bristol; also in the neighbourhood of Paris, and in Sicily, from which last place we obtain the finest specimens.

CELLEPO'RA. Animals belonging to the class Vermes, order Zoophytes. Generic characters:— Animal an hydra or polype; coral somewhat membranaceous, composed of cells. Species:—The principal species are the Cellepora pumicosa, annulata, spongites, &c.—*Crabb.*

CELLEPO'RITE. A fossil cellepora. Seven species of celleporites have been determined by Goldfuss as belonging to the organic remains of the cretaceous group: two other species are found occurring in the oolite, and two distinct species in the grauwacke. Fossil celleporæ may be distinguished from fossil flustræ, by their cells being urceolated and irregularly placed, and by the constricted appearance of their mouths.

CE'LLULAR MEMBRANE. In anatomy, that tissue of filmy meshes which connects the minute component parts of most of the structures of the body.

CE'LLLUAR INTEGUMENT. In botany, the succulent pulpy substance situated immediately under the cuticle; the seat of colour, mostly green, particularly in the leaves or branches. Leaves consist almost entirely of this substance, covered on each side by the cuticle; the stems and branches of both annual and perennial plants are invested with it. This tissue is membranous and transparent: in its simplest state, it appears like a mass of globules or vesicles, crowded together; these, by pressing against each other, assume a six-sided, or hexagonal, form. This cellular tissue allows of the passage of fluids, and is consequently porous, but no pores or openings have been discovered in it. Cellular integument is in itself colourless, but its vesicles contain that colouring matter which gives to the corolla its brilliant colours, and to the herbaceous parts of plants their green.

CEMENTA'TION. (*cementation*, Fr.) 1. The act of uniting by means of cement. 2. A chemical process by which iron is converted into steel; glass into porcelain, &c. This process is effected by surrounding the body to be acted on with some other substance, as iron with charcoal, and subjecting it to the action of fire in a closed vessel.

CEMOR'IA. A genus of patelliform univalve shells, separated by M. Leach from Fissurella, on account of its having the fissure placed behind the apex, which is produced and incurved. C. novehina has been found in pleistocene marls at Dalmuir, and in Sweden and Norway.—*Lycett*.

CENTIFO'LIOUS. (from *centum* and *folium*, Lat.) Having a hundred leaves.

CENTRE OF GRAVITY. A point in every body, which if supported, the body will remain at rest, in whatever position it may be placed. Above that point, all the parts exactly balance one another. The centre of gravity of the solar system is within the body of the sun, because his mass is much greater than the masses of all the planets and satellites added together.

CENTRI'FUGAL. (*centrifugo*, It. *centrifuge*, Fr. *Terme de Physique. Un corps qui se meut en rond, à une force centrifuge.*) That power which bodies in rapid rotatory motion acquire of flying off from the centre. The force with which a revolving body tends to fly from the centre of motion: a sling tends to fly from the hand in consequence of the centrifugal force. The centrifugal force arising from the velocity of the moon in her orbit, balances the attraction of the earth. The dimensions of the earth being known, as well as the time of its rotation, it is easy thence to calculate the exact amount of the centrifugal force, which, at the equator, appears to be $=\frac{1}{289}$th part of the force or weight by which all bodies, whether solid or liquid, tend to fall towards the earth.

CENTRI'PETAL. (*centripète*, Fr. *Qui tende approcher d'un centre. Les planètes ont une force centripète vers le soleil.*) The contrary power to centrifugal, for while the one drives, as it were, the surrounding bodies from the centre, the other, the centripetal, attracts and holds them to it.

CEPHALAS'PIAN. Relating to the cephalaspis; resembling the cephalaspis. "The remains of a large cuirassed fish of the cephalaspian type."—*Hugh Miller*.

CEPHALA'SPIS. (from κεφαλή, a head, and ἀσπίς, a buckler, buckler-head; pl. cephalaspides). A fossil fish of the cornstone formation, so named by Agassiz, from its head being covered by a sort of shield, having the bones united into one osseous case: in form, this fish bears a resemblance to the elongated trilobites of the transition rocks. Of the genus cephalaspis Agassiz established four species, namely, C. Lyellii, C. rostratus, C. Lewissii, and C. Lloydii. Of these four species, the first is found most abundantly in the red sandstone, both of England and Scotland, and is constituted the type of the genus. An entire fish of the cephalaspis Lyellii has been found in the old red sandstone of Forfarshire, the head of this fish is very large in proportion to the body, and occupies very nearly one-third of its whole length. The outline is rounded, in the form of a crescent, the lateral horns inclining slightly towards each other, while the anterior and central parts project much. The middle of the head, including the region of the eyes, the cranium, and the occipital crest, is elevated, whilst the sides and anterior edge are considerably dilated and horizontally extended. The eyes are placed in the middle of the shield and near each other, but a little nearer to the end of the snout than the occipital crest; they appear to have been directed straight upwards. The posterior and middle portion of the head is nearly square, and is edged by the first series of scales. The C. Lloydii has a head the least resembling that of a fish, it occurs in the cornstone of Shropshire, Worcestershire, Herefordshire, &c. Cephalaspis appears to have formed the connecting link between crustaceans and fishes; no existence of the present creation at all resembles the cephalaspis.

CEPHALOPHORA. The first class of the sub-kingdom Mullusca. This class has usually been called Cephalopoda, but, according to the classification of Pictet, it is now divided into six orders, namely, Dibranchiata, Tetrabranchiata, Pulmonata, Pteropoda, Gasteropoda Diœcia, and Gasteropoda, Monœcia.

CEPHA'LOPOD. ⎫ (from κεφαλὴ, head,
CEPHALO'PODA. ⎬ and πόδες, feet,
CEPHALO'PODES. ⎭ Gr.) A term applied by Cuvier to a large family of molluscous animals, from their having the feet placed around their heads, and walking with their heads downwards. The feet are lined internally with ranges of horny cups, or suckers, by which the animal seizes on its prey, and adheres to extraneous bodies. The mouth, both in form and substance, resembles a parrot's beak, and is surrounded by the feet. It is now well established that the living species of cephalopods which possess no external shell, are protected from their enemies by a peculiar internal provision, consisting of a bladder-shaped sac, containing a black and viscid fluid, or ink, the ejection of which, by rendering the surrounding water opaque, conceals and defends them. The sepia vulgaris and loligo afford us familiar examples.—*Buckland.*

According to Prof. Phillips, the following are the genera of Cephalopoda: Bellerophon, Orthoceras, Belemnite, Nautilus, Ammonite, Edomite, Scaphite, Baculite, Nummulite. The only living species belong to the genus Nautilus, of which there are two.

The Cephalopoda, in the arrangement of Cuvier, form the first class mollusca, and comprise the following genera, which he divided according to the nature of the shell, Sepia, Nautilus, Belemnites, Ammonites, and Nummulites.

CERAU'NII LA'PIDES. (from κεραυνὸς, Gr.) A name formerly given to fossil echinites, from a supposition that they were formed in the air.

CERATOPHY'TA. In Linnæus's arrangement, an order, the 6th, of the class, Zoophytes, or animal plants, comprising Gorgonia, Corallium, Pennatula, &c. They have a horny axis, covered with a fleshy substance, from the cavities of which polypi occasionally appear.

CEREBE′LLUM. (Lat.) Dim. of cerebrum; the little brain situate behind the brain, or cerebrum.

CE′REBRUM. (Lat.) The brain.

CE′REBRAL. } (cerebral, Fr. cerebrale,
CE′REBRINE. } It.) Belonging to the brain: relating to the brain.

CERIOP′ORA. The name assigned to a genus of zoophytes.

CERI′OPORITE. A fossil ceriopora. Twenty-one species have been determined by Goldfuss, as occuring in the cretaceous group; nine species have been found in the oolite, and six in the grauwacke group.

CE′RITE. The siliceous oxide of cerium.

CERI′THIUM. A turreted or turriculated univalve, with an oblique opening. Lamarck has discovered sixty fossil species of the genus Cerithium in the neighbourhood of Paris. The recent cerithium is found at depths varying to seventeen fathoms, and it is stated that so tenacious of life is at least one species, the cerithium telescopium, that a specimen sent from Calcutta in sea water, lived out of water in a tin-box for more than a week. The recent cerithium has a veil on its head, with two separated tentacula.

CE′RITHITE. The fossil cerithium.

CE′RIUM. A metallic substance discovered by Berzelius and Hissinger in 1804. It was obtained from a mineral called cerite, which was formerly supposed to be an ore of tungsten; it is also found in allanite.

CE′RVIX. (Lat.) The neck.

CERVI′CAL. (cervicalis, Lat. cervical, Fr. cervicale, It.) Belonging to the neck, as the cervical vertebræ, the cervical muscles, the cervical arteries.

CERVI′COBRANCHIATA. In the conchological system of De Blainville, the name given to an order of shells, comprising the two families Retifera and Branchifera, and the genera Patella, Fissurella, Emarginula, and Parmophorus.

CESTRA′CION. } In Agassiz's Ta-
CESTRA′CIONTS. PL. } bular View of the Genealogy of fishes, the Cestracionts, and they only, sweep across the entire geological scale. With this family, so far as is yet known, ichthyic existence first began.—*Hugh Miller.*

The Cestracions constitute a family of fishes of the placoid order. "It does not appear that on the globe we inhabit there was ever an ocean, tenanted by living creatures, that had not its cestracion".—*Hugh Miller.*

The first and oldest sub-family of sharks. The cestracionts have only one living representative, the Cestracion Philippi, or Port Jackson Shark. The character of this sub-family of sharks is marked by the presence of large polygonal obtuse enamelled teeth, covering the interior of the mouth with a kind of tesselated pavement. In some species, not fewer than sixty of these teeth occupied each jaw. They are rarely found connected together in a fossil state, in consequence of the perishable nature of the cartilaginous bones to which they are attached. They are found abundantly dispersed throughout all the strata, from the carboniferous to the most recent chalk series.—*Buckland.*

CESTRA′CIONT. Resembling a cestracion; pertaining to a cestracion. "This fish belongs to the *cestraciont* family, of the placoid order."—*Hugh Miller.*

CETA′CEA. Vertebral, warm-blooded animals, living in the sea; they have no gills; there is an orifice on the top of the head through which they breathe, and eject water; and they have a flat horizontal tail. The cetacea breathe by means of lungs, and this compels them to rise frequently to the surface of the water for the purpose of respiration; they also sleep on the surface. The cetacea both bring forth their young alive, and suckle them. Cetacea constituted the ninth order of the class Mammalia, and is divided into six families, namely Zeuglodontidœ, Delphinidœ, Monodontidœ, Heterodontidœ, Physeteridœ, and Balœnidœ.

CETIOSAURUS. A genus of extinct reptiles, bearing some resemblance to the cetaceans. The vertebræ are almost circular at the ends, and have a very short body; the front articulating surface is nearly flat, and the other concave, in the dorsal vertebræ, but in the caudal, both ends are deeply hollowed; they are often found eight inches in diameter. The Cetiosauri must have equalled the whales in bulk.

CETOSIS. A genus of fossil multilocular shells having a stellated opening at the pointed termination.

CETOTOLITE. (from κῆτος, a whale, ὠς, an ear, and λιθός, a stone, Gr.) The name assigned by Prof. Owen to a fossil ear-bone of the whale. "I have proposed to call the bodies in question *cetotolites*, as they consist of portions of the ear-bones of large cetacea."—*Prof. Owen*.

CEYLANITE. The pleonaste of Haüy; zeylanite of Werner; ceylanit of La Metherie; candite of Bournon. A dark green or black variety of dodecahedral corundum. It occurs in the sand of the rivers of Ceylon, from which island it obtains its name. From its being found in the rivers and alluvial country around Candy in the isle of Ceylon, Bournon named it candite. Some mineralogists place Ceylanite in the members of the ruby family By some it is regarded as a variety of spinel, but its specific gravity is higher, and it differs from spinel as well in colour as in composition. Ceylanite is found sometimes in rounded grains, is often crystallized in octahedrons, sometimes with truncated edges; also in dodecahedrons with rhombic faces, of which eight solid angles are sometimes truncated. Specific gravity, according to Haüy, is from 3·76 to 3·79. It is of sufficient hardness to scratch quartz. Structure indistinctly foliated; fracture shining and conchoidal, with large smooth cavities.

It is nearly or quite opaque, its common colour being of a dark blue or nearly black, but its fragments transmit a dark greenish light.

It is met with in the drusy cavities of certain lavas of Somma and Vesuvius. According to the analysis of Descotils, it consists of alumina, 68·0, silica 2·0, magnesia 12·0, oxide of iron 16·0. Laugier discovered 2·0 of lime in what he analysed.

When met with in grains it may be confounded with *tourmaline*, but it may be distinguished from *tourmaline* by its resinous lustre, greater specific gravity, and by its not becoming electric when heated. La Metherie first established it as a distinct species, and as such it was subsequently acknowledged by Haüy and Werner.

CHA'BASITE. Rhombohedral zeolite, The chabasie of Haüy, and schabasit of Werner. A mineral of a white colour, with sometimes a rosy tinge. It occurs crystallized and massive. It is composed of silica, alumina, lime, potass, soda, and water, silica forming about fifty per cent. of the whole.

CHALA'ZA. (χάλαζα, Gr.) In botany, a small swelling on the outside of the seeds of some plants, it is sometimes coloured: the lemon and orange afford examples of the chalaza.

CHALCE'DONY. See *Calcedony*.

CHALK. (*kalk*, Germ., the craie blanche of the French.) A white earthy limestone, composed of lime and carbonic acid; a variety of carbonate of lime. According to M. Berthier, the chalk of Meudon, when the sand disseminated in it was separated by washing, contained in 100 parts,—carbonate of lime 98, magnesia and a little iron 1, alumina 1. It has an earthy fracture, is meagre to the touch, and adheres to the tongue; it is dull, opaque, soft, and light; its specific gravity being from 2.30 to 2.80. It contains an inconsiderable proportion of silex and iron. The harder varieties of this substance were formerly used for building, and, when protected from the influence of the atmosphere by a thin casing of limestone or flint, proved very durable. The ruins of the Priory of St. Pancras, near Lewes, which have stood nearly 800 years, prove this.—*Dr Mantell*.

The rock commonly known as chalk preserves its peculiar mineral character throughout a considerable area in Europe, but it is rarely of such thickness as in many parts of the south east of England, where horizontally stratified masses about one thousand feet thick are composed of it. In proportion as we depart from the great central deposit of Europe, we find the chalk greatly varying in its texture; in some parts becoming oolitic.

Various conjectures have been offered respecting the probable origin of chalk, and the mode of its formation. Patrin supposed that it was the production of three different causes. 1. Animal earth, proceeding from the decomposition of organic bodies. 2. Calcareous lava, ejected by submarine volcanos. 3. Detritus of calcareous mountains. Delamétherie imagined it to have been deposited by water in a state of great agitation. Dr. Mantell observes that it may have been precipitated from water holding lime in solution, from which an excess of carbonic acid was expelled. Mr. Bakewell says, "according to Ferrare, streams of liquid chalk, or chalk in the state of mud, were ejected from the mud volcano of Macaluba, in Sicily, in 1777. If then we allow submarine aqueous eruption of calcareous matter, and siliceous solutions from thermal waters, to have been poured into a deep ancient ocean, we shall have all the circumstances required to form thick beds of chalk, interspersed with nodules of flint. Mr. Lonsdale, on examining some pieces of chalk from various parts, found what appear to the eye simply white grains, were, in fact, well preserved fossils. From each pound in weight of chalk, he obtained about 1,000 of these.

CHALK FORMATION. This term is applied in the nomenclature of geology to a group of deposits very dissimilar in their lithological compositions, but agreeing in the character of the organic remains which they contain, and referrible to the same epoch of formation, or series of strata, of great depth, which are spread over a large portion of the eastern and south-eastern counties of England, northern France, Germany, Denmark, Sweden, European and Asiatic Russia, and the United States of North America. It consists of strata that have been accumulated in the depths of a sea of vast extent, and affords a striking illustration of the character of oceanic deposits. Scarcely a trace of chalk can be found in Scotland or Wales, but it occurs in Ireland on the north coast. The chalk formation is composed of six divisions, namely, 1. The Maestricht beds; 2. The upper chalk with flints; 3. The lower chalk without flints; 4. The Upper green sand; 5. The gault; 6. The lower green sand. Some authors divide the chalk formation differently, constituting the grey marl, which lies between the chalk without flints, above, and the firestone, or upper green sand, below, a distinct member. Others separate the green sand formation from the chalk formation, making the chalk marl the lowest member of the chalk formation, and placing the upper green sand, or firestone, the gault, and the lower green sand, in the green sand formation. The whole of these are marine deposits. It must however be kept in mind that this order is far from constant. The members of the cretaceous group are ranked as the last of the secondary period; and, in the order of superposition, are placed above the wealden, and below the earliest of the tertiary period, or eocene. "In general, an interval seems to have taken place between the completion of this formation, and the deposition of those which repose upon it; and the surface of the chalk, at the line of junction, usually bears marks of having undergone, during that period, a partial destruction subsequently to its consolidation; a bed of debris being spread over it, consisting chiefly of flints washed out of its mass, and the surface being irregularly worn into frequent cavities, many of them of considerable depth, filled with similar debris. On this debris rests the plastic clay: here, therefore, the transition from the chalk to the more recent formations appears to have been abrupt, not gradual; in a few instances, however, a bed of intermediate character, a cretaceous marle, is interposed at the junction, which may seem to countenance this idea,—that where the series of deposits was permitted, from the circumstances under which they have been formed, to proceed quietly, such a gradation may have taken place."—(*Conybeare.*) The greatest thickness of the chalk strata in England may be estimated at from 600 to 1,000 feet. The organic remains in the chalk formation are exclusively marine. The nodules and veins of flint which occur in the chalk, show that water holding silex in solution must have been very abundant at the cretaceous period, although we are ignorant by what means silex may be dissolved in water. Mammalia are not known in the cretaceous rocks. The testacea hitherto obtained from the various members of the cretaceous group amount to about one thousand. The plants found in the chalk of England and France are principally marine. Sulphuret of iron is the only metallic substance, or metalliferous ore, found in either the upper or lower chalk.

The chalk hills of England are bounded by a line which stretches from south-west to north-east, and they form three principal mountain ranges. The first, leaving Berkshire, runs north through Bucks, Bedfordshire, and Hertfordshire, to Gogmagog hills, near Cambridge. The second, passing from Berkshire eastward, stretches through Surrey, where it forms the Hog's Back, a beautiful ridge extending from **Farnham** to Guildford, and then appears at Boxhill. This branch forms the hilly country and the Downs north of Reigate, Bletchingly, and Godstone. It enters Kent to the north of Westerham, and extends to Folkstone and Dover. One division of this ridge is continued to the north coast of Kent, and terminates at the North Foreland. The third range, leaving Wilts and Berks, enters Hants, and to the south passes round Petersfield, then stretching to the east, forms a barrier against the sea along the coast from Chichester, constituting the South Downs, ranging from Mapledurham to Beachy-head. For a description of the organic remains of the chalk formation, the reader is referred to the article *Cretaceous group.*

CHALK MARL. (Craie Tufau, Fr. Kreide mergel, Germ.) The beds of chalk marl, says Mr. Conybeare, which occur immediately beneath the chalk, graduate into the lowest strata of that subtance, in such a manner that very often no distinct line of separation can be traced. On the other hand, the lowest beds of the chalk marl often pass into those of the green sand, and it is sometimes difficult to catch any precise characters for a formation which is thus rather intermediate between two others, than possessed of independent features. The composition of these beds consist apparently of three ingredients, intimately blended, but in various proportions; cretaceous matter; argillaceous matter; and sand. In the upper beds, the cretaceous matter is the most abundant; and these commonly have the appearance of chalky beds, but distinguishable from the true chalk by a **mottled** or greyish character, by a more laminated texture, and by falling to pieces when dried, after having been wetted. When the argillaceous matter greatly prevails, a tenacious bluish-grey marl is the result. When the sand prevails, a fine-grained grey-coloured sandstone, having a loose texture, is produced; and this last is found graduating into the green sand. The chalk **marls attain a** thickness, in some situations, **of from** 300 to 500 feet. The inclination of the strata is conformable to that of the superjacent chalk, being **usually** nearly horizontal, but where, **as in some** situations, the chalk has been tilted up, as in the Isles of Wight and Purbeck, there also will the chalk marl be found in conformable position. The chalk marl has been called by some authors grey marl. For further particulars see *Grey marl.*

CHA′LICO-THERIUM. An extinct animal, belonging to the order of Mammalia, allied to the tapir, and referrible to the miocene period.

CHAMA. *(Animal a chiton.)* A genus of inequivalved adhering bivalves, with unequal incurvated beaks. It is placed both by De Blainville and Lamarck in the family Chamaceæ, together with Diceras, Etheria, &c. Bruguiere limits this genus to those shells possessing a single hinge-tooth only. Many species have been found fossil, more particularly in the neighbourhood of Paris. The shells of this genus are inhabitants of the ocean, and live in deep water. Twenty-five species have been described; one only of these has been discovered in our seas, namely, **the** Chama Cor.

CHAMA′CEÆ. A family of bivalves placed by Lamarck in the order Dimyaria, and by De Blainville in the order Lamelli-branchiata. It comprises the genera chama, diceras, **etheria,** isocardium, trigonia, &c.

CHA′MA GIGAS. (The Tridacna of Lamarck.) A species of chama inhabiting the Indian ocean; it is the largest and heaviest shell yet discovered, being sometimes of the enormous weight of 530 pounds, and its occupant so large as to furnish one hundred and twenty men with a meal; it is said **to be** very palatable. Fossil shells of the chama gigas are collected by the inhabitants of the East Indian archipelago, for the purpose of being formed into armlets and bracelets.

CHA'MBERED. (*chambré*, Fr.) Divided into compartments by septa: the chambered shells have also been called multilocular. The fossil chambered shells are exceedingly numerous, and afford proofs of not only having performed the office of ordinary shells, as a defence for the body of their inhabitants, but also, of having been hydraulic instruments of nice operation and delicate adjustment, constructed to act in subordination to those universal and unchanging laws, which appear to have ever regulated the movements of fluids. The history of chambered shells illustrates also some of those phenomena of fossil conchology, which relate to the limitation of species to particular geological formations; and affords striking proofs of the curious fact, that many genera, and even whole families, have been called into existence, and again totally annihilated, at various and successive periods, during the progress of the construction of the crust of our globe.—*Prof. Buckland*.

CHAMITE. The fossil shell of the genus chama. The Hon. Mr. Strangways, in his description of the Geology of the environs of St. Petersburg, says, "There occurs a bed of yellowish-white sand containing organic remains. These are only one large species of *chamite*, in very good preservation, usually of a brownish colour and retaining the original polish of the shell. The two beds immediately above this consist also of a sandstone containing chamites. The lowest is usually of a reddish or even rosy colour, and contains vast quantities of *chamites*, or rather fragments of them, strewed throughout its mass, in the direction of the planes of stratification. The upper bed is equally filled with chamites, some of which have a tendency to iridescence or metallic lustre. Chamites are found in the supracretaceous deposits, in the cretaceous group, and in the oolite formation.

CHARA. A genus of aquatic plants found both recent and fossil. Fossil charæ occur in formations of different eras. Charæ are often of considerable importance to the geologist in characterizing entire groups of strata. The seed-vessel of these plants, says Sir C. Lyell, is remarkably tough and hard, and consists of a membranous nut covered by an integument, both of which are spirally striated or ribbed. The integument is composed of five spiral valves, of a quadrangular form. The stems of charæ are found fossil in great abundance in the Scotch marl or travertin, they are striated, and while the striæ of the stems turn, like the worm of a screw, from right to left, those of the seed-vessel wind from left to right. When first these seed-vessels were discovered, in a fossil state, they were supposed to be the shells of an unknown species of mollusk, and a genus was formed for their reception, and termed gyrogonites, or twisted stones, a name by which they are still often designated.

CHA'RACEÆ. In the vegetable kingdom, the fourth order of Thalogens, comprising the chara, &c.

CHARA HISPIDA. A species of the genus chara, above described. The stems of this species are longitudinally striated, possessing at the same time a disposition to spirality. In this, and other, species of chara, the living plant contains so large a portion of carbonate of lime in its structure, that when dried it effervesces strongly with acids.

CHATOYANT. } (Fr. Il se dit d'un objet
CHATOYANTE. } dont la couleur varie, suivant la direction de la lumiere, qui le frappe.) A term used to signify that changeable play of light observable in certain minerals,

as in moonstone and chrysoberyl, the opal, labradorite, &c., &c.

CHEIRACANTHUS. (from χείρ, the hand, and ἄκανθα, a thorn or spine, Gr.) The name assigned to a genus of ichthyolites found in the lower old red sandstone. The cheiracanthus, says Hugh Miller, must have been an eminently handsome little fish, slim, tapering, and described in all its outlines, whether of the body or the fins, by gracefully-waved lines. The body was covered with small angular scales, brightly enamelled, and delicately fretted into parallel ridges that run longitudinally along the upper half of the scale, and leave the posterior portion of it a glittering smooth surface. The tail had the unequal-sided character common to the formation. There is a single dorsal fin placed about two thirds down the back. But it is rather in the construction of the fins than from their position, that the peculiarities of the cheiracanthus are most marked. The anterior edge of each, as in the pectorals of the existing genera Cestracion and Chimæra, is formed of a strong large spine. In the Cheiracanthus, each fin seems to consist of but a single spine, with an angular membrane fixed to it by one of its sides, and attached to the creature's body on the other. Its fins are masts and sails, the spine representing the mast, and the membrane the sail; and it is a curious characteristic of the order, that the membrane, like the body of the ichthyolite, is thickly covered with minute scales.

CHEIROLE'PIS. (from χείρ, a hand or fin, and λεπίς, a scale, Gr.) A genus of ichthyolites of the old red sandstone. In this genus we find a union of the cartilaginous with the osseous skeleton. The external skull, the great shoulder-bone, and the rays of the fins, are all un- equivocally osseous, while the internal skeleton is cartilaginous. The cheirolepis is found from four to fourteen inches in length. Five species have been enumerated.— *Hugh Miller.*

CHEIRO'PTER. (from χείρ, a hand, and πτερόν, a wing, Gr.) An animal having the fingers elongated, for the expansion of membranes which act a wings, as in the Vespertilio or bat. The Cheiroptera form the tenth order of mammalia, which is subdivided into Ch. Frugivora, and Ch. Insectivora.

CHEIRO'PTEROUS. Furnished with elongated fingers, or toes, for the expansion of membranes which serve as wings; belonging to the family Cheiroptera.

CHELI'FEROUS. (from χηλή, Gr. a claw, and *fero*, Lat.) Furnished with claws; armed with claws.

CHELO'NIA. (from χελώνη, testudo.) The tortoise tribe. Chelonia, the first order of Reptilia, is divided into four families, namely, land tortoises, testudo; pond tortoises, emys; river tortoises, trionyx; and sea tortoises, or turtles, chelonia.

CHE'LONICTHYS. The name assigned by Agassiz, in the first instance, to an ichthyolite of old red sandstone; this name has subsequently given place to Asterolepis.

CHE'LONITE. A name given to some fossil echinites, from their resemblance, in their sutures, to the shells of the tortoise. The chelonite belongs to the family Cidaris, class Anocysti.

CHELY'OPHORUS. A genus of icthyolites of the old red sandstone, described by Agassiz in his *Poissons Fossiles.*

CHŒROPO'TAMUS. } An extinct genus
CHEROPO'TAMUS. } in the order Pachydermata, or animals having thick skins. The cheropotamus was an animal most nearly allied to the hog; forming a link between the

Anoplotherium and the Peccary.—*Buckland*.

The remains of the cheropotamus have been found in the gypsum of Paris, and in the eocene limestone of the Isle of Wight. From an examination of these, Cuvier considered that this genus more nearly approximated to the genus porcus than the anoplotherium or palæotherium, and yet did not resemble the living swine. In no other part of Great Britain but in the Isle of Wight, has any portion of the cheropotamus been found.

CHERT. (Dr. Johnson deduces chert from *quartz*.) A variety of hornstone. Chert is also, by some, called horn-stone. A siliceous stone, resembling flint, but less splintery in the fracture, and fusible; which latter property is probably owing to some admixture of calcareous matter. A gradual passage from chert to limestone is not uncommon. Although the words flint and chert are frequently used indiscriminately, they are not to be regarded as synonymous. Chert is not generally to be found in distinct globular masses as flint is, but rather in continuous layers, separating thicker strata of rocks.

CHIA'STOLITE. (from χιαστός, *decussationis formam habens*, and λίθος.) A mineral whose crystals are arranged in four-sided nearly rectangular prisms, presenting a black cross in their transverse section: it has obtained its name from being marked with the form of an X, in dark lines, visible on the summits of the crystals. Its constituent parts are, silica 60·49, alumina 30·17, magnesia 4·12, oxide of iron 2·7, water 0.27. It is the Holspath of Werner, and the Macle of Haüy. It is found in Cumberland and Argyleshire, occurring in clay-slate.

CHIASTOLI'TIC. Composed of chiastolite; containing chiastolite. A mass of chiastolitic and hornblendic slates forms the base of the clay-slate system of Cumberland.

CHILA'GNATHA. An order of the class Myriapoda, which see.

CHILO'PODA. An order of the class Myriapoda, the centipede is an example.

CHIMÆ'RA. (from χίμαιρα, Gr.) A genus of animals, placed in Cuvier's arrangement in the order Sturiones, or Chondropterygii Branchiis Liberis, class Pisces. Professor Buckland observes, "The Chimæra is one of the most remarkable among living fishes, as a link in the family of Chondropterygians; and the discovery of a similar link, in the geological epochs of the oolitic and cretaceous formations, shows that the duration of this curious genus has extended through a greater range of geological epochs, than that of any other genus of fishes yet ascertained by Professor Agassiz. The jaws of four extinct species of fossil fishes of the genus Chimæra have been discovered, and Dr. Mantell states that the jaw, or mandible, of a Chimæra, has been found in the Kentish Rag. The only known species is the Chimæra monstrosa, or Arctic chimæra, two or three feet in length, of a silvery colour, and spotted with brown. This species has the first ray of the dorsal fin enlarged into a strong bony spine, armed with sharp hooks, and placed over the pectorals; like the Icthyodorulite of the earliest fossil sharks. It produces large coriaceous eggs with flattened and hairy borders.

CHINA CLAY. Called also porcelain earth. The name given to disintegrated protogine; decomposed granite. A clay found in some parts of Cornwall, in pits of the depth, sometimes, of twenty feet. This clay when first raised from the pits has the appearance and consistence of mortar: it contains numerous grains of quartz, which are disseminated through it in the same manner as in granite. In some parts the clay is stained of a rusty colour, from the presence of veins, and imbedded portions of shorl and quartz. By a process of washing and mixing with water, the china clay is freed from all extraneous matters, and is then allowed to subside; it is afterwards cut into pieces, thoroughly dried, packed in casks, and shipped for the potteries. The quantity exported from Cornwall exceeds 7,000 tons a-year. China clay contains no alkali and is consequently wholly without felspar.

CHINA STONE. A kind of granite, (containing talc, but wholly destitute of mica), the felspar of which has undergone a partial decomposition. It has obtained its name from being employed in the manufacture of English china. Many thousand tons are annually shipped from Cornwall for the potteries.

CHINE. A narrow ravine with vertical sides. These are numerous in the Isle of Wight, and are objects of curiosity and admiration, being sometimes of great depth. The word *chine* appears to be synonymous with *gully*. In Hampshire, and in the Isle of Wight, the chasms through which the rivulets empty themselves into the sea, are commonly termed *chines*; thus we have Shanklin Chine, Blackgang Chine, &c.

CHIROTHE'RIUM. A name proposed to be given by Professor Kaup to the great unknown animal whose footsteps have been discovered in beds of red sandstone. These footsteps are beautifully figured in Professor Buckland's Bridgewater Treatise. The name proposed by Kaup is on account of a distant resemblance, both of the fore and hind feet, to the impression of a human hand.

These impressions of feet are partly hollow, and partly in relief; all the depressions are upon the upper surfaces of slabs or sandstone, while the reliefs are only upon the lower surfaces, covering those which bear the depressions. These footsteps follow one another in pairs, at intervals of fourteen inches, from pair to pair, each pair being in the same line. Both large and small steps have the great-toes alternately on the left and right side; each has the print of five toes, and the first, or great-toe, is bent inwards like a thumb. The fore and hind foot resemble each other in form, though they differ greatly in size.

CHI'TON. (from χιτών, Gr.) An oval convex, multivalved shell, having eight arcuated valves, partly lying over each other in a row across the back of the animal. The *chiton* is found both fossil and recent; recent, attached to rocks in the southern seas; fossil, at Grignon. The animal inhabiting the shell, a Doris. In Turton's Linné, twenty-eight species of chitons are described, seven of which have been found in the seas of our coasts.

CHLAMY'PHORUS. (from χλαμύς, and φέρω, Gr.) The name it possesses has been given to this animal from its being cased in a coat of armour. The chlamyphorus and armadillo are the only known animals that have a compact coat of plated armour. The chlamyphorus lives almost entirely in burrows beneath the surface of sandy plains; its scales are of a dense substance, resembling hard leather.

CHLORINE. (from χλωρός, green, Gr.) A yellowish-green coloured gas, of a pungent smell, and most injurious to respiration. Chlorine was discovered by Scheele in 1774. It is not permanent over water, which absorbs twice its volume. Specific gravity 2·47, 100 cubic inches weighing 75·67 grains. United with an equal volume of hydrogen gas it gives two volumes of hydrochloric or muriatic acid. In nature, chlorine exists most abundantly in chloride of sodium or common salt. Its equivalent number is 35·42; its symbol Cl. Chlorine has never been decomposed.

CHLO'RITE (from χλωρὸς, green, Gr.) A mineral, consisting of silica, 27·43, alumina 17·9, lime 0·50, oxide of iron 30·63, magnesia 14·56, potash 1·56, water 6·92. It is a dark green variety of talc; has a glistening lustre; minutely foliated structure; is soft and unctuous to the feel; and has obtained its name from its colour. There are several varieties of talc having a dark green colour, and these are known as compact chlorite, earthy chlorite, chlorite slate, foliated chlorite, &c. Chlorite and talc pass by insensible gradations into each other, and in this state they supply the place of mica, in most of the granitic rocks in the vicinity of Mont Blanc.

CHLORITE ROCK. A name proposed, by Dr. Boase, to be given to a genus of rocks found in Cornwall. The term is synonymous with Chlorolite, which see.

CHLO'RITE-SCHIST. A metamorphic rock, of a green slaty character, abounding with chlorite.

CHLORI'TIC-SAND. Sand coloured green by an admixture of chlorite.—*Lyell.*

CHLORI'TIC GRANITE. Granite containing particles of chlorite.

CHLOROLITE. (from χλωρὸς, green, and λίθος, a stone, Gr.) A compound of granular compact-felspar and a mineral resembling chlorite. Three species are enumerated by Dr. Boase, namely, compact, lamellar, and schistose chlorolite. Chlorolite abounds in quartz veins containing compact and crystalline chlorite, and which are often very rich in metallic minerals, particularly in copper ores. Chlorolite is found in Cornwall and is a primary rock.

CHLOROPHŒITE. A mineral discovered by Dr. Mc Culloch. It was found imbedded in the amydaloids of the cliffs of Scuirmore in the Isle of Rum, and has since been brought from Iceland.

CHŒROPOTAMUS. ⎫ The name assigned
CHEROPOTAMUS. ⎭ by Cuvier to a genus of fossil quadrupeds of the hog tribe, the earliest, according to Professor Owen, of this tribe introduced upon our planet. This quadruped, says Owen, must have resembled the Peccari, but was about one-third larger. The only locality in Great Britain where any portion of the Chœropotamus has been discovered is in the Isle of Wight, in the eocene limestone, a locality corresponding with the Paris basin in mineral character, as well as in date of origin.

CHO'ANITE. A zoophyte of the chalk formation, intermediate between Alcyonia and Ventriculites. Dr. Mantell, in his "Wonders of Geology," states, "the choanite, called petrified sea-anemone by lapidaries, bears a close resemblance to the recent Alcyonia. In the choanite, crucial spines, resembling those in the recent Alcyonia, may be detected. The choanite is of a sub-cylindrical form, with root-like processes, and having a cavity or sac, which is deep and small in comparison to the bulk of the animal. The inner surface is studded with pores, which are the terminal openings of tubes, disposed in a radiating manner, and ramifying through the mass." The beautiful pebbles found on the shores of Bognor and Worthing owe their markings to the internal structure of the choanite, and these are worked into a variety of ornaments, as brooches, buckles, earrings, &c.

CHOKE-DAMP. A name given by miners to carbonic acid.

CHO′NDRODITE. Hemiprismatic chrysolite. Another name for brucite; a mineral composed of magnesia 54, silica 32, fluoric acid 4, oxide of iron 2, potass 1, and water 1.

CHONDROPTER′YGII. One of the two great divisions in the classification of fishes, comprising all the cartilaginous genera. The cartilaginous fishes—Chondropterygii of Cuvier. —*Hugh Miller.*

CHONDROPTER′YGIAN. Belonging to the division of fishes named Chondropterygii.

The cartilaginous or *chondropterygian* fishes. The long-fronted *chondropterygian* series of Cuvier. —*Hugh Miller.*

CHROMATE. A mineral containing chromic acid with a base, as chromate of lead.

CHRI′STIANITE. An earthy mineral so named by Monticelli, after Prince Christian of Denmark. See Anorthite.

CHROME. } (from χρωέα, colour,
CHRO′MIUM. } Gr.) This mineral is said to have obtained its name from the property it possesses of imparting colour to other bodies in a very remarkable degree. Chromium was first discovered by Vauquelin in 1797, after a variety of discordant analyses made by Macquart, Bindheim, and others. Specific gravity about 5·0. Its principal ore is found in Siberia, and is a salt of lead, formed by an acid oxide of chromium. To the presence of chrome the emerald and the ruby owe their hues. It is used in tinting glass of an emerald green.

CHRYSOBE′RYL. (from χρύσος, gold, and βηρύλλιον, gemma.) The cymophane of Haüy. Chrysopal of Delametherie. Werner first made the chrysoberyl a distinct species, and gave it the name which it now bears. Colour, a light yellowish or asparagus green. This gem is found in the Brazils, in Ceylon, in America, and in Siberia. It occurs in the alluvial deposits of rivers, and consequently in rolled and frequently much worn masses; also chrystallized. It consists of alumina 76·75, glucina 17·79, and protoxide of iron 4·50,—according to Thomson; Klaproth's analysis gives alumina 71·5, lime 6·0, silica 18·0, oxide of iron 1·5. Its specific gravity 3·6 to 3·8, hardness = 8·5. The primary form of its crystal is a right rectangular prism. By friction it becomes electric. Alone it is infusible before the blow-pipe, but with borax it melts slowly, forming a transparent glass. It displays a beautifully changeable play of light. When free from flaws it is a handsome gem, but not of the first order in value.

CHRY′SOLITE. (from χρύσος gold, and λίθος, a stone, *chrysolite*, Fr.) The Peridot of Haüy, and Krisolith and Olivin of Werner. The term chrysolite was applied, without any regard to distinction, or any discrimination, to a great variety of precious stones, till Werner defined it accurately, and confined it to that stone which the French mineralogists distinguish by the appellation of Peridot (sorte de pierre précieuse, peu recherchée, qui tire un peu sur le vert.) Chrysolite contains a very large proportion of magnesia, according to some authorities more than half its weight, but agreeably to the analysis of others from forty to fifty per cent. Chrysolite is of a green colour, inclining to yellow; its texture is foliated; fracture conchoidal. It causes double refraction. It is infusible at 150°, but at that temperature loses its transparency, and becomes of a dark grey. With borax it melts, without effervescence, into a transparent glass of a light green colour. The chrysolites of commerce come from Upper Egypt and the Brazils; they are also found in Ceylon, in South America, and in Bohemia. The variety called Olivine is met with in Scotland; of this the colour is olive-green. According to the analysis of Klaproth, chrysolite consists of magnesia 43, silica 38, oxide of iron 19.

CHRY'SOPRASE. (from χρύσος, gold, and *prasinus*, green; *chrysoprase*, Fr. *Pierre précieuse d'un vert clair mêlé d'une nuance de jaune.*) A precious stone of an apple-green colour. It is a variety or subspecies of calcedony. It owes its colour to the presence of the metals nickel and iron, in small quantities. It is found in different parts of Germany, particularly in Silesia. It is always amorphous, and possesses but little lustre. It consists of 96 per cent. silica, 1 per cent. oxide of nickel, with a trace of iron, alumina, lime, and magnesia.

CHUSITE. A mineral discovered by Saussure in the cavities of porphyritic rocks near Limbourg, and in Switzerland.

CI'CATRICE. ⎫ (*cicatrix*, Lat. *cicatrix*,
CI'CATRIX. ⎭ Fr. *cicatrice*, It.
1. The scar remaining after a wound.
2. In conchology, the glossy impression in the inside of the valves, to which the muscles of the animal have been affixed.

CI'DARIS. (*cidaris*, Lat.) A family of echinites, characterized by being hemispherical, globular, or suboval; with porous ambulacra, diverging equally on all sides, from the vent to the mouth; vent vertical; mouth beneath, and central. The name cidaris has been given to them from their supposed resemblance to turbans.

From other characters, derived from their spines, they have obtained the names of sea-urchins, sea-hedgehogs, sea-thistles, &c., and those in a petrified state have obtained various names, according to the particular, fanciful, and erroneous notions which have been entertained respecting their origin. Thus, they were called ombria, from ὄμβρος, Gr. signifying the heavy rain, in which it was supposed they fell; brontia, from βροντή, from an idea that they were thrown to the earth by thunder; ceraunii lapides, from κεραυνός, under an impression that they were formed in the air and generated by lightning; chelonites, from their resemblance to the shells of the tortoise; and ova anguina, from their being supposed to be the eggs of serpents.—*Parkinson*.

The species are numerous; eighteen species are described as occurring in the oolite, and nine in the chalk formation.

CI'LIA. (from *cilium*, Lat.)
1. The eye-lash.
2. Hair-like vibratile organs. The organs of motion in the radiated animals. The cilia resemble very minute hairs, and are only visible with the microscope. In the simpler forms of animals, the cilia are the organs for motion, respiration, and the obtaining of food. Dr. Grant has calculated four hundred millions of them on a single flustra foliacea.

CI'LIATED. Fringed, or edged, with parallel hair, bristles, or appendages; occupied with short stiff hairs.

CIMOLIO'RNIS. (from κιμωλία, chalk, and ὄρνις, a bird, Gr.) The name assigned to a fossil bird found in the chalk formation.

CIMOLITE. (So named from its abounding in the island of Cimolia, now called Argentiera, in the Mediterranean.) A clay of a light greyish-white, which, by exposure to the air, becomes of a reddish colour. It occurs massive, and exhibits a somewhat slaty texture. It consists of silex 63, alumine 23, water 12, and oxide of iron 1·25. It was highly prized as a medicine by the ancients. It is opaque, dull, and has an earthy fracture. Though somewhat difficult to break, it receives an impression from the nail. It adheres to the tongue. Specific gravity 2.

CI′NNABAR. (from κιννάβαρι, Gr. *cinabre*, Fr. *cinabro*. It.) The mercure sulphuré of Haüy. Native cinnabar is a red, heavy, sulphureous ore of quicksilver, the principal mines of which are at Idria in Carinthia, and at Almaden in Spain. Cinnabar is called "ore of mercury," since from it mercury is obtained. Before the blow-pipe, it is infusible without addition, but loses its color, and becomes opaque. Specific gravity from 2·60 to 3·25.

CI′NNAMON-STONE. (The Kanelstein, or Kaneelstein, of Werner; Hiacint of Mohs; Essonite of Haüy.) So named from the resemblance of its colour to that of cinnamon. Cinnamon-stone has been hitherto found in masses only, full of fissures, and in grains in the mud of rivers. It is transparent, or only translucent; scratches quartz with difficulty. Fracture imperfectly conchoidal, with small cavities, and shining. Specific gravity 3·60. Before the blow-pipe, it fuses into a brownish-black enamel. When cut, it has rather a greasy feel. The Romanzovite of Nordenskiold is considered to be a variety of cinnamon-stone. A blood-red, or hyacinth-red, variety of the dodecahedral garnet. It consists of silica, alumina, lime, and oxide of iron. The finest specimens are brought from Ceylon, where it is found in the sand of the rivers. It is also called Essonite.

CIPOLI′NO. A granular limestone, containing mica.

CI′RRHOPOD. (from *cirrus*, Lat. and ποῦς, Gr.) The cirrhopods, or cirrhopoda, like the entomostraceous crustacea, are articulated animals, enclosed in shells like those of mollusca, so that they present both forms of the skeleton. The cirrhopods are almost always enclosed in multivalve shells, secreted from the outer surface of a fleshy, thin, enveloping, mantle, and are attached to submarine bodies either directly, by their base, or by means of a fleshy tubular peduncle. The barnacle is an example of the cirrhopoda. In Cuvier's arrangement the cirrhopoda form the sixth class of Mollusca. Linnæus comprised them all in one genus, Lepas; they have since been divided into two, and again, by others, subdivided. Prof. Grant places the cirrhopoda in the sub-kingdom diplo-neura or helminthoida. He thus describes them, " aquatic, sub-articulated, diploneurose animals, with numerous lateral articulated cirrhi, distinct branchiæ for respiration, a pulsating dorsal vessel for circulation, body covered with a fleshy mantle and fixed inverted in a sessile or pedunculated multivalve shell."

CI′RRIPEDE. An annulose, articulated, animal without jointed feet. Cirripedes are found only in the upper secondary, and in tertiary deposits.

CIRRIPED′IA. The sixth order of the class crustacea.

CI′RRUS. (*cirrus*, Lat.) A genus of fossil spiral shells of the chalk deposit. This genus bears great resemblance to trochus, from which, however, it may be distinguished by its deep funnel-shaped umbilicus.

CLADO′DUS. The name assigned by Agassiz to an ichthyolite of the old red sandstone formation.

CLATHRA′RIA LYE′LLII. (described in the Geological Transactions as Clathraria anomala.) A fossil plant discovered by Dr. Mantell, and thus named by him in honour of Sir Charles Lyell. The following description is extracted from Dr. Mantell's works. The Clathraria Lyellii bears an analogy to the yucca, and dracæna or dragon-blood plant. Stems, with the markings of the bases of the leaves, point out the relation of this vegetable to the arborescent ferns, while its internal structure is essentially different. The clathraria has only been found in the quarries in Tilgate Forest. This vegetable appears to have possessed a thick epidermis, or false bark, formed by the union of the bases of the leaves, and covered externally with distinct rhomboidal scales, each scale being surrounded by an elevated ridge. The form of the leaves is not positively known, although, from some imperfect traces on the stone in a specimen bearing the impressions of the cicatrices of the bases of the leaves, there is reason to conclude that they were of a lineari-lance olate form. The axis, or interior part of the trunk, originally enclosed by the bark, occurs in the state of solid subcylindrical blocks of sandstone, attenuated at their base, the surfaces of which are marked with longitudinal interrupted ridges, and, in some instances, are deeply imbricated; they are generally of a dark-brown colour.

CLAVAGE'LLA. A genus of bivalves, of which only one species has been found recent, in the Sicilian seas. It has two irregular, flattish valves, one of which is clasped by the tube, the other being left free. Mr. Sowerby observes, "The shells composing this genus are found in stones, madrepores, &c. and appear to form the connecting link between Aspergillum, which has both valves cemented into the tube, and Fistulana, which has both free."

CLA'VICLE. (*clavicula*, Lat. *clavicule*, Fr. *clavicula*, It.) The collar-bone.

CLAY. When clay is quite pure and unmixed (and in this state it is one of the rarest substances in the mineral kingdom) it is termed alumina, but under the term clay is comprehended an extensive class of compounds, of which silex is a principal constituent. Clay, then, may be defined an unctuous and tenacious earth, capable of being moulded into form; any earth which possesses sufficient ductility, when kneaded with water, to be fashioned like paste, by the hand, or by the potter's lathe. Clays are firmly coherent, weighty, compact, and hard when dry, but stiff, viscid, and ductile when moist; being smooth also and unctuous to the touch. Besides alumina and silica, clays often contain carbonate of lime, magnesia, barytes, oxide of iron, &c. When clay is breathed on, it yields a peculiar smell; it has also a strong affinity for moisture, which is shown by its sticking to the tongue, when applied to it. The purest clay is kaolin, or porcelain clay. All clays appear to be mechanical deposits, not one of them occurring crystallized, or with a crystalline structure; some are found slaty.

CLAY-SLATE. (The argillite of Kirwan.) An indurated clay or shale, common to the fossiliferous and metamorphic series. Clay-slate is opaque, of various shades of colour, and of different degress of hardness, but easily scratched by iron. It is composed of about fifty per cent. of silex, twenty-five of alumine, and ten or twelve of oxide of iron. Some varieties are used as whetstones. The common hone is a variety of slate containing a smaller proportion of alumine and some lime.

CLA'YSTONE. An earthy stone resembling indurated clay, and generally of a colour approaching to purple; it is a variety of prismatic felspar.

CLEA'VAGE. A peculiar fracture, impressed by nature, which is sometimes mistaken for stratification. This is prettily described by Dr. Mantell: "If I take a flint and break it at random, it still preserves a conchoidal fracture, a sharp cutting edge; and sub-divide it as I may, it still retains the same character. If I shiver to pieces calcareous spar, every fragment presents, more or less distinctly, a rhomboidal form; so true is the remark, that we cannot break a stone but in one of nature's joinings."

The regular partings or cleavages in many slate rocks which intersect the beds, nearly at right angles to their dip or inclination, have often been mistaken for strata seams, and have led geologists of some eminence to draw very erroneous conclusions.

"Cleavage means the superinduced fissile structure of clay-slate; cleavage pays no regard to the laminæ of deposition, or original bedding of the rock, but frequently cuts right across it."—*Jukes*.

Where only part of a bed is exposed, it is often difficult to distinguish the lines of cleavage from the true planes of stratification; but the doubt may be cleared by observing the upper and under surface of the bed at the line of its junction with its superstratum and substratum, especially if these be of a different substance.—*Conybeare and Phillips*.

CLEAE'VLANDITE. A mineral, to which this name has been given after Professor Cleaveland; it has been also called albite.

CLIMATIUS RETICULATUS. An ichthyolite of the old red sandstone, described by Agassiz in his *Poissons Fossiles*.

CLI'NKSTONE. (So named from its yielding a metallic sound when struck.) Called also phonolite, a felspathic rock of the trap family. In basalt or wacké, when the felspar greatly prevails, and the texture becomes nearly compact, basalt passes into clinkstone; again, when clinkstone has a more earthy texture, it passes into claystone. Clinkstone often contains imbedded crystals of felspar, and then becomes a trap-porphyry, varying in colour according to the prevailing ingredients of its base. The colour of clinkstone is grey, of various shades. According to Gmelin, natron and potash characterize *clinkstone*; iron and magnesia, basalt.

CLINO'METER. (from κλίνω and μέτρον, Gr.) An instrument, invented by R. Griffith, Esq., for measuring the dip of mineral strata. The following description of the clinometer is extracted from a paper by Lord Webb Seymour, presented to the Geological Society. The clinometer consists of two parts, the plate and the quadrant. The plate is circular and of brass; it is supported by three feet placed at equal distances, and made of wood, with their ends flat and broad. The clinometer may be used to determine the position of any plane surface to which the plate may be applied so as to admit of observation with the quadrant.

CLO'VATE. In conchology, thicker towards the top, elongated towards the base.

CLO'VEN. In botany, leaves are called cloven, when the margins of the segments and fissures are straight.

CLUNCH. A provincial term for a sort of indurated clay which is found dividing the coal seams. The clunch yields those infusible kinds of clay which are adapted for fire-bricks; it varies in hardness, and is black, grey, yellow, white, &c. In the coal series it is generally found immediately beneath each bed of coal, and where it crops out at the surface becomes soft clay.

CLYME′NIA. A genus of Ammonites established by Count Munster, (the Endosiphonites of Ansted) distinguished by having the siphon placed on the ventral margin or close to the body whorls. Morris's catalogue records seven English species from the Devonian system in Cornwall.—*Lycett.*

CLY′PEIFORM. Of a shield—like shape or form.

CLY′PEUS. (Lat.) A division of the first class of echinites. The fossil echinites of the second division of anocysti are distinguished as clypei, from their similitude in form to the round bucklers of the ancient foot-soldiers.

COAL. (*col,* Sax. *kol,* Germ. *kole,* Dutch.) A substance of vegetable origin, composed of charcoal, bitumen, and earthy matter; the latter forms the ashes which remain after combustion. Common coal is a black, solid, and compact substance, generally of a foliated, or rather laminated, structure, which necessarily directs its fracture. Its specific gravity is 1·25 to 1·37. It cakes into cinders during combustion in proportion to its degree of purity, and the nature of the earths which enter into its composition. Coal has obtained various names from varieties of appearance, hardness, situation whence obtained, &c., &c. The very great improbability of finding good coal above the chalk is now acknowledged by all who have even the smallest acquaintance with the geological relations of the English coal mines. The analysis of coal shews it to consist principally of carbon, hydrogen, oxygen, and nitrogen: a specimen of Newcastle coal yielded carbon 75·28, hydrogen 4·18, oxygen 4·58, nitrogen 15·96.

COAL FORMATION. The carboniferous group succeeds the grauwacke in the ascending series of Europe, and is so called because the great mass of European coal is included among the rocks of which it is composed. Considered in its greatest generality, and with reference to where the masses appear in the greatest simplicity, the carboniferous system consists of three formations, namely, the *coal formation,* a mass 1000 yards or more in thickness, consisting of indefinite alternations of shales and sandstones of different kinds, with about fifty feet of coal in many beds, some ironstone layers, and (very rarely) thin layers of limestone; *mountain limestone,* a mass of calcareous rocks, from 500 to 1500 feet in thickness; and *old red sandstone,* a mass of arenaceous and argillaceous rocks, varying in thickness from 100 to 10,000 feet. The coal formation may be understood as applied to the great and principal formation of that mineral, interposed between the newer red or saliferous sandstone, and the great carboniferous limestone and older sandstone formations, or, where these are absent, resting on transition rocks. This is the deposit distinguished by the Wernerians as the independent coal formation. The total thickness of coal existing in the English and Scotch fields is generally about 50 or 60 feet, divided into 20 or more beds, of a thickness of from six feet to a few inches, alternating with from twenty to fifty or one hundred times as great a thickness of shales and sandstones. Every coal district has its peculiar series of strata, unconnected with any other. A district with its peculiar series of strata is called a coal-field. Coal-fields are of limited extent, and the strata frequently dip to a common centre, being often arranged in basin-shaped concavities, which appear to have been originally detached lakes, that were gradually filled up by repeated depositions of carbonaceous and mineral matter. In some of the larger coal-fields, the original form of the lake cannot be traced, but in the smaller ones it is distinctly preserved. The stratum lying over a bed of coal is called its roof, and the stratum under it, the floor. On the eastern side of England, the coal strata generally dip to the south-east point: on the western side, the strata are more frequently thrown into different and opposite directions, by, what are termed, faults and dykes. A fault is a break or intersection of strata, by which they are commonly either suddenly raised or depressed, so that in working a coal mine, the miners come suddenly to its apparent termination. A dyke is a wall of mineral matter which, from igneous or volcanic action, has been forced upwards through the strata, cutting them in a direction nearly vertical. In these cases, sometimes the coal is reduced to a cinder for some distance on either side of the wall or dyke. One of the green-stone dykes of Ireland, passing through a bed of coal, has reduced it to a cinder for the space of nine feet on each side. Our ancient *coal formation* has not been found in Italy, Spain, Sicily, or in any of the more southern countries in Europe. Coal is now universally admitted to be of vegetable origin, a question which was long disputed. It is not uncommon to find among the cinders beneath our grates, traces of fossil plants, whose cavities having been filled with silt, at the time of their deposition in the vegetable mass, that gave origin to the coal, have left the impression of their forms upon clay and sand enclosed within them, sharp as those received by a cast from the interior of a mould. Mr. Hutton has recently discovered the most decisive and indisputable proof of the vegetable origin even of the most bituminous coal; he has ascertained that if any of the three varieties of coal found near Newcastle be cut into very thin slices, and submitted to the microscope, more or less of vegetable structure can be recognised. He says, "each of these three kinds of coal, beside the fine distinct reticulation of the original vegetable texture, exhibits other cells, which are filled with a light wine-yellow-coloured matter, apparently of a bituminous nature, and which is so volatile as to be entirely expelled by heat, before any change is effected in the other constituents of the coal." The plants of the carboniferous group are by no means confined to the simplest forms of vegetation, as to cryptogamic plants; but, on the contrary, belong to all the leading divisions of the vegetable kingdom; some of the more fully developed forms, both of the dicotyledonous and monocotyledonous class, having been already discovered, in the first three or four hundred species brought to light. If violence had attended the transport of the plants now converted into coal, or discovered fossil in the associated beds, the appearance of those in the latter would not be as we now find them; instead of appearing as if spread out by the botanist for examination, we should have had them crushed and disfigured. Moreover, tranquillity seems requisite to explain the condition of those vertical, or nearly vertical, stems of plants discovered in the coal measures of different situations, where they have been gradually enveloped by different beds of sandstone or shale through which they appear to pierce. The alternations of limestones containing marine remains, and of sandstones, shales, and coal-beds, with no trace of a marine animal in them, are exceedingly remarkable, and seem difficult of explanation, without calling in the aid of oscillations of the solid surface of the earth, by which very gradual risings and depressions are effected.

It is, however, the opinion of some authors that the alternations of freshwater shells with marine remains, do not prove as many relative changes of sea and land; but that the coal-measures were deposited in an estuary, into which flowed a considerable river, subject to occasional freshes; and this opinion is supported by the fact of frequent alternations of coarse sandstones and conglomerates with beds of clay or shale, containing the remains of the plants brought down by the river. The quality of coal, even in the same district, very much depends on the nature of the bed which immediately covers the coal stratum; for when the superincumbent stratum is sandstone, the coal is greatly deteriorated, being, more or less, mixed with iron pyrites: on the other hand, if the coal lie immediately beneath argillaceous shale, its quality is much better. For economical purposes, coal may be deemed more or less valuable, in proportion to the quantity of bitumen it contains.

The study of the more ancient coal deposits has yielded the most extraordinary evidence of an extremely hot climate; for it appears from the fossils of that period that the flora consisted almost exclusively of large vascular cryptogamic plants. M. Ad. Brongniart states that there existed at that epoch equiseta upwards of ten feet high, and from five to six inches in diameter; tree-ferns, or plants allied to them, of from forty to fifty feet in height, and arborescent lycopodiaceæ, of from sixty to seventy feet high, exceeding in their development those now found in the hottest parts of the globe. The Newcastle coal-field is supplying rich materials to the fossil flora of Great Britain. The finest example of distinctly preserved vegetable remains is that witnessed in the coal-mines of Bohemia. "The most elaborate imitations," says Prof. Buckland, "of living foliage upon the painted ceilings of Italian palaces, bear no comparison with the beauteous profusion of extinct vegetable forms, with which the galleries of these instructive coal-mines are overhung. The roof is covered as with a canopy of gorgeous tapestry, enriched with festoons of most graceful foliage, flung in wild, irregular profusion over every portion of its surface. The spectator feels himself transported, as if by enchantment, into the forests of another world; he beholds trees, of forms and characters now unknown upon the surface of the earth, presented to his senses almost in the beauty and vigour of their primeval life; their scaly stems, and bending branches, with their delicate apparatus of foliage, are all spread forth before him; little impaired by the lapse of countless ages, and bearing faithful records of extinct systems of vegetation, which began and terminated in times of which these relics are the infallible historians." I can hardly conclude this article better, than by again drawing on the composition of the above quoted elegant and eloquent author, in transferring to my page from his delightful work on Geology and Mineralogy the following beautiful passage. "The important uses of coal and iron in administering to the supply of our daily wants, give to every individual amongst us, in almost every moment of our lives, a personal concern in the geological events of these very distant eras. We are all brought into immediate connexion with the vegetation that clothed the ancient earth, before one-half of its actual surface had yet been formed. The trees of the primeval forests have not, like modern trees, undergone decay, yielding back their elements to the soil and atmosphere by which they had been nourished; but, treasured up in subterranean storehouses, have been transformed into enduring beds of coal, which, in these later ages, have become to man the sources of heat, and light, and wealth. My fire now burns with fuel, and my lamp is shining with the light of gas, derived from coal which has been buried for countless ages in the deep and dark recesses of the earth. We prepare our food, and maintain our forges and furnaces, and the power of our steam-engines, with the remains of plants of ancient forms and extinct species, which were swept from the earth ere the formation of the transition series was completed. Thus, from the wreck of forests that waved upon the surface of the primeval lands, and from ferruginous mud that was lodged at the bottom of the primeval waters, we derive our chief supplies of coal and iron; those two fundamental elements of art and industry, which contribute more than any other mineral production of the earth, to increase the riches, and multiply the comforts, and ameliorate the condition of mankind.

COAL MEASURES. (The Terrain Houiller of the French, the Steinkohlengebirge of the German geologists.) The name given to one division of the carboniferous group. The coal measures consist of beds of coal, sandstone, and shale, irregularly interstratified, sometimes mixed with conglomerates. "The organic remains discovered in the coal measures are principally terrestrial plants; with these are a few freshwater shells, and certain marine exuviæ, which, for the most part, would rather appear to occur in beds alternating with the coal beds and their accompanying shales and sandstones, than mingled with the terrestrial remains." Thirty-five genera and three hundred and ten species of plants have been discovered: ten genera and fourteen species of conchifers: four genera and fourteen species of mollusks: and two genera and three species of fishes.—*De la Beche.—Manual of Geology.*

COA'RCTATE. (*coarctatus*, Lat.) Pressed together. A term used in entomology, to express that state wherein the larva is.

CO'BALT. (The word cobalt seems to be derived from *cobalus*, or *kobold*, the name of a spirit, or goblin, that, according to the superstitious notions of the times, haunted mines, destroyed the works of the miners, and often gave them much unnecessary trouble. It was once customary in Germany to introduce into the church service a prayer that God would preserve miners and their works from kobalts and spirits.)

This metal is of a gray colour, with a shade of red, with but little lustre; its texture is fibrous; specific gravity 8·6, or according to some 7·8. Fusible only at a temperature of 16·677 of Fahrenheit. When heated, cobalt is partly malleable; it is permanently magnetic. The fine blue mineral called zaffre is an impure oxide of this metal. The colour of this oxide is so intense that a single grain of it will impart a full blue to 240 grains of glass. An oxide of cobalt, dissolved in muriatic acid, forms a sympathetic ink;

the characters written with it being invisible when cold, but on exposure to heat assuming a bright green colour, which on cooling they again lose. The principal use of cobalt is to give to glass and porcelain a beautiful blue colour.

Co′bble. } A pebble. This word is given by
Co′bble-stone. } Ray as belonging to the northern counties. Cobble has the same signification as boulder.

Cocci′ferous. (from κόκκος, a berry, and *fero*, to bear.) Any plant or tree bearing berries.

Co′ccolite. (from κόκκος, a grain, and λίθος, a stone, Gr.) A mineral of a green colour, a variety of augite: called also Granular Augite.

Cocco′steus. (from κόκκος, a berry, and οστέον, a bone, Gr.) An ichthyolite of the lower old red sandstone. Hugh Miller says, "the figure of the coccosteus I would compare to a boy's kite. There is a rounded head, a triangular body, a long tail attached to the apex of the triangle, and arms thin and rounded where they attach to the body. The manner in which the plates are arranged on the head is peculiarly beautiful. There are two marked peculiarities in the jaws of the coccosteus. The teeth, instead of being fixed in sockets, like those of quadrupeds and reptiles, or merely placed on the bone, like those of fish of the common variety, seem to have been cut out of the solid, like the teeth of the saw, or the teeth in the mandibles of the beetle, or in the nippers of the lobster. The position of the jaws is vertical not horizontal, and yet the creature belonged to the vertebrata. Four species are established. The average length of the coccosteus somewhat exceeds a foot.

Co′ccygal. } Pertaining to the coccyx,
Coccyge′al. } or terminating bones of the spinal column.

Coccyx. The terminating bones of the spinal column. In the human subject the coccyx consists of four caudal or coccygal vertebræ, which, however, often unite into one undivided portion. In most mammalia, their number is far greater.

Co′chleate. } Twisted like a screw,
Co′chleated. } or the shell of a snail; of a screwed or turbinated form.

Co′ckle. A term applied by the Cornish miners to either schorl or hornblende.

Co′cos. Petrifactions resembling nuts of that genus.

Cœlentera′ta. A sub-kingdom of the kingdom animalia; comprising two classes, Actinozoa and Hydrozoa.

Cœ′liac. Relating or pertaining to the abdominal cavity, or belly.

Coleo′ptera. (from κολεός, a sheath, and πτερόν, Gr. a wing.) An order of insects, having four wings, the two upper being crustaceous, and forming a shield. The second order of the class insecta. This order is divided into five sub-orders, containing many families and tribes.

Coleo′pterous. Belonging to the order Coleoptera; having a horny hollow case under which the wings are folded. Coleopterous insects have four wings, the two superior resembling horizontal scales, and joining in a straight line along the inner margin; the inferior wings are merely folded transversely, and covered with cases, commonly called elytra.

Co′llyrite. The name of a hydrated silicate of alumina, in which there is one equivalent of silica to two of alumina.

Co′lolite. (from κῶλον, and λίθος, Gr.) The name given to the fossil intestines of fishes by M. Agassiz.

Co′lon. (κῶλον, Gr. *colon*, Lat. *colon*, Fr.) One of the large intestines, and by much the longest. The colon commences in the cœcum, and terminates in the rectum.

COLO'PHONITE. (from κολοφωνία, Gr. resin.) So named from its resin colour. The grenat resinite of Haüy. Found near Pitigliano, in Italy, and in Norway. A brown or red variety of dodecahedral garnet, having a resino-adamantine lustre; it is chiefly found at Arendal, in Norway. It consists of silica 37, alumina 13·6, lime 29, oxide of iron 7·4, magnesia 6·5, oxide of manganese 4·7, oxide of titanium 0·5, water 1·0. Specific gravity 3·5.

COLU'MBITE. A mineral ore, the ore of columbium.

COLU'MBIUM. A metal first discovered in 1801 by Mr. Hatchett in a mineral brought from North America, from which it received its name. It is of a dark grey colour, very dense, and difficult of fusion. This metal takes fire when heated in contact with air, and burns into columbic acid, which consists of 85 columbium and 24 oxygen.

COLUMBE'LLA. A genus of spiral univalve shells; they are thick, oval or angular; spire shut, aperture long and narrow, terminating in an anterior canal; the outer lip is thick and dentated, inner lip crenulated. The C. sulcata is recorded from the Crag of Walton on Naze.—*Lycett*.

COLUME'LLA. (Lat.) In conchology, the upright pillar in the centre of most of the univalve shells.

CO'LUMN. In botany the central point of union of the partitions of the seed-vessel, (that is in a capsule containing many cells) to which the seeds are usually attached.

COLU'MNAR. Formed in columns; having the form of columns; having the circumference always circular, but the thickness indeterminate. In this last character *columnar* differs from cylindrical;

cylindrical bodies being equally thick throughout.

CO'MATE. (*comatus*, Lat.) Hairy. In entomology, having the upper part of the head, or vertex, alone covered with long hairs.

COMA'TULA. An existing genus of radiaria of the family of Crinoïdea. The comatula presents a conformity of structure with that of the pentacrinite, almost perfect in every essential part, except that the column is either wanting, or at least reduced to a single plate. Peron states that the comatula suspends itself by its side arms from fuci, and in this position watches for its prey, and obtains it by its spreading arms and fingers. —*Miller*. Four species of the genus comatula discovered at Solenhofen, in the oolite, have been determined by Goldfuss, namely, C. filiformis, C. pectinata, C. piunata, and C. tenella.

COMB.   } These words, thus differently
COMBE.  } written, appear to be of
COOMB. } Saxon origin. Ray gives the second as a south and east country word, and defines it to be a valley, "*vallis utrinque collibus insita*." Lyell states it to be a provincial name for a valley on the declivity of a hill, and which is generally without water. Buckland says, "the term Combe is usually applied to that unwatered portion of a valley, which forms its continuation beyond, and above the most elevated spring that issues into it; at this point, or springhead, the valley ends and the *combe* begins." A narrow undulating ravine.

COMBU'STION. (*combustion*, Fr. *combustione*, It.) Consumption by fire; the disengagement of light and heat which accompanies chemical combination.

CO'MMISSURE. (*commissura*, Lat.) A joint, seam, or suture.

**Common opal.** The Gemeiner opal of Werner. L'Opale commune of Brochant. Quartz resinite commun of Haüy. A subspecies or variety of opal. *See Opal.*

**Common garnet.** (The Gemeiner granat of Werner.) An earthy mineral of a reddish-yellowish, or blackish-brown color, differing from precious garnet only in being opake or translucent. It occurs in granular masses, and crystallized in dodecahedrons. *See Garnet.*

**Compa′ct.** A term used in mineralogy when no particular or distinct parts are to be discerned in a mineral; a compact mineral cannot be cleaved or divided into regular or parallel portions. It is sometimes confounded with the term massive.—*Phillips.*

**Compact feldspar.** The name assigned to one of the feldspathic trappean rocks.

**Compre′ssed.** (*compressus*, Lat.)
1. In botany, leaves are so termed when flattened laterally.
2. In conchology, having one valve flatter than the other.

**Co′mptonite.** A mineral thus named after Lord Compton, who first brought it to England, it is found in the erupted matter of Vesuvius.

**Concamera′tion.** (*concameratio*, Lat.) An arched chamber. In conchology, concamerations are those small chambers into which multilocular shells are divided by transverse septa, as in the nautilus, ammonite, &c.

**Conce′ntric.** (*concentrique*, Fr. *concentrico*, It.) Having one common centre, as the coats of the onion; running to a centre. A term applied to the direction taken by the lines of growth in spiral bodies.

**Conce′ntric lame′llar.** A term used to describe the appearance of a body which being of a spherical form has received successive coverings or depositions.

**Co′nchifer.** A class of mollusca, the constructors and inhabitants of bivalves. All turbinated and simple shells are constructed by molluscs of a higher order than the conchifers, which construct bivalves; the former have heads and eyes; conchifers are without either, and possess but a low degree of any other sense than touch and taste. Thus the whelk is an animal of a higher order than the muscle or oyster.—*Buckland.*

**Conchi′fera.** Constitues the second class of Mollusca. It comprises the cockle, oyster, mussel, and all ordinary bivalve shells.

**Co′nchite.** } (*conchytes*, Fr. *coquilles*
**Co′nchyte.** } *pétrifées.*) A petrified, or fossil, shell.

**Co′nchoid.** In geometry, the name given to the curve invented by Nicomedes.

**Conchoi′dal.** Shelly; shell-like. The fracture of flint is said to be conchoidal, that is to resemble a shell, having convex elevations and concave depressions.

The surface of fracture is termed *conchoidal* when it more or less resembles the appearance of a shell; thus, there are the perfect, imperfect, large, small, and flat conchoidal.

**Conchoidal hornstone.** The Muschlicher Hornstein of Werner. A subspecies of Hornstone, occurring in metalliferous and agate veins, also in striped jasper, and in pitchstone porphyry. *See Hornstone.*

**Conchole′pas.** A genus of oval, vaulted, univalvular mollusks; one species only is known, the concholepas Peruviana, brought from Peru.

**Concho′logy.** (from κόγχη, concha, and λόγος, Gr. *conchyliogogie*, Fr.) That branch of natural history which treats of testaceous animals, or animals having a testaceous covering, whether they inhabit the ocean, or fresh water, or the land.

It is upon the exclusive shape of the shell, and not the animal inhabitant, that the arrangement of conchology is founded. In early periods, naturalists hesitated whether to construct the arrangement from the animal or the shell; it was, however, very wisely determined that it should be from the latter. The greater part of shells are found without the animal in them, and all fossil shells can only be determined by their form. The Linnæan arrangement of shells consists of three orders, namely, Univalves, Bivalves. and Multivales. Univalves consist of shells complete in one piece, as the cyprea, bulla, buccinum, &c. Bivalves are shells of two parts, or valves, generally connected by a cartilage, or ligament, as the oyster, muscle, cockle, &c. Multivalves are shells consisting of more parts than two, as chiton, lepas, and pholas. Every part of a shell which is connected by a cartilage, ligament, hinge, or teeth, is called a valve of such shell. Of the three orders of shells, the univalves are the most numerous, both in genera and species.

CONCHYLIOLI'THUS. } (from κόγχη, a
CONCHY'LIOLITE. } shell, and λίθος, a stone, Gr.) A fossil shell.

CO'NDYLE. (κόνδυλος, Gr. condylus, Lat. condyle, Fr.) The condyles are bony projections, or eminences, at the ends of bones, as the condyles of the shoulder-bone at the elbow; the condyles of the thigh-bone at the knee.

CO'NDYLOID. (from κόνδυλος and εἶδος, form, Gr.) An apophysis of a bone; resembling a Condyle.

CONE. (κῶνος, Gr. conus, Lat. cône, Fr. cono, It.)
1. A solid figure having a circle for its base, and terminating in a point; a figure resembling a sugar-loaf.
2. The fruit of the fir-tree; a catkin hardened, and enlarged into a seed-vessel.

CONFE'RVA. A genus of plants, class Cryptogamia, order Algæ.

CONFERVITES. Fossil remains of plants belonging to the genus Conferva.

CONFIGURA'TION. (Fr. Forme extérieure, ou surface qui borne les corps, et leur donne une figure particulière.)
1. The form of a body in relation to its various parts, and their mutual adaptation.
2. The conjunction, or mutual aspect of the planets.

CONFO'RMABLE. (conforme Fr. conforme, It.) A term used in geology to express parallel strata lying upon each other, or when their general planes are parallel to each other: thus, when several horizontal strata are deposited one upon another, they are said to be in a conformable position, but when horizontal are placed over vertical strata, they are said to be unconformable, so far as regards the horizontal in relation to the vertical strata.

CONFO'RMABLY. In agreement with one another. Horizontal strata placed on parallel strata lie conformably; when placed on vertical strata, or strata having an inclination, or dip, they rest unconformably.

CONFORMA'TION. (conformatio, Lat. conformation, Fr. conformazione, It.) The form, shape, or structure of a body, as regards the disposition of the various parts, and their relation to each other.

CONGE'NEROUS. (congenereux, Fr.) Of the same kind or nature.
1. In anatomy, muscles which act together to produce the same movement are called congenerous.
2. In botany, plants of the same genera.

CONGE'RIES. (Lat.) A collection of many particles into one mass; an aggregate, or mass, of particles.

CONGLO'BATE. (conglobatus, Lat.) Gathered together in a round ball; conglobate glands are such as are smooth in their surface, and seem to be made up of one continued surface.

CONGLO'MERATE. (*conglomeratus*, Lat.) This in geology has the same meaning as breccia, and puddingstone. A mass of fragments united by some cement. Geological writers have chosen to define the term variously, and oppositely, to one another; thus Lyell states a conglomerate to be "rounded water-worn fragments of rock or pebbles, cemented together by another mineral substance." Simply, gravel bound together by a cement. Mantell defines it "fragments cemented together." Bakewell "large fragments of stone, whether rounded or angular, and imbedded in clay or sandstone." Ure "a compound mineral mass, in which angular fragments of rock are imbedded. The Italian word breccia has the same meaning." Mantell in his "Wonders of Geology," p. 417, has "the most interesting beds of these conglomerates, or breccia, in this country."

CONGLO'MERATE GLAND. A gland composed of several glomerate glands, whose excretory ducts unite into one common duct: the liver, kidneys, pancreas, &c. are all conglomerate glands.

CONI'FERÆ. (from *conus* and *fero*, Lat.) An order of trees bearing cones or tops, containing the seeds; the fifteenth order in Linnæus's Fragmenta Methodi Naturalis, and the fifty-first of his natural orders. The Coniferæ are plants whose female flowers, placed at a distance from the male, either on the same or distinct roots, are formed into a cone.

"The Coniferæ," says Professor Buckland, "form a large and very important tribe among living plants, which are characterised not only by peculiarities in their fructification, (having their seeds originally naked, and not enclosed within an ovary; for which reason they have been arranged in a distinct order, as Gymnospermous Phanegoramiæ,) but also by certain remarkable arrangements in the structure of their wood, whereby the smallest fragment may be identified. The recognition of these peculiar characters in the structure of the stem, is especially important to the geological botanist, because the stems of plants are often the only parts which are found preserved in a fossil state. A transverse section of any coniferous wood, in addition to the radiating and concentric lines, exhibits under the microscope a system of reticulations by which coniferæ are distinguishable from other plants. It appears that the coniferæ are common to all fossiliferous strata of all periods; they are least abundant in the transition series, more numerous in the secondary, and most frequent in the tertiary series. All the trees of this order secrete resin, have branched trunks, and linear, rigid, entire leaves: species are found in the coldest as well as in the hottest regions."

CO'NILITE. A genus of molluscous univalves, placed both by Lamarck and De Blainville in the family Orthocerata. It is conical, straight, or slightly curved.

"The difference between conilites and baculites, is that the external sheath of the latter is thin, and not filled up with solid matter, from the points of the alveole to the apex, as in the former."—*Sowerby*.

CONI'STON FLAGSTONE. } The names
CONI'STON LIMESTONE. } assigned to two divisions of the Upper Cambrian rocks. They answer to the Upper Bala rocks, and Bala limestone.

CO′NITE. An ash-coloured mineral, becoming brown by exposure to the atmosphere.

CO′NJUGATE. (*conjugatus*, Lat.) A pinnate leaf having only one pair of leaflets; leaves that consist of one pair of pinnæ or leaflets.

CO′NNATE. (*connatum*, Lat.) Applied to leaves, when two leaves are so united at their base as to have the appearance of one leaf.

CO′NOID. (from κῶνος and εἶδος, Gr.) Resembling a cone in form; sugar-loaf shaped.

CONO′VULUS. A genus of fossil ovate pyramidal univalves occurring in the Suffolk Crag, and formerly attributed to Auricula.

CONTEMPORANE′ITY. (*contemporanéité*, Fr.) The state of being contemporary with. "It becomes a very curious problem to determine what are the lines of *contemporaneity* in the oolitic system."—*Phillips*.

CONULA′RIA. A genus of orthocerata, of a conical shape, and polythalamous, the transverse septa being imperforate. The conularia has no siphon, and in this character differs from orthoceras.

CO′NULUS. A genus of echinites; in it are contained those which rise from a circular base into a cone, (from which form they obtain their name,) with an acute or obtuse vertex, from which five pairs of punctated or crenulated lines, or ambulacra, pass; dividing the shell into five large and five small areæ, that in which the anus is placed being rather the largest. All the species which constitute the genus are known only as fossils, and are distinguished by the modification of their form.

CO′NUS. (κῶνος, Gr. *conus*, Lat.) Animal, a Limax; shell univalve, convolute, turbinate; aperture effuse, longitudinal, linear, without teeth, entire at the base; pillar smooth. This genus is divided by some into five families. The recent conus is an inhabitant of the ocean, and is generally found on rocky shores. Some of the shells are very beautiful, and are both rare and valuable; one species, the cedo nulli, is valued at one hundred guineas. The conus does not inhabit our seas.

COOMB. } See *Comb*.
COOMBE. }

CO′PPER. (*cuprum*, Lat. *kupfer*, Germ. *koper*, Dutch. The word is derived from the island of Cyprus, where it was first wrought.) When pure, copper is of a red colour; its specific gravity is from 8·6 to 8·9, or nearly nine times as heavy as water. Copper is found in primary and secondary rocks, and is often native, i. e. in a pure metallic state; it is also found crystallized. In smell and taste copper is excessively nauseous. It is very malleable, next so in degree after gold and silver, and can be hammered out into extremely thin leaves, so thin as to be blown about by the slightest breeze. In ductility it ranks after gold, silver, platinum, and iron; while in tenacity it yields only to iron. A copper wire one-tenth of an inch in diameter will sustain a weight of 385 lbs. Copper is the most sonorous of all metals: its fusing point is 1450 Fah., and it can be volatilized by an increased temperature; when allowed to cool slowly, it assumes a crystalline form. At common temperatures, copper is not acted on by water, but, if long exposed to the action of the atmosphere and moisture, it oxidizes; as it does in the air alone, if heated to redness. It combines with oxygen in two proportions. Copper admits of a greater degree of condensation by hammering than any other metal. Copper has been known from the earliest ages. As stated before, it occurs frequently in the native state, either in masses, grains, or crystallized in cubes and octohedrons. The most abundant, and most generally diffused ore, and that from which the metal is chiefly obtained, is the sulphuret of copper, termed copper pyrites, composed of copper, sulphur, and a small portion of iron. Copper has never been combined with carbon, hydrogen, or azote; but it combines readily with sulphur and phosphorus, forming with them compounds called sulphuret and phosphuret of copper. Copper, having the property of increasing the hardness of gold without injuring its colour, is used in the making of gold coin; that of Great Britain is an alloy of 11 parts of gold and 1 of copper.

COPEPO'DA. The fourth order of the class crustacea.

CO'PPERAS. (*copparosa*, It. *couperose*, Fr. *kupferwasser*, Germ.) Sulphate of iron; green vitriol. Sulphate of iron has a fine green colour; its crystals are transparent rhomboidal prisms, the faces of which are rhombs, with angles of 79° 50′ and 100° 10′ inclined to each other at angles of 98° 37′ and 81° 23′. It has a strong styptic taste, and reddens vegetable blues. It is prepared by moistening the sulphurets of iron, which are found native in abundance, and exposing them to the open air. These are slowly covered with a crust of sulphate of iron, which is first dissolved in water and, subsequently, by means of evaporation, obtained in crystals.

CO'PPLE-STONES. Boulders; cobble-stones, *which see*.

CO'PROLITE. The petrified fæcal matter of carnivorous reptiles; the petrified fæcal remains of certain fishes. The following description of coprolites is taken from a memoir on the subject, by Professor Buckland, published in the transactions of the Geological Society, as well as from his splendid Bridgewater Treatise :—" In variety of size and external form, the coprolites resemble oblong pebbles or kidney potatoes. They, for the most part, vary from two to four inches in length, and from one to two inches in diameter. Some few are much larger, and bear a due proportion to the gigantic calibre of the largest ichthyosauri; some are flat and amorphous, as if the substance had been voided in a semifluid state; others are flattened by pressure of the shale. Their usual colour is ash-grey, sometimes interspersed with black, and sometimes wholly black. Their substance is of a compact earthy texture, resembling indurated clay, and having a conchoidal and glassy fracture. Their structure is in most cases tortuous, but the number of coils is very unequal; the most common number is three. Some coprolites, especially the small ones, shew no traces of contortion. The sections of these fæcal balls, show their interior to to be arranged in a folded plate, wrapped spirally round from the centre outwards, like the whorls of a turbinated shell; their exterior also retains the corrugations and minute impressions, which, in their plastic state, they may have received from the intestines of the living animals. Dispersed irregularly throughout the petrified faces, are the scales, and occasionally the teeth and bones, of fishes, that seem to have passed undigested through the bodies of the saurians; just as the enamel of teeth, and sometimes fragments of bones, are found undigested both in the recent and fossil *album græcum* of hyænas." On the shore at Lyme Regis, in Dorsetshire, coprolites are found in great abundance, lying scattered in the ground like potatoes. The true character and real nature of the coprolite was long misunderstood, having formerly been called Juli, and believed to be fossil fir cones. The animal origin of coprolites had previously been suggested by M. König, but it is to the investigation of these substances by Professor Buckland, and to his sagacity, that we owe our present knowledge of their true nature. Coprolites are found in all strata which contain the remains of carnivorous reptiles. The real origin of these coprolites is placed beyond all doubt, by their being found frequently within the intestinal canal of fossil skeletons of ichthyosauri. The preservation of such fæcal matter, and its lapidification, result from the imperishable nature of the phosphate of lime, one of the constituents of bony matter.

COPROLY'TIC. Composed of coprolites; resembling coprolites; containing coprolites.

COQUILLA'CEOUS. Containing shells. A term applied by some authors to strata abounding in shelly remains.

CORACOID. (from κόραξ, a crow, and εἶδος, Gr.) Resembling the beak of a crow. A name given to the upper anterior point or process of the scapula.

CO'RAL. (κοράλλιον, Gr. *corallium*, Lat. *corail*, Fr. *coralla*, It. It is somewhat marvellous to find Todd following Johnson in his description of coral, and stating it to be a plant.) The red coral is a branched zoophyte, somewhat resembling in miniature a tree deprived of its leaves and twigs. It seldom exceeds one foot in height, and is attached to the rocks by a broad expansion or base. It consists of a bright red, stony axis, invested with a fleshy, or gelatinous substance, of a pale blue colour, which is studded over with stellular polypi. Coral is composed of carbonate of lime and animal matter. The powers of the organic creation, says Lyell, in modifying the form and structure of the earth's crust, which may be said to be undergoing repair, or where new rock formations are continually in progress, are most conspicuously displayed in the labours of the coral animals. We may compare the operations of these zoophytes in the sea to the effects produced on a smaller scale upon the land, by the plants which generate peat. In corals, the more durable materials of the generation that has passed away serve as the foundation on which living animals are continuing to rear a similar structure. Of the numerous species of zoophytes which are engaged in the production of coral banks, some of the most common belong to the genera meandrina, caryophyllia, millepora, and astrea, but especially the latter. It has been asked, "From whence do these innumerable zoophytes and molluscous animals procure the lime, which, mixed with a small quantity of animal matter, forms the solid covering by which they are protected? Have they the power of separating it from other substances, or the still more extraordinary faculty of producing it from simple elements? The latter I consider the more probable; for the polypi which accumulate rocks of coral have no power of locomotion; their growth is rapid, and the quantity of calcareous matter they produce, in a short space of time, can scarcely be supposed to exist in the waters of the ocean to which they have access, as sea-water contains but a minute portion of lime." Le Sueur, who observed them in the West Indies, describes these polypes, when expanded in calm weather at the bottom of the sea, as covering their stony receptacles with a continuous sheet of most brilliant colours. Ehrenberg, the distinguished German naturalist, was so struck by the splendid

P

spectacle presented by living polyparia covering every portion of the bottom of the Red Sea, that he is said to have exclaimed. "Where is the paradise of flowers that can rival in variety and beauty these living wonders of the ocean.—*Lyell. Mantell. Buckland. Bakewell.*

CORALLI'FERI. An order of polypi, embracing those species which were so long considered to be marine plants.

CO'RALLINE. Belonging to the class Zoophyta, order Eschara, each polypus being contained in a calcareous or horny shell, without any central axis. The animal which secretes and inhabits coral. Fossil corallines abound among the radiata of the transition series, proving that this family had entered thus early upon the important geological functions of adding their calcareous habitations to the solid material of the strata of the globe.

CO'RAL-RAG. (So named from an abundance of fossil corals generally found in it.) A member of the middle division of oolite, of the thickness of about forty feet, in the Bath district. "The coral-rag of England, and analogous zoophytic limestones of the oolitic period in different parts of Europe, bear a resemblance to the coralline formations now in progress in the seas of warmer latitudes."—*Lyell*. The coral-rag comprises a series of beds, occupying in some places a thickness of from one to two hundred feet.

CO'RAL REEF.
CO'RAL ISLAND. } It is a curious, but indisputable fact, that a considerable portion of the earth's surface is the result of organic secretion, and the same process is still going on extensively in the Pacific and Indian seas, where innumerable coral islands rise above, and innumerable reefs and shoals lie just below, the surface of the waves. The observations of modern voyagers have thrown much light on the formation of coral islands and reefs; they concur in the opinion that these reefs and islands do not rise from the depth even of many hundred yards, but commence on the summit of some volcanic elevations, or other submarine ridges and rocks, not far below the surface of the sea. M. M. Quoy and Gaimard observe that the species which form the most extensive banks belong to the genera Meandrina, Caryophyllia, and Astrea, but especially to the latter; and that these genera are not found at depths exceeding a few fathoms. The calcareous masses usually termed coral reefs are by no means exclusively composed of zoophytes; a great variety of shells, and among them some of the largest and heaviest of known species, contributing to augment the mass. The reefs, which just raise themselves above the level of the sea, are usually of a circular or oval form, and surrounded by a deep, and often unfathomable ocean. In the centre of each, there is usually a comparatively shallow lagoon, where there is still water, and where the smaller and more delicate kind of zoophytes find a tranquil abode, while the stronger species live on the exterior margin of the isle. When the reef is of such a height that it remains almost dry at low water, the corals leave off building. Fragments of coral limestone are thrown up by the waves, until the ridge becomes so high, that it is covered only during some seasons of the year by the high tides. The heat of the sun often penetrates the mass when it is dry, and splits it. The force of the waves subsequently separates blocks of the coral and throws them upon the reef. Afterwards the calcareous sand, removed from the action of the waves, lies undisturbed, and offers to the seeds of

trees and plants, cast upon it by the waves, a soil upon which they rapidly vegetate. Wherever circumstances are compatible with vegetable life there we find plants arise. Islands formed by coral reefs, which have risen above the level of the ocean, become, in a short time, covered with verdure. The slightest crevice or irregularity is sufficient to arrest the invisible germs that are always floating in the air, and affords the means of sustenance to diminutive races of lichens and mosses. These soon overspread the surface, and are followed, in the course of a few years, by successive tribes of plants, gradually increasing in size and strength, till, at length, the island is converted into a natural and luxuriant garden. Entire trunks of trees, carried by the rivers from other countries, find here a resting-place: with these come small animals, such as insects, lizards, &c., as the first inhabitants. Even before the trees form a wood, the sea-birds nestle here; strayed land-birds take refuge in the bushes: and at a much later period, man appears, and builds his hut on the fruitful soil.—*Phillips. Lyell. Kotzebue. Bakewell.*

CORALLI'GENOUS. Producing coral. The depth at which the coralligenous zoophyta commence their labour is said not to exceed fifteen or twenty fathoms.

CORALLINE CRAG. The crag of Suffolk consists of two portions, namely, Red crag and Coralline crag. The coralline crag is the older of the two, and is a mass of soft marly sands of a white colour.

CO'RALLOID. } (from *coral* and εἶδος, CORALLOI'DAL. } Gr.) Resembling coral; having the form of coral.

CORALLO'IDES. (*coralloides,* Fr. *seme del corallo bianco,* It.) Coral-wort; the clavaria coralloides of Linnæus.

CORBIS. A bivalve genus of shells belonging to the family Nymphacea of Lamark; it is equivalve, nearly equilateral, oval, thick, ventricose, the hinge has two lateral and two cardinal teeth in each valve. The only recent species, C. fimbriata, is an inhabitant of the Indian ocean; several fossil species are recorded from the eocene strata at Crignon, and upwards of six are known in the oolitic rocks of England.—*Lycett.*

CO'RBULA. (*corbula,* Lat.) A genus of bivalves belonging to the family Corbulacea in Lamarck's arrangement, and to that of Conchacea in De Blainville's. The corbula is a marine animal, found at depths varying to thirteen fathoms, in sandy mud. Some authors place the genus corbula in Solen, others in Mya. Corbulæ are found both fossil and recent. Fossil corbulæ occur in the London clay, calcaire grossier, and Norfolk crag. They are also found in the Shanklin sand, at Parham, and elsewhere.

CORBULA'CEA. A family of bivalves in Lamarck's system, belonging to the order Dimyaria, and comprising the two genera Pandora and Corbula.

CORDI'ERITE. A mineral, so named by Leonhard after Cordier. It is better known as Iolite, or prismatic quartz, *which see.*

COR-MARINUM. A genus of echinites, characterized by the bilabiated mouth being in the third region of the axis of the base, and the anus in the side of the truncated extremity. In this genus, or, as he terms it, family, Leske, with Muller, includes spatangus, spatagoides, brissus, and brissoides, not considering the absence of the groove to be a generic distinction.—*Parkinson.*

CORI'NDON. Another name for corundum or spinel.

CO′RNEA. (from *cornu*, Lat. *cornée*, Fr. *cornea dell'occhio*, It.) The anterior transparent portion of the ball of the eye, or that portion of the front of the eye which allows the rays of light to pass through, and permits objects to be reflected on the retina at the back.

CO′RNEAN. A name applied by De la Beche to designate a variety of Trappean rocks met with in Pembrokeshire, which rocks may be divided into felspathic, quartzose, and hornblendic, as those minerals prevail in the mass.

CO′RNEANITE. The name assigned by Dr. Boase to a genus of rocks having a basis of compact felspar combined with hornblende, which latter is generally in a smaller proportion than in greenstone. Corneanite sometimes contains particles, granules, and minute veins of calcspar: sometimes it abounds in granules and nodules of quartz: sometimes it contains veins of antimony and lead; but it more abounds in manganese.

CORNE′LIAN. For an account of this sub-species of calcedony, see *Carnelian*.

CO′RNEAN. The name given to a felspathic trappean rock.

CORNEO′CALCITE. The name proposed by Dr. Boase to be given to a dark limestone, abundantly occurring in Cornwall. It is composed of carbonate of lime, with hornblende and compact-felspar. Of this genus Dr. Boase enumerates six species.

CO′RNBRASH. A coarse shelly limestone; a provincial term. Cornbrash is a marine deposit, a member of the oolite; it occurs in Wiltshire.

CO′RNSTONE. A mottled, red and green limestone, occurring in the old red sandstone. The name of this and of the preceding word may be considered as provincial, and given to them from their presumed utility in producing fertile corn-land. Sir R. Murchison has divided what he terms the old red system into three parts, the central of these is the Cornstone Formation. He says "the central masses of this system are chiefly composed of red and green argillaceous spotted marls, affording, on decomposition, the soil of the richest tracts of the counties in which it occurs." There is no district in which the nature and relations of the *cornstone* can be better studied than to the north of Ludlow, where this formation occupies a distinct range of hills, rising to the height of four or five hundred feet above the low country, and presenting escarpments to the valley of Corvedale. The spotted marls can never be distinguished from those of the new red sandstone, except, perhaps, when they are separated from each other by beds of hard, micaceous, sandstone. Wherever the order of superposition is not apparent, the fragments of fossil fishes which occur in abundance throughout the cornstones, and which were first detected by Dr. Lloyd, of Ludlow, constitute the best distinction between this formation and the lower new red sandstone, which it so much resembles. These fishes are of very peculiar forms, and their fragments being often of brilliant purple and blue colours, are excellent points of attraction for the eye of the geologist; presenting a striking contrast to the surrounding dull red and green matrix in which they are enveloped. The cephalaspis appears to be a characteristic fossil of the cornstone.

CORNUB′IANITE. (from Cornubia, the Latin name for Cornwall.) Dr. Boase proposes to apply the name of Cornubianite to the rock hitherto distinguished by the name of Killas: he says that the killas appears to constitute a rock *sui generis*; which ought to be distinguished by a peculiar denomination; and as it is exceedingly rich in tin and copper ores (being the principal seat of the Cornish mines) the name of *cornubianite* might be adopted. Cornubianite consists of a basis of compact felspar, coloured by a dark mineral resembling mica, with which it is not only intimately combined, but also contains this mineral in distinct granules, or scales, variously disposed. This rock always occurs in contact with granite; by numerous beds and veins of which, it is frequently intersected. Dr. Boase enumerates six species of cornubianite, namely, compact, quartzose, lamellar, striped, micaceous, and schistose.

CORO'LLA. (Lat.) The corolla consists of the delicate petal, or petals, forming, what, in common language, are termed the blossoms; and in polypetalous flowers, the petals are usually called the leaves of the flower. The corolla constitutes the beauty of the flower, and the odour and fragrance of the plant frequently reside therein, as in the rose, jessamine, violet, &c. The *corolla* has a diversity of forms, as well as of colour, being found of every shade and variety except black. It includes two parts, the petals and the nectary; the latter is sometimes a part of the former, and sometimes separate from it. The leaves of the *corolla* are called petals, and these are either distinct, when the corolla is termed polypetalous, as in the rose, ranunculus, &c. or they are united by their edges, in which case the *corolla* is said to be monopetalous, as in the honey-suckle, convolvulus, &c. The *corolla* is either regular or irregular; when the petals are all alike in size and form, the corolla having a symmetrical appearance, it is called regular; but when the petals are unequal, or unlike each other, it is termed irregular, as in the pelargonium, violet, &c. A *papilionaceous corolla* consists of five petals of particular forms, of which the uppermost is turned back, and is called the vexillum or standard; the two next resemble each other, but differ from the first; they have their faces towards each other; they are called the alæ, or wings: the remaining two, which are placed below the others, also resemble each other, but differ from the three already mentioned; they are usually united by their lower edge, and form a figure resembling the keel of a boat, whence they obtain the name of carina, or keel. This *corolla* is the characteristic of the leguminosæ, a very large order of plants, of which the broom, lupin, sweet-pea, vetch, &c. are examples.

In some plants the *corolla* has one or more of its petals spurred, as in the violet.

In the orchideæ, the *corolla* consists of three pieces, one differing very greatly in form and size from the other two; it is called the *labellum* or little lip, and is often spurred. In many species, this resembles an insect.

The lower part of the single petal of a *corolla*, by which it is fixed to the receptacle, is named the claw.

The cruciferous plants have four petals, and these are so arranged as to resemble a cross, from which circumstance they have been named Cruciferæ. The stock, radish, cabbage, mustard, &c. are examples.

The outer part of the heads of many composite flowers is formed of the ligulate *corollas* of the exterior florets, and these are commonly white, blue, or yellow, as in the aster, daisy, &c.; this part of the head is termed the *ray*, the central part being called the *disk*, which disk is composed of florets, with regular corollas.

A *corolla* with two lips is called *bilabiate:* when the two lips present an appearance resembling the mouth of an animal, the *corolla* is called *ringent*.

The petals of all corollas are placed alternately with the sepals of the calyx.

Coro'lliflorae. In botany, the fourth order of the class Exogens.

Coro'na. (Lat.) In botany, an appendage of the corolla or perianth.

Co'ronoid. See *Coracoid*.

Coro'nula. A regular subrotund, or subconical shell, divided into twelve areae, with an opening both in the superior and inferior part; that in the superior closed by a four-valved operculum.

Co'ronated. (*coronatus*, Lat.) In conchology, crowned, or girt towards the apex, with a single row of eminences.

Co'rpuscle. } (*corpusculum*, Lat.
Corpu'scule. } *corpuscule*, Fr. *corpuscolo*, It.) A minute particle of a body; an atom.

Coru'ndum. (The Korund and Demant-spath of Werner: the Corindon-harmophane of Haüy: Corindon adamantin of Brongniart: Adamantine spar of Kirwan.) A genus of gems comprising four species.
1. Spinel, or dodecahedral corundum.
2. Automolite, or octahedral corundum.
3. Sapphire, or rhombohedral corundum.
4. Chrysoberyl, or prismatic corundum.

These will all be described under their several names.

Some mineralogists constitute corundum a species comprising sapphire, corundum-stone, and emery; others place corundum amongst the members of the ruby family, while some consider it to be a sub-species of sapphire. The colour of corundum is greenish-white, sometimes nearly colourless, passing into greenish-grey, occasionally reddish; sometimes it possesses a Berlin or azure blue, at others it is of a cochineal or crimson red. Its colours are usually weakened by exposure to heat. Before the common blow-pipe it does not yield but with borax, but before the compound blow-pipe, it fuses into a grey globule. The form of the primitive of corundum is a slightly acute rhomboid. It occurs in crystals as well as in amorphous masses of a moderate size, sometimes rolled. Lustre of the cross fracture shining and glistening. Fracture perfect, foliated, with a four-fold cleavage. Its infusibility and hardness serve to distinguish it from all minerals which it resembles in its external characters. Specific gravity from 3·710 to 3·873. It consists of alumina 91, silica 5, iron 1·5. Corundum is found in India, Malabar, the Carnatic, and other eastern parts, and in Italy. It occurs imbedded in primary rocks, having scales of mica and felspar frequently adhering to its surface. It is employed in polishing gems and other hard substances.

Co'rymb. (*corymbe*, Fr. *corymbus* Lat.) A kind of efflorescence. A raceme. A spike of flowers, whose partial peduncles take their rise from different heights upon the common stalk, but the lower peduncles being longer than the upper ones, they all form nearly a level surface at the top.

Cory'mbiated. Garnished with bunches of berries or blossoms, in the form of corymbs.

Corymbi'ferous. (from *corymbus* and *fero*, Lat.) Bearing berries or blossoms in the form of corymbs.

CORY'PHODON. (from κορυφή, a point, and ὀδούς, a tooth, Gr.) The name assigned by Prof. Owen to a sub-genus of extinct fossil tapiroid, one species of which, Coryphodon Eocænus, has been discovered in the eocene clay, on the coast of Essex.

COSMACANTHUS MALCOLMSONI. A fossil fish of the old red sandstone, described by Agassiz in his *Poissons Fossiles*.

COSMO'GONY. (κοσμογένεια, Gr. *cosmogonie*, Fr.) The science of the formation of the universe.

COSMO'GRAPHER. (from *cosmographe*, Fr. *cosmografo*, It. κόσμος and γράφω, κοσμογράφος, Gr.) One who describes the several parts of the creation by writing.

COSMO'GRAPHY. (*cosmographie*, Fr. *cosmografia*, It. κοσμογραφία, Gr.) The science which describes the several parts of the creation, delineating them according to their number, positions, motions, magnitudes, figures, &c.

COSMOLO'GICAL. (*cosmologique*, Fr. κοσμολογικὸς, Gr.) Pertaining to the science of cosmology.

COSMO'LOGIST. A pursuer of the science of cosmology; one who describes the several parts of creation.

COSMO'LOGY. (*cosmologie*, Fr. κοσμολογία, Gr.) The science which treats of the general laws by which the physical world is governed; the study of the world in general.

CO'STA. (*costa*, Lat. plural costæ.) A rib.

CO'STAL. (*costal, costale*, Fr.) Belonging to the ribs.

CO'STATE. } (from *costatus*, Lat.)
CO'STATED. } Ribbed, or having ribs.

CO'TYLE. (from κοτύλη, Gr. *cotyl*, Fr. *cavité d'un os dans laquelle un autre os s'articule*.) The cavity or socket of a bone which receives another bone in articulation, as the socket of the hip which receives the head of the femur, or thigh-bone.

COTYLE'DON. (κοτυληδὼν, Gr. *cotyledon*, Fr.) The side lobe, or seed-lobe of seeds, furnishing nourishment and protection to the corculum, and forming the chief bulk of the seeds: these lobes swell and expand in the ground, and as the stem ascends they are usually raised out of the ground, assume a green colour, and perform the functions of leaves until the young leaves unfold, when they generally wither. The cotyledon is found at the point of union of the radicle and plumule. The most essential difference in the structure, mode of growth, and character of the plants growing from the seeds, is found connected with the number or position of the cotyledons. Those plants, the seeds of which have only one cotyledon, or if more, these alternate on the embryo, are called monocotyledonous. All monocotyledonous plants can be recognized without any difficulty, by a characteristic feature of the leaf, the veins of the leaf being parallel, and not reticulated; all the palms, the tulip, lily, aloe, &c., are instances. Those plants which have two cotyledons, and those opposite, are called dicotyledonous; all dicotyledonous plants have the veins of their leaves reticulated.

COTYLE'DONOUS. Having cotyledons.

COU'ZERANITE. A mineral so named by Leonhard from its having been found in the country called des Couzerans.

CO'WRY. The common or familiar name for shells belonging to the genus Cypræa.

CRAG. A tertiary deposit of the older pliocene period, which has obtained this name from a provincial term signifying gravel. The crag is chiefly developed in the eastern parts of Norfolk and Suffolk, extending thence into Essex; it is seen to rest on the chalk and on the London clay, but generally on the chalk. By some the crag has been divided into two groups, the lower, or coralline, which is, in some places, fifty feet or more in thickness, and the upper, or red crag, thus named from its ferruginous colour. The fossils of the crag are very numerous. From an examination of a collection of shells of the crag made by M. Deshayes, it appears that out of 111 species, 66 were extinct or unknown, and 45 recent, the last, with one exception, being inhabitants of the German Ocean. From this result Sir C. Lyell concludes that the crag belongs to the olden Pliocene period.

The sands of this formation vary in colour from white, through different shades of yellow, up to orange-red: the colour proceeding partly from a ferruginous stain, and partly from the intermixture of yellow oxide of iron.

CRA'NIA. (from *cranium*, Lat. a skull, in consequence of a supposed resemblance of the interior of the shells to a skull, arising from some deep muscular impressions.) A regular inequivalved bivalve; the upper valve very convex, patelliform, with the umbo near the centre, the lower valve flat, and nearly round, and pierced internally with three unequal and oblique holes. The arms of the animal are ciliated. Cuvier places *crania* in the class Brachiopoda, division Mollusca. By Lamarck this genus is placed in the family Rudistes, order Monomyaria; and by De Blainville in the order Palliobranchiata. Craniæ are found attached to stones and shells, and are brought up, probably from great depths, by cod-lines, off the coast of Shetland, and with corals in the Mediterranean. Several species of fossil craniæ are found in the chalk formation.

CRA'NIUM. (Lat.) The skull.

CRASSATE'LLA. (from *crassus*, thick, Lat.) A genus of equivalved inequilateral close bivlaves. The hinge teeth two, with an adjoining fossa; the lateral teeth obsolete. The cartilage inserted in a pit formed in the hinge.

As the crassatella advances in age, the valves become very greatly thickened, from which circumstance it obtains its name. Cuvier places this genus in the family Mytilacia, order Acephela; Lamarck, in the family Mactracea; and De Blainville, in the family Conchacea.

Recent crassatellæ inhabit sandy mud at depths varying from eight to twelve fathoms. Some species of fossil crassatellæ have been found in the tertiary formations.

CRA'SSINA. The name assigned by Lamarck to a genus of shells. See *Astarte*.

CRE'NATE. } (*crenatus*, Lat.) Notched
CRE'NATED. } at the margin; scolloped; indented.

1. It is applied to leaves when the notches or teeth on the borders are rounded, and the notches not directed to either end of the leaf.

2. In entomology, a margin with indentations, not sufficient to be called teeth, the exterior whereof is rounded.

CRE'NATURE. The notch or indentation of a leaf.

CREN'ATULA. (from *crenatus*, Lat.) This name has been given to a genus of bivalves from the hinge showing a row of roundish or oval pits, making it appear as if crenulated. An irregularly formed flat bivalve; closed, not giving passage to any byssus; the hinge linear, excavated, and crenulated; umbones terminal. It is found in sponges, and moored to coral-lines, &c. Parkinson, in describing the crenatula, says, "there are very few among the fossil shells of this or of any other country, which, at first sight, are more dissimilar from any of the recent shells, than the fossil *crenatula*." It is very rarely found.

CRENELLA. A genus of bivalve conchifera belonging to Lamarck's family Arcacea, and established by Brown.

CRE'NULATE. } (*crenélé*, **Fr.**) Indented round the margin with small notches. The fine saw-like edge of the shell of the cockle, which so nicely fits into the opposite shell, is a familiar example of a crenulated margin.

CRETA'CEOUS GROUP. This group comprises the different strata from the chalk of Maestricht to the lower green-sand inclusive. In Lyell's Principles, they are thus arranged: 1. Maestricht beds; 2. Chalk with flints; 3. Chalk without flints; 4. Upper green sand; 5. Gault; 6. Lower green-sand. The whole of these formations are marine.

The cretaceous group are also divided into Upper Cretaceous, comprising the Maestricht and Faxoe beds; white chalk, with flints; white chalk, without flints; chalk marl; upper green-sand; and the gault; and lower cretaceous or Neocomian, comprising the lower greensand, the Speeton clay, and the Wealden Beds.

CRI'CHTONITE. The name given to a black, opaque, shining mineral, after Dr. Crichton.

CRIO'CERATES. A genus of ammonites, having the whorls disconnected.

CRINO'IDAL. Containing fossil crinoïdean remains. The Derbyshire encrinital marble is composed principally of the fossilized remains of crinoïdea, cemented together by carbonate of lime.

Although the representatives of crinoïdeans in our modern seas are of rare occurrence, this family was of vast numerical importance among the earliest inhabitants of the ancient deep. The extensive range which it formerly occupied among the earliest inhabitants of our planet, may be estimated from the fact, that the crinoïdeans already discovered have been arranged in four divisions, comprising nine genera, most of them containing several species, and each individual exhibiting, in every one of its many thousand component little bones, or ossicula, a mechanism which shows them all to have formed parts of a well-contrived and delicate mechanical instrument. —*Prof. Buckland.*

CRINOÏDEA. (from κρίνον and εἶδος, Gr.) "I have derived the name of this family," says Miller, "from the Greek τὰ ζῶα κρινοείδεα, the lily-shaped animals, and have used this word, with another distinguishing term prefixed, to form the name of the genera." Lily-shaped zoophytes. In the most modern classification, Crinoïdea constitutes the fifth order of the class Echinodermata, and is divided into seven families. A name given to the whole class of encrinites and pentacrinites, from their resemblance to the head of the lily. "Of more than thirty species of Crinoïdeans," says Prof. Buckland, "that prevailed to such enormous extent in the transition period, nearly all became extinct before the deposition of the Lias. We may judge of the degree to which the individuals of these species multiplied among the first inhabitants of the sea, from the countless myriads of their petrified remains which fill so many limestone beds of the transition formations, and compose vast strata of entrochal marble, extending over large tracts of country in Europe and America."

The fossil remains of this order have been long known by the name of stone lilies, or encrinites, and have lately been classed under a separate order by the name of Crinoïdea. This order comprehends many genera and species, and is ranged by Cuvier after the asteriæ, in the division of zoophytes. The skeleton of the crinoïdea is composed of numerous ossicula, the number of bones in one skeleton being computed at upwards of thirty thousand. Mr. Miller, in his work, entitled "a Natural History of the Crinoïdea," thus defines them: "An animal with a round, oval, or angular column, composed of numerous articulating joints, supporting at its summit a series of plates, or joints, which form a cuplike body, containing the viscera, from whose upper rim proceed five articulated arms, divided into tentaculated fingers, more or less numerous, surrounding the aperture of the mouth, situated in the centre of a plated integument, which extends over the abdominal cavity, and is capable of being contracted into a conical or proboscal shape."

The existence and preservation of the muscular portion of the Crinoïdea, have been proved by Parkinson, who placed well preserved portions of columns in diluted acid, which gradually removed the calcareous matter, and left the fine animal pellicle behind.

The detached ossicula of the crinoïdea occur in myriads in the mountain limestone and transition rocks, forming successions of strata, each many feet in thickness, and miles in extent; showing how largely the bodies of animals have contributed by their remains, to increase the mass of materials which compose the mineral world. If we imagine a star-fish to possess a long flexible column, the base of which is attached to a rock, we shall have a correct idea of the general character of the crinoïdea, or lily-shaped animals; which are so called from their fancied resemblance, when in a state of repose, to a closed lily. The columns and columnar joints of the crinoïdea, by their frequent occurrence and remarkable figure, attracted the attention of naturalists at an early period. The round columns and their depressed single perforated joints, marked upon the upper and lower surfaces with radiating striæ, have acquired names founded on superstitious ideas, their resemblance to other bodies, and the use they were applied to; as rosary beads, giant's tears, fairy stones, wheel stones, trochites, entrochites, &c. The angular columns, being generally star-shaped, received the names, star-stones, asteriæ, &c.

The essentially distinguishing character of the family of Crinoïdea, is the column formed of numerous joints which separates them from the Polypi, whilst the arms and fingers surrounding the mouth, prove their affinity to them and the Stelleridæ.

Miller establishes four divisions of the family of Crinoïdea, namely, Articulata, comprising the genera Apiocrinites, Encrinites, and Pentacrinites; Semiarticulata, genus Poteriocrinites; Inarticulata, genera Cyathocrinites, Actinocrinites, Rhodocrinites, and Platycrinites; and thirdly Coadunata, genus Eugeniacrinites.—*Buckland. Mantell. Miller.*

CRINOÏDE'AN Belonging to the order Cirnoïdea.

CRIO'CERAS. A genus of Ammonites; proposed by M. Léveillé, with disconnected whorls; the Tropæum of Mr. J. Sowerby. The C. Bowerbankii belongs to the lower green sand of the Isle of Wight; the C. Duvalii and the C. Plicatilis to the Speeton clay of Yorkshire.—*Lycett.*

CRO'CODILE. (κροκόδειλος, Gr. *crocodilus*, Lat. *croccodillo*, It. *crocodile*, Fr.) An amphibious voracious animal of the order Crocodilia, family Reptilia. It is covered with very hard scales, which can be pierced with great difficulty, except under the belly. It has four feet, and a tail, with five toes on each of the fore, and four toes on each of the hind feet, of which only the three internal ones on each foot are armed with nails. It has a wide throat, with several rows of teeth. The fossil remains of crocodiles are common and abundant. Crocodiles are omnivorous. The living species of the crocodile family are twelve, one Gavial, three Alligators, and eight true Crocodiles. Crocodiles, it is said, continue to grow throughout the whole of their existence, and Buckland states their increase to be no less than four hundred times their original bulk, between the period at which they leave the egg and their full maturity. Crocodiles are furnished with a frequent succession of teeth, in order to maintain a duly proportioned supply during every period of their life. The vertebræ of the neck rest on each other through the medium of small false ribs, whereby all lateral motion is rendered difficult, and the crocodile is unable to deviate suddenly from his course; this renders escape from them facile, by either running round them, or pursuing a zigzag course. The eggs of the crocodile are as large as those of the goose. They inhabit fresh water, but they cannot swallow their food under water. The remains of crocodiles occur in all the secondary formations of England, from the lias to the chalk inclusive, as well as in the tertiary formations. The fossil species are numerous, differing greatly both from each other, and from existing species.

CROP-OUT. A term used by miners to express the rising up at the surface of one or more strata. A stratum rising to the surface from beneath another stratum is said to crop out. Beds are said to crop-out when they make their appearance on the surface from beneath others.

CROSS-STONE. Called also Staurolite, and Harmotome; it is the Paratomer Kuphonspath of Mohs, and the Kreutzstein of Werner. Colours white and grey; occasionally it is found with a reddish and yellowish cast. It is composed of 47 parts silica, 21 baryta, 15 alumina, 0·88 potash, 0·10 lime, 15 water. It occurs in small quadrangular prisms terminated by four rhombic planes, crossing each other. The surface of the smaller lateral planes is doubly plumosely streaked. It is found in galena veins and agate balls in the mines of Strontian, in Argyleshire, and in other parts of Scotland; also at Andreasburg, in the Hartz, and in Norway.

CROWS'NEST. The common name given to certain fossil cycadeous plants of the genus Mantellia, from an idea that they were formerly built by crows in the fossil trees, which have become silicified. The largest specimens are about two feet high and three feet in circumference.

CROWSTONE. A hard argillaceous rock, sometimes found forming the floor of the coal-beds: it may be considered to be a highly indurated variety of clunch.

CROY'LSTONE. Crystallized cauk. In this the crystals are small.

CRUCI'FEROUS. (from *crux* and *fero*, Lat.) The name given to a large order of plants, whose petals, four in number, are so arranged as to resemble a cross. The radish, cabbage, stock, &c., are *cruciferous* plants.

CRU'CIFORM. (from *crux* and *forma*, Lat.) Cross-shaped; in the form of a cross. In botany, polypetalous flowers are so called, when the petals are placed in the form of a cross; this is particularly the case in a very large order of plants, which have four petals, so arranged as to resemble a cross.

CRU'CIBLE. (*crucibulum*, Lat.) A vessel, or melting-pot, made of earth, so named, according to some, from its having been formerly made in the shape of a cross; but, according to others, from the metals being tortured in it by fire to compel them to become gold.

CRU'RA. (The plural of *crus*, Lat.) Applied to parts from their resemblance to legs; the legs.

CRU'RAL (*crural*, Fr. *cruralis*, Lat.) Belonging to the leg.

CRUST. (*kruste*, Germ. *crusta*, Lat. *croûte*, Fr. *crosta*, It.) Any shell, hard coat, or external covering. That portion of our globe which is accessible to our inspection and observation is called by geologists, the earth's crust. It is this crust which offers proper occupation to the geologist. The greatest depth to which he has been hitherto able to extend his observations, from the uppermost strata to the lowest beds, is from eight to ten miles; a thickness which, compared with the bulk of the earth, does not exceed that of the thickness of the paper which covers a globe a foot in diameter. The inequalities and crevices in the varnish applied over the surface of such a globe would fairly represent, and be in proportion to, the highest mountains and deepest valleys of the world. The mean density of the earth's mineral crust has generally been taken at 2·5: according to De la Beche 2·6 would be a nearer approximation.

"The term is not used with the intention of conveying an opinion that the earth consists only of a crust, or that its centre is hollow; for of this we know nothing."—*Phillips.*

Mr. W. Hopkins gives 800 miles as the minimum thickness of the solid external crust of the earth.

CRUSTA'CEA. } (from *crusta*, Lat.)
CRUSTA'CEANS. } The crustacea possess a hard external covering, and numerous articulated limbs; antennæ, and palpi; a heart, with circulating vessels and gills, and a nervous system. The crab, lobster, sea urchin, shrimp, &c., are examples. Crustaceous animals possess the most solid form of the skeleton met with in the articulated classes. It is found in the larger decapods to contain nearly half its weight of carbonate of lime, and there is also a considerable proportion of phosphate of lime, with traces of magnesia, iron, and soda. These substances are exuded from the surface of the true skin, along with a tough coagulable animal gluten, which connects all their particles, and forms a thin varnish on the surface. The Rev. J. Williams suggests,—"instead of supposing these animals to secrete the calcareous coverings which they inhabit, say that they emit or secrete a gluten, to which the calcareous particles adhere, and thus the shells are formed." The colouring matter is generally beneath this varnish, and on the exterior surface of the calcareous deposit, but sometimes it pervades the whole substance of the shell.—*Dr. Rob. Grant. Professors Buckland and Fyfe.*

The crustaceæ respire by means of branchiæ; these branchiæ, sometimes situated at the bottom of the feet, at others on the inferior abdominal appendages, either form pyramids composed of laminæ in piles, or bristled with setæ; and in some cases consist seemingly wholly of hairs. The crustaceæ differ from the testaceæ in one most striking point of view: lobsters, crabs, &c., cast their shell or covering every year, whereas the testaceous animals retain theirs as long as they exist. The shells of crustaceous animals appear to grow all at once, whereas those of testaceous animals are evidently formed by the animal adding gradually to them, either annually or periodically, and they are all composed of layers.

Fossil crustaceans are by no means rare both in the most recent, as well as in the most ancient strata, but they are rarely found in a state of complete preservation. Crustacea constituted the fourth class of the sub-kingdom Annulosa, and is divided into nine orders, two of which, Trilobita and Eurypterida, are found fossil only.

CRUSTA'CEAN. ⎫ (*crustacée*, Fr. *crust-*
CRUSTA'CEOUS. ⎭ *aceo*, It.) Shelly, with joints. The crustaceous animals possess a hard shelly covering divided into parts by joints, while the testaceous have a continued uninterrupted shell. The crustaceous animals are the spiders of the sea.

CRUSTACITE. The name given by some authors to any fossil crustacean.

CRY'OLITE. (from κρύος and λίθος, Gr.) Ice-stone. A rare mineral of a white, brown, or red colour, hitherto found only in Greenland, at the arm of the sea named Arksut, where it occurs in gneiss, associated with iron-pyrites and galena. It consists of fluoric acid 44, soda 32, alumina 24.

CRYPTOGA'MIA. (from κρύπτος, concealed, and γάμος, nuptials, Gr.) The 24th class of plants in the Linnæan artificial system, comprehending those whose fructifications are concealed, either through minuteness, or within the fruit. According to more modern classification, the cryptogamia include the thalogens, the anogens, and the acrogens. The carboniferous era abounded in the vascular cryptogamia to a degree unexampled at the present time; the plants belong to species and genera now extinct, but allied to existing types by common principles of organization. The numerical preponderance of the cryptogamia in the coal is such, that while in the present order of nature they are to the whole number of known plants as one to thirty, at that epoch they were in the proportion of twenty-five to thirty. In the saliferous system, about fifty species have been ascertained, some of which differ from any observed in the coal measures. The class cryptogamia contains the ferns, mosses, funguses, and sea-weeds: in all of which the parts of the flowers are either little known, or too minute to be evident.

CRYPTOGA'MIC. ⎫ A term applied to
CRYPTOGA'MOUS. ⎭ plants not bearing flowers with stamens and ovarium visible; belonging to the class Cryptogamia. Ferns, mosses, fungi, &c., are cryptogamic plants. In the transition rocks, about thirteen species of cryptogamic plants, four of which are algæ, and the remainder ferns, comprise all that is known of the vegetable kingdom, anterior to the carboniferous system.

CRY'STAL. (from κρύσταλλος, Gr. *crystallus*, Lat. *crystal*, Fr. *cristallo*, It. *krystall*, Germ.) A crystal may be defined as a more or less symmetrical, geometrical solid, commonly bounded by plane surfaces, which, in mineralogical language, are termed planes, or faces. There are many mineral, or inorganic, substances, which assume certain regular forms when becoming solid from a fluid state, or when, after being dissolved in a fluid, this fluid is evaporated. These regular figures are termed crystals. The cause of a body's possessing this power, or property, is unknown, but it is supposed to be connected with the form of the molecules of which it is composed. Crystals are symmetrical forms. "It has been said of crystals," says the Abbé Haüy, "that they are the flowers of minerals;" an observation concealing a very just idea beneath the air of a comparison which appears to be only ingenious. The importance of their form will become more evident if, in pursuing our enquiries into the niceties of the mechanism of structure, we conceive all these crystals as the assemblages of integrant molecules perfectly resembling each other, and subject to the laws of regular arrangement. Thus although, by a superficial notice of crystals, we might adjudge them to be only the sports of nature, a more intimate acquaintance with them leads to this conclusion,—that the Deity, whose power and wisdom prescribed the unerring laws of the planetary motions, has also established those, which are obeyed with the same fidelity, by the molecules composing the various substances concealed in the recesses of the earth. There are six primitive forms of crystals.

1. The regular tetrahedron, having four equilateral triangles for its faces.
2. The regular cube of six squares for its faces.
3. A dodecahedron, or solid of twelve faces, each being a rhombus.
4. The octohedron, having eight triangles for its faces.
5. A six-sided prism.
6. A parallelopiped, or a solid of six faces, each two of which are parallel and equal, as a cube, a rhomboid, &c. From these six primitive forms of crystals every variety may be supposed to be produced, by cutting away its angles or edges in various manners; or by additions supposed to be made on its faces. The regularity of the figure will be influenced by the rapidity of the evaporation, as when the evaporation is hurried the crystals will be confused, and wanting in regularity; sometimes the evaporation must be spontaneous, or not assisted by the addition of heat, for procuring regular and large crystals. It must not be supposed that every mineral crystallizes naturally in, or can be cut into, all the forms which might be deduced from its primitive form; but it never occurs that the same mineral is found assuming a form, which cannot be shown on these principles to be related to its primitive, or in which primitive it either is occasionally found, or to which the other forms in which it occurs may not be reduced.— *Min. and Metals.*

When bodies dissolved in any fluid are separated by crystallization, they are always found to retain a part of the fluid. The water thus retained by saline crystals is called the water of crystallization. This water appears to be essential to the transparent crystalline form of salts. Most salts may be deprived of their water of crystallization by heat; some lose it in the common temperature of the atmosphere, and fall into a pulverulent mass; others attract moisture so strongly that they, from exposure to the atmosphere, deliquesce.

CRY'STALLINE HUMOUR. } ($\kappa\rho\upsilon\sigma\tau\acute{\alpha}\lambda\iota\nu o\varsigma$,
CRY'STALLINE LENS. } Gr. *crystallinus,* Lat.) A solid body of a lenticular form, being a part of the eye. It appears most absurd ever to have given to this solid body the name of humour. The crystalline lens is situated behind the aqueous humour, opposite to the pupil, and its posterior portion is received into a depression on the fore-part of the vitreous humour. It has two convex surfaces, like a common lens, the anterior being the less convex; the two being formed of

segments of spheres of unequal size.

CRYSTALIZA'TION. } *(crystallisation,*
CRYSTALLIZA'TION. } Fr. *cristallisazione,* It.) A methodical arrangement of the particles of matter according to fixed laws; congelation into crystals.

CRYSTALLIZED QUARTZ. See *Quartz.*

CTENA'CANTHUS. A genus of ichthyolites of the old red sandstone, two species of which have been described by Agassiz; C. Ornatus, **and** C. Serrulatus.

CTEN'ODUS. A genus of ichthyolites of the old red sandstone; Aggassiz enumerates six species.

CTENOI'DEAN. Belonging to the third order of fishes, according to the arrangement of M. Agassiz.

CTENOI'DIA. (from κτείς, *pecten,* a comb; and είδος, Gr.) The third order of fishes in the arrangement of M. Agassiz. The ctenoidians have their scales jagged on the posterior margin, resembling the teeth of a comb, from which circumstance they derive their name; the perch is an example. The ctenoidians first appear at the commencement of the cretaceous formations, succeeding the placoidean and ganoidian orders.—*Prof. Buckland.*

CTENOPTI'CHIUS. **A genus of icthyolites of the old red sandstone.**

CUBE. (from κύβος, Gr. *cubus,* Lat. *cube,* Fr. *cubo,* It.) A regular solid body consisting of six square and equal faces, with right, and therefore equal, angles: a die is a small cube; a prism contained by six equal squares.

CUBE-ORE. A name given to the mineral hexahedral olivenite.

CU'BIT.
1. A measure, according to Dr. Arbuthnot, equal to one foot nine inches, and 888 decimal parts.—*Horne.*
2. That part of **the arm** which extends from the elbow to the wrist.

CU'BIZITE. A name given by Werner to analcime.

CUBOI'DES. A bone of the foot, in shape somewhat resembling a cube; it is placed at the fore and outer part of the tarsus.

CUCULLÆA. A genus of bivalve equilateral, **deep**; hinge straight, shells; subquadrate, equivalve, subwith a series of angular teeth small near the umbones, larger, and more oblique towards the extremities; the umbones are separated by a large flat area; the anterior muscular impression is bordered by a raised, sharp-edged plate or ledge, projecting from the side of the shell. The recent species of Cucullæa are few, but the fossil are very numerous; upwards of 30 are recorded in Mr. Morris's catalogue from the cretaceous, oolitic, carboniferous, and Devonian systems of rocks, and additional species are known, though not described.—*Lycett.*

CUCU'LLATE. } *(cucullatus,*Lat.) Hood-
CUCU'LLATED } ed; having the shape of a hood. Applied to leaves when their edges meet in the lower and expand towards the upper part.

CUCUMERI'NA. (from *cucumer,* Lat.) A species of fossil spine belonging to the echinus, and possessing something of the form of a cucumber, whence its name is derived. There are several varieties.

CUCU'MITES. The name given by Mr. Bowerbank to a genus of fossil fruits found in the London clay. The generic characters are thus given. "Pepo succulent, one celled, many seeded. Seeds ovate, enveloped in a thin membranous arillus." All the parts of these fruits so closely resemble those of various members of the recent genus Cucumis, both in their outward form and their internal structure, that no reasonable doubt can remain of their being true Cucurbitacea.

CULM. (Welsh.)
1. A kind of fossil coal, of indifferent quality, burning with little flame, and emitting a disagreeable smell.

Culm is also called stone coal. Between culm or stone coal and bituminous or common coal there is no *geological* difference. They are mere mineral varieties, which occur in formations accumulated at the same period. The coal of the greater part of the basin of South Wales is stone-coal or culm. Culm generally presents a pure, clean, and polished fracture. Plants common in other coal-fields occur not only in the shale, but in the culm itself. A numerous collection of fossil plants from the culm measures has been submitted to Professor Lindley, who considers that they all occur in other coal-fields. They consist of various Lepidodendra and Calamites together with Neuropteris gigantea, Pecopteris conchitica and Pecopteris nervosa, Iphenoyhyllum Schlotheimii, Stigmaria ficoides, &c., &c. Prof. Lindley, after examining a collection of plants from the Devonshire culm-measures, states, "I have looked over them carefully, and I do not see one single species which might not have been met with at Newcastle, with the exception of two round compressed bean-like bodies, which, if of vegetable origin, are unknown to me."

2. An herbaceous stem peculiar to grasses, rushes, and some other plants allied to them. Culms are either hollow or solid, jointed or without joints, round or triangular, rough or smooth, hairy or downy, and bear both leaves and flowers.

CUL'MIFEROUS. Containing culm; such are the *culmiferous* rocks of Devonshire, South Wales, &c.

CU'MBRIAN SYSTEM. } The word Cumbrian means the rocks of
CU'MBRIAN.
CA'MBRIAN.
Cumberland; the word Cambrian the rocks of Wales. The Cambrian rocks have been divided into Upper and Lower; the upper consisting of the Coniston Flagstone, the Coniston Limestone, and Slates and Porphyry; the lower comprises the Skiddaw slates. The Cumbrian or slate system, as described by Professor Sedgwick, extends over a large portion of Cumberland, Lancashire, and Westmoreland, attaining an elevation in some places of upwards of three thousand feet, and affording the splendid scenery of North Wales and of the lakes. The strata are of great, but unknown, thickness, possessing a slaty character, and nearly destitute of organic remains. The Cumbrian, or, as it has been also called, Grauwacké system, includes the Plynlymmon rocks, the Bala limestone, and the Snowdon rocks.

CU'NEIFORM. } Having the form of a
CU'NIFORM. } wedge. Three bones of the foot have obtained the name of cuneiform bones from their wedge-like shape; they are situated at the fore part of the tarsus and inner side of the os cuboides, and are applied to each other like the stones of an arch.

CUPANÓIDES. The name given by Mr. Bowerbank to a genus of fossil fruits found in the London clay, from their resemblance to the pericarp of Cupania Americana. Eight species are figured and described in his admirable work on the *Fossil Fruits and Seeds of the London Clay.*

CUPRESSINÍTES. The name given by Mr. Bowerbank to a genus of fossil fruits found in the London clay, thirteen species of which he has described. He says, "the fruits forming this group are evidently members of the natural order *Cupressinæ*. I have thought it advisable to place the whole of them together, and to term them *Cupressinites*, which will allow of our uniting under one designation, a greater number of these evidently very nearly allied fruits, than could have been done had I attempted to refer them to M. Adolphe Brongniart's genus *Cupressites*." Mr. Bowerbank has separated this genus into four divisions, comprising thirteen species; for a description, the reader is referred to his admirable work on the *Fossil Fruits of the London Clay*.

Cu′pule. (*capula*, Lat.) The cup of the acorn and of similar fruits.

Curl. The name given by the miners to a variety of argillaceous limestone, found in connexion with the iron-stone. Werner gave it the name of *Dutenmergel* or *funnel marl*. It has been thus described by the Rev. James Yates. "The name evidently alludes to the convoluted form of its distinct concretions, each of which is either itself a complete and regular cone, or is wrapped round part of a cone, which serves as the basis of its structure. Each distinct concretion, on being parted from the conical surface to which it has been attached, presents on its concavity a series of wrinkles, regularly indented, and always parallel to the base of the cone. The convex surface is longitudinally striated. The circular bases of the cones project a little one beyond another, and thus give to the external surface of the mass the appearance of leaves folded over one another. Hence has arisen the conjecture, that the mineral in question is a petrified palm, or lotus. The fracture of this mineral is splintery; its colour greyish-black. The masses, which are exposed to the weather, soon acquire externally a yellow rusty aspect, from the combination of oxygen with the iron which they contain.—*Geological Transactions*.

Curragh. The name given in some parts, as in the Isle of Man, to any tract of peat bog.

Curso′res. The sixth order of the class aves, comprising the ostrich, emu, bustard, &c.

Cu′spated. (from *cuspis*, Lat.) Pointed; terminating in a point, as the leaves of the thistle.

Cu′spidate.  }
Cu′spidated. }
1. A botanical term, applied to leaves terminating in sharp ridged spines.
2. In entomology, having a pointed process much extended, and nearly setiform.

Cuta′neous. (*cutanés*, Fr. *cutaneo*, It.) Pertaining to the skin.

Cu′ticle. (*cuticula*, Lat. *cuticule*, Fr.)
1. The scarf-skin; the outermost skin. The cuticle is a thin, greyish, semi-transparent, insensible membrane, which covers the skin, and adheres to it by small vascular filaments. It is this which is separated by the application of blisters.
2. In botany, the outward covering of plants. Every plant is covered by a cuticular expansion, analogous to the scarf-skin that covers animal bodies. The cuticle, or epidermis, of plants varies in thickness, being extremely delicate on some parts of a flower, and very thick, hard, and coarse on the trunks of many trees.

Cu′tis. (Lat.) The skin, dermis, or true skin, as distinguished from the cuticle or scarf-skin. It lies immediately under the corpus mucosum, and gives a covering to the whole body. It is formed of fibres intimately interwoven, and running in every direction, like the hairs in the felt of a hat, and is so plentifully supplied with nerves and blood-vessels, that the smallest puncture cannot be made in any part of it, without occasioning pain and a discharge of blood. It is that part of quadrupeds of which leather is made. The cutis can be entirely dissolved by the action of boiling water, and consists chiefly of gelatin, from which circumstance it is a principal article in the manufacture of glue.

CU'TILE. } The sepia of Linnæus.
CU'TTLE-FISH. } A species of Cephalopoda, genus Mollusca. The bone of the sepia (which is an internal bone, flat and broad, somewhat resembling a sole in its appearance,) is found, commonly, washed up on our coasts, and when ground into fine powder is used as pounce, and is sometimes employed in the making of tooth-powder. The sepia attains to an immense size in the seas of India and China, and it is said that its arms, which are eight in number, are sometimes several fathoms long, so that it will, by throwing them around a boat, endanger the safety of the boat's crew, and that it is usual to keep on board a hatchet for the purpose of severing them on such occasions. The cuttle-fish has no external shell, but is protected from its enemies by a peculiar internal provision, consisting of a bladder-shaped sac, containing a black and viscid ink, soluble in water, the ejection of which, by rendering the surrounding water opaque, conceals and defends the animal. The sepia has its feet around its head, and walks along the bottom of the sea with its head downwards. The feet are lined internally with little round serrated cups, or suckers, by which the animal both seizes its prey and adheres to other bodies. The mouth, which resembles a parrot's beak, or the bill of a hawk, is placed in the centre of the arms. The ink of the cuttle-fish is said to form an ingredient in the composition of Indian ink.

Professor Buckland states, in describing the ink found in a fossil ink-bag of the cuttle-fish, "So completely are the characters and qualities of the ink retained in its fossil state, that when, in 1826, I submitted a portion of it to my friend Sir Francis Chantrey, requesting him to try its power as a pigment, and he had a drawing prepared with a triturated portion of this fossil substance, the drawing was shown to a celebrated painter, without any information as to its origin, and he immediately pronounced it to be tinted with *sepia* of excellent quality."

The common sepia used in drawing is from the ink-bag of an oriental species of cuttle-fish.

CY'ANITE. (from κύανος, Gr. *color cæruleus*, or sky-coloured.) Called also Kyanite, and by Saussure, Sappare, is a mineral of a grey, blue, and blueish-green colour. It occurs regularly crystallized, as well as massive and disseminated; the form of the primitive crystal is an oblique prism. Its texture is foliated; laminæ long; fragments splintery. It feels somewhat greasy. Before the blow-pipe it becomes almost perfectly white, but it does not melt. Its constituent parts are, alumina 64·30, silica 34·33, with a trace of oxide of iron and a very small portion of lime.

CY'ANOGEN. (from κύανος, blue colour, and γεννάω, to produce, Gr.) A colourless gas which burns with a purple-blue flame. It is the essential ingredient in Prussian Blue.

CYA'THIFORM. (from *cyathus* and *forma*, Lat.) In the form of a cup, or drinking-vessel; cup-shaped.

CYATHOCRINI'TES. (from κύαθος, a cup, and κρίνον, a lily, Gr.) Cup-like, lily-shaped animal. The name assigned by Miller to a genus of crinoidea. There are many species. Miller thus describes the generic characters. "A crinoidal animal, with a round or pentagonal column formed of numerous joints, having side arms proceeding irregularly from it. On the summit adheres a saucer-shaped pelvis of fine pieces, on which are placed in successive series, five costal plates, five scapulæ, and an intersecting plate. From each scapula proceeds one arm, having two hands. The several species of Cyathocrinites occur in the mountain limestone and transition strata; no recent specimen has hitherto been discovered. One species, C. rugosus, has been mistaken for a species of Marsupite, but the marsupite possessed no column, whereas the Cyathocrinite has one.

CYATHOPHY'LLOUS. (from κύαθος, and φύλλον, Gr.) Having cup-shaped leaves.

CYCA'DEA. (from κύκας, cycas, Gr.) A genus of plants. The cycadeæ hold an intermediate place between the palms, ferns, and coniferæ. "That curious tribe," says Lindley, "that stands on the very limits of Monocotyledons and Dicotyledons, and of flowering and flowerless shrubs." Some species are very short, as the zamia; others attain a height of thirty feet and upwards. This beautiful family of plants in their external habit resemble that of palms, whilst their internal structure approximates to that of coniferæ. The cycadeæ are natives of warm climates, mostly tropical, though some are found at the Cape of Good Hope. Leaves of cycadeæ are of frequent occurrence in the shale of the oolitic formation near Scarborough, and they have been found in the Stonesfield slate. Cycadeæ have been found in the coal formation of Bohemia. The trunk of the cycadeæ has no true bark, but it is surrounded by a dense case, composed of persistent scales, which have formed the bases of fallen leaves; these, together with other abortive scales, constitute a compact covering that supplies the place of bark. The prevalence of cycadeæ gives a distinctive character to the flora of the upper secondary formations. The stems found in the Isle of Portland, and the leaves and fruits in the oolitic formations of Yorkshire, show considerable analogy to the existing forms of the tribe at the Cape of Good Hope, in India, and Australia.

CY'CADITES. A name applied to some fossil species of cycas. Our fossil cycadites are closely allied by many remarkable characters of structure to existing cycadeæ.—*Buckland.*

CYCA'DEOIDEÆ. The name given by Prof. Buckland to the petrified remains of certain plants allied to the natural family of Cycadeæ, and resembling the existing genera Zamia and Cycas, though still distinct from both. These fossil remains were obtained from the Isle of Portland; they are now converted into silex, their substance varying from a coarse granular chert to imperfect calcedony: everything seems to favour the supposition, that the plants thus petrified, like those of the analogous recent genera, were the inhabitants of a climate much warmer than that of this country at the present day. M. Adolphe Brongniart has assigned the name Mantellia to this new genus.

CY'CAS. (κύκας, Gr.) The term Cycas was first applied by Theophrastus to a palm tree; it is now used to distinguish a natural order of vegetables, introduced by botanists and phytologists as a connecting link between the ferns and the palms. A genus of plants belonging to the first natural order Palmæ, according to the first arrangement of Linnæus, but subsequently placed among the ferns.

CY′CLAS. (pl. *cyclades*.) A genus of lacustrines, or fresh-water bivalves. The calciferous grit near Hastings is full of cyclades, and several species of cyclas occur, in myriads, in the shales and clays of the Wealden formation.—*Mantell*.

The clycas is an ovato-transverse bivalve, not inflected on the fore part; the hinge with three hinge-teeth and two lateral teeth, compressed and rather remote.—*Parkinson*.

While the clycas of Europe is described as small, thin, and horny, abounding in ditches, ponds, and slow streams, that of Asia is stated to be very large. The cyclas is viviparous.

CY′CLE. (from κύκλος, Gr. *cycle*, Fr. *cielo*, It.) A round of years which go on from first to last, and then return to the same order as before; a space in which the same revolutions begin again.

CY′CLOID. (from κύκλος, and εἶδος, Gr. *cycloïde*, Fr.) A geometrical curve; a figure made by the upper end of the diameter of a circle, turning about a right line.

CYCLOI′DIANS. (from κύκλος, Gr.) The fourth order of fishes, according to the arrangement of M. Agassiz. Families of this order have their scales smooth and simple at their margin, and often ornamented with various figures at the upper surface. The salmon and herring are examples.

CYCLOI′DEAN. Belonging to the fourth order of fishes, according to the arrangement of M. Agassiz. The cycloidean and ctenoidean orders succeeded the placoidean and ganoidean.

CY′CLOLITE. (from κύκλος and λίθος, Gr.) Another name for madrepore.

CY′CLOPITE. A sort of zeolite, resembling analcime, found in the pores of the lava of Etna. The pores of the lava are sometimes coated, or entirely filled, with carbonate of lime, and with a zeolite resembling analcime, which has been called *cyclopite*.—*Lyell, Principles of Geology*.

CYCLOP′TERIS. A genus of plants of the coal series.

CY′LINDER. (κύλινδρος, Gr. *cylindrus*, Lat. *cylindre*, Fr. *cilindro*, It.) A solid formed by the revolution of a rectangular parallelogram about one of its sides, so that it is extended in length equally round, and its ends or extremities are equal circles.

CYLI′NDROID. A solid, in many respects resembling a cylinder, but having elliptical instead of circular extremities, yet parallel and equal.

CYLINDRICO′DON. The name given to a genus of oviparous quadrupeds.

Under this name, Dr. Jæger, of Stutgard, has described the remains of a fossil reptile, of which almost the entire upper jaw, with the teeth, has been discovered by him in the Keuper formation of Germany, near Wurtemburg.

CYME. (*cyma*, Lat. κῦμα, Gr.)
1. A form of inflorescence, the general appearance of which resembles an umbel, and agrees with it in this respect, that its common stalks all spring from one centre; but differs in having those stalks alternately and variously divided. The oleander and elder are examples.
2. A sprout, as of a cabbage.

CYPERA′CEÆ. A tribe of plants answering to the English sedges; they are distinguished from grasses by their stems being solid and generally triangular, instead of being hollow and round. Together with gramineæ, they constitute what writers on botanical geography often call glumaceæ.—*Lyell, Principles of Geology*.

CYMOPHANE. A name given by Haüy to the chrysoberyl, *which see.*

CYMO'SÆ. Plants whose inflorescence is disposed in the form of a cyme; the sixty-third natural order of Linnæus.

CYPRÆ'A. (The cowry.) Animal a slug; shell univalve, oval, or oblong, involute, smooth, obtuse at each end: aperture long, narrow, extending the whole length of the shell, and dentated on each side. The mantle sufficiently ample to fold over and envelope the shell, which at a certain age it covers with a layer of another colour. The genus cypræa consists of beautifully coloured shells very highly polished. They live in sand at the bottom of the ocean; the animal is provided with a membrane, which it throws over its shell, which not only preserves the fine polish, but prevents testaceæ from fixing on it. One hundred and twenty species have been described, one only of which belongs to our seas; the rest are all tropical. In some parts the shell of this animal is used in the place of money, and passes current. By some it is thought that the cypræa casts its shell annually.

CYPRI'FEROUS. Containing shells of the genus Cypris. Entire layers of stone are sometimes composed of the consolidated remains of the cypris; these shells occur in the Hastings sand and sandstone, in the Sussex marble, and in the Purbeck limestone. The cypris contains many species of a genus of crustaceous animalcules formerly called monoculus, from its single eye.

CY'PRIS. A genus of animals, enclosed within two flat valves, like those of a bivalve shell, inhabiting the waters of lakes and marshes. The cypris throws off its integuments every year, which the conchiferous molluscs do not. This circumstance serves to explain the presence in certain places of the countless myriads of the shells of the cypris. The cypris is a microscopic crustacean, with which certain clay beds of the Wealden are so abundantly charged, that the surfaces of many laminæ, into which this clay is easily divided, are often entirely covered with them, as with small seeds. The Sussex marble abounds in the shells of the cypris.

The cypris has two antennæ terminated by a pencil of hairs; one eye and four legs; the head concealed, and the tail small. It inhabits fresh-water only. Three or four species of cypris have been discovered in the Wealden group, but the cypris taba is the most abundant.

CY'PRINA. An equivalve, inequilateral, sub-orbicular, marine bivalve; living in sandy mud. Fossil species occur in the tertiary deposits.

CY'PRINE. Cupreous idocrase. See *Idocrase* and *Vesuvian.*

CYPRICA'RDIA. A genus of bivalve shells belong to the family Cardiacea of Lamarck. Fourteen species are established as occurring in our English deposits, namely, five in the Silurian, three in the Devonian, four in the carboniferous, one in the oolitic, and one in the tertiary system.

CYRE'NA. A genus of small bivalve fluviatile shells, eight species of which are recorded as found in the tertiary series of rocks.

CYRTO'CERAS. (from κυρτός, carved, and κέρας, a horn, Gr.) A genus of fossil chambered shells established by Goldfuss, the form is bent, arched or partially convoluted, the free end being sometimes elongated and straight, the siphuncle is sub-dorsal or marginal, the aperture nearly orbicular. Cyrtoceras has been found only in the Silurian and Devonian systems of rocks; 13

English species are recorded in Mr. Morris's catalogue, one only of which, the C. læve, belongs to the Silurian system.—*Lycett.*

CYSTIDEA. The seventh order of the class Echinodermata. All the genera of this class are extinct, and found fossil only.

CYSTIPHYLLUM. (from κύστις, vesica, and φύλλον, folium, Gr.) A genus of corals found in the Silurian rocks, and thus named by Mr. Lonsdale. Externally, they are striated; internally, composed of small bladder-like cells. From this internal structure, and from the absence of a distinct centre, Mr. Lonsdale has named the genus, separating it from the Cyathophylla of Goldfuss.

CYTHERÆ'A. A marine bivalve; equivalve, lenticular, oval; hinge with two cardinal teeth; one anterior lateral tooth in each valve, which distinguishes this genus from Venus. It is found in depths of the ocean varying to fifty fathoms, in mud and coarse sands. Several species have been found fossil in the tertiary deposits. Cytheræa nitidula is mentioned by Dr. Mantell as occurring in the London clay, and cytheræa convexa in the Plastic clay.

# D

DACHSTEIN BEDS. Beds of the triassic period. The Dachstein beds attain in some parts of Germany a thickness of 2,000 feet, they are of white or greyish limestone. Below, these beds are unfossiliferous, but above, they contain beds made up of corals.

DACTYLOPO'RA. A genus of lapideous free polypifers, of a cylindrically elevated form, with a perforation in the narrower extremity. The surface reticulated with rhomboidal meshes, the network itself porous.

DA'OURITE. The siberite of Lermina. A variety of the red shorl of Siberia, called also rubellite. This stone is found in Siberia mixed with white quartz. It is composed of silica 56, alumina 36, with some oxide of manganese, and oxide of iron. Daourite is another term for rubellite; it is in fact a variety of tourmaline, of a red colour. It has obtained a variety of names, as, rubellite, siberite, daourite, tourmaline, apyre, red schorl of Siberia, &c. See *Tourmaline.*

DA'SYPUS. (δασύπους, from δασύς and πούς, Gr.) The armadillo, *which see*

DASYU'RUS. An animal of the marsupial order. The dasyurus is said to be the largest of the carnivorous marsupial animals. The head of a species of dasyurus has been discovered in the Eocene freshwater limestone of Auvergne.

Dasyurus ursinus is a very ferocious creature about the size of a badger, its actions and habits much resemble those of a bear.

DA'THOLITE. } The Dystom-spath of
DA'TOLITE. } Mohs. A sort of spar-stone; the siliceous borate of lime. According to Menil, it is a combination of silica 38·50, lime 35·60, boracic acid 21·30, water 4·60. Its varieties are named Botryolite, Earthy Botryoidal Datolite, and Common Datolite. It has been found principally in Norway, in beds of magnetic iron-ore.

DA'VYNE. A earthy mineral, described by Monticelli and Covelli in their Podromo della Mineralogia Vesuviana, and by them named after Sir H. Davy. It is of a white, or yellowish colour; transparent; translucent, or opaque. Specific gravity 2·4; hardness 5 to 5·5.

DEBA'CLE. (*Debacle*, Fr. *Amas de glaçons qui arrivent avec impétuosité, dans un dégel subit, après qu'une rivière a été prise long-temps.*) A violent torrent or rush of waters, which, overcoming all opposing barriers, carries with it stones, rocks, and other fragments, spreading them in all directions.

DEBOU'CHE. (*débouché*, Fr. *L'extrémité d'un défilé, d'un col de montagnes.*) The outlet of a narrow pass.

DEBR'IS. (*débris*, Fr.) The fragments of rocks; the ruins of strata; the rubbish, sand, grit, &c., brought down by torrents.

DECAHE'DRAL. (from δέκα, and ἕδρα, Gr.) Having ten sides.

DECAHE'DRON. A figure which hath ten sides.

DECA'NDRIA. (from δέκα, and ἀνὴρ, Gr.) A class of plants characterized by having ten stamens; it includes cassia, ruta, saxifraga, &c.

DECA'NDRIAN. Belonging to the class Decandria; having ten stamens.

DECAPHY'LLOUS. (from δέκα, and φύλλον, Gr.) A calyx which hath ten leaves.

DECA'PODA. (from δέκα, ten, and πούς, foot.) The first order of crustacea. Having the antenniferous region of head confluent with the thorax. This order includes lobster, crab, craw-fish, shrimp, &c.

DECA'PODAL. Belonging to the order Decapoda; having ten feet. Synonymous with decempedal.

DECE'MFID. (from *decem* and *fissus*, Lat.) Ten-cleft; in botany, a term for a calyx cleft, or divided, into ten parts.

DECEMLO'CULAR. (from *decem* and *loculus*, Lat.) Ten-celled; in botany, an epithet for a pericarp divided into ten loculi or cells.

DECI'DUOUS. (*deciduus*, Lat.)
1. In botany, falling off; plants which lose their leaves in autumn are called deciduous; applied also to stipules falling in the autumn; to calyces falling soon after the expansion of the corolla; and to the corolla when falling with the stamens.
2. In conchology, to shells having a tendency in the apex of the spire to fall off; to crustaceans, annually casting their shells.

DE'COMPOSE. (*decomposer*, Fr. *Réduire un corps à ses principes, ou séparer les parties dont il est composé.*) To resolve a body into its constituent elements; to overcome the power of affinity, and thereby to separate elementary particles.

DE'COMPOUND. Doubly compound. Leaves are so called when the petioles, instead of bearing leaflets, branch out into other petioles to which the leaflets are attached.

DE'CREMENT. (*decrementum*, Lat.) Gradual waste, or wearing away, as of rocks by the action of water; gradual diminution. In mineralogy, decrement is considered as of two kinds, single and compound. When in crystals the planes decrease equally to a point, they are said to arise from *simple decrement*; but when, as in the pentagonal dodecahedron, the planes do not decrease equally on all sides, the decrement is termed *compound*.

DECRE'SCENT. (*decrescens*, Lat.) Gradually becoming less.

DECU'RRENT. (from *decurro*, Lat.) Running downwards. Applied to sessile leaves when the base runs down the stem and forms a border or wing; applied also to stipules when extending downwards along the stem. In some plants, as in some of the thistles, the margins of sessile leaves run down on each side of the stem, so as to appear to be of one piece with it; these leaves are called *decurrent*.

DECU'RSIVELY PINNATE. Applied to leaves having their leaflets decurrent, or running along the petiole.

DECU'SSATE. (*decusso*, Lat.) To intersect at acute angles; to cross each other at right angles. Applied to branches growing in pairs, and alternately crossing each other at right angles; applied also to leaves alternately opposite. In conchology, applied to striæ, crossing or intersecting each other at acute angles.

DEFLE'XED. (*deflexus*, Lat.) In entomology, having the sharp edge bent downwards.

DEGRADATION. (*degradation*, Fr. *Il signifie dépérissement, etat de décadence, de ruine.* This term is used by geologists to signify the lessening or wearing away of rocks, strata, &c., by the action of water, or other causes.

DEHI'SCENT. (*dehiscens*, Lat.) In botany, fruits which open when ripe, so as to enable the seeds to escape, are termed dehiscent. Gaping; opening.

DELPHI'NULA. (from *delphinus*, Lat.) A turbinated, subdiscoidal, umbilicated univalve. The aperture round and pearly; operculum horny. The delphinula creeps on rocks and sea-weeds. This genus is formed of shells formerly included by Linnæus in his genus Turbo. Lamarck places delphinula in the family Scalariana. The fossil delphinula occurs in the tertiary deposits; it is also recent.

DE'LTA. A term applied by geologists to the alluvial deposits formed at the mouths of rivers. It has obtained its name from a supposed resemblance to the Greek letter Δ. Deltas are occasionally of immense size, and they are divided into lacustrine, mediterranean, and oceanic, the first being those formed in lakes, as the delta at the mouth of the Rhone, at the upper end of the lake of Geneva; the second, or mediterranean, are those formed in inland seas, as that at the mouth of the Rhone, where it enters the Mediterranean; the third, or oceanic, are those formed on the borders of the ocean, as the delta of the Ganges.

DELTHY'RIS LIMESTONE. A shaly limestone, met with in the Heidelberg group.

DE'LTOID. (from *delta*, the fourth letter of the Greek alphabet.) The name of a muscle of the shoulder, from its supposed resemblance to the Greek letter Δ; triangular.

DE'NDRACHATE. (from δένδρον and ἀχάτης, Gr.) An agate with delineations of trees, ferns, moss, &c. Some of these are exceedingly beautiful, and are so elegantly depicted that they have been erroneously taken for real plants, whence their name. These pebbles are found abundantly on the shore from Bognor to Brighton, and, when cut and polished, are made into very beautiful necklaces, brooches, snuff boxes, &c., &c.

DE'NDRITE. (δενδρίτις, Gr.) The same as dendrachate.

DENDR'ITE AGATE. A species of agate, thus named from its containing in its interior brown, reddish-brown, or blackish delineations of leafless trees, shrubs, &c. For a description see *Mocha stone*.

DENDRI'TICAL. Containing the resemblance of trees, ferns, or mosses.

DENDR'ODUS. In the transverse section of these reptile teeth the cancelli are found to radiate from the open centre towards the circumference, like the spokes of a wheel from the nave; and each spoke seems as if it had sprouted into branch and blossom, presenting the appearance of a well-trained wall-tree: hence the generic name Dendrodus, assigned by Professor

Owen. The name Dendrodus, appears to have been supplanted by that of Cricodus. The Dendrodus is an icthyolite of the old red sandstone, of which six species have been particularized by Agassiz.

DE'NDROITE. A fossil resembling the branch of a tree.

DE'NDROLITE. (from δένδρον and λίθος, Gr.) Fossil wood; the fossil branch of a tree.

DE'NSITY. (*densitas*, Lat. *densité*, Fr. *densità*, It.) Closeness; compactness; that property directly opposite to rarity, whereby bodies contain such a quantity of matter in such a bulk. The densities of bodies are proportional to their masses, divided by their volumes. Hence if the sun and planets be assumed to be spheres, their volumes will be as the cubes of their diameters. The strata of the terrestrial spheroid are not only concentric and elliptical, but the lunar inequalities show that they increase in density from the surface of the earth to its centre.

The absolute density of, or the quantity of matter contained in, the earth, compared with an equal bulk of any known substance, may be nearly determined by the attractive force which any given mass of matter exerts upon a plummet, when suspended in its vicinity, to draw it from a vertical line. By this method it has been found that the mean densisty of the earth is about five times greater than that of water, or nearly twice the average density of the rocks and stones on the surface. The mean density of the ocean is only about one-fifth part of the mean density of the earth. More recent observations, by the astronomer royal, on the pendulum at the surface and at the bottom of deep mines, give a mean of 6·809 for the earth's specific gravity, while those made by the ordnance survey, on the deflection of the plumb-line, give it as 5·14.

DE'NTAL. } (from *dens*, Lat.) A
DENTA'LIUM. } shell-fish belonging, according to Linnæus, to the class Vermes, order Testaca. The shell consists of one tubulous arcuated cone, open at both ends. There are many species, distinguished by the angles, striæ, &c., of their shells.

The observations of Deshayes lead to the conclusion that the genus Dentalium approaches very closely to the molluscs, if, indeed, it does not belong to them. The dentalia are found in deep water, frequently near the shore, inhabiting the ocean only; they are solitary. Captain Vidal drew up *dentalia* from the mud of Galway Bay from a depth of 240 fathoms. The animal is a terebella. The shells are known commonly by the name of tooth-shells, or sea teeth. Twenty-two species have been described, seven of which inhabit our coasts.

DE'NTALITE. } (from *dens*, a tooth,
DE'NTALITHE. } and λίθος, a stone.) A fossil dentalium found in the tertiary formations, in the galt and in the lower green sand. Of these there are many species; as the Dentalium planum, D. striatum, D. ellipticum, D. decussatum, &c. See *Ditrupa*.

DENTA'TA. A name given to the second vertebra of the spinal column, from a tooth-like process which it possesses.

DE'NTATE. } (*dentatus*, Lat.) Indent-
DE'NTATED. } ed; jagged; notched; toothed. In botany, leaves are called dentated, when the border is beset with horizontal projecting points or teeth, with rather a distant space between each, and of the same consistence as the substance of the leaf itself: applied also to stipules having spreading teeth about the margin, remote from each other.

8

DE'NTATURE. Pertaining to the teeth of an animal, to their structure and character.

DE'NTED. (*denté*, Fr. *decoupé en pointes serrées les unes contre les autres.*) Notched; indented.

DE'NTICLE. } (*denticule*, Fr.) A small
DE'NTICULE. } tooth or projecting point.

DE'NTILE. A small tooth, as that of a saw: a term used in conchology.

DENTI'CULATED. (*denticulatus*, Lat.) Set with small teeth, as in the area.

DE'NTOID. (from *dens* and εἶδος, Gr.) Of the shape, or form, of a tooth.

DENUDA'TION. (*denudatio*, Lat. *dénudation*, Fr.) The laying bare; the act of divesting of its covering; the uncovering of strata by the washing away of their covering; the stripping off the superstrata.

DEO'XYDATE. } To reduce from the
DEO'XYDIZE. } state of an oxyd by depriving it of its oxygen.

DEO'XIDIZED. } Deprived of oxygen;
DEO'XYDIZED. } disunited, or sepa-
DEO'XIDATED. } rated from the oxy-
DEO'XYDATED. } gen with which it was previously joined.

DEPO'SIT. Matter laid or thrown down; that which having been suspended or carried along in a medium lighter than itself at length subsides, as mud, gravel, stones, detritus, organic remains, &c.

DEPRE'SSED. (*depressus*, Lat.) Pressed down; low; shallow; flat. In botany, leaves are called depressed when flattened vertically: radical leaves are thus called when they are pressed close to the ground.

DEPRE'SSION. (*depressio*, Lat. *dépression*, Fr. *depressióne*, It.) The sinking, or falling in, of a surface.

DEPRE'SSOR. The name given to such muscles as have the power of depressing, as the depressor anguli oris, &c.

DEPURA'TION. (*dépuration*, Fr. *depurazióne*, It. *depuratio*, Lat.) The action of freeing from impurities, of cleansing.

DERA'CINATE. (*déraciner*, Fr. *tirer de terre, arracher de terre un arbre.*) To tear up by the roots; to extirpate.

DE'RBYSHIRE SPAR. This beautiful substance is fluate of lime, a combination of calcareous earth with fluoric acid; it occurs in nodular masses, and in crystals. It is found in great beauty and abundance in Derbyshire, whence it has obtained its name, but it is also plentiful in other parts of England. It is also called flour-spar and blue-john, *which latter see*.

DE'RMAL. (from δέρμα, Gr.) Belonging to the skin; composed of skin. Thus we read of the dermal fringe of the iguana; the dermal bones of the hylæosaurus.

DE'RMA. } (δέρμα, Gr.) The true
DE'RMIS. } skin, as distinguished from the cuticle, epidermis, or scarf-skin.

DE'RMOID. (from δέρμα, and εἶδος, Gr.) Belonging to the skin; resembling the skin.

DE'SMINE. A mineral found in the lava of extinct volcanoes accompanying spinellane; its form of crystallization is in small silken tufts.

DETRI'TION. The act of wearing away.

DETRI'TAL. Composed of detritus; consisting of the disintegrated materials of rocks.

DETRI'TUS. (Lat.) The worn off, or rubbed off, materials of rocks. "Beneath the whole series of stratified rocks," says Professor Buckland, "that appear on the surface of the globe, there probably exists a foundation of unstratified crystalline rocks, bearing an irregular surface, from the detritus of which the materials of stratified rocks have in great measure been derived."

DEUTO'XIDE. } (from δεύτερος, Gr. and
DEUTO'XYDE. } oxyd.) Called also

Binoxide. A substance in the second degree of oxidation, or containing two prime proportions of oxygen: a protoxide is in the first or smallest degree; a tritoxide denotes a third proportion, and a peroxide has the greatest degree of oxidation.

DEVE'XITY. (*devexitas*, Lat.) Declivity; a bending downwards.

DEVOLU'TION. (*devolutio*, Lat. *dévolution*, Fr. *devoluzióne*, It.) The act of rolling down, as the removal of earth or strata into a valley.

DEVO'LVE. (*devolvo*, Lat.) To roll down, as "every headlong stream devolves its winding waters to the main." In this sense, however, the word is not modernly used: in its common acceptation, at the present day, it signifies to pass by succession from one person to another.

DEVONIAN SYSTEM. A term assigned by Sir R. Murchison to a series of strata largely developed in Devonshire and Cornwall, and belonging to the Old Red Sandstone. "Though the term Old Red Sandstone, when designating great groups of rocks like the Cornish killas and Devonian slates, should involve no error of classification, still it would, mineralogically, be most inappropriate. We purpose therefore, for the future, to designate these groups collectively by the name *Devonian system*, as involving no hypothesis, and being agreeable to analogy. Thus the terms Carboniferous system, Devonian system, Silurian system, and Cambrian system, will represent a vast and apparently uninterrupted sequence of deposits." —*Sedgwick and Murchison.*

DEW. A considerable refrigeration of the surface of the ground below the temperature of the air resting upon it, amounting to 10 or 20 degrees, occurs every calm and clear night, and is caused by the radiation of heat from the earth into space. On becoming colder than the air above, the ground will condense the moisture of the air in contact with it, and be covered with dew. The air, however clear, is never destitute of watery vapour, and the quantity of vapour which air can retain depends on its temperature; air at $32°$ being capable of retaining 1-150th of its volume of vapour, while at $52°$ it can retain as much as 1-86th. That the deposition of dew depends entirely on radiation is fully established by the following circumstances: 1st. It is on clear and calm nights only that dew is observed to fall: when the sky is overcast with clouds, no dew falls, for then the heat which radiates from the earth is returned by the clouds above, and prevented from radiating into space, so that the ground does not become colder than the air. 2nd. The slightest screen, such as a cambric handkerchief, stretched between pins, at the height of several inches from the ground, is sufficient to protect the objects below it from this chilling effect of radiation, and prevent the formation of dew or hoar-frost upon them. Plants derive a great part of their nourishment from this source; and as each possesses a power of radiation peculiar to itself, they are capable of procuring a sufficient supply for their wants.

DEW-LAP. The loose skin which hangs down under the throat of the cow and other animals, and thus called from its licking or lapping the dew when grazing.

DEX'TER. } (Latin.) The right, as
DE'XTRAL. } opposed to the left. In conchology, shells are divided into dextral and sinistral. The more common turn of shells is with the apparent motion of the sun, or as the index or hand of a clock moves. On the contrary, a reversed, or sinistral, shell, when

placed in a perpendicular position, has its spriral volutions in an opposite direction to the motion of the index of a clock, and resembles what is called a sinistral, or left-handed screw. The sinistral shells are sometimes termed heteroclitical, and heterostrophe shells. There has been considerable confusion amongst conchological writers in describing the position in which shells should be held, to ascertain the right from the left side, &c. Perhaps, the most simple plan is, to place the apex of any spiral shell towards the eye with the mouth downwards; dextral shells will then be found to have their aperture on the right side of the *axis*; sinistral shells, on the contrary, will have theirs on the left of the *axis*.

DI'ABASE. A crystalline granular, sometimes porphyritic, or even a slaty, mixture of augite and labradorite or oligoclase, mostly with some chlorite.—*Jukes*.

DIADE'LPHIA. (from δίς and ἀδελφος, Gr.) The seventeeth class of plants in the artificial system of Linnæus. The stamens are united into two parcels at the base. This class has papilionaceous flowers and leguminous fruits. Familiar specimens will be found in the garden pea, bean, &c. &c.

DIADE'LPHOUS. Having its stamens united into two parcels at the base; belonging to the class Diadelphia.

DIA'GONAL. (διαγώνιος, Gr. *ab angulo ad angulum perductus: diagonius*, Lat. *diagonal*, Fr. *diagonále*, It.) A line reaching from one angle to another, so as to divide a parallellogram into equal parts. Diagonals principally belong to quadrilateral figures.

DIA'GONALLY. (*diagonalement*, Fr *diagonalemente*, It.) In a diagonal direction.

DI'ALLAGE. Schiller spar; a variety of augite or crystallized serpentine. The colour of diallage is dark-green.

DI'AMOND. (*diamant*, Fr. *diamánte*, It. ἀδάμας, Gr. *adamas*, Lat.) The hardest and most valuable of all the precious stones. Some mineralogists form a family of the various sorts of diamond, under the title Diamond Family, placing it in the class of earthy minerals: others place the diamond amongst those minerals termed combustible, of which the basis is either sulphur or carbon. Strange as it may appear, diamond consists of pure carbon. If the best charcoal be burnt in oxygen, carbonic acid gas is formed, the weight of which is nearly equal to that of the charcoal and the oxygen, there being a small residuum of earthy ashes left after the combustion; but if, in like manner, a diamond be burnt in oxygen, carbonic acid gas is equally the result, though, in the latter case, there is no residuum, and the carbonic acid gas obtained is precisely equal in weight to the two elements, the oxygen and the diamond. Why, or how, it is that the same elementary substance can, with little or no addition, form two such excessively dissimilar bodies as diamond and charcoal,— the former the hardest and clearest body in nature, the latter a mere black soft, brittle mass,—is a mystery beyond our finite powers to comprehend. The primitive crystal of the diamond is the regular octahedron, each triangular facet of which is sometimes replaced by six secondary triangles bounded by curved lines; so that the crystal becomes spheroidal, with 48 facets. When rubbed, the diamond shews positive electricity. It reflects all the light falling on its posterior surface at an angle of incidence greater than 24° 13′, whence its great brilliancy is derived. Its lustre is brilliant adamantine; fracture conchoidal. Neither acids nor alkalies produce any effect upon it. The diamond is hard in the highest **degree, and** scratches all other known minerals. Specific gravity from 3·5 to 3·6. It burns when **heated** to 14° of Wedgwood's pyrometer, a point just below that at which silver fuses. It is the natural edge of the diamond only that has the property of cutting glass, all artificially formed edges will only tear or scratch it. Diamonds are found of nearly every shade of colour; those which are colourless are deemed the most valuable. The weight and value of diamonds is estimated in carats, one carat being equal to four grains, and the difference between the price of one diamond and another, all other matters being equal, is as the squares of their respective weights, thus the value of three diamonds of one, two and three carats weight, is as one, four, and nine. To estimate the value of a wrought diamond, ascertain its weight in carats, multiply this by two, then multiply this product into itself, and lastly multiply this latter sum by £2. Thus a wrought diamond of one carat is worth £8, one of two carats is worth £32, one of three carats £72, one of four carats £128; and so on. Zircon is sometimes substituted for diamond. The largest diamond known is said to be that which belonged to the late Emperor of the Brazils; it is uncut, and weighs 1680 carats, or 11 ounces 96 grains. This magnificent gem would be worth, supposing the table of rates to be applicable to stones above a certain size, £5,645,000, but the highest price that has ever been given for a single diamond is £150,000. A diamond in the possession of the Great Mogul is of the size of half a hen's egg. The Pitt diamond, now the property of the king of the French, was sold for £100,000.; it weighs 136 carats, or nearly one ounce.

Brazil and Hindostan are the localities where diamonds are principally obtained. Those of Brazil are generally less large, but of the finest water. Diamonds are cut and manufactured into what are termed brilliants and rose-diamonds, The former being mostly made out of octahedral crystals, the latter from the spheroidal varieties. In the formation of either of these, so much is cut away, that the weight of the polished gem does not exceed the half of the rough crystal from which it was wrought, wherefore the value by weight of a cut diamond is twice that of a rough diamond, exclusive of the expence of workmanship.

DI'AMOND SHAPED. Leaves are so called when approaching to a square, having four sides, of which those opposite are equal: the four angles are generally, two obtuse, and two acute.

DIA'NCHORA. A genus of attached inequivalved bivalves; the attached valve having an opening instead of a beak, the other beaked and eared; the hinge toothless.

DIA'NDRIA. (from δίς, and ἀνήρ, Gr.) The second class of plants in Linnæus' artificial arrangement; they have two stamens. This is a very numerous class, consisting of three orders, and comprehends all hermaphrodite flowers having two stamens.

DIA'NDRIAN. Having two stamens; belonging to the class Diandria.

DIA'PHANOUS. (διαφανὴς, pellucid, from διαφαίνω, Gr. *diaphane*, Fr.) Which may be seen through; transparent; pellucid. That which allows a passage to the rays of light.

DI'APHRAGM. (διάφραγμα, Gr. *diaphragma*, Lat. *diaphragme*, Fr. *diafragma*, It.) A large transverse muscle, which separates the chest from the belly; the midriff.

DIA'SPORE. (from διασπείρω, Gr. to disperse.) This name has been assigned to the mineral from the peculiarity that when exposed to heat it decrepitates violently, and is *dispersed* in numerous small spangles. It is a rare mineral and but little known: it is a combination of alumina and water, often mixed with hydrate of iron. Specific gravity 3·43.

DIATOMOUS SCHILLER SPAR. (from διά, through, and τέμνω, to cut, Gr. from being easily cleavable in one direction. The schillerstein of Werner; diatomer schiller-spath of Mohs; diallage chatoyante of Brochant.) An earthy mineral, a combination of bisilicate of magnesia, protoxide of lime and iron, with hydrate of magnesia. It is prismatic. Cleavage in two directions, forming together an angle of about 135°; one cleavage perfect and easily obtained, the other appearing only in traces. Hardness = 3·5, — 4, Specific gravity = 2·6 — 2·8. Its colours are green, grey, and brown. Occurs in granular concretions; disseminated, and seldom massive. Lustre shining or splendent, and metallic pearly. Opaque, and yields to the knife. Streak greyish or yellowish white, dull. Before the blow-pipe it becomes hard, and assumes a metallic appearance. Occurs in serpentine and greenstone and in secondary trap rocks. — *Jameson. Phillips.*

DIBRA'NCHIATA. The first order in the class Cephalopora. This order, called by Pictet Cephalopoda Acetabularia, has been divided by him into two sub-orders, Octoheda, and Docanida.

DI'CERAS. (from δίς and κέρας, Gr.) A genus of fossil shells discovered in oolitic rocks. At Mount Saléve, near Geneva, two species have been discovered, but the genus has not been met with in England, it is thus named from possessing two prominent spiral umbones, which resemble two twisted horns.

DICHOBUNE. (from δίχα, bipartitio, and βουνός, collis, Gr.) This genus was proposed by Cuvier instead of Anoplotherium minus. It is closely allied to the Anoplotherioid genus Xiphodon, the dental formula is the same, only there is a slight interval between the canine and the first premolar in both jaws: the first three premolars are sub-compressed, sub-trenchant, but less elongated from behind forwards than in Xiphodon.—*Owen.* The *dichobune* has points, arranged in pairs, on the back molars of the lower jaw. It has been found fossil only, and in strata more recent than the chalk.

DICHO'TOMOUS. (from δίχα and τέμνω, Gr. *dichotome*, Fr. *dicotomo*, It.) Forked; regularly and continually divided by pairs from the top to the bottom: applied to stems dividing into two parts; example, the misletoe.

DIC'HODON. The name of a genus of extinct Artiodactyle Mammals of the Eocene period. A portion of the upper jaw of a species Dichodon Cuspidatus is in the British Museum, as also a portion of a lower jaw.

DICH'ROISM. A term used in mineralogy to express a property possessed by some bodies of exhibiting different colours, when examined by transmitted light, in determinate directions. Iolite, tourmaline, and mica possess this property.

DI'CHROIT. ⎫ (Of Greek etymology,
DI'CHROITE. ⎭ implying double colour, because its crystals present a very deep blue when viewed in a

direction parallel to their axis, while they appear of a brownish colour, when viewed in a direction perpendicular to this axis. Leonhard named this mineral Cordierite, after Cordier, but Cordier himself gave it the name of Dichroite.) A mineral, called also iolite. The prismatic quartz of Mohs; iolithe of Haüy. Dichroite is of a blue colour, shining lustre, and conchoidal fracture. It consists of nearly 50 per cent. of silica, alumina 30, magnesia 11, oxide of iron 5, with a trace of oxide of manganese. It occurs in granite and gneiss.

DICO'CCOUS. (from δὶς and κόκκον, Gr.) A capsule which consists of two cohering grains, or cells, with one seed in each.

DICOTYLE'DON. (from δὶς and κοτυληδών.) A plant that has two cotyledons or seminal leaves.

DICOTYLE'DONOUS. Every plant the embryo of whose seed is made up of two lobes, or which possesses two cotyledons, or seminal leaves, is included in this great division of the vegetable kingdom; or is a dicotyledonous plant. The stems of dicotyledonous plants are all exogenous, that is, they increase externally by the addition of concentric layers from without; these concentric additions being made annually, a vertical section of a tree of this division will show, at once, its age; the number of rings or circles marking its number of years. Dicotyledonous plants may always be distinguished from monocotyledonous by their leaves: monocotyledonous plants have the veins of their leaves parallel and not reticulated, while all dicotyledonous plants have the veins of their leaves reticulated.

DICTYO'GENÆ. An order of plants, belonging to the class Exogens, comprising the yam and smilax.

DIDA'CTYLE. (εἰδάκτυλος, Gr.) An animal having two toes only.

DIDA'CTYLOUS. Two-toed; having two toes only.

DIDE'LPHIS. } (from δὶς and δελφύς,
DIDE'LPHYS. } Greek, having two wombs.) A genus of animals, belonging to the class Mammalia, order Feræ. All the animals of this genus are marsupial, that is, possess an external abdominal pouch, marsupium, or sac, in which the fœtus is placed after a very short period of uterine gestation, and where it remains suspended to the nipple by its mouth, until sufficiently matured to come forth to the external air. The opossum and kangaroo are examples. The didelphys afford the only known example of mammalian remains in the secondary formations.

DIDE'LPHOID. Belonging to the genus didelphys.

DIDY'MIUM. One of the sixty simple or elementary bodies. Its symbol is D.

DIDYNA'MIA. (from δὶς and δύναμις, Gr.) The name given to the 14th class in Linnæus's artificial arrangement: it has four stamens, two long and two short. This class is easily distinguished from the 4th class, Tetrandria, which has also four stamens. The flowers of this class are generally labiate; corolla monopetalous. It is divided into two orders: Gymnospermia, with four naked seeds in the bottom of the calyx, and Angiospermia, the seeds numerous and contained in a seed-vessel. In the first order, with the naked seeds, the plants are mostly aromatic and wholesome, including the mint, lavender, &c. In the second, where the seeds are contained in a seed-vessel, we find digitalis, and other poisonous plants.

DIDYNA'MIC. } Belonging to the class
DIDYNA'MOUS. } Didynamia. Plants having four stamens, two of which are shorter than the others, are called *didynamous*.

DI′ELECTIC. (from διὰ, through, and ἤλεκτρον, Gr.) Any body through which the electric fluid may be transmitted.

DIGA′STRIC. (from δὶς and γαστήρ, Gr.) Having two bellies.

DI′GITAGRADA. In Cuvier's arrangement, the second tribe of carnivora. The name *digitagrada* has been applied to them from the circumstance of the animals which compose this tribe walking on the ends of their toes. The lion, wolf, and hyæna are examples.

DI′GITATE. }
DI′GITATED. } (*digitatus*, Lat.) A sort of compound leaf, composed of two or more leaflets. Botanists include under the name digitate, binate and ternate leaves, as well as those having more than five leaflets, as the horse-chesnut, which has seven leaflets.

DIGY′NIA. (from δὶς and γυνή, Gr.) The second order in Linnæus's artificial system, comprehending such plants as have two styles, or pistils.

DIGY′NIAN. } Having two styles, or
DIGY′NIOUS. } pistils; belonging to the order Digynia.

DIKE. }
DYKE. } Sax. die, Germ. *deich*, D. *dyk*.

1. A ditch; a channel to receive water.

2. A mound; defence; wall; fortification.

3. Geologists use the word *dike* to express a wall of mineral matter, cutting through strata in nearly a vertical direction. Trap dykes are to be found in all parts of the world, the composition of the rock varying materially, even in the dyke itself. Dykes are often of great extent; one of the longest with which we are acquainted has been described by Prof. Sedgwick, reaching from High Teesdale to the confines of the eastern coast, a distance exceeding sixty miles. The trap dykes of Ireland have been traced upwards of sixty miles, and they are found cutting through all the stratified rocks, from the gneiss to the carboniferous limestone inclusive. They are known as greenstone, porphyry, basaltic or other dykes, according to the kinds of rock of which they are composed. Lyell observes, "That it is not easy to draw the line between dikes and veins; the former are generally of larger dimensions, and have their sides parallel for considerable distances; while veins have generally many ramifications, and these often thin away into slender threads."

In the coal districts, the dykes are an endless source of difficulty and expence to the coal-owner, throwing the seams out of their levels, and filling the mines with water and fire-damp. At the same time they are not without their use; when veins are filled, as is often the case, with stiff clay, numerous springs are dammed up and brought to the surface; and by means of downcast dikes valuable beds of coal are preserved, which would otherwise have cropped out and been lost altogether. Whatever be the throw or difference of level occasioned in the coal measures by these dikes, it never happens, as might be expected, that a precipitous face of rock is left on the elevated side; or that the lower side is covered by an alluvial deposit, which connects the inequality of the beds that are in situ; but the surface of the ground covering the vein is rendered level by the absolute removal of the rocky strata on the elevated side. The coal in contact with the dykes is sometimes charred, and resembles exactly the coke obtained by baking coal in close iron cylinders, in the process of distilling coal-tar. In some parts the coal is deteriorated to the distance of twenty yards from the dike; in others, to that of three or four yards only. Sometimes the coal first becomes sooty, and at length assumes the appearance of coke.

DILU'VIAL. (*diluvialis*, Lat.) Relating to the deluge. A term introduced by Professor Buckland to distinguish accumulations consequent on the deluge. "It is always," says Dr. Mantell, "in *diluvial* beds spread over the surface of plains, or accumulated in the bottoms of valleys, that the teeth and bones of mammalia have been discovered in various parts of England."

DILU'VIAL DEPOSITS. "Next in order to the alluvial deposits," says the Rev. Dean Conybeare, "we find a mantle, as it were, of sand and gravel indifferently covering all the solid strata, and evidently derived from some convulsion which has lacerated and partially broken up those strata, inasmuch as its materials are demonstratively fragments of the subjacent rocks, rounded by attrition. The fragmented rocks constituting these gravel deposits are heaped confusedly together, but still in such a manner that the fragments of any particular rock will be found most abundantly in the gravel of those districts, where the parent rock itself appears *in situ* among the strata. In these deposits, and almost in these alone, the remains of numerous land animals are found, many of them belonging to extinct species, and many others no longer indigenous to the countries where their remains are thus discovered."

DILU'VIALIST. One who attributes certain effects, denied by others, as consequent on the Noachian deluge.

DILUVIAN. (Lat.) A name applied by Professor Buckland to the superficial beds of gravel, clay, and sand which he considers to have been produced by the Noachian deluge; loose and water-worn strata not at all consolidated, and deposited by an inundation of water.

DILU'VION. } (*diluvium* Lat.)
DILU'VIUM. }

The term Diluvium has been applied to the general covering of debris indiscriminately thrown together from all the strata by an inundation which must have swept over them universally, which inundation, known as the Noachian deluge, was the last great geological change to which the surface of our planet appears to have been exposed. By this name it is proposed to distinguish it from the partial debris produced by causes now in operation, and to which the term Alluvium is assigned.

Sir R. Murchison observes "those coarse and sometimes far transported fragments, to which some geologists apply the term *diluvium*, to avoid misconstruction, I call *drift*. See *Drift*."

The Hon. Mr. Strangways observes "I need not say what is intended by the term Diluvium; it is meant to express that superficial deposit which covers everything, and is composed of almost everything. Its composition and thickness are very various: the latter amounting to thirty, forty, or even fifty feet. In some instances it seems to form entire hills. It is usually thickest in the vallies, or on the flat summits of some of the hills. Its composition, much as it differs in different places, may be considered twofold; first, as the debris of rocks upon which, or in the neighbourhood of which, it is accumulated; secondly, as that of a set of rocks totally foreign to the country, the analogies of some of which have been recognised *in situ* at a vast distance, while others remain yet to be identified."

For the purpose of impressing

more strongly the distinction between diluvial and postdiluvial deposits, it will be convenient, says Prof. Buckland, if geologists will consent to restrict the term *diluvium*, to the superficial gravel beds produced by the last universal deluge; and designate by the term *alluvium* those local accumulations that have been formed since that period.

"All that transported matter commonly termed *diluvium*," says Sir H. De la Beche, "requires severe and detailed examination. At the present time, there would appear to be three principal opinions connected with the subject. One, supposing the transport to have been effected at one and the same period; another, that several catastrophes have produced these superficial gravels; while a third would seem to refer them to a long continuance of the same intensity of natural forces as that which we now witness. These different opinions, though they cannot each be correct in the explanation of all the observed facts, may each be so in part; and it were to be **wished** that **the phenomena were examined without the control of a preconceived theory.**"

DIMEROCRINI'TES. (from διμερής, bipartite, and κρίνον, a lily, Gr.) The name assigned by Professor Phillips to a new genus, or subgenus, of encrinites belonging to the Silurian **rocks**, he says "**the** two following species, D. decadactylus and D. icosidactylus, appear to me really different, generically, from Actinocrinites, both by the character of the intercostal plates and the exact bifurcation of the hands and arms.—*Murchison's Silurian System.*

DIMO'RPHISM. In mineralogy, the property which some substances have of crystallizing in two different forms belonging to two different systems of crystallization.

DIMO'RPHOUS. (from δις, twice, and μορφή, form, **Gr.**) Having two forms.

DI'NGLE. A dale; a narrow valley between hills; a hollow.

DINOSAU'RIA. The fourth order of the class Reptilia; this order which comprises Megalosaurus, Hylæosaurus, Iguanodon, Pelorosaurus, Regnosaurus, and Plateosaurus, is altogether extinct.

DINOTHE'RIUM. (from δεινός, and θηρίον, Gr.) An extinct genus of terrestrial mammalia. The dinotherium may be considered to have been the largest of terrestrial mammalia. The most abundant **fossil** remains of this genus have been found at Epplesheim, in **Germany**, where, in 1836, an entire head of this animal was discovered, measuring about four feet in length by three feet in breadth. In various parts of the south of France, large molar teeth and osseous fragments of dinotheria have been found occasionally, and these were referred by Cuvier to a gigantic species of tapir, and named by him Tapir giganteus.

Subsequent discoveries have enabled Prof. **Kaup** to place the dinotherium in a new genus, and to establish the fact that it was an herbivorous aquatic animal, inhabiting marshes and lakes, and that one species the D. giganteum, sometimes attained the length of eighteen feet. The dinotherium holds an intermediate place between the tapir and the mastodon, supplying a link between the cetacea and pachydermata. The scapula, or shoulder-blade, is the most remarkable bone hitherto discovered, belonging to this animal; it resembles that of the mole, and seems to indicate that the fore-leg was adapted for digging up the earth. It appears also certain that this huge creature was furnished with a proboscis, by means of which it carried to its mouth the vegetable food collected by its tusks and claws. The dinotherium is referrible to the miocene period.

A very remarkable peculiarity of formation in the dinotherium consisted in the possession of two immense tusks, which were curved downwards, and resembled those of the upper jaw of the walrus, but in the dinotherium these tusks, **nearly four feet** in length, were placed at the anterior extremity of the lower jaw. These tusks are supposed to have served as instruments for raking and grubbing up by the roots large aquatic vegetables. It is also thought that these tusks might have been **used by the** animal for the purpose of hooking itself, as it were, to the bank, and thereby enabling it to keep its head above water during sleep; as well as for means of defence.

Dɪ'ODON. In Cuvier's arrangement, a genus of fishes belonging to the family Gymnodontes, and thus named in consequence of their jaws being undivided, and forming one piece only above and one below. Their skin is in all parts so armed with spines, that they resemble the case of the fruit of the horse chesnut. Teeth supposed to belong to *diodon histrix* have been found in the chalk.

"One day," says Mr. Darwin, "I was amused by watching the habits of a Diodon which we caught. This fish is well known to possess the singular power of distending itself into a nearly spherical form. Cuvier doubts whether the Diodon in this state is able to swim, but not only can it thus move forward in a straight line, but likewise it can turn round to either side. This Diodon possessed several means of defence. It could give a severe bite, and could eject water from its mouth to some **distance**. By the inflation of its body, the papillæ, with which the skin is covered, become erect and pointed. But the most curious circumstance was, that it emitted from the skin of its belly, when handled, a most beautiful carmine-red and fibrous secretion, which stained ivory and paper in so permanent a manner, that the tint is retained with all its brightness to the present day."

Mr. Darwin adds "I am quite ignorant of the nature and use of this secretion."

Dɪœ'cɪA. (from δὶς and οἶκος, Gr.) The twenty-second class of plants in Linnæus's artificial system. The stamens and pistils are in separate flowers, and situated on two separate plants. The orders in this class depend on the circumstances of their male flowers.

Dɪo'PSIDE. (from διοψις, Gr. *transpectus*, in reference to its transparency.) A mineral known also as alalite, baikalite, and musite. It is a white or pale-green variety of augite. It occurs massive, disseminated, and crystallized. It is found, generally, imbedded in Serpentine. It consists of more than half silica, lime, about eighteen per cent. magnesia, with a trace of alumina and protoxide of iron.

The Abbé Haüy states the primitive crystal of Diopside to be an acute rhomboidal prism. The measurements of the angles by the reflecting goniometer are 87° 5′ and 92° 55′. Diopside is considered by Haüy to be a variety of Pyroxene.

Dɪo'PTASE. (The Cuivre Dioptase of Haüy.) Emerald copper-ore, a very rare mineral of an emerald-green colour, consisting of oxide of copper and silica in nearly equal proportions, with about eleven per cent. of water.

Dɪ'ORITE. A variety of greenstone, composed of hornblende and albite.

Dɪp. In geology, the downward

inclination of strata. The point of the compass towards which strata incline is called their dip, and the angle of such inclination with the horizon is termed the dip, or angle, of inclination. It sometimes happens that a stratum, without varying its direction, may be so bent as to dip two ways in the same mountain, like the sloping sides of the roof of a house, or the letter V reversed Λ.

DIPE′TALOUS. (from εἰς and πέταλον, Gr.) Having two flower-leaves or petals.

DIP′LACANTHUS. (from ἐιπλέος, double, and ἄκανθα, a spine, Gr.) A genus of ichthyolites or fossil fishes, found in the old red sandstone by Hugh Miller and thus described by him, "though the smallest ichthyolite of the formation yet known, it is by no means the least curious. The length from head to tail in some of my specimens does not exceed three inches; the largest fall a little short of five. The scales, which are of such extreme minuteness that their peculiarities can be detected by only a powerful glass, resemble those of the Cheiracanthus. There are two dorsals, the one rising immediately from the shoulder, the other directly opposite the anal fin. The ventrals are placed near the middle of the body. But the best marked characteristic is furnished by the spines of its fins, which are of singular beauty. Each spine resembles a bundle of rods, or rather the sculptured semblance of a bundle of rods, which finely diminish towards a point, sharp and tapering as that of a rush. The rest of the fin presents the appearance of a mere scaly membrane, and no part of the internal skeleton appears. The spines run deep into the body, as a ship's masts run deep into her hulk." Agassiz has enumerated four species; D. crassispinus, D. longispinus, D. striatulus, and D. striatus.

DIPLO-NEURA. See *Helminthoida*.

DIPLO′PTEROUS. The name given by M. Agassiz, to a genus of ichthyolites, belonging to the family of Sauroid fishes, and found in the old red sandstone. "Like the genus dipterus, it has two dorsal opposite two anal fins, but the caudal fin is of a very peculiar form; the throat is very large, and the jaws are armed with large conical teeth."—*Murchison. Silurian System.*

DI′PTERA. (from εἰς and πτερὸν, Gr.) The sixth order of insects, or insects having two wings. The musca, or common fly; the culex, or gnat; and the œstrus, or gadfly, are familiar examples. In Cuvier's arrangement diptera forms the twelfth class of insecta, and their distinguishing characters are said to be possessing six feet, and two membranous extended wings, accompanied generally by two moveable bodies, called halteres, which are placed behind the wings; the organs of manducation are a sucker composed of squamous, sectaceous pieces, varying in number from two to six, enclosed in an inarticulated sheath, most frequently in the form of a proboscis. In these divisions almost every entomologist is disposed to make alterations, and the systems proposed are far too numerous to be recounted.

DI′PTEROUS. Two-winged insects; belonging to the order Diptera.

DIPTERUS. (from εἰς, twice, and πτερὸν, a wing, Gr. Double wing.) A genus of ichthyolites, belonging to the old red sandstone, established by Cuvier. This genus, says Sir R. Murchison, was at first separated by Valenciennes and Pentland into four species, but after an attentive examination of a great variety of specimens, M. Agassiz has concluded that although the genus Dipterus ought to be retained, the supposed four species are only differently modified forms of the same animal. The generic character of the Dipterus, as now confirmed by M. Agassiz, consists in "two dorsal fins opposite to two similar anal fins, with a caudal fin conforming to that of the genus Palæoniscus, in having the vertebral column prolonged into the extremity of the tail." The dipterus is a characteristic fossil of the tilestone. The dipterus differs from the osteolepis chiefly in the position of its fins, which are opposite, not alternate; the double dorsals exactly fronting the anal and ventral fins.

DIPY′RE. (from δυο two, and πυρ, fire, indicating the double effect of fire to produce fire and phosphorescence in this mineral. It is generally classed with scapolite.) The schmelzstein of Werner, A mineral, a variety of scapolite found in the Pyrenees, thus named by Haüy. It consists of silica 60·0, alumina 24·0, lime 10·0, and some water and loss.

DIRECTION. In geology, a term applied to the course which strata take at right angles to their line of dip.—*De la Beche.*

DISC. } (*discus*, Lat.)
DISK. } 1. In conchology, the middle part of the valves, or that which lies between the umbo and the margin; the convex centre of a valve, or most prominent part of the valve, supposing it to lie with its inside undermost.
2. In botany, the central florets of a compound flower; the whole surface of a leaf.

DI′SCINA. A genus of recent ovate, unequal, roundish bivalves.

DISCO′PORA. A genus of cellepora, found recent only, and differing from tubulipora in the cells being more sunk and less free; and from cellepora in having no lobated, convoluted, or ramose expansions.

DISCO′BOLI. The name given to a family of fishes, in Cuvier's arrangement, from the disk formed by their ventrals. The fishes of this family form two genera.

DI′SCOID. } 1. In the form of a disc.
DISCOI′DAL. } In botany, plants, the petals of whose flowers are set so closely and evenly as to make the surface plain and flat like a dish.
2. In conchology, when the whorls are so horizontal as to form a flattened spire.

DISCOI′DES. A genus of fossil echinus, one species only of which has been found, namely discoides subuculus.

DISC′OPHORA. The third order of the class annulata.

DISCO′RBIS. A genus of microscopic spiral discoidal univalves.
These have been thought by some authors, and the number includes Lamarck, to be found in a fossil state only; but Parkinson states that they are found recent on our coasts.

DISCORBI′TES. Fossil shells of the genus discorbis.

DISCO′PORA. The name given by Lamarck to a genus of fossil corals. Three species are described as belonging to the Wenlock limestone, namely, D. antiqua; D. squamata; and D. favosa. The two last have been thus named by Mr. Lonsdale.

DISI′NTEGRATED. Separated into integrant parts by mechanical division.

DISINTEGRA′TION. The separation of a body into its integrant parts by mechanical division; the wearing down of rocks; utter separation of particles.

DISK. See *Disc.*

DISLOCA′TION. (*dislocation*, Fr. *dislogazióne*, It.) The state of being displaced; displacement of portions of the earth's crust. According to the theory of M. De Beaumont, the principal dislocations of the earth's crust of the same geological age range in lines parallel to one and the same great circle of the sphere; those of different ages are parallel to different circles. The geological era, consequently, of the elevation of mountains, may be ascertained from the direction of their axes of movement.

DISPE'RMOUS. (from δὶς, and σπέρμα, Gr.) Two-seeded; an epithet for fruit containing two seeds only; stellate and umbellate plants are thus termed.

DISSE'PIMENT. (*dissepimentum*, Lat.) In botany, the partition which divides a capsule into cells. In many plants the *dissepiments* do not reach to the axis or centre, in some plants the dissepiments are not formed, or subsequently disappear, and leave the placenta in the centre of the ovarium, like a column, with the seeds adhering to it.

DI'STHENE.  ⎫ Disthene is the
DI'STHEN-SPATH. ⎭ name given by Haüy, and Disthen-spath by Mohs, to the mineral Kyanite or Cyanite, *which last, see*.

DISTICHOPORA. A genus of foraminated polypifers, established by Lamarck. Mr. Parkinson thus describes it: "A stony, solid, fixed, ramose, and rather compressed polypifer. The pores unequal and marginal, placed on the two opposite edges, in longitudinal rows, and in the form of sutures; stelliform, wart-like, projections are scattered on the surface of the branches."

DI'STICHOUS. (δίστιχος, Gr. *distichum*, Lat.). Two-ranked; applied to leaves occupying two sides of a branch, but not regularly opposite at their insertion, as the fir, yew, &c.; applied also to branches when they spread into two horizontal directions; and to flowers, placed in two opposite ranks.

DI'TRUPA PLA'NA. A species of pteropodous mollusk, formerly called Dentalium planum; found in the lowest tertiary strata, especially on the western coast of Sussex.

DIVE'RGING. In botany, applied to the position of leaves during sleep, signifying that the leaflets approach at their base, and are open at their summits.

DODE'CAGON. (from δώδεκα and γωνία, Gr. *dodécagone*, Fr. *dodecagono*, It.) A regular polygon having twelve equal sides and angles.

DODECAHE'DRON. (from δώδεκα, twelve, and ἕδρα, base, Gr.) A geometrical solid, comprehended under twelve equal sides, each whereof is a pentagon. It is one of the regular or plutonic bodies.

DODECAHE'DRAL. Having twelve equal sides; relating to a dodecahedron.

DODECAHE'DRAL CORUNDUM. Called also Spinel; the Spinelle and Pleonaste of Haüy. There are two varieties, the Ceylanite and Spinel Ruby. Colours red, blue, brown, black, green, and white. It consists of alumina 74, silica 16, magnesia 8, oxide of iron one and a half, and lime 0·75 per cent.

DODECAHE'DRAL GARNET. A species of garnet containing ten sub-species or varieties; these are the Grossullaire, or asparagus-green variety; the Pyrenaite, or greyish-black variety; the Colophonite, or red variety in granular concretions; the Precious Garnet, or highly crystallized and transparent red variety; the Topazolite, or yellow variety; the Melanite, or velvet-black opaque variety; the Allochroite, or brown, green, and grey massive variety; the Pyrope, or deep blood-red variety; the Essonite, or hyacinthine and orange-yellow variety; the Common Garnet, or brown and green variety, in granular concretions and translucent.

DODECAHE'DRAL MERCURY. Called also native amalgam, the Mercuré Argental of Haüy. A mixture of mercury and silver in the proportions of nearly three-fourths of the former, and rather more than one-fourth of the latter. It is found in quicksilver mines together with cinnabar. It is of the colour of silver, and regularly crystallized.

DODECA'NDRIA. (from δώδεκα and ἀνήρ, Gr.) The eleventh class of plants in Linnæus's artificial system. The plants in this class have from twelve to nineteen stamens; the common houseleek will illustrate it.

DODECA'NDRIAN. Belonging to the class Dodecandria; having from twelve to nineteen stamens.

DO'DO. A genus of birds belonging to the order of gallinæ. The bill is contracted in the middle by two transverse rugæ; each mandible is inflected at the point; and the face is bare behind the eyes. The dodo is a case in point serving strongly to illustrate the views and opinions of those who argue for the extinction of species, even in the present day. Lyell says, "The most striking example of the loss, even within the last two centuries, of a remarkable species, is that of the dodo, a bird first seen by the Dutch, when they landed on the Isle of France, at that time uninhabited, immediately after the discovery of the passage to the East Indies by the Cape of Good Hope. It was of a large size, and singular form; its wings short like those of an ostrich, and wholly incapable of sustaining its heavy body, even for a short flight. In its general appearance it differed from the ostrich, cassowary, or any known bird. Many naturalists gave figures of the dodo after the commencement of the seventeenth century; and there is a painting of it in the British Museum, which is said to have been taken from a living individual. Beneath the painting is a leg, in a fine state of preservation, which ornithologists are agreed cannot have belonged to any other known bird. In the museum at Oxford, also, there is a foot and a head." "The dodo," as Dr. Mantell observes, "has been annihilated, and become a denizen of the fossil kingdom, almost before our eyes. The bones of the dodo have been found in a tufaceous deposit, beneath a bed of lava, in the Isle of France; so that if the very few remains of the recent bird, above alluded to, had not been preserved, these fossil relics would have constituted the only record that such a creature had ever existed on our planet. Nevertheless, two centuries since, the dodo formed the principal food of the inhabitants of the Isle of France." No living dodo has been seen since the year 1691.

DOLABE'LLA. A genus of univalvular molluscs, the known species of which are found in the Indian ocean and in the Mediterranean. They differ from Aplysiæ only in the position of their branchiæ and their surrounding envelope.

DOLA'BRIFORM. (from *dolabra* and *forma*, Lat.) Hatchet-shaped; a term more commonly applied to leaves, cylindrical at the base and having the upper part dilated, thick on one edge and cutting on the other.

DO'LERITE. A crystalline, granular, distinct mixture of labradorite and augite, with some titaniferous magnetic iron ore, and also with some carbonate of iron and carbonate of lime. General colour, dark grey; an augitic lava.—*Jukes*.

DO'LIUM. (*dolium*, Lat. a tub, a tun.) A subglobular ventricose univalve, spirally ribbed in the direction of the whorls; the inferior whorl ample and ventricose; outer lip crenated, or dentated, throughout its whole length. Aperture oblong, ample, and notched; epidermis light and horny.

DO'LOMITE. A variety, or modification,

of limestone, consisting of magnesian earth 48 parts, and calcareous earth 52 parts. It derives its name from Dolomieu, a French geologist. There are three sub-species. Von Buch maintains that limestone has been converted into dolomite by its proximity to porphyry in fusion, and that the magnesia has been transferred from magnesian minerals in the porphyry to the limestone; the magnesia being reduced to vapour or gas.—*Bakewell.*

The name Zetchstein (from *zeche* and *stein*, Germ.) has also been given to dolomite or magnesian limestone. This is a calcareous deposit, of a somewhat variable aspect; it is fossiliferous. The zechstein has not yet afforded any remains of trilobites. It does not appear to be a deposit widely spread over the European area. As yet, it is principally known in Germany and England. Dolomite is generally of a light fawn or yellow colour, and in some parts of a crystalline, in others of a concretionary character. It is included in the new red sandstone group, its position being immediately above the coal measures. It is frequently traversed by veins of carbonate of lime, and there are sometimes met with enclosed in it hollow geodes of calcareous spar, with sulphate of strontian and sulphate of barytes.

Do'LOMITE MA'RBLE. A variety of dolomite of a white colour, occurring in small granular concretions; these concretions are frequently so loosely united as to fall apart by the slightest pressure.

Do'MITE. A variety of trachyte, and thus named from being found in the Puy de Dome, in Auvergne, in France. It has the appearance and gritty feel of sandy chalk.

Do'NAX. (*donax*, Lat. δόναξ, Gr.) Animal a tethys; an equivalved inequilateral bivalve, with a crenulate margin, the frontal margin obtuse; hinge with two cardinal teeth in one valve, one in the other; the lateral teeth one or two, rather distant. The shells of this genus are in general triangular, inequilateral, flattened, truncated before, and wedge-shaped. It is found in sands and sandy mud, at depths varying to ten fathoms. Nineteen species have been described, six of which have been found in our seas. Several fossil species occur in the neighbourhood of Paris.

Do'RIS, In the Linnæan arrangement, a genus of gasteropoda, belonging to the class Mollusca. An animal inhabiting a shell; body, creeping, oblong, and flat beneath; mouth below, on the fore part; vent behind, on the dorsum, surrounded by a fringe; feelers two and four, retractile, and placed on the front of the upper part of the body. They are all marine, and are found in every sea. In Cuvier's arrangement *doris* is placed in the order Nudibranchiata.

Do'RSAL. (*dorsal*, Fr. *dorsâle*, It. from *dorsum*, Lat.) Appertaining to the back, as the dorsal fin, the dorsal ligaments, &c.

Do'RSUM. (*dorsum*, Lat.) The back: the ridge of a hill is sometimes called the dorsum. In conchology, it generally means the upper surface of the body of the shell, when laid upon the aperture or opening. In the genera of patella and haliotis, the dorsum means the upper convex surface.

DRE'ISSIMA. A genus of bivalve shells separated from mytilus by Dr. Vanbeneden; it is regular, equivalve, inequilateral; umbo with a septum in its interior; there are three muscular impressions. The great depth of the anterior side, together with the nearly semilunar figure, distinguish it from mytilus. It has been found fossil in the secondary and tertiary rocks.—*Lycett.*

DRIFT. "All those coarse and sometimes far transported materials, to which some geologists apply the word *diluvium*, I," says Sir R. Murchison, "to avoid misconstruction, designate *drift*. Diluvium, as used by M. Elie de Beaumont and the modern foreign geologists, means precisely what I term drift. Geologists having now completely ascertained that each region of the earth has its own superficial diluvia, produced by distinct and separate action, the unambiguous word *drift* is proposed, which when preceded by the name of the tract whence the materials were derived, expresses at once the intended meaning. Hence Silurian drift, Northern drift, Scandinavian drift, &c., &c." *Sir R. Murchison. Silurian System.*

DRUPE. (*drupæ*, Lat. ὁρυπεπῆς, Gr.) A pulpy pericarp, or seed-vessel, containing a single hard and bony nut, to which it is attached: the epicarp and sarcocarp separable from each other, and from the endocarp, which is stony; the nectarine, peach, apricot, &c., furnish us with familiar examples.

DRUPA'CEOUS.
1. Having the characters of a drupe, as drupaceous fruit.
2. Bearing drupes, as drupaceous trees.

DRUSE. A hollow space in veins of ore, generally lined with crystals.

DRUSY. This word says Phillips, "has been adopted from the German term *drusen*, for which we have no English word. The surface of a mineral is said to be drusy when composed of small prominent crystals, nearly equal in size; it is often seen in iron pyrites."

DUCT. (*ductus*, Lat.) A tube, canal, or passage through which anything is conveyed.
1. In anatomy the ducts are very numerous; thus we have the cystic duct, the hepatic duct, the nasal duct, &c., &c.
2. In botany, ducts are membranous tubes, having their sides dotted or barred; they are large enough to be visible to the naked eye, and are plainly seen when a cane, or vine-branch, is cut across.

DU'CTILE. (*ductilis*, Lat. *ductile*, Fr. *duttile*, It.) That may be drawn out into greater length without breaking. The term is applied to metals only, and is sometimes confounded with malleable, whereas the two have very different significations; thus copper is both malleable and ductile, but lead is only malleable and not ductile; some metals are neither malleable nor ductile, but brittle, as antimony, manganese, tellurium, &c., &c.

DUCTI'LITY. (*ductilité*, Fr. *duttilità*, It.) That property which metals possess of being drawn out into greater length with diminished thickness, without separation of parts. The French use the word ductilité to express malleability; but we do not. "La ductilité est un synonyme de malléabilité."— *Dict. De L'Acad. Françoise.*

DUDLEY ROCKS. Called also Wenlock Rocks. A marine formation, composing one of the divisions of the Silurian system. The Dudley rocks have been sub-divided into Dudley limestone, a mass of highly concretionary grey and blue subcrystalline limestone; and Dudley shale, an argillaceous shale, of a liver or dark grey colour, rarely micaceous, with nodules of earthy limestones. Amongst the organic remains contained in the Dudley rocks may be enumerated corals and crinoidea, in great abundance; trilobites, &c., &c.

DUDLEY FOSSIL. } Names given to
DUDLEY TRILOBITE. } that species of Calymene named by M. Ad. Brongniart C. Blumenbachii. This species has obtained the name of

Dudley fossil in consequence of its being found in such great abundance in the neighbourhood of Dudley; it is however by no means confined to that locality, but is met with over a considerable area through England, Germany, Sweden, and North America.

Dug. The teat or nipple.

Du'gong. A species of phytophagous, or herbivorous, cetacea.

Dune. By geological writers, this word is used to signify a low hill, or bank, of drifted sand, and in no respect is synonymous with down, as might be inferred from Todd and Webster. The downs, both north and south, are very extensive ranges of chalk hills, principally covered with short grass, affording excellent herbage for sheep, whereas dunes are banks of drifted sand, scarcely of sufficient heights to be ranked as hills.

Duode'num. (*duodenum*, Lat. *duodenum*, Fr.) The first of the small intestines, immediately adjoining the stomach, and called duodenum from its length, supposed to be twelve inches.

Dust. In botany, the pollen of the anther. The pollen or dust is contained in the anther. In dry and warm weather the anther contracts and bursts, when the pollen is thrown out. It is found, from microscopic examination, that each particle of dust is generally a membranous bag, either round or angular, smooth or rough, which on meeting with any moisture instantly bursts and discharges a subtile vapour. To the perfecting the seeds of plants, it is necessary that the pistil, or female organ, be impregnated by the pollen of the anther; the fluid contained in the pollen, when the anther bursts, penetrates the stigma, and is conveyed to the seeds, whereby they are rendered fertile, or endued with the property of growing, and producing a plant resembling the parent one.

Dyke. See *Dike*.

Dy'namics. (from δύναμις, force, power, Gr.) That branch of mechanical science which treats of moving powers, and of the action of forces on solid bodies, when the result of that action is motion.

Dy'namics geological. These include the nature and mode of operation of all kinds of physical agents, that have at any time and in any manner, affected the surface and interior of the earth.—*Buckland's Bridgewater Treatise.*

Dyso'dile. (from ενσώδης, *graviter olens, fœtidus*, Gr.) A mineral of a greenish colour found near Syracuse, which burns like coal, but gives out during its combustion a most intolerable odour.

# E

Ea'gle-stone. Called also ætites. A variety of argillaceous iron ore, of a nodular form, something resembling a kidney in shape, and containing a sort of loose kernel. It obtained its name from a supposition that it was either found in, or had dropped from, the nests of eagles.

Earth's crust. That portion of our planet which is accessible to our observation and inspection.

"It comprises," says Sir C. Lyell, "not merely all of which the structure is laid open in mountain precipices, or in the cliffs overhanging a river, or the sea, or whatever the miner may reveal in artificial excavation; but the whole of that outer covering of the planet on which we are enabled to reason by observations made at or near the surface."

It has been concluded, both from astronomical and geodesical observations, that the figure of the earth is a spheroid. This spheroid has been considered as one of rotation, or such figure as a fluid body would assume if possessed of rotatory motion in space. The amount of the flattening of the poles, or the difference of the diameter of the earth from pole to pole, and its diameter at the equator, has been variously estimated, the commonly received opinion is, that the polar axis is to the equatorial diameter as 304 to 305, the difference in favour of the equatorial diameter amounting to 26 miles. As regards the density of the earth, various opinions have been formed; it however appears certain, that the internal exceeds that of the solid superficial density. Daubuisson infers from the observations of Maskelyne, Cavendish, and Playfair, that the mean density of the earth is above five times greater than that of water, and consequently, about twice that of the earth's mineral crust. Laplace calculated the mean density to be 1·55, the solid surface being 1. Baily, in his astronomical tables, states the mean density of the earth to be 3·9326 times greater than that of the sun, and to that of water as 11 to 2.

A very great proportion of the earth's surface bears the most incontestible evidence of having been deposited under water; of having been, at one time, not only submerged by the sea, but *gradually* accumulated at its bottom. Nearly all the strata contain the remains of shells in great profusion, the bones and scales of fishes, fragments or entire trunks of timber, fragments also of stone, very commonly in the form of pebbles, evidently worn and rounded by friction under water.

The term "earth's crust" relates only to the comparative extent of our knowledge beneath its surface, and is not used with the intention of conveying an opinion that the earth consists only of a crust, or that its centre is hollow; for of this we know nothing. The nature of the earth's crust is most readily studied in mountains, because their masses are obvious; and also because, as they are the chief depositories of metalliferous ores, the operations of the miner tend greatly to facilitate their study.

EBOU'LEMENT. (Fr.) Fall of any detached rock. The fall of parts of mountains is so common an occurrence in the Alps, that it is expressively called an éboulement, from the verb ébouler; tomber en ruine.

EBU'RNA. (from *eburnus*, Lat. ivory.) An oval or elongated univalve with a deeply umbilicated columella; the aperture oblong, and notched at the bottom. The recent eburna lives in sandy mud. Fossil eburnæ are rarely met with. Parkinson states that Lamarck does not notice them among the Paris fossils, but that a shell exists among the Essex fossils which he names Eburna glabrata. Dr. Mantell gives eburna as a fossil of the chalk marl, but affixes a note of interrogation to it, as though doubtful.

ECHI'NIDEA. An order of the class Echinodermata, and divided into three families, viz., Spatangoidea, Clypeasteroidea, and Cidaridæ.

ECHIDNIS. A fossil resembling an orthoceratite, its specific character being the alternate circular risings and depressions on its surface.

ECHI'NATE.  }  (*echinatus*, Latin.)
E'CHINATED. }  Bristled like a hedge-hog; set with spines; having sharp points or spines.

ECHI'NIDAN. A fossil belonging to the class Echinoderms. Professor Buckland states that the family of echinidans appears to have extended through all formations, from the epoch of the transition series to the present time.

ECHI'NITE. The fossil echinus, or sea-urchin. Echinites vary greatly both in form and structure, and are arranged accordingly into many sub-genera; they are all marine. The chalk formation abounds with these fossil shells, some of which are exceedingly beautiful from their elegant and minute decorations. The Ananchytes cretosus, a sub-genus, is found in some places in shoals, and in every condition from the youngest to the oldest age. The Spatangus cor-marinum, another sub-genus, silicified, is frequently found on our shores and in our gravel-pits, and the spines of the different sub-genera, detached from the shells, are very numerously dispersed throughout the chalk.

The various opinions entertained of the real nature of echinites are curious and amusing. Rumphius believed that they, as well as belemnites, fell from the sky. Monnius supposed that they were the petrified eggs of serpents. The Romans imagined they were the eggs of toads, or petrified toads themselves, or that they fell from the clouds during heavy rains and thunder. Some authors supposed that they were figured stones, to which nature had assigned their peculiar form, and these and various other absurd notions were entertained until the time of Aldrovandus, who asserted the true origin of these fossil bodies.

ECHI'NUS. (*echinus*, Lat. ἐχῖνος, Gr.) The sea-urchin, or egg. The echinus is included in the order Echinodermata, being covered by a hard and coriaceous skin. The shell is spherical, and composed of an immense number of polygonal plates, closely fitted to each other, and has attached to it many spines or prickles, which serve as instruments of motion. The mouth of the echinus is placed beneath, and is provided with several triangular teeth. Some species of the echinus are edible, more especially the E. Esculentus. The echinus feeds principally on small shell-fish, which it seizes with its tentacula.

ECHI'NODERMS. Simple aquatic animals, with a radiated, globular, or elongated body, covered with a spiny shell or coriaceous skin. They are entirely marine, slow-moving or fixed, predaceous, and commonly provided with a distinct nervous, muscular, sanguiferous and respiratory system, and organs of sense. They are termed echinoderms, from the surface of their skin being covered generally with calcareous spines, whether the surface be calcified or calcareous.—*Prof. Grant.*

ECHINODERMATA. The seventh class of the sub kingdom Annulosa. This class comprises seven orders, namely Holothuridæ, Echinidea, Ophiuridæ, Asteridea, Crinoidea, *Blastoidea* and *Cystidea*; the two last, in italics, are entirely extinct.

ECHINA'NTHUS. The name given to a section of Catocysti by Leske. The genus echinanthus comprises all the echinites of this section. Of this genus there are many species. The echinanthus was named Scutum by Klein. The shells are of an irregular figure, resembling a buckler. On the base, which is concave, five grooves pass from the margin, and terminate at the mouth in the centre. Five rays ornament the upper part. The mouth is placed in the centre of the base, and is of a pentagonal form. The whole of the surface is marked with very small depressions, of a circular form, with central tubercles.—*Parkinson.*

ECHINARA'CHNIUS. (from ἐχῖνος and ἀράχνη, Gr.) A genus of echini belonging to the class Pleurocysti. To this genus Klein gave the name Arachnoides.

ECHINOCO'RYS. A genus of fossil echinites, thus named by Leske, belonging to the class Catocysti. This genus comprises all those echinites which Klein divided into Galeæ and Galeolæ. There are several species.

ECHINODI'SCUS. The name given by Breynius to a section of echini belonging to the class Catocysti. The echinodiscus is of a depressed discoidal figure, whence it has its name, nearly flat on both sides. Echinodisci are found both recent and fossil. There are many species.

ECHINOPHO'RA. A genus of stony polypifers, recent, and found in the sea of New Holland. The *Echinophora* is fixed, flat, expanded in a rounded membrane, free, and bearing the form of a leaf, finely striated on both sides.

ECPYRO'SIS. The sect of Stoics taught that catastrophes were of two kinds; the *cataclysm*, or destruction by deluge, which sweeps away the whole human race, and annihilates all the animal and vegetable productions of nature; and the *ecpyrosis*, or conflagration, which dissolves the globe itself.—*Lyell.*

E'CLOGITE. A greenstone rock, composed of green smaragdite and red garnet.

EDENTA'TA. The twelfth order of the class mammalia, divisible into four families; Tardigradæ, Gravigradæ, Dasypidæ, and Myrmecophagidæ.

EDENTA'TED. (*edentatus*, Lat. *edenté*, *edentato*, It.) Without teeth.

EDI'NGTONITE. A mineral, resembling some varieties of felspar and prehnite.

EDMO'NDIA. A genus of fossil bivalve shells, established by M. Koninck; he describes it as convex, equivalve, inequilateral, transverse, sub-oval, or rotund, the lunuli gaping, with no cardinal teeth, with an internal transvere hinge, the ligament internal, and very small. One species, the E. unioniformis, is recorded from the carboniferous limestone of Holland.—*Lycett.*

EDRYOPHTHA'LMIA. The second order of crustacea, comprising Læmodipoda, Amphiboda, and Isopoda.

EFFLORE'SCENCE. (*effloresco*, Lat. *efflorescence*, Fr. *efflorescenza*, It.)
1. Production of flowers.
2. Excrescences in the form of flowers.
3. The pulverescence of crystals on exposure to the atmosphere.

It is applied, says Phillips, to such minerals as are found in extremely minute fibres on old walls, &c.

EFFLORE'SCENT.
1. Shooting out in the form of flowers.
2. Becoming pulverulent on exposure to the atmosphere; the reverse of deliquescent.

EFFO'SSION. (from *effodio*, Lat.) The digging out of the earth, as of fossils, &c.

EFFU'SE. In conchology, a term applied to shells where the aperture is not whole behind, but the lips are separated by a gap.

E'GERAN. A variety of idocrase or vesuvian, occuring near Eger, in Bohemia, whence its name.

EGE'RIA. A genus of fossil bivalves belonging to the tertiary formation.

EGYPTIAN JASPER. Called also Egyptian pebble. The Egyptischer jaspe of Werner. Jaspe Egyptien of Brongniart and Brochant. Quartz agathe onyx of Haüy. A subspecies of jasper, characterized by its globular or spheroidal form, sometimes flattened, and by the arrangement of its colours. These colours are brown, of different shades, yellow and grey, always arranged in zones, or bands, more or less concentric, sometimes intermixed with spots and dentritic marks. According to Dr. Clarke, Egyptian jasper is found in great abundance scattered over the surface of the sandy desert eastward of Grand Cairo, in the sand near Suez in Egypt, and in the adjoining deserts; it forms a constituent part of extensive beds of a siliceous breccia, which, by their decomposition furnish these pebbles in a loose state.—*Cleaveland.*—*Phillips.*

ELA′OLITE. (from ἐλαια, an olive, and λίθος, a stone.) A sub-species of pyramidal felspar, known also as fettstein, or fatstone, a name given to it from its greasy feel.

ELASMOTH′ERIUM. The name given by Fischer to what he considered an extinct genus of fossil pachydermata.

"What," says Griffith, "distinguishes the elasmotherium from all known animals is that the laminæ of the teeth form a very elevated shaft, which grows like that of the horse, preserving a long time its prismatic form, and that they descend vertically through the entire extent of this shaft, not dividing into roots until after a considerable time, while, in other animals, they unite promptly into a single osseous body which is itself speedily divided into roots; and also that their section has its edges festooned like those of the transversal bands of the molars of the Indian elephant."

The only relic hitherto found occurred in Siberia.

ELA′STIC. (from ἐλάω, Gr. *élastique,* Fr.) Having the power of returning to the form from which it is distorted or withheld; springy. It is applied to such minerals as being bent have the property of springing back to their original form, and therein differ from those that are merely flexible; thus, talc is only flexible, mica is elastic.

ELA′STIC MINERAL PITCH. Called also elaterite and mineral caoutchouc; a brown, massive, elastic variety of bitumen: it consists of about 52 per cent. of carbon, 40 per cent of oxygen, 0·15 of nitrogen, and 8 per cent. of hydrogen.

ELA′TERITE. Another name for elastic mineral pitch.

ELE′CTRUM. Argentiferous gold ore, a variety of hexahedral gold, of a pale brass-yellow colour. Pliny informs us that it was a mixture of gold and silver, and thus writes, " Omni auro inest argentum vario pondere. Ubicunque quinta argenti portio est, electrum vocatur." It has been attempted to prove that platinum is the electrum of the ancients, but such is not the case.

ELEME′NTARY. (*elementarius,* Lat *élémentaire,* Fr. *elementále,* It.) Uncompounded; uncombined; simple; primary.

ELEME′NTARY SUBSTANCES. There are about sixty simple, or elementary substances at present known, that is, substances, which, under the conditions yet applied to them, are found to be incapable of further analysis, and are therefore called simple, or elementary substances.

ELEVA′TION. The question of the elevation and subsidence of the earth's surface is one which long gave rise to controversy, and various were the arguments adduced in support of, and in opposition to, opinions which now are unhesitatingly and universally received, and on which the vitality, as it were, of geology depends. It may not, however, be amiss to quote here some of the views of our best and soundest geologists, on a point of so great importance, and one which, to the Neophyte, seems often so startling. The fact of great and frequent alteration in the relative level of the sea and land is so well established, that the only remaining questions regard the mode in which these alterations have been effected. The evidence in proof of great and frequent movements of the land itself, both by protusion and subsidence, and of the connection of these movements with the operations of volcanoes, is so various and so strong, derived from so many different quarters on the surface of the globe, and every day so much extended by recent inquiry, as almost to demonstrate that these have been the causes by which those great revolutions were effected; and that although the action of the inward forces which protrude the land has varied greatly in different countries, and at different periods, they are now, and ever have been, incessantly at work in operating present change, and preparing the way for future alteration in the exterior of our globe.—*Dr. Fitton.*

Sir C. Lyell says, "We may regard the doctrine of the sudden elevation of whole continents by paroxysmal eruptions as invalidated. In 1822, the coast of Chili was visited by a most destructive earthquake; when the district round Valparaiso was examined on the following day, the whole line of coast, for the distance of above 100 miles was found raised above its former level. The area over which this permanent alteration of level extended was estimated at 100,000 square miles; the rise upon the coast was from two to four feet, inland it was from five to seven feet." The following extracts are from Mr. Bakewell's Introduction to Geology: "The granite-beds in the Alps were not elevated till a late geological epoch, after the deposition of the oolites and chalk. M. Elie de Beaumont has proved, that whole mountain-chains have been elevated at **one** geological period, that great physical regions have partaken of the same movement at the same time, and that these paroxysms of elevatory force have come into action at many successive periods. I agree with Professor Sedgwick, and M. Elie de Beaumont, that the elevation of mountain-ranges, where the beds are nearly vertical, was effected by a sudden and violent upheaving, yet I am persuaded that the elevation of continents, or extensive tracts of country, was (as Sir C. Lyell observes) a long continued process, and that these operations were distinct from each other. The elevation of large continents and islands, was not effected by the same operation, which upraised the primary rocks. I consider it probable, that all large tracts of country or continents emerged slowly from the ocean, forming at first mountainous islands, before the lower countries were raised above the level of the sea. In the Wealden beds the strata have been upheaved and submerged more than once. All the coal-basins were either formed in inland marshes or lakes, or were surrounded by dry land; but a great submergence of the land took place, and they were covered in many parts by thick depositions of marine limestone. At a subsequent period they again emerged from the ocean with a covering of marine secondary strata. The elevations of limited portions of the earth's surface, at a distance from any known volcanic agency, are not uncommon. Loose stones, or shingles of an ancient sea-beach, are found at heights considerably above the present level of the sea in many parts of England. The elevation of extensive islands or continents was probably always

accompanied by the depression of other portions of the earth's crust."
M. Elie de Beaumont has discovered probable evidence of no less than twelve periods of elevation, affecting the strata of Europe. The Isle of Portland affords us an admirable example of alternate elevations and submersions of strata.

1. We have evidence of the rise of Portland stone, till it reached the surface of the sea, wherein it was formed.

2. This surface became, for a time, dry land, covered by a temporary forest, during an interval which is indicated by the thickness of a bed of black mould, called the dirt-bed, and by the rings of annual growth in large petrified trunks of prostrate trees, whose roots had grown in this mould.

3. We find this forest to have been gradually submerged, first beneath the waters of a fresh water lake, next of an estuary, and afterwards beneath those of a deep sea, in which cretaceous and tertiary strata were deposited.

4. The whole of these have been elevated by subterranean violence.—*Prof. Buckland.*

It is now clearly ascertained that the whole country from Frederickshall, in Sweden, to Abo, in Finland, is slowly and visibly rising, while the coast of Greenland is being gradually depressed. Certain parts of Sweden are being gradually elevated at the rate of two or three feet in a century.

ELY'TRA. (from ἔλυτρον, Gr.) The hard cases which cover the wings of coleopterous insects; the wing-sheaths, or upper crustaceous membranes, which cover the true membranous wings of insects of the beetle tribe.

EL'VAN. The name given to a stone which frequently occurs in the mines of Cornwall: it is various in its appearance and composition as well as in its relative situation, and in its apparent effects on metallic veins and lodes. It occurs in Cornwall in inclined strata, which are scarcely sufficiently horizontal to be called beds: the miners call these channels or courses, Elvan courses.

ELV'AN COURSE. The Elvans all along the coast of Cornwall occur in beds and veins of every possible thickness, from forty feet to half an inch, sometimes overlying, but more frequently traversing the killas in various directions, under such circumstances as are apparently irreconcilable with any other theory, than that which supposes them to be of contemporaneous formation with the rock containing them; the result of some play of affinities which allowed a part of the mass to assume a crystalline texture, while its coarser and more abundant portions were left to arrange themselves in the slaty or tortuous form which characterizes the killas.

ELVE'NITE. Quartziferous porphyry.

EMA'RGINATE. ⎫
EMA'RGINATED. ⎬ (*emargino*, Lat.)

1. In botany, applied to leaves terminating in a small acute notch at the summit.

2. In conchology, to shells having no margin; or when the edges, instead of being level, are hollowed out.

3. In mineralogy, to minerals having all the edges of the primitive form truncated, each by one face.

4. In entomology, when the end has an obtuse incision.

EMARGIN'ULA. A genus of obliquely conical univalves, the vertex inclined, and the posterior margin notched. Found both recent and fossil.

E'MBRYO. (ἔμβρυον, Gr. *embryon*, Lat. *embryon*, Fr.)

1. In botany, the germ, or most essential part of a seed, and without which no seed is perfect, or capable of re-production. The embryo is usually placed within the substance of the seed, either central, excentral out of the centre, or external; its direction is curved or straight, and in some instances spiral.—*Flora Medica*.

2. The offspring yet enclosed in the uterus, and in the early stage only of uterogestation; it is afterwards called the fœtus.

EMBOUCHU'RE. (Fr.) The mouth of a river, or that part where it enters the sea.

E'MERALD. (*emeraude*, Fr. *smeráldo*, It. μάραγδος, Gr. *smaragdus*, Lat.) A precious stone of a green colour, found crystallized. Under the genus emerald are comprised two species, the first, the prismatic emerald, or euclase of Werner and Haüy, and prismatischer smaragd of Mohs; the second, the rhombohedral emerald, or rhomboedrischer smaragd of Mohs. This last species contains two varieties, the precious emerald and the beryl, or common emerald. Prof. Jameson places the emerald amongst the members of the schorl family, dividing it into sub-species, namely, emerald and beryl. He observes, "its characteristic, and, we may almost say, its only colour, is emerald green, of all degrees of intensity from deep to pale. The deep sometimes inclines a little to verdigris green, and oftener to grass-green: the pale varieties sometimes nearly pass into greenish-white." Prof. Phillips says "the only important difference between emerald and beryl is in their colours; which, since they present an uninterrupted series, is altogether insufficient for a division of the species. The emerald and beryl are crystallized compounds of an earth called glucina, with silex, alumine, lime, and oxide of iron the splendid green of the emerald is attributed to the presence of oxide of chromium. The finest emeralds are brought from Peru. Vauquelin, in analysing the emerald, first discovered the earth which he called glycina, or glucina. Cleaveland says "the emerald is always crystallized;" Jameson states, "it is said to occur masive, and in rolled pieces," but of such Werner has seen no specimens; Phillips says "it occurs in rolled masses in secondary depositories." The primitive form, of which Haüy has described six modifications, is a regular hexahedral prism, whose sides are squares. The integrant particles are triangular prisms. The emerald yields readily to cleavage, parallel to all the planes of its primary form. Before the blow-pipe it fuses into a white, and somewhat vesicular, glass. Specific gravity from 2·60 to 2·77. Hardness from 7·5 to 8·. It scratches glass easily, quartz with difficulty.

E'MERY. (*emeri*, Fr. *pierre ferrugineuse fort dure, dont on se sert pour polir les métaux et les pierres*.) The schmiergel of Werner; emeril of Brongniart; corindon granulaire of Haüy. A massive, nearly opaque, greyish-black variety of rhombohedral corundum, consisting of alumina 86, silica 3, oxide of iron 4. Emery sometimes occurs in fine granular distinct concretions. Lustre glistening; fracture uneven, and sometimes splintery. It is sufficiently hard to scratch quartz. Specific gravity from 3·6 to 4. It is found in Europe, Asia and America. Emery powder is used for the purpose of polishing metals and hard stones, and also for domestic purposes, sprinkled upon, and fastened to, brown paper; then called emery-paper.

E'MYS. (from ἠμύω, Gr.) Emydes,

pl. The fresh-water turtle or tortoise. This has five nails to the fore feet, and four to the hind ones. Most of them feed on insects, small fishes, &c. Their envelope is generally more flattened than that of the land tortoises. In fresh-water tortoises all the toes are nearly equal, and of moderate length; in land tortoises the toes are also nearly equal, but they are short; in the marine tortoise, or turtle, the toes are all long, and the middle toe of the fore paddle is considerably longer than the rest. Fossil species of the emys have been discovered in the Wealden, as well as in lacustrine deposits of the tertiary period.

ENA'LIOSAUR. An extinct genus of saurians of the old red sandstone.

ENCE'PHALON. (ἐγκέφαλος, Gr.) The brain.

E'NCRINAL. Pertaining to encrinites; composed of encrinites; containing encrinites.

E'NCRINITE. (from κρίνον, Gr. *lilium*, Lat; a lily.) A fossil encrinus. A genus of the order Crinoïdea, known by the name of stone-lily. Mr. Miller gives the following description of the genus Encrinites, or true lily-shaped animal. "A crinoïdal animal with a column formed of numerous round depressed joints, adhering by a radiating grooved surface, and becoming sub-pentangular near the pelvis, which is composed of five pieces, giving a lateral insertion to the first series of costal plates, to which the second series and scapulæ succeed, whence the tentaculated arms or fingers proceed, formed by double series of joints. Dr. Mantell observes, "there are some kinds of star-fish which, instead of the five flat rays of the common species, have jointed rays, which surround the body and mouth, like the tentacula of the polypus. These arms are composed of thousands of little bones, or ossicula, and the whole are inclosed in the common integument or skin. The asterias is a free animal, floating at liberty in the water. Now, if we imagine a star-fish, like that which I have described, to possess a long flexible column, the base of which is attached to a rock, we shall have a correct idea of the general character of the crinoïdea, or lily-shaped animals." Prof. Buckland states, "successions of strata, each many feet in thickness, and many miles in extent, are often half made up of the calcareous skeletons of encrinites." The encrinite differs from the pentacrinite, another genus of the same order, in having the bones of its column circular, or elliptical, whereas those of the pentacrinite are angular or pentagonal. In the encrinites moniliformis, a species of encrinite, Mr. Parkinson states the upper part of the skeleton to consist of nearly 27,000 ossicula, or small bones. Fossil encrinites are so various that they have been divided into several subgenera, according to the formation of the central body.

E'NCRINITES MONILIFORMIS. The lily encrinite, or bead columned true lily-shaped animal. A species of the genus encrinites. Its specific characters are thus described by Miller: "A crinoïdal animal with a column formed of numerous round joints, alternately, as they approach the pelvis, larger and smaller, becoming subpentangular when nearly in contact with it. On the pelvis, formed of five pieces, adhere laterally the first series of costæ, on which the second series of costæ is placed, succeeded by the scapulæ from which the ten tentaculated arms or fingers proceed. Animal permanently affixed by exuded indurated matter. The encrinites moniliformis has hitherto been found only in the muschelkalk of the new red sandstone group, and principally in Lower Saxony.

ENCRINI'TAL. Containing the remains of encrinites. The Derbyshire *encrinital* marble is formed of the fossilized remains of the crinoïdea, cemented together by carbonate of lime.

ENCRI'NUS. A genus of the order Pedicellata, class Echinodermata, in Cuvier's arrangement.

"The characters of this genus," says Pidgeon, "may thus be given: a stelliform or radiated body, composed of five principal rays, subdivided into three or four articulated branches, pinnated in their entire length, presenting, at the upper concave surface, a series of pores. The body is supported at its extremity by a long stem, vertical, polygonous, and articulated, and furnished in its length with a variable number of verticillæ, composed of five small simple branches, equally articulated, and probably adherent to submarine bodies."

E'NDOCARP. (from ἔνδον, within, and καρπὸς, fruit, Gr.) The stone or shell of certain fruits is called the *endocarp*, as in the peach, cherry, &c.; the outer skin the *epicarp*; the fleshy substance, the *sarcocarp*.

ENDOGE'NS. The fourth class of the vegetable kingdom, comprising, Glumiferæ, Petaloideæ, and Dictyogenæ.

ENDO'GENOUS. Plants are called endogenous (from two Greek words, ἔνδον and γίνομαι) the growth of whose stems takes place by addition from within, while those whose growth takes place by addition from without are named exogenous. The ferns and equisetaceæ are endogenous plants.

ENDOGENI'TES ECHINA'TUS. The name assigned by M. Brongniart to the fossil trunk of a tree, nearly four feet in diameter, obtained from the calcaire grossier at Vaillet, near Soissons.

ENDOGENI'TES ERO'SA. A fossil plant discovered by Dr. Fitton at Hastlings, imbedded in clay. The stems, when cut and polished, exhibit the monocotyledonous structure, and were considered related to the palms. It occurs in the strata of Tilgate forest. A small specimen exhibiting that very peculiar eroded appearance of the exterior, which its specific name denotes, is beautifully figured in Dr. Mantell's Geology of the South-East of England.

Dr. Fitton thus describes some specimens of this fossil, "all the specimens lay with their longer diameter and their flatter surfaces in the horizontal position. Their appearance, when first uncovered by the removal of the rock above, was that of elongated and flattened elliptical bodies, tapering at both extremities. They consist of two distinct portions; a stony nucleus, of a dark brownish grey colour, with a very slight tinge of purple; and a crust or case, in the state of lignite, which has externally a nearly uniform surface, and varies in thickness from about one tenth to half an inch. The size of the different specimens varies considerably. The largest that I saw in its place must have been in the whole full nine feet long The width in the middle was 12 inches, and the greatest thickness four inches. The original form of this vegetable was probably cylindrical; and that shape is still retained in a large specimen of a nucleus from Tilgate Forest, now in the British Museum." *Geological Events.*

ENDOSI'PHONITE. (from ἔνδον, and σίφων, Gr.) A cephalopod, found in the Cambrian rocks. The siphuncle is ventral, differing therein from the ammonite, in which it is dorsal, and from the nautilus, in which it is central.

ENDOSKE'LETAL. Having its skeleton within.

E'NNEAGON. (from ἐννέα, nine, and γωνία, angle, Gr.) A polygon with nine faces.

ENNEAPE'TALOUS. (from ἐννέα, nine, and πέταλον, a petal, Gr.) In botany, a corolla having nine petals.

E'NSIFORM. (*ensiformis*, Lat.) Sword-shaped; two-edged; tapering towards the point like a sabre. In botany, applied to two-edged leaves, slightly convex on both surfaces, and gradually tapering to a point from the base to the apex.

ENTI'RE. (*entier*, Fr. *intéro*, It.) Whole; undivided; complete in all its parts. In botany, a term applied to leaves when the margins are devoid of notches, serrations, or incisions. In conchology, when a shell is whole and undivided, neither interrupted nor intermarginated, it is termed entire.

ENTOMO'IDA. The third sub-kingdom of the animal kingdom, called also Diplo-gangliata. "This sub-kingdom," says Prof. Grant, "is chiefly composed of articulated animals with articulated members, the *insects* of Linnæus, which have an elongated, segmented form of the trunk, with tubular jointed organs of motion symmetrically disposed along its sides. Their exterior covering is more consolidated, and generally contains phosphate of lime. Some respire by branchiæ, others by ramified tracheæ, and others by pulmonary sacs. Most are active, carnivorous, and predaceous." This sub-kingdom comprises the following classes, namely, Myriapoda, Insecta, Arachnida, and Crustacea.

ENTO'MOLITE. (from ἔντομα, an insect, ans λίθος, a stone, Gr.) A fossil insect; a **petrified** insect. These are found either in amber or in fossil stones.

ENTOMOLI'THUS PARADO'XUS. The name given, erroneously, at one time to fossil trilobites. Fossil trilobites were long confounded with insects, under the name of entomolithus paradoxus; after many disputes, their place is now established in a separate section of the class Crustacea.

ENTOMO'LOGY. (from ἔντομα, and λογος, Gr.) That part of the science of zoology which treats exclusively of insects, of their history and habits; that branch of natural history which treats of insects. The object of entomology is, to investigate the nature of insects; its design is to show how the insect is organised and formed, and why it was obliged to adopt this particular conformation and internal structure; and, when this is accomplished, it proceeds to the generalisation and development of the various vital phenomena observable in the class. Its view, however, is not limited to show the mere general form of the body of the insect, but it also displays how this general form varies in the several orders of insects, and how far this general transformation and change may extend, without destruction to its identification.

ENTOMOSTO'MATA. In the conchological system of De Blainville, the entomostomata form the second family of Siphobranchiata, and include many genera, as the buccinum, dolium, cerithium, eburna, and other univalves.

ENTOMO'STRACON. } (from ἔντομα,
ENTOMO'STRACA PL. } an insect, and ὄστρακον, a shell, Gr.) Shelled insects. In Cuvier's arrangement the entomostraca form the second section of Crustacea. Entomos-

traca are both dentated and edentated; they are mostly microscopic, they are without exception aquatic, and they mostly, though not without exceptions, inhabit fresh water. This order of crustaceans has the head rarely, if ever, distinct from the thorax, but provided with antennæ. Feet always distinct. Animals undergoing metamorphosis. Mr. Rupert Jones first noticed the occurrence of Entomostraca in the Permian system of England. During the Permian period, the prevailing forms of Entomostraca seem to have belonged to two groups, to Bairdia, and to an undetermined genus. The list of Permian entomostraca is now rather extensive; the Permians of Durham possess a list of 21 species, 13 of which are peculiar to them. Five species are peculiar to Germany, and six to Russia.

ENTOMO´TOMY. (from ἔντομα, an insect, and τεμνω, to cut, Gr.) The dissection of insects, by which we learn their internal construction, and become acquainted with the form and texture of their organs.

ENTOZO´A. (from ἐντὸς and ζωή, Gr.) Intestinal worms.

E´NTROCHAL. (from *entrochite*.) Resembling an entrochite; containing entrochites.

E´NTROCHITE. (from ἐν and τροχὸς, Gr.) Wheel-stone; a name given to the broken stems of fossil encrinites. Some beds of mountain limestone are almost entirely composed of broken stems and branches of encrinites, frequently called entrochites. The detached vertebræ of the radiaria are known by the name of trochitæ; and when several are united together, so as to form part of a column, the series is termed an entrochite. The perforations in the centre of the vertebræ afford a facility for stringing them as beads, from which, in ancient times, they were used as rosaries, and in the northern parts of England they still continue to be known under the name of St. Cuthbert's beads.

E´OCENE. (from ἠὼς, *aurora*, and καινὸς, *recens*, because, as Sir C. Lyell observes, the very small proportion of living species contained in these strata indicates what may be considered the *dawn*, or first commencement, of the existing state of the animate creation.) M. Deshayes and Sir C. Lyell, have proposed a fourfold division of the marine formations of the tertiary series, founded on the proportions which their fossil shells bear to marine shells of existing species. To these divisions Sir C. Lyell has, with the soundest judgment, applied the terms Eocene, Miocene, Older Pliocene, and Newer Pliocene, and well would it be for the advancement of geology, if its nomenclature were, in all instances, derived from some universal language. In fully explaining the meaning of these terms, I shall borrow largely from Sir C. Lyell's Principles of Geology. In proportion as geological investigations have been extended over a larger area, it has become necessary to intercalate new groups of an age intermediate between those first examined; and we have every reason to believe that, as the science advances, new links in the chain will be supplied, and that the passage from one period to another will become less abrupt. All those geological monuments are by Sir C. Lyell called tertiary, which are newer than the secondary formations, and which, on the other hand, cannot be proved to have originated since the earth was inhabited by man. All formations, whether igneous or aqueous, which can be shewn, by any proofs to be of a date posterior to the introduction of man will be called recent. The European strata may be referred to four successive periods, each characterised by containing a very different proportion of fossil shells of recent species. These four periods will be called, Newer Pliocene, Older Pliocene, Miocene, and Eocene. In the older groups we find an extremely small number of fossils identifiable with species now living; but as we approach the superior and newer sets, we find the traces of recent testacea in abundance. The latest of the four periods before alluded to, is that which immediately preceded the recent era. To this more modern period may be referred a portion of the strata of Sicily, the district round Naples, and several others. They are characterised by a great preponderance of fossil shells referable to species still living, and may be called the Newer Pliocene strata.

Out of 226 fossil species brought from beds belonging to this division, M. Deshayes found that not fewer than 216 were of species still living, ten only being of extinct or unknown species. Nevertheless, the antiquity of some Newer Pliocene strata of Sicily, as contrasted with our most remote historical eras, must be very great, embracing perhaps myriads of years. There are no data for supposing that there is any break, or strongly marked line of demarcation, between the strata of this and the recent epoch; but, on the contrary, the monuments of the one seem to pass insensibly into those of the other.

The Older Pliocene strata contain among their fossil shells a large proportion of recent species, amounting to nearly one-half. Thus out of 569 species examined from Older Pliocene strata in Italy, 238 were found to be still belonging to living, and 331 to extinct, or unknown, species.

The next division of the marine formations of the tertiary period is the Miocene, from μείων, minor, and καινὸς, recens. In this division a small minority, less than eighteen per centum, of fossil shells being referable to living species. From an examination of 1021 shells of the Miocene period, M. Deshayes found 176 only to be recent. As there are some fossil species which are exclusively confined to the Pliocene, so are there many shells equally characteristic of the Miocene period. The Miocene strata are largely developed in Touraine, and in the South of France, near Bourdeaux; in Piedmont; in the basin of Vienna, and other localities.

The oldest division of the marine formations of the tertiary period is the Eocene, the derivation of which term is given at the commencement of this article. To this era the formations first called tertiary, of the Paris and London basins, are referrible. The total number of fossil shells of this period known when the tables of M. Deshayes were constructed, was 1238, of which number 42 only are living species, being at the rate of three and a half per centum. Of fossil species, not known as recent, forty-two were found to be common to the Eocene and Miocene epochs. Of the present geographical distribution of those recent species which are found fossil, in formations of such high antiquity as those of the London and Paris basins, there is much of great interest and importance. Of the forty-two Eocene species, which occur fossil in England, France, and Belgium, and which are still living, about one-half now inhabit the seas within, or near the tropics, and almost all the rest are inhabitants of the more Southern parts of Europe.

The heat of European latitudes during the Eocene period, does not seem to have been superior, if equal, to that now experienced between the tropics.

The English Eocene **deposits are** generally conformable **to the chalk,** being horizontal where **the beds of** chalk are horizontal, **and** vertical where they are **vertical; so that** both series of **rocks** appear to have participated in nearly the same movements.

As a summary of the preceding, the numerical proportion of recent to extinct species of fossil shells, in the four different tertiary periods, is as follows:—
Newer Pliocene period 90 to 95
Older Pliocene period 35 to 50
Miocene period . . 18
Eocene period . . $3\frac{1}{3}$
per centum of recent fossils.

In the British Islands, strata belonging to the Eocene system are found only in the South Eastern parts of England, namely, in the country round London, in the south of Hants, and in the north of the Isle of Wight. The Eocene system consists of the following groups, Lower Eocene, containing the Plastic Clay, the Woolwich Beds, Thanet Sands, London Clay, and Bognor Rock, all fresh-water. The Middle Eocene, the Bagshot and Bracklesham Sands, the Barton Clays, marine, and the Headon, fluvio-marine and fresh-water. The Upper Eocene, containing the St. Helen's Sands, the Bembridge Sands, and the Hampstead series, principally fresh-water. Of the Pleiocene group there is a sub-division, named Pleistocene, or the maximum proportion of the recent, containing from 90 to 95 per cent. of the recent, and after these has been added the "Post Pleiocene, or Quarternary," in which all the fossil shells are still found living, though not perhaps in the immediate neighbourhood of the places where they are found fossil.

EPHE′MERA. (ἐφημερία, ex. ἐπί et ἡμέρα, **Gr.**) Insects, so called from their short term of life in their perfect state. Their body is extremely soft, long, tapering, and terminated posteriorly by two **or** three long and articulated setæ. The antennæ **are** very small and composed **of** three joints, the last of which **is** very long, and in the form **of a** conical thread. The ephemera usually appear at sunset, in **fine** weather, in summer and autumn, along the banks of rivers and lakes. The continuation of their species **is** the only function those animals have to perform, for they take no food, and frequently die on the day of their metamorphosis. In another condition, as larvæ, their **existence is much** longer, extending **from two to** three years. **In this first state** they live in **water.**

E′PICARP. (from ἐπί, upon, and καρπός, fruit, Gr.) In botany, the outer skin of fruits is called the *epicarp*; the fleshy substance, or edible portion, is termed the *sarcocarp*, and the stone is called the *endocarp*.

EPIDE′RMAL. ) Composed of epidermis;
EPIDE′RMIC. } relating to the epidermis; resembling the epidermis.

EPIDE′RMIS. (ἐπιδερμίς, Gr. *epidermis*, Lat. *epiderme*, Fr. *epidermide*, It.) The scarf-skin, or cuticle, of animals. In conchology, the outer skin or cuticle, with which the exterior surface of many of the univalve and bivalve shells is covered. It is membranaceous, and resembles the periosteum which covers the bones of animals. The skin seems to be formed entirely by the animal, and is always met with in some species, and never in others; those shells with a ragged surface have almost always an epidermis. In some it is laminated, velvety, fibrous, or rough; in others it is thin and pellucid, allowing the colours of the shell to show through it. It often falls off of its own accord, and without any injury to the surface of the shell; the beauty of many shells is hidden by this outer coat. —In botany, the outward covering of plants: every plant is covered by a skin, or membrane, analogous to the scarf-skin that covers animal bodies; this epidermis varies in thickness, being extremely delicate and diaphanous on some parts of a flower, and very thick, hard, and coarse, on the trunks of many trees.

EPIDI'DYMIS. (ἐπιδιδυμίς, from ἐπί, and δίδυμος, Gr.) A body principally composed of minute, tender, elastic tubes, intricately convoluted, termed tubuli seminiferi, and placed at the outer and back part of the testis.

E'PIDOTE. The Prismatoidischer Augitspath of Mohs, and Pistazit or Pistacite of Werner; the thallite of Lemetherie; akanticone of Dandrada; delphinite of Saussure; glassyactinolite of Kirwan; arendalit of Karstein; glassiger strahlstein of Emmerling; and la rayonnante vitreuse of Brochant. The above long array of names assigned to one mineral by various mineralogists, affords a striking illustration of the evils arising from ill-arranged nomenclature, and adds greatly to the perplexity of the student. A mineral of a green or grey colour; passing on the one hand into blackish-green, on the other, into dark olive-green, oil-green, and siskin-green. It occurs regularly crystallized, in granular, prismatic, and fibrous concretions, and is said to derive its name from the Greek word ἐπιδίδωμι, from an enlargement of the base of the prism in one direction : it is also found massive and granular. Its lustre is shining externally, and vitreous; internally, glistening and resinous. It is hard, brittle, frangible, and scratches glass. Specific gravity 3.42. Hardness = 6.0—7.0. The primary crystal is a right oblique angled prism, of about 115°30′ and 64°30′. It cleaves with brilliant surfaces, parallel with the lateral planes of the prism. Before the blow-pipe it melts into a dark brown or blackish scoria; and this property, according to Saussure, is very characteristic. Two varieties of epidote have been separated; one has been named zoisite by Jameson, after Baron Von Zois, its discoverer; the other, skorza, by Brochant. It is found, principally, in primary rocks, and in many parts of Scotland, as well as in England, Norway, France, &c. It consists of silica 37·0, alumina 27·0, lime 14·0, oxide of iron 17·0, oxide of manganese 1·5. There are many varieties.

EPIGA'STRIC. (from ἐπί, above, and γαστήρ, the belly, Gr. épigastrique, Fr.) Belonging to the upper part of the abdomen, or epigastric region.

EPIGA'STRIUM. (ἐπιγάστριον, Gr. epigastrium, Lat. épigastre, Fr.) The upper part of the abdomen or belly.

EPIGLO'TTIS. (epiglottis, Lat. épiglotte, Fr. ἐπιγλωσσίς, vel ἐπιγλωττίς, Gr. membrana cartilaginosa rotunditatis oblongæ gutturi claudendo et reserando.) One of the five cartilages of the larynx, situated above the glottis, whose use is to close the glottis during the act of swallowing, and thereby to prevent the passage of food into the trachea, or windpipe.

EPIPHYLLOSPE'RMOUS. (from ἐπί, φύλλον and σπέρμα, Gr.) A term in botany, applied to plants bearing their seeds on the back part of their leaves; ex. the ferns.

EPIOLI'TIC. A term proposed to be given to certain rocks by Professor Catullo. Sir R. Murchison says: "I hold that the rocks which Professor Catullo has termed epiolitic, when separated from the neocomian, are simply the representatives of the Oxfordian group." Professor Catullo divides the Epiolitic group, which consists of lime stones, into the upper and lower, the former composed of the red lime stones of the Jura range, the lower, the Epiolitic lime stone of the Venetian Alps.

EPI'PHYSIS. (ἐπίφυσίς, from ἐπιφύω, Gr.) A process of bone attached to a bone, but not being a part of the same bone, as in the case of apophysis.

EPI'PLOON. (ἐπίπλοον, from ἐπιπλέω, Gr. *epiploon*, Fr.) The omentum, or caul; that membranous expansion which hangs from the bottom of the stomach and covers the intestines.

EPIZOO'TIC. (from ἐπὶ and ζῶον, Gr.) Containing animal remains, as epizootic hills, or epizootic strata.

E'POCH. (ἐποχή, Gr. *epocha*, Lat. *epoque*, Fr. *epoca*, It.) A term literally signifying a stop, a fixed point of time, from which succeeding years are numbered; the period at which a new computation, or reckoning, is begun.

EQUA'TOR. (*equateur*, Fr. *æquator*, Lat. *equatóre*, It.) A great circle of the sphere, equally distant from the two poles of the world, or having the same poles with those of the world. It is called the equator, because when the sun is in it, the days and nights are equal; whence also it is called the equinoctial. Every point of the equator is a quadrant's distance from the poles of the world; whence it follows, that the equator divides the spheres into two hemispheres, in one of which is the northern, and in the other the southern pole.

EQUATO'RIAL. Pertaining to the equator : the equatorial diameter of our planet exceds its polar diameter by about 26 miles; the length of the equatorial diameter being 7927 miles, that of the polar 7900.

EQUIA'NGULAR. (from *æquus* and *angulus*, Lat. *équiangule*, Fr. *equiángolo*, It.) A figure whose angles are all equal; consisting of equal angles; having equal angles.

EQUISETA'CEÆ. (from *equisetum*, horse-tail.) These plants are known in this country as the horse-tail of our ditches. Equisetaceæ are found fossil and recent. M. Ad. Brongniart has, in his "Histoire des Végétaux Fossiles," divided fossil equisetaceæ into two genera; the one exhibits the characters of living equiseta, and as a fossil is rare; the other differs greatly in its form, frequently attaining an immense magnitude; these last have been arranged under the distinct genus Calamites. Equisetaceæ are found from Lapland to the Torrid Zone; the species are most abundant in the temperate zone: as we approach a more frigid temperature, they diminish in size and abundance, and in the warm and humid regions of the tropics they acquire their greatest magnitude.

EQUISETUM. (Lat. A genus of the order Filices, belonging to the Cryptogamia class of plants.) Horse-tail. Of this genus there are numerous species. The equisetum fluviatile of our marshes is the largest of all the species, growing sometimes to the height of three feet, and nearly an inch in diameter. It has a succulent, erect, jointed stem, with attenuated foliage surrounding the joints in whorls. In the coal measures, remains of the equiseta are in great abundance, and occur of a magnitude quite unknown at the present day, some of the stems being fourteen inches in diameter. M. Ad. Brongniart enumerates twelve species of calamites and two of equiseta found in strata of the carboniferous series. Equiseta occasionally occur in the Wealden strata, and where they are found they are abundant.

EQUISE'TUM LYELLII. The name given by Dr. Mantell to a distinct species of equisetum, found in the grey and blue grit and limestone at Pounceford, in honour of Sir C. Lyell. When perfect, it probably attained a height of two feet or more.

EQUI'VALENT. (from *æquus* and *valens*, Lat.) In geology, where one bed supplies the place of another which, in that situation, is wanting, such bed is called the *equivalent* of the wanting bed. When a stratum suddenly terminates, and its place is supplied by a stratum of a different character, the latter is called the *equivalent* of the former. The equivalents of compound substances are the sums of those of their elements; thus the equivalent of water is $(8 + 1 = 9)$.

E'QUIVALVE. (from *æquus* and *valva*, Lat.) In conchology, when the shells of bivalves are formed exactly alike, as regards their length, width, depth, &c. The shells of mya, solen, tellina, &c., are generally of the kind called *equivalve*, while those of ostrea, pinna, &c., are inequivalve.

E'RA. (*æra*, Lat. Written frequently *era*.) A particular account and reckoning of time and years, from some remarkable event. Webster quoting from some encyclopædia, says, "it differs from epoch in this; era is a point of time fixed by some nation or denomination of men; epoch is a point fixed by historians and chronologists. The Christian era began at the epoch of the birth of Christ."

ERE'CTILE. (from *erigo*, Lat.) A tissue peculiar to certain parts of the body, as the nipple, &c.

ERE'CT. (*erectus*, Lat.) In botany, leaves are so called when they form a very acute angle with the stem. The term is also applied to branches rising in an upright direction, to petioles rising nearly perpendicularly; and to flowers and pedicles rising perpendicularly.

E'RINITE. A name given to a species of native arseniate of copper, from its having been discovered in Ireland. It is of an emerald-green colour; its constituent parts are oxide of copper, arsenic acid, alumina, and water.

ERO'SE. } (*erosus*, Lat.) Jagged;
ERO'SUS. } applied to leaves very irregularly cut or notched, and having the appearance of being gnawed or eaten by insects.

ERO'TYLUS. A genus of insects, belonging to the Vivalpi, or the seventh family of the Tetramera. In the Erotyli the intermediate joints of the antennæ are almost cylindrical, and the club, formed by the last ones, is oblong; the interior and corneous division of their maxillæ is terminated by two teeth. They are peculiar to South America.

ERINA'CEUS. (Lat.) The hedgehog.

ERPETO'LOGY. (from ἐρπετός and λόγος, Gr.) That branch of natural history which treats of the structure, habits, &c., of reptiles.

ERRA'TIC BLOCK GROUP. One of the sub-divisions of detrital deposits. Professor Phillips observes, "In the British islands, very considerable tracts of country have been traversed, since the land had its present general aspect of hill and dale and was inhabited by large quadrupeds, by currents of water due to some unknown cause, which transported rock masses with so great a degree of force, to points so elevated, in such directions, and at such distances, that we cannot avoid feeling extreme astonishment, and look around in disappointment on the physical processes now at work on the earth, for anything similar. But it is only in particular tracts that the magnitude of the transported rocks is such as to deserve the name of *erratic blocks*.

It appears to be certain that, in the dispersion of boulders, the present physical configuration of the neighbouring regions had great influence; they are found to descend from the Cumbrian mountains northward in the vale of Eden to Carlisle, eastward to the foot of the Penine chain, southward by the Lune and the Kent to the narrow tract between Bolland Forest and the bay of Morecambe; and from the vicinity of Lancaster they are traced at intervals through the comparatively low country of Preston and Manchester, lying between the sea and the Yorkshire and Derbyshire hills, to the valley of the Trent, the plains of Cheshire and Staffordshire, and the vale of the Severn, where they occur of great magnitude."—*Phillips' Treatise on Geology.*

ERUCTA′TION. (*eructatio*, Lat. *éructation*, Fr.) A violent belching forth of wind or other matter, as from a volcano or geyser.

ERU′GINOUS. (from *ærugo*, Eat. *erugineux*, Fr. *ruginoso*, It.)
1. Of the nature of copper.
2. Of a bright green colour, inclining to blue.

ERY′CINA. An equivalved, inequilateral, transverse bivalve. The hinge-teeth two, diverging upwards, with a small intermediate pit; the lateral teeth compressed and oblong. The cartilage inserted in the hinge-pit. Lamarck is of opinion that the shells of this genus exist only fossil, and enumerates eleven species found in the environs of Paris. He places them in the family Mactracea.

ESCA′LOP. ⎫ Commonly called scollop.
ESCA′LLOP. ⎬ A bivalve, whose shell is regularly indented.

E′SCAR. The name given to certain glacial deposits, chiefly consisting of pebbles of carboniferous limestones, heaped sometimes into ridges from 40 to 80 feet high, and from one to twenty miles long, these are called *Escars*.

ESCA′RPMENT. (*escarpement*, Fr.) The steep face of a ridge of high land; the escarpment of a mountain range is generally on that side which is nearest to the sea.

ES′CHARA, (*eschara*, Lat.)
1. Fishes which are said to chew the cud.
2. In Linnæus's arrangement, eschara forms the fifth order of Zoophytes, each polypus being contained in a calcareous or horny shell, without any central axis.
3. A genus of zoophytes, ten species of which have been named by Goldfuss as occurring in the cretaceous group, in the neighbourhood of Maestricht.

ESCHARI′NA. A genus of corals thus named by Milne Edwards.

ESO′PHAGUS. (from ὄιω and φαγεῖν, Gr. *esophage*, Fr. *esófago*, It.) The canal, or passage, leading from the pharynx to the stomach, and through which the food passes from the mouth to the stomach. It is also written œsophagus.

E′SOX (Lat.) The pike, a genus of fishes of the order Abdominales. The esox has small intermaxillaries furnished with little pointed teeth in the middle of the upper jaw, of which they form the two-thirds, those on the sides of the jaw being edentated. The vomer, palatines, tongue, pharyngeals, and rays of the branchiæ, bristled with teeth resembling those of a carp. The dorsal fin is exactly opposite to the anal.

ESSENTIAL CHARACTER. That single circumstance which serves to distinguish a genus from every other genus.

E′SSONITE. Another name for cinnamon-stone. A variety of dodecahedral garnet, of an orange yellow, or hyacinth colour. The finest are brought from Ceylon. See *Cinnamon stone.*

ETHE´RIA. A genus of large inequivalve molluscs belonging to the family Ostracea. They differ from the ostreæ in having two elongated muscular impressions in each valve, which are united by a slender palleal impression. The animal is not known to produce a byssus.

E´THMOID. (from ἠθμὸς, a sieve, and εἶδος, like, Gr.) A bone of the nose to which the name ethmoid has been given from its being cribriform, or perforated like a sieve, for the passage of the olfactory nerves.

E´TITE. See *Ætites*.

E´TYUS. A genus of crustaceans, some species of which have been discovered in the galt.

EUCHYSI´DERITE. A mineral occurring crystallized. Primary form, an oblique rhombic prism of the same cleavage and measurements as pyroxene. Colour, brownish-black. Lustre, vitreous; nearly opaque. Specific gravity 3·34, hardness 6 to 6·5. Streak, yellowish-grey. Fracture, imperfect conchoidal. Found in Norway.

EU´CLASE. (from εὖ, well, and κλάω, to break, Gr.) The Prismatischer Smaragd of Mohs; Prismatic Emerald. This stone has obtained its name from the ease with which it is broken. It is a rare and beautiful mineral, and was brought first from Peru by Dombey; it was at first confounded with the emerald, in consequence of its green colour. The primitive form of its crystals is a rectangular prism, whose bases are squares. It is of sufficient hardness to scratch quartz. Its constituents are silica, alumina, glucina, and the oxides of iron and tin. Euclase has hitherto only been found crystallised. Specific gravity 3·06, hardness = 7·5. It is transparent, and possesses strong double refraction. Before the blow-pipe it first loses its transparency, and then fuses into a white enamel.

EUDI´ALITE. A mineral of a brownish-red colour, having an octohedral cleavage.

E´UDOSIPHONITES. The same assigned by Professor Ansted to a genus of ammonites. See *Aymenia*.

EUGE´NIA CRINITES. (So named from the Eugenia caryophyllata, or clove tree, the unripe fruit of which it resembles, and κρίνον, a lily.) A genus of fossil crinoïdea, six species of which have been determined by Goldfuss and Munster as occurring in the oolitic group.

EUKAI´RITE. Cupreous seleniuret of silver, consisting of silver 39, selenium 26, copper 23, alumina 8.

EU´LISITE. A mixture of olivine-like oxide of iron, green augite, and brownish-red garnet.—*Jukes*.

EUO´MPHALUS. A univalve unchambered fossil shell, found in the mountain lime-stone.

EU´PHOTIDE. }
EU´PHOTITE. } Names given by the French mineralogists to Saussurite. A green stone in which the hornblendic mineral is diallage, and the feldspar labradorite.

EU´PODA. The name given in Cuvier's "Règne Animale" to the fifth family of Tetramerous Coleoptera; Eupoda comprises two tribes, Sagrides and Criocerides.

EURYPTE´RIDA. An order of crustaceans, comprising eurypterus and pterygotus, found fossil only.

EU´RITE. White-stone, the weissstein of Werner. A variety of granite in which feldspar predominates, and named Eurite by the French mineralogists. It occurs in beds, in common granite, in Cornwall. In its most compact form, it becomes a porphyry, and is closely allied to volcanic rocks in Auvergne; felspathic granite.—*Bakewell*.

EXCE´RN. (*excerno*, Lat.) To excrete;

to separate and emit through the pores.

E'xogen. Exogens are plants which have a pith in the centre of their stems, not descending into the roots; or having their woody system separated from the cellular, and arranged in concentric zones. They increase by additions to the outside of their wood, as the name implies.

Exo'genous. (from ἔξω and γεννάω, Gr.) Plants in which the growth takes place by additions from without, or by external increase. Exogenous trees augment, both in height and diameter, by the successive application, externally, of cone upon cone of new ligneous matter, so that if we make a transverse section near the base of the trunk, we intersect a much greater number of layers than nearer to the summit. We can ascertain the age of an oak or a pine, by counting the number of concentric rings of annual growth, seen in a transverse section near the base, so that we may know the date at which the seedling began to vegetate. The Baobab-tree of Senegal is supposed to exceed almost every other in longevity. Adanson was of opinion that one which he measured, the diameter of which was thirty feet, had attained an age of 5150 years.

Exo'ssated. (*exossatus*, Lat.) Deprived of bones.

Exo'sseous. (from *ex* and *ossa*, Lat.) Destitute of bones; animals not possessing bones.

Exosto'sis. (from ἐξ and ὀστέον, Gr.) A diseased growth of bone.

Exo'tic. (*exoticus*, Lat, ἐξωτικὸς, Gr. *exotique*, Fr. *esotico*, It.) In botany, plants not natives of the countries in which they are cultivated.

Explana'ria. A genus of stony polypifers, fixed, expanded in a free, foliaceous, undulated, or convoluted and sublobated membrane, with one stelliferous surface.

Exte'nsor. (from *extendo*, Lat. *extenseur*, Fr.) The name of such muscles as extend or straighten the parts, and serve as antagonist muscles to the flexors.

Exu'viæ. (Lat.) Cast shells; cast skins; organic remains.

Exu'viable. That may be cast or thrown off, as the skeletons of articulated animals.

Eyed agate. The name given to such specimens of agate as have their coloured zones arranged in concentric circles. The lapidary, by cutting and rounding these agates in a particular manner, is able to produce a striking resemblance to the eyes of certain animals.

# F

Fab'aceous. (*fabaceus*, Lat.) Of the nature of a bean; resembling a bean.

Faboidea. The name given to a genus of fossil seeds found in the London clay, and resembling our garden beans, whence the name. Twenty-five species are described by Mr. Bowerbank.

Face. (*face*, Fr. *faccia*, It. *facies*, Lat.) One of the figures which compose the superficies of a body; the surface which presents itself to the sight. Polyhedrons have several faces; a cube has six faces.

Fa'cet. (*facetta*, It. *facette*, Fr. *l'un des côtés d'un corps qui a plusieurs petits côtés.*) A superficies cut into several angles.

FA'CIAL. (from *facies*, Lat. *facial*, Fr.) Belonging to the face, as the facial nerves, &c.

FACIAL ANGLE. An angle composed of two lines, one drawn in the direction of the base of the skull, from the ear to the roots of the superior incisores, the other from that point to the superciliary ridge of the frontal bone. The facial angle of Camper was obtained by drawing a line from the most prominent part of the forehead to the edge of the upper incisors, or front teeth, and then, by making a basilar line from the external aperture of the ears to the lower edge of the aperture of the nostrils, so as to bisect the previous line. Camper states this angle to be fifty-eight degrees in the young Orang, seventy degrees in the young Negro, and eighty degrees in the European. In consequence, however, of some variations in the relative position of the parts above mentioned, Cuvier proposed, as a more certain mode of ascertaining the facial angle, to draw a basilar line parallel to the floor of the nostrils, the angle formed with which, by a facial line drawn from the anterior convexity of the forehead to the greatest prominence of the sockets of the front teeth, he states to be sixty-seven degrees in the young Orang, seventy degrees in the adult Negro, eighty-five degrees in the adult European, and ninety degrees in the European child.

FÆ'CES. (*fæx*, Lat. used plurally only.) Excrement; sediment. The fossil fæces of certain fishes are called coprolites; the excrement of dogs and wolves, album græcum; of mice, album nigrum.

FA'HLUNITE. (from *Fahlun*, in Sweden, where it is found.) An earthy mineral, called also Tricklasite; it occurs in masses, and in thin layers, and is of a dark red-brown colour and opaque. It consists of silex 46·74, alumina 26·73, magnesia 2·97, oxide of iron 5·11, and water 12·5.

FAIRY-RING. In meadows and grasslands, circles of a different hue from the surrounding grass are often seen; these are commonly called fairy-rings, from a vulgar saying that at night fairies dance thereon. The true cause of these appearances, which have excited the astonishment of many, is as follows: they are external indications of the centrifugal growth of the subterranean stems of certain agarics, which, originally springing from a common point, continually spread outwards upon the same place, the centres, or first formed parts, perishing as the circumference, or last formed parts, develope themselves.

FAIRY-STONE. A name sometimes given to the echinite.

FA'LCATE. A figure formed by two curves bending the same way, and meeting in a point at the apex, the base terminating in a straight margin, resembling a sickle.

FALLING-STONE. } A meteoric body,
FALLING-STAR. } commonly called an aërolite.

FA'LUN. (*falunière*, Fr. *assemblage de coquilles brisées, qu'on trouve en masse à une certaine profondeur de terre.*) A provincial name given to some shelly strata in the neighbourhood of the Loire, and which resemble, in their lithological characters, what is denominated the crag. The faluns, or marls of Tourraine and the Loire, constitute an extensive formation of marl beds, which are now admitted to be of later date than the most recent of the fresh-water beds in the Paris basin. They are regular depositions, formed during an epoch of tranquillity, and subjected to laws of which the action is continued on the present shores. The great mass of fossil shells which these beds contain, differ from those of the Paris basin: in nearly 400 species, there are only about 20 identical with the Paris fossils. The terrestrial and river shells are in the same state of mineralization as the marine shells. The bones of the mastodon, rhinoceros, and hippopotamus, are in the same state of preservation as those of whales, and other cetaceous animals, with which they are intermixed. They are coated with marine polypi and serpulæ, which proves that they were long covered by a tranquil and stationary sea. These *faluns* are distinct from the tertiary beds of the Seine, and more recent than any of them; but they are themselves the lowest term of a new system, more important, and more extensive, than the formations of the Paris or London basins, and which has been continued to the present epoch, during all the numerous up-heavings of the ground, the changes in the relative level of seas and continents, and the successive modifications of organic beings.

FARCILITE. Farcilite is the prevailing rock about Manresa in Spain.

FARI'NA. (Lat.) Meal; flour: in botany, the pollen, or dust of the anther. The pollen, or farina, is contained in the anther. In dry and warm weather the anther contracts and bursts, when the pollen is thrown out. From microscopic observation we find each particle of dust to be generally a membranous bag, either round or angular, smooth or rough, which on meeting with any moisture instantly bursts with great force, and discharges a subtile vapour.

FA'RINOSE. In entomology, having the surface covered with dust, resembling flour, which the slightest touch will remove.

FA'SCIA. (Lat.) The tendinous expansion of a muscle, inclosing others like a band.

FA'SCIATED. Filleted, or enclosed with a band.

FA'SCICLE. (*fasiculus*, a little bundle, Lat.) A bundle, or little bundle: applied to flowers on small stalks, when many spring from one point, and are collected into a close and level bundle at the top; as the sweet-william.

FASCI'CULAR. (*fascicularis*, Lat.) United, or growing together, in a cluster, or tuft, as the larch, and some species of pine; applied also to roots, when many tubes proceed from the same centre, shooting forth in an elongated form.

FASI'CULATED. Arranged in small bundles. In mineralogy when the crystals are collected, as it were, into bundles.

FASCIO'LA. The fluke-worm. A genus of internal worm belonging to the order Parenchymata, family Trematodea. There are many species; they are furnished underneath the body, or at its extremity, with organs resembling cupping-glasses, by which they adhere to the viscera. In this genus is included the Distoma hepatica, or Fasciola Hepatica of Linnæus, which so infests, and is so common in, the hepatic vessels of sheep.

FASCIOLA'RIA. A subfusiform univalve, channelled at its base, without any projecting sutures, and having two or three very oblique folds on the columella.—*Parkinson.*

FASCIOLI'TE. A subcylindrical, shelly, or bony body, about half an inch in length, rather tapering at the ends, and formed by the spiral arrangement of perpendicular, concamerated tubes, the tapering end of which is obliquely and transversely folded on that of the preceding one. The tubes are seen to be distinct, and, where the outer surface has been removed, the concamerations are perceived, resulting from the interposition of very numerous and minute septa, transversely disposed. The tubes are placed perpendicularly round the centre, and it appears that round the first formed tube, or chamber, successive increasing columnar tubes were disposed, folding over each other at their ends. Whether these several tubes were internally connected with each other or not, or whether the chambers communicated, or not, with each other, by a siphuncle, are questions not yet clearly ascertained. Like some of the nummulites, this body, when polished, has more the appearance of bone than of shell, and from this and other circumstances, it seems to approximate nearer to the nummulite than to any other fossil.

FA'SSAITE. (from *Fassa* in the Tyrol.) Called also Pyrgom. A mineral, a dark-green variety of augite; it is also found in Scotland and Ireland, in beds of primitive trap, limestone, and magnetic ore.

FASTI'GIATE. } *(fastigiatus*, Latin.)
FASTI'GIATED. } Pointed; a term applied to a stem, peduncles, umbel, &c.

FAT QUARTZ. } (The Quartz hyalin
FETID QUARTZ. } gras of Haüy.) Quartz, both crystallised and massive, sometimes exhibits, when fractured, a greasy polish on the surface, equal to that which would be produced by rubbing it with oil: it sometimes, though not always, gives out a fetid odour when struck; from these two circumstances the above names have been given to this variety.

FAULT. *(faute*, Fr.) A break or intersection of strata; interruption of the continuity of strata, with displacement; the sudden interruption of the continuity of strata, in the same plane, accompanied by a crack or fissure, varying in width from a mere line to several feet, such fissure being generally filled with fragments, &c. Although the two sides of a fault often come into close contact, there is very frequently a clayey substance interposed which is impervious to water; and it rarely happens that water on one side of a fault passes to the other side. On the contrary, the water is usually discharged along the line of the fissure, particularly on mountain sides, in the shape of springs. When a *fault* occurs in strata they are generally either elevated or depressed, so that in working a bed or vein there appears to be a sudden termination of it. *Faults* consist of fissures traversing the strata, extending often for several miles, and penetrating to a depth, in very few instances ascertained; they are accompanied by a subsidence of the strata on one side of their line, or, which amounts to the same thing, an elevation of them on the other; so that it appears, that the same force which has rent the rocks thus assunder, has caused one side of the fractured mass to rise, or the other to sink. Of the extent of displacement of strata some idea may be formed from the following statement; the old red sandstone of the Fans, situated in South Wales, is proved to have been upcast to an extent of 700 feet. Mr. Bald mentions that the great south slip in the Clackmannanshire coal field throws down the strata 1230 feet. If we suppose a thick sheet of ice to be broken into fragments of irregular area, and these fragments again united, after receiving a slight degree of irregular inclination to the plane of the original sheet, the re-united fragments of ice will represent the appearance of the component portions of the broken masses, while the intervening portion of more recent ice represents the clay and rubbish that fill the faults. In the coal-fields, these *faults* operate as coffer-dams, and are of the greatest possible advantage. Faults are of two kinds, true faults, and Symon faults. When a stratum of coal tapers away and disappears amid the shales and sandstones, it is locally termed a "Symon" fault. See *Symon Fault*.

FAU'NA. (*fauni*, Lat.) As the plants peculiar to a country constitute its *flora*, so do the animals constitute its *fauna*; the zoology of a country.

FAUX. (Fr.) That portion of the cavity of the first chamber of a shell which may be seen by looking in at the aperture.

FAV'OSITES. A genus of foraminated polypifers, resembling the honeycomb in appearance, from which circumstances the name has been applied.

FAVULA'RIA. A genus of fossil plants. Stem-furrowed; scars of leaves small, square, and of a breadth with the ridges of the stem. In the favularia, the trunk was entirely covered with a mass of densely imbricated foliage, the bases of the leaves are nearly square, and the rows of leaves separated by intermediate grooves. The genus is believed to be extinct, but is found fossil in the coal formation.

FEA'THERY. Plumose; applied to plants furnished with lateral hairs.

FECU'LA. (from *fæx* Lat. *fécule*, Fr.) 1. The sediment or grounds of any liquid. The word *fecula*, says Dr. Paris, originally meant to imply any substance which was derived by spontaneous subsidence from a liquid.
2. The green matter of plants.

FEE'LERS. In conchology, those crenated arms, evolved from the side of the Lepas anatifera, and other shells of the second division of Lepas. While the animal is in the water it continually moves its feelers, evidently for the purpose of entangling minute marine insects as food.—*Brown*.

FE'LDSPAR. } A mineral which enters
FE'LSPAR. } into the composition and, next to quartz, constitutes the chief material of many rocks. There are many species and sub-species, or varieties of this mineral, though all agree nearly in their chemical composition, and all are found both crystallised and massive. Feldspar is lamellar in its structure, but not in so great a degree as mica; it scratches glass, and is nearly opaque. It is composed of silex 64, alumina 18, potash 13, lime 3, and some oxide of iron. Common feldspar is perhaps the most generally diffused mineral, next to quartz and iron. It is one of the components of granite, gneiss, and some other primary rocks; and granite owes its variety of appearance and colour principally to the abundance, or otherwise, of the feldspar it contains. In some kinds of granite the feldspar is in large whitish crystals of irregular forms, occasionally of one or two inches in length. From the liability of feldspar to be decomposed by atmospheric action, granite containing large crystals of it is less durable than that which is finer grained, and it is said that Waterloo-bridge, being unfortunately built of granite containing large crystals of feldspar, will be less durable than could be wished for. Felspar forms, in general, more than half of the mass of modern lavas. When it is in great excess, lavas are called trachytic; when augite (or pyroxene) predominates, they are termed basaltic.

Felspar assumes a considerable variety of forms, which differ so greatly from each other, that a novice finds it difficult to recognise in them the same substance. In an earthy, vitreous, or compact state, it forms the basis of all lavas, and of the quarter number of trap rocks. Associated with augite, and usually in a vitreous form, it constitutes some of the well-known modern volcanic basalts, in which the greater or less preponderance of the latter mineral confers the more or less black, dense, and ferruginous character, which they often **assume.** Mixed with hornblende, **it** forms a large class of ancient rocks, also called basalt when the minerals are intimately blended, or greenstone when each is distinguishable. In another condition, felspar, in a glossy but loosely aggregated state, composes a rock of porous, rough, and earthy aspect, called trachyte. In a compact state, the same mineral is the base of many of the porphyries; and **in** a more or less crystalline form, associated with quartz, mica, and other minerals, it composes the great class of granite rocks. As felspar is not found in any of the aqueous sedimentary deposits, except in a decomposed or regenerated state, it may therefore be considered the most characteristic ingredient of all igneous rocks. Professor Jameson divides felspar into five species, namely, 1. Rhombohedral Feldspar, or Nepheline. 2. Prismatic Feldspar, or Common Feldspar. 3. Tetarto-prismatic Feldspar. or Scapolite. 4. Polychromatic or Labrador Feldspar. 5. Pyramidal Feldspar, or Scapolite.

1. The rhombohedral feldspar, or Nepheline of Haüy and Werner, is of a white or grey colour, and occurs both massive and crystallised; it is externally splendent, internally vitreous and shining. Cleavage fourfold. Fracture conchoidal, melts with difficulty before the blow-pipe. Its crystals form druses. It occurs in drusy cavities. Its constituent parts are, according to Gmelin, silica 43·46, alumina 33·49, soda 13·36, potass 7·13. Other authors, however, give a different analysis, stating lime and oxide of iron to form a portion of its constituents.

2. Prismatic feldspar, or common feldspar. The prismatischer feldspath of Mohs. Potash feldspar. Of this there are many varieties, namely, adularia, or moonstone, a transparent variety with a silvery or pearly opalescence; glassy feldspar, a translucent variety, with various shades of colour, such as white and red, which from its abundance has obtained its name; amazon-stone, a blue or green variety; Norwegian Labrador feldspar, a dark-green variety with a beautiful changeableness of colour, obtained from Frederickswarn, in Norway; compact feldspar, a feebly translucent variety, with a splintery fracture; slaty feldspar, or clinkstone, a slaty variety; porcelain earth, earthy feldspar, and claystone, varieties, in a comparatively loose state of aggregation, without lustre or transparency, and varying in their degree of compactness.

3. Tetarto-prismatic feldspar, or albite. See *Albite*.

4. Polychromatic or Labrador feldspar. Lime feldspar. The polychromatischer feldspath of Mohs. This beautiful mineral was first discovered on the coast of Labrador, as a constituent part of syenite. When light falls on it in certain directions it exhibits the most beautiful changeability of colour. It occurs massive and disseminated. Cleavage splendent. Fracture glistening. It has been subsequently found in different parts of Europe. It contains about eleven per cent. of lime and four of soda. It breaks into rhomboidal fragments. In its changeability of colour, it exhibits patches

of blue, green, yellow, red, and grey colour.

5. Pyramidal feldspar, or scapolite. Meionite. Pyramidaler feldspath of Mohs. Of this species of feldspar there are many varieties, namely, Meionite, Scapolite, Paranthine, Wernerite, Dipyre, and Schmelzstein. These will be all described under their several heads.

FE′LDSPARITE. } Names assigned by
FE′LSPAR ROCK. } Dr. Boase to a genus of primary rocks found in Cornwall. "All the varieties of this genus, have, in common, a basis of compact felspar, (i.e. a compound of felspar and quartz) which assumes various characters, passing gradually from crystalline felspar into hornstone, and even into quartz, according as one or other of its constituents predominates.

FE′LDSPATH. } See *Feldspar*.
FE′LSPATH. }

FELSPA′THIC. } Any mineral in which
FELSPA′THOSE. } feldspar greatly predominates; of the nature of feldspar.

FE′LDSTEIN. } The names assigned to
FE′LSTONE. } a feldspathic granite rock. A compact, smooth, hard flinty looking rock. It is probably a mixture of a feldspar with silica, in a state of paste. It has two principal varieties.

FE′LIS. (Lat.) A genus of quadrupeds belonging to the order of Feræ, the characters of which are these:—The fore-teeth are equal; the molares, or grinders, have three points; the tongue is furnished with rough sharp prickles, pointing backward; the claws are sheathed and retractile, and being raised perpendicularly, and hidden between the toes when at rest, by the action of an elastic ligament, lose neither point nor edge. The species of this genus are very numerous, and various with regard to size and colour, though they are all similar with respect to form.

FE′MORAL, *(femoralis,* Lat.) Belonging to the thigh.

FE′MUR. (Lat.) The thigh bone; the thigh.

FENESTE′LLA. The name given by Hugh Miller to a genus of fossil corals of the Dudley rocks and mountain limestone. The fenestella is a stony coral, fixed at the base, and composed of branches, which unite by growth and form a cup. "Externally the branches anastomose, or regularly bifurcate; internally they form a net-work, the intervals being generally oval. One row of pores on each of the branches externally, the openings being circular, and projecting when perfect." Mr. Lonsdale has distinguished four species, namely, F. antiqua; F. Milleri; F. prisca; and F. reticulata, all of which are figured and described in Sir R. Murchison's splendid work, the Silurian System.

FE′RGUSONITE. A brownish-black ore, occurring in quartz; thus named after Mr. Ferguson of Raith.

FERN. (Sax. *fearn.*) Ferns are distinguishable from all other vegetables by the peculiar division and distribution of the veins of the leaves; and in arborescent species by their cylindrical stems without branches, and by the regular disposition and shape of the scars left upon the stem, at the point from which the petioles, or leafstalks, have fallen off. The brake, or fern, of our commons and waste lands, is a familiar example of this remarkable and numerous family of plants, distinguished by the peculiar distribution of the seed-vessels. The family of ferns, both in the living and fossil flora, is the most numerous of vascular cryptogamous plants. The total number of living species of ferns is about 1500. The large tree ferns are confined almost exclusively to the tropics; an elevated and uniform temperature and great humidity being the conditions most favourable to their development. The existence of immense fossil arborescent ferns, from thirty to forty feet in height, in the coal formation, is one of the strongest possible evidences of the great diminution of temperature and change of climate which the earth has undergone. In the coal formation there are not fewer than 130 known species of ferns, nearly all of which belong to the tribe of Polypodiaceæ. An arborescent fern, forty-five feet high, from Silhet in Bengal, may be seen in the stair-case of the British Museum. In the strata of the secondary series there is a considerable diminution in the absolute and relative number of ferns; and in the strata of the tertiary series the ferns seem to bear nearly the same proportion to other vegetables as in the temperate regions of the earth at the present day.—*Buckland. Lyell. Mantell.*

FER'RUGINOUS QUARTZ. (Called also Iron-flint.) The Eisenkeisel of Werner; Quartz Rubigineux of Haüy. A subspecies of quartz, opaque, or translucent at the edges only. Fracture uneven, more or less conchoidal, but imperfect, shining, and nearly vitreous. It occurs massive and crystallized. Some authors divide ferruginous quartz into yellow and green varieties. Cleaveland mentions two varieties, yellow and red. Ferruginous quartz consists of silica 93, oxide of iron 5, oxide of manganese 1.

FE'TID LIMESTONE. The Stinkstein of the Germans. A limestone which, when struck with the hammer, gives off a fetid smell, like that of sulphuretted hydrogen gas.

FETUS. *(fœtus,* Lat.) Commonly written *fœtus.* Of viviparous animals, the young in utero; of oviparous, the young in the shell: in the earliest stages of utero-gestation, the young is usually called the embryo, and when fully formed, or after a certain period, the fetus.

FI'BER. } *(fibra,* Lat. *fibre,* Fr. *fibra,*
FI'BRE. } It.) A filament or thread, whether of animal, vegetable, or mineral structure.

FI'BRIL. *(fibrille,* Fr. *petite fibre, fibrilla,* It.) A small fibre; the diminutive of fibre.

FI'BROUS. *(fibreaux,* Fr. *fibroso,* It.) Composed of fibres; containing fibres.

In botany, a fibrous root consists of numerous fibres, either simple or branched; these are the most simple of all roots, conveying nourishment directly to the stem, or leaves.

FIBROUS QUARTZ. See *Cat's Eye.*

FI'BROLITE. (from *fibra,* Lat. and λίθος, Gr.) A mineral of a white or grey colour, occurring with corundum. Cleavage imperfect. Hardness more considerable than that of quartz. Consists of alumina 46, silica 33, oxide of iron 13. It is composed of minute fibres, from which circumstance it obtains its name, some of which appear to be rhomboidal prisms. It is found in China and in the Carnatic.

FI'BULA. (Lat.)
1. The small bone of the leg, thus named, according to some authors, from being placed opposite to the part where the knee-buckle, or clasp, was formerly used.
2. A fossil echinite, resembling, not a buckle, but a button. By some oryctologists these have been termed Bufonitæ and Scolopendritæ, and by others Pilei; and, by the English, Capstones.—*Parkinson.*

FIBULA'RIA. A genus of echinites, subglobular, ovoid and nearly round, with no determinate margin; the ambulacra forming petaloidal, short, narrow, and circumscribed figures, the mouth beneath, central; the vent near to the mouth. Many species have been described.

FI′GURE STONE. Agalmatolite, a variety of talc-mica, of a grey, green, white, red, or brown colour. The finest are brought from China.

FI′LAMENT. *(filamenta*, Lat. *filament*, Fr. *filaménto*, It.
1. A long thread or fibre; a slender thread-like process.
2. In botany, the long thread-like part that supports the anther; the filament is not essential, being sometimes wanting; the form is various, being sometimes short and thick, or long and slender, or forked, one point only supporting the anther; generally smooth, sometimes hairy; the number varies from one to many. Most filaments are simple, some are bifid; others tricuspidate or broad, and trifid at the extremity.

FILAMENTOUS. *(filamenteux*, Fr. *filamentoso*. It.) Composed of fine threads or fibres.

FILA′RIA. A genus of nematoidea, belonging to the class Entozoa.

FILE. A name given by the chalk-diggers to the striated and prolonged cucurmerine claviculæ of echinites.

FILI′CES. *(filix*, Lat.) Ferns, the first order of Cryptogamia, in Linnæus's artificial system; the first tribe of acotyledonous plants. In modern classification, filices constitute the fourth order of Acrogens.

FILICITE. A fossil fern.

FILICOIDE′Æ. (from *filix*, Lat. and εἶδος, Gr.) Fern-like plants.

FI′LIFORM. (from *filum*, a thread, and *forma*, form, Lat.) Thread-like; thread-shaped; slender, and of equal thickness. In botany, applied to peduncles when very fine, resembling threads; applied also to the tube of monopetalous flowers when of a thread-like form; and also to aments.

FIN. (Sax.) The organ in fishes by which they steady and keep upright their bodies in the water; the cau-

dal fin alone assists in progressive motion. The fin consists of a membrane supported by rays, or little bony or cartilaginous ossicles.

FIN-FOOTED. Palmipedous; having palmated feet, or feet with membranes between the toes, connecting them with each other.

FI′ORITE. A siliceous incrustation deposited by the thermal waters of Ischia, first noticed by Dr. Thompson; called also Pearl-sinter. By some mineralogists, this is considered to be merely a variety of siliceous sinter; others constitute it a distinct sub-species. It occurs in stalactitical, botryoidal, globular, and cylindrical masses of a milk-white, yellowish-white, pearl-grey, and yellowish-grey colours. It is less hard than quartz, but sufficiently hard to scratch glass. It is infusible before the blow-pipe, without addition. It consists of silica 96, or, according to some, 94, alumina 2, lime 4. It is regarded by some authors as a volcanic product.

FIRE-DAMP. Choke-damp. Carburetted hydrogen gas. This is sometimes very abundantly evolved in coal mines, and is productive of the most dreadful results, occasionally nearly all employed in the mines perishing from its combustion. When carburetted hydrogen gas constitutes more than one-thirteenth of the volume of the atmosphere of pits and mines, the whole become explosive whenever a flame is brought into contact with it; to prevent the disastrous consequences which were so frequently resulting, Sir H. Davy invented a *safety-lamp*, which being formed of wire-gauze, in the form of a cylinder, consumes, but does not explode, the explosive mixture. It has been well and truly observed, "if the genius of Davy had merely produced his safety-lamp, it would alone have entitled him to the applause and thanks of mankind." *Fire-damp*, or carburetted hydrogen gas, appears to be generated by the decomposition of iron pyrites in coal, and may often be heard issuing from the fissures in coal-beds with a bubbling noise, as it forces the water out with it. In Messrs. Conybeare and Phillips' Geology of England and Wales, it is stated that the pit-men occasionally open with their picks, crevices in the coal or shale, which emit 700 hogsheads of fire-damp in a minute. These blowers, as they are termed, continue in a state of activity for many months together. The afterdamp, or stythe, which follows the explosions, is a mixture of the carbonic acid and azotic gases, resulting from the combustion of the carburetted hydrogen in atmospheric air, and more lives are destroyed by this than by the violence of the fire-damp explosions. Sir H. De La Beche observes, "it appears very remarkable that in the coal districts of the British Isles, where such a large amount of carburetted hydrogen is annually produced, means have not been adopted for making an economical use of this gas, both as repects light and heat."

FIRE OPAL. Called also Girasole. The Feur Opal of Klaproth; Quartz Resinite Girasol of Haüy. A subspecies or variety of opal, according to some authors, of a hyacinth-red colour, but according to others presenting bright hyacinth-red and yellow colours, when turned towards the light. It has been hitherto obtained chiefly from Mexico and the Faroe Islands. See *Girasole* and *Opal*.

FIRE-STONE. (The craie chloritée, ou glauconie crayeuse, of the French geologists.) Known also as the Merstham beds; and Upper Green sand. From the application of the stone found in some parts of this deposit in the construction of furnaces, ovens, &c., it has obtained the name of fire-stone. An arenaceo-argillaceous deposit of a greyish green colour, composed of marl and grains of silicate of iron; in some places, in a state of sand; in others, forming a stone sufficiently hard for building. The transition from the marl to the *fire-stone* is in many localities so gradual, and the sandy particles are so sparingly distributed, that the chalk-marl may be said to repose immediately on the galt; in others, however, the characters of the *fire-stone* are very peculiar, and some geologists have deemed them of sufficient importance to rank this deposit as an independent formation. The fire-stone contains the same fossils as the grey-marl, and a few species not found in any other bed.—*Dr. Mantell.*

FI'SSILE. (*fissilis*, Lat.) Capable of being split, or divided, in the direction of the grain or cleavage.

FISSIRO'STRES. A family of birds, numerically small, but very distinct from all others in the beak, which is short, broad, horizontally flattened, slightly hooked, unemarginate, and with an extended commissure, so that the opening of the mouth is very large, which enables them to swallow with ease the insects they capture while on the wing: the swallow belongs to this family, and is an example.

FISSURE'LLA. A gasteropod; a genus of the order Scutibranchiata. A buckler-formed univalve, without spire: the vertex perforated by a small ovate or oblong orifice, which affords a passage to the water required for respiration; this orifice penetrates into the cavity of the branchiæ, which are situated on the fore part of the back. The fissurella has been found in the Essex cliffs.

FISTULA'NA. A genus of the family

Inclusa, class Acephala, division Mollusca. Nearly all of the family Inclusa live buried in sand, stones, ooze, or wood. The external tube of fistulana is entirely closed at its larger end, and is more or less like a bottle or club. The fistulanæ are sometimes found buried in submerged fragments of wood, or in fruits, and the animal like the teredo, has two small valves, and as many palettes. Recent specimens are only obtained from the Indian ocean, but *fistulanæ* are found fossil in the Shanklin sand, where, in some instances, the wood is studded with the remains of a small species of fistulana, of a pyriform shape, about one-third of an inch long, to which the name of Fistulana pyriformis has been given. Fistulanæ personatæ are found in the chalk formation, and in the arenaceous limestone, or sandstone, of Bognor; and Fistulana pyriformis, at the junction of the Galt and Shanklin sand, imbedded in wood.

FI'STULIFORM. (from *fistula* and *forma*, Lat.) In round hollow columns.

FI'STULOUS. (*fistuleux*, Fr. *infistolito*, It.) Hollow; tube-like.

FI'XITY. (*fixité*, Fr. *propriété qu'ont quelques corps de n'être point dissipés par l'action du feu.*) Coherence of parts: that property which some bodies possess of resisting dissipation by heat.

FLAGSTONE. When sandstones are very thin-bedded, or the beds are easily split along the lines of lamination, they are called *flagstones*. Flagstones are not exclusively arenaceous, but may be argillaceous, or even calcareous.—*Jukes*.

FLAMMI'VOMOUS. (from *flamma* and *vomo*, Lat.) That vomits forth flames; volcanoes are *flammivomous*.

FLE'XIBLE. (*flexibilis*, Lat. *flexibile*, Fr. *flessibile*, It.) That can be bent; not brittle; pliable. That substance is said to be *flexible* which, being bent, does not itself resume its former shape; but continues in the form forcibly given to it. Substances which, being bent forcibly, spring back to their former position, are termed elastic.

FLINT. (Sax.) Siliceous earth, nearly pure. Flint is the commonest form in which quartz exhibits itself; it is rather harder than quartz, and contains a minute portion of alumine, lime, and oxide of iron; 98 per cent. being pure silex. A remarkable circumstance attending flint is, that it is found in masses, dispersed in regular parallel beds, in chalk-rocks. This is elucidated, and partly explained, in a beautiful manner in the manufacturing of porcelain. Porcelain is made of flint and clay, pounded extremely fine, and mingled together with water so perfectly, as to form a smooth fluid, of the consistence and colour of cream; if this fluid be left a long time tranquil, the flint separates from the clay, and collects in small masses, in a manner analagous to that in which the natural masses occur in the chalk. When flint is first extracted from the quarry it is much more brittle, and requires a much lighter blow to break it, than flint that has been long exposed. This may perhaps be owing to the moisture or water belonging to the flint in its natural state, but which it loses in great measure by the joint action of the sun and air. It has a conchoidal fracture, and feeble lustre; thin fragments are translucent. Specific gravity 2·594. According to Klaproth's analysis, it consists of silex 98, lime 0·5, alumine 0·25, oxide of iron 0·25, water 1. Before the blow-pipe, flint per se is infusible, but it whitens and becomes opaque. When two pieces of flint are rubbed together, they emit a peculiar smell, and phosphorise greatly. The constant occurence of flint in the upper chalk, and the apparent conversion of animal remains into flint, has given rise to much speculation respecting its origin; and it was at one time maintained, that flint and chalk were convertible, or capable of undergoing a mutual transmutation. One thing appears very certain, that the veins of flint so freely distributed throughout the chalk, are invariably confined to that formation. I propose to submit a few observations from the pens of our first writers on this interesting and intricate subject, for after having considered the matter in every point of view; after having carefully read the opinions of others, and again and again examined strata of flint nodules and tabular flints, flints horizontally and diagonally distributed throughout the numerous chalks pits in my neighbourhood; after having observed their crushed but not disordered condition, and having commonly found flints imbedded in flints, I am totally unable to arrive at anything approaching to a legitimate deduction from the various phenomena.

"That the beds of chalk and flint were deposited periodically," says Dr. Mantell, "cannot admit of the slightest doubt. Specimens are not unusual, in which angular fragments of black flint, that could not possibly have been originally formed in their present state, are imbedded in chalk. Sir Henry Englefield was the first who directed the attention of geologists to the subject of the shattered condition of the flints found in certain strata. In a paper read before the Linnæan Society, he notices several beds of shattered flints, which occur in a chalk-pit at Carisbrook, in the Isle of Wight; and, after describing their situation and appearance, proceeds to offer some conjectures upon the probable cause of their destruction. This he supposed might have been occasioned by some sudden shock or convulsion, which in an instant shivered the flints, though their resistance stopped the incipient motion; for the flints, though crushed, are not displaced, which must have been the case, had the beds slid sensibly. Chalcedony is often found occupying the hollows of flints, and on this subject it has been remarked that although in the present compact state of the matter of flint, it is not easy, though possible, to force a fluid slowly through its pores, yet it is probable that before its consolidation was complete, it was permeable to a fluid whose particles were finer than its own; and that the particles of chalcedony, whilst yet in a fluid state, being finer than those of common flint, did thus pass through the outer crust to the inner station they now occupy; where *they* also allowed a passage through their own interstices to the still purer siliceous matter, which is often crystallized, in the form of quartz, in the centre of the chalcedony, and is so entirely surrounded by it, that it could have no access to its present place, except through the substance of the chalcedony, and the flint enclosing it." In Professor Buckland's Bridgewater Treatise we find the following: "We may in like manner refer the origin of those large quantities of silex, which constitute the chert and flint beds of stratified formations, to the waters of hot springs, holding siliceous earth in solution, and depositing it on exposure to reduced degrees of temperature and pressure, as silex is deposited by the hot waters that issue from the geysers of Iceland." Again Dr. Mantell, "the nodular masses of flint are very irregular in form, and variable in magnitude; some of them scarcely exceeding the size of a bullet, while others are several feet in circumference. Although thickly distributed in horizontal beds and layers, they are never in contact with each other, but every nodule is completely surrounded by chalk. Flints so commonly enclose the remains of sponges, alcyonia, and other zoophytes, that some geologists are of opinion that the nucleus of every nodule was originally an organic body, and Townsend states, 'so far as my observation goes, zoophytes appear universally to have formed the nuclei of nodulated and coated flints. The nodules of flint frequently exhibit the internal structure of the enclosed zoophyte most beautifully and delicately preserved." A theory offered by Professor Buckland is to this effect: "It does not appear possible that flints could have been formed by infiltration into pre-existing cavities, like the regularly disseminated geodes of the trap rocks. Assuming that the mass which is now separated into beds of chalk and flint, was, previously to its consolidation, a compound pulpy fluid, and that the organic bodies now enveloped in the strata were lodged in the matter of the rock, before the separation of its calcareous from its siliceous ingredients, the bodies thus dispersed throughout the mass would afford nuclei, to which the flint, in separating from the chalk, would, upon the principle of chemical affinity, have a tendency to attach itself. The chalk and flint proceeded through a contemporaneous process of consolidation; the separation of the siliceous from the calcareous ingredients being modified by attractions, which drew to certain centres the particles of the siliceous nodules, as they were in the act of separation from the original compound mass. The distances of the siliceous strata must have been regulated by the intervals of precipitation of the matter from which they are derived; each new mass, as it was discharged, forming a bed of pulpy fluid at the bottom of their existing ocean, which being more recent than the bed produced by the last preceding precipitate, would rest upon it as a foundation similar in substance to itself, but of which the consolidation was sufficiently advanced to prevent the ingredients of the last deposit, from penetrating or disturbing the productions of that which preceded it." The present shape of many chalk flints being that of organic bodies, demonstrates the latter to have existed before the consolidation of the former; for the fidelity with which the site has often copied the organization, and even the accidents and irregularities of the bodies enveloped is so accurate, that it is impossible to attribute the form of the flint to any other cause than that of the body on which it was deposited. Sometimes the organization is so delicately retained that it seems not to have undergone the smallest derangement before the siliceous cast was taken, and the model is thus permanently preserved. The organic bodies that afforded nuclei to these nascent flints, appear to have been dispersed pretty uniformly through the original compound mass, which is now divided into beds of chalk and flints, but it is not easy to determine what cause it was that regulated the distances at which the beds of flint have been disposed, or to say why we sometimes find organic bodies preserved in flint, at other times enveloped and filled only by pure chalk. In many silicified sponges and alcyonia, the outer crust being composed of flint in its common state, represents rudely the outline of the body inclosed. But the internal structure retains traces of all its tubes and fibres, most delicately preserved in a reddish caldedony. Sir H. Davy found pure flint in the cuticle of many grasses; it is also found in the hollow stems of bamboo; the ashes of wheat straw are also found to contain it.

FLI'NTY SLATE. Flinty slate differs from common slate, in containing a large portion of siliceous earth. Slate and flinty slate not only pass into each other, but frequently alternate. When flinty slate ceases to have the slaty structure, it becomes hornstone, or, what the French geologists term petrosilex. If it contains crystals of felspar, it becomes hornstone porphyry.—*Bakewell.*

FLO'ATSTONE. The white and grey porous varieties of rhombohedral quartz. In consequence of their extreme pourousness they swim on the surface of water, and have therefrom been named floatstone, or spongiform quartz.

FLOETZ. *(flötz,* Germ.) The name given by Werner to certain rocks which were flat, horizontal, and parallel to each other. Werner employed this term, in lieu of secondary, for the rocks reposing on his transition series, from the belief that they generally were stratified in planes nearly horizontal, while those of the older strata were inclined to the horizon in considerable angles. This however, holds good only as regards the structure of countries comparatively low; in mountain ranges Werner's floetz formations are highly inclined. As therefore vertical floetz formations imply a manifest contradiction of terms, the use of the word floetz in the place of secondary is inadmissable.

FLO'RA. (Lat.) As the animals peculiar to any country constitute its *fauna,* so do the trees and plants its *flora;* the botany of a country.

FLO'RAL. *(floralis,* Lat.) An epithet for a bud or leaf; pertaining to flowers; belonging to the flower. The calyx is the outer set of the *floral* envelopes.

FLO'RET. *(fleurette, petite fleur,* Fr.) A floret is a small monopetalous flower, many of which, enclosed in one calyx or perianthium, and placed sessile on a common undivided receptacle, form a species of compound flower.

FLU'ATE. A compound of fluoric acid with a salifiable base.

FLU'CAN. A provincial name for a fault or dam; particularly used by the Cornish miners. 2. A term applied by the Cornish miners to either a white or a greenish clay, without regard to its composition.

FLU'OR. } (Lat.) Octahedral
FLU'OR SPAR. } fluor. Octaedrisches Flus-Haloide of Mohs. Chaux Fluaté of Haüy. Fluate of lime; consisting of 67·75 lime and 32·25 fluoric acid. If a cube of fluor spar be split with a knife and a hammer, it will yield only in the direction of the solid angles, and if the division be pursued the result will be an octohedron. There are three varieties of fluor spar; the first, with even fracture and feeble lustre, is called compact fluor; the second, in which the cleavage is distinct, foliated fluor; the third, which occurs incrusting other minerals, earthy fluor. See also *Blue-john* and *Derbyshire spar.*

FLUOR'IC ACID. } An acid first procured by Gay
FLUO'RINE. } Lussac, or by Margraff, and called fluorine by Sir H. Davy. It may be obtained by putting a quantity of fluor-spar in powder into a leaden retort, pouring over it an equal quantity of sulphuric acid, and then applying a very gentle heat. From its exceedingly de-

structive properties it has been called *phtore*, from φθόριος, Gr., by M. Ampere. It destroys the skin, almost immediately, if applied to it, producing very painful wounds. The most singular property which it possesses is that of corroding glass and siliceous bodies, especially when hot, and the thickest glass vessel can only withstand its action for a short time.

Fluorine enters into the composition of some minerals which form constituent portions of great masses of rocks. Fluoric acid is found in mica and hornblende, two minerals of very great importance, as component parts of many rocks. Fifteen analyses of mica, from various parts of the world, by Klaproth, Vauquelin, Rose, and Bendant, afford as a mean, 1·09 per cent. of fluoric acid; and Bousdorf's analysis of hornblende, gives 1.5 per cent. of the same substance. Calculation affords us 0·36 of fluoric acid in gneiss with mica, 0·54 in mica slate, 0·75 per cent. in hornblende rock and greenstone, 0·18 in granite with mica, 0·5 of the same substance in sienite, 0·5 per cent. in porphyritic greenstone. Fluor spar is, however, the mineral in which the greatest relative amount of fluorine is detected.—*De la Beche.* The symbol of fluorine is F.

FLU'STRA. A genus of polyparia, class Vermes, order Zoophyta. If we carefully observe the patches of white calcareous matter, called flustræ, that may be seen on every sea-weed or shell on the shore, appearing like delicate lace, we shall discover that these apparently mere specks of earthy substance belong to the animal kingdom. The flustra, when taken fresh and alive out of the water, presents to the naked eye the appearance of fine network, coated over with a glossy varnish. With a glass of moderate powers, it is discovered to be full of pores, disposed with much regularity. If a powerful lens be employed, while the flustra is immersed in sea-water, very different phenomena appear; the surface is seen to be invested with a fleshy, or gelatinous, substance, and every pore to be the opening of a cell or cavity, whence issues a tube with several long feelers or tentacula; these expand, then suddenly close, withdraw into the cells, and again issue forth; and the whole surface of the flustra is studded with these hydra-like forms, sporting about in all the energy and activity of life. For a more full account, see Dr. Mantell's Wonders of Geology, whence the above is taken.

FLUX. *(fluxus,* Lat. *flux,* Fr. *mouvement réglé de la mer vers le rivage à certaines heures du jour.)*

1. The flow of the tidal wave: the flux is the rise; the reflux, the ebb of the tide.

2. Any substance added to facilitate the fusion of metals or minerals.

FŒ'CAL. See *Fecal.*
FŒ'CES. See *Feces.*
FO'LIATED. *(foliatus,* Lat.)

1. In botany, leaved or having leaves.

2. In conchology, in laminæ or leaves, as when the edges of the shelly layers are not compact, but seem to separate from each other. This may easily be seen in the large coarse oyster shell.

3. In mineralogy, the term foliated was used by Werner to express the structure of such minerals as may be divided or cleaved regularly, and were therefore said by him to consist of folia or leaves.

FOLIA'TION.

1. In botany, vernation or leafing of trees, &c.

2. In mineralogy, the act of beating into thin leaves.

3. Mr. Darwin proposes to give the term foliation to the tendency in

certain rocks to split, *foliate*, or cleave in a certain given direction; by foliation is meant, a separation into layers of different chemical composition, while cleavage means only a tendency to split, in a mass of the same composition.

FO'LKSTONE MARL. A stiff marl, varying in colour from a light grey to a dark blue, more generally known under the provincial term Galt. The thickness of this bed is in some places, in the South of Sussex, not less than between two and three hundred feet. It is a member of the cretaceous group, lying between the upper and lower green-sand. Where the Folkstone marl is exposed, and forms the surface of the country, the soil is exceedingly tenacious, and ranks amongst the finest and most productive. The Folkstome marl abounds in fossils.

FO'LLICLE. (*folliculus*, Lat. *follicule*, Fr. *follicola*, It.)
1. In botany, a univalvular pericarp, opening on one side longitudinally, and having the seeds loose in it; a membranous seed-vessel of one valve and one cell, bursting lengthwise, and having no apparent suture to which the seeds are attached.
2. In anatomy, a small secreting cavity.

FONTANE'L. (*fontanelle*, Fr. *fontanélla*, It.) An opening left in the skull at birth, which is subsequently closed by osseous deposit; there are two fontanels.

FORA'MEN. (Lat.) A hole; an opening; generally, by which nerves or blood vessels obtain a passage through bones. In botany, the opening in the ovulum. When the foramen is visible on the seed, as is the case in the bean and pea, it is called the *micropyle*.

FORA'MINATED. } (from *foramino*, Lat.)
FORA'MINOUS. } Pierced with small openings; full of small holes; porous.

FORAMINI'FERA. The second order of the class Astomata, comprising Rotalia, Nodosaria, Nummulites, and Orbitolites. This order has been constituted a class by Pictet and D'Orbigny, and divided into seven families.

FORAMINI'FEROUS. Belonging to the order foraminifera.

FO'RCIPATED. (*forcipatus*, Lat.) Hooked, or furnished with pincers, as the claws of a lobster, crab, &c.

FORMA'TION. (*formatio*, Lat. *formation*, Fr. *formazione*, It.) Any assemblage of rocks possessing some character in common, either as regards their age, origin, or composition. When a series of strata of a similar rock are arranged with occasional strata intervening, of rocks of another kind, which recur in different parts of the series, they are regarded as having been all formed nearly at the same epoch, and under similar circumstances; and such series are called by geologists *formations*. Thus, the strata of shale, sand-stone, and iron-stone, that accompany beds of coal, are called the coal *formation*. Strata of different kinds, in which a gradation is observed into each other, and which contain similar species of organic remains, constitute a *geological formation*. The chalk with flints above, the lower chalk without flints, the chalk-marl, and the green-sand under the chalk, are all regarded as members of the chalk *formation*.—*Bakewell*.

FO'RNICATED. (*fornicatus*, Lat.) Concave within, and convex without; vaulted; arched.

FO'RNIX. (Lat.) In conchology, the excavated part under the *umbo*. It likewise signifies the upper, or convex shell in the ostrea.

FO'RTIFICATION AGATE. The name given to a variety of agate which, when cut and polished, presents lines running in such directions as to resemble a fortification, the centre being often amethyst, with jasper, quartz, &c., surrounding. In composition it resembles other agates; it occurs in irregular nodules, commonly in amygdaloid, as in that of Scotland, called Scotch pebble.

Fo'ssil. } *(fossilis,* from *fodio,* Lat.
Fo'ssile. } *fossile,* Fr. *fossile,* It.)
Dug out of the earth, as fossil shells, fossil bones, fossil coal, &c. The adjective is frequently spelt *fossile.*

Fo'ssil. A substance dug out of the earth. At the present day, the word fossil is used by geologists to express only the remains of animal, or vegetable, substances found buried in the earth's crust.

Fossil shells of forms such as now abound in the sea, are met with far inland, both near the surface and at all depths below it, as far as the miner can penetrate. They occur at all heights above the level of the ocean, having been observed at an elevation of from 8,000 to 9,000 feet in the Alps and Pyrenees, more than 13,000 feet in the Andes, and above 15,000 feet in the Himalayas.—*Lyell.*

Mr. W. Smith was the first to notice that certain fossils are peculiar to, and are only found lodged in, particular strata. The rapid advances made in geology have taught the following facts respecting fossils: that exactly similar fossils are found in distant parts of the same stratum, not only where it traverses this island, but where it appears again on the opposite coast; that in strata of considerable comparative depth, fossils are found, which are not discovered in any of the superincumbent beds: that some fossils, which abound in the lower, are found in diminishing numbers through several of the superincumbent, and are entirely wanting in the uppermost strata: that some fossils, occurring in considerable numbers in one stratum, become very rare in the adjacent portion of the next superincumbent stratum, and afterwards are lost: that most of the remains which are abundant in the superior strata, are not at all found in the lower.—*Geological Transactions.*

Fra'gmentary. Composed of fragments. Dr. Johnson says, "a word not elegant, nor in use:" in elegance or euphony it may, or may not, be deficient, but, at the present day, it is in use by geologists.

Frangibi'lity. (In mineralogy, one of the physical characters of minerals.) Capability of being broken. The *degree* of frangibility consists in the ease with which a thing may be broken. This quality varies greatly in different substances, ranging through all the intermediate degrees, from very brittle to very tough. Frangibility, strictly speaking, ought not to be considered as connected with the ease or difficulty with which minerals yield in directions parallel to their natural joints; it is rather applicable to their property of yielding to mechanical force in other directions.

Free-stone. Any kind of stone, the texture of which is so free or loose that it may be easily worked.

Fr'eshet. A flood or overflowing of a river, by means of heavy rains or melted snow; an inundation of fresh-water. Freshets take place, more or less, in all rivers, greatly augmenting their velocities and transporting power, and carrying forward substances that could not have been moved under ordinary circumstances.

Frond. *(frons,* Lat.)
1. In botany, implies peculiar union of the fructification with the leaf and stem, namely, the flowers

and fruit are produced from the leaf itself.

2. The herbaceous parts of flowerless plants, resembling leaves, are called fronds; they differ from true leaves in their structure in many respects.

FRONT. *(front,* Fr. *fronte,* It.) In conchology, when the aperture in univalves is turned towards the observer.

FRO'NTAL. *(frontale,* Lat. *frontal,* Fr. *frontale,* It.) Appertaining to the forehead.

FRUCTIFICA'TION. *(fructification,* Fr. *fruttificazióne,* It.)

1. The temporary part of a vegetable appropriated for generation, terminating the old vegetable, and beginning the new. It consists of the following parts; namely, the calyx, corolla, stamen, pistillum, pericarpium, semen, and receptaculum.

2. The act of bearing fruit; fertility; fecundation.

FUCHSITE. A variety of mica containing chrome.

FU'COID. (from φῦκος and εἶδος, Gr.) A species of fucus. Fucoids are very abundant in many of the strata, occurring in the transition strata of North America in numerous thin layers. An account of these has been published by Dr. Harlan, of America, and by Mr. R. C. Taylor, in Loudon's Magazine of Natural History. Fucoids are found in great abundance in the grauwacke slate of the Maritime Alps, in the lias, and in the chalk. There is one species, the Fucoides targionii, that abounds in the upper green-sand. To a fine species, discovered in the chalk by Dr. Mantell, he has given the name Fucoides Brongniarti.

FU'CUS. *(fucus,* Lat. φῦκος, Gr.) A genus of the order of Algæ, belonging to the class Cryptogamia. This genus comprehends most of those plants commonly called seaweed: pl; fuci.

FU'LGORITE. *(fulguritus,* Lat.) Anything struck by lightning. Rocks, and the tops of mountains, often bear the marks of fusion from the action of lightning; and occasionally vitreous tubes, descending many feet into banks of sand, mark the path of the electric fluid. Some years ago, Dr. Fiedler exhibited several of the *fulgorites* in London, which had been dug out of the sandy plains of Silesia and Eastern Prussia.—*Mrs. Somerville.*

"In a broad land of sand-hillocks," says Mr. Darwin, "which separate the Laguna del Potrero from the shores of the Plata, at the distance of a few miles from Maldonado, I found a group of those vitrified, siliceous tubes, which are generally supposed to have been formed by lightning entering the loose sand. These tubes resemble in every particular those from Drigg in Cumberland, described in the second vol. of the Geological Transactions. The internal surface is completely vitrified, glossy, and smooth. A small fragment examined under the microscope, appeared, from the number of minute entangled air, or, perhaps, steam bubbles, like an assay fused before the blow-pipe. The sand is entirely, or in greater part, siliceous; but some points are of a black colour, and from their glossy surface possess a metallic lustre. The thickness of the wall of the tube varies from a thirtieth to a twentieth part of an inch, and occasionally even equals a tenth. On the outside, the grains of sand are rounded, and have a slightly glazed appearance: the tubes are generally compressed, and have deep longitudinal furrows, so as closely to resemble a shrivelled vegetable stalk, or the bark of the elm or cork tree. Their circumference is about two inches, but in some fragments which are cylindrical and without any furrows, it is double, or four inches. The compression from the surrounding loose sand, acting while the tube was still softened from the effects of the intense heat, has evidently caused the creases or furrows. Judging from the uncompressed fragments, the measure or bore of the lightning must have been about one inch and a quarter. At Paris, M. M. Hachette and Beaudant succeeded in making tubes, in most respect similar to these fulgorites, by passing very strong shocks of galvanism through finely powdered glass: when salt was added, so as to increase its fusibility, the tubes were larger in every dimension. They failed both with powdered felspar and quartz. One tube formed with pounded glass, was very nearly an inch long, namely, ·982, and had an internal diameter of ·019. When we hear that the strongest battery in Paris was used, and that the effect on a substance of such easy fusibility as glass, was to form tubes so diminutive, we must feel greatly astonished at the power of a shock of lightning, which, striking the sand in several places, has formed a cylinder, in one instance, of at least thirty feet long, and having an internal bore, where not compressed, of full an inch and a half; and this is in a material so extraordinarily refractory as quartz!" Specimens of vitreous sand tubes or fulgorites are contained in the Museum of the Geological Society, and the fifth volume of the Society's transactions contains an engraving of them.

FULI'GINOUS. *(fuliginosus,* (Lat.) *fuligineux,-euse,* Fr. *fulliginóso,* It.) Sooty; dark; smoky; dusky; of the colour of soot.

FU'LLER'S-EARTH. A marl of a close texture, soft and unctuous, containing about 25 per cent. of alumina. It derives its name from its being used by fullers to take the grease out of cloth before they apply soap. Any clay having its particles of silica very fine, may be considered as fuller's earth; for it is the alumina alone which acts upon the cloth, on account of its strong affinity for greasy substances.

FUNGIA. A genus of stony, free polypifers, simple, orbicular, or oblong; convex and lamellated in the upper part with an oblong central groove, concave and rough beneath. Many species have been identified.

FU'NGIFORM. Having its termination resembling the head of a fungus: certain substances are occasionally found of this shape, as calcareous stalactites, &c.

FU'NGUS. (Lat. *fungus.*) One of the orders of the class Cryptogamia, according to the artificial system of Linnæus. A mushroom; an excrescence from trees or plants not naturally belonging to them; any morbid sponge-like excrescence.

FU'NNEL-SHAPED. In botany, applied to a monopetalous corolla, having a conical border placed upon a tube. A form which gradually increases in thickness towards its apex, and scooped out, or hollowed, at its apical margin.

FU'RCULA. (Lat.) A fork, a peculiar formation of bone in birds, of a fork-like shape. The *furcula,* commonly known as the merry-thought-bone, is seldom wanting in birds. It is in form like a V, common to both shoulders, and joined by its point to the most prominent part of the crista of the sternum, while the other extremities are connected to the humeral end of the clavicles, and the point of the scapulæ, where these two bones are articulated with each other, and with the os humeri. The furcula serves to keep the wings at a proper distance in flying, and is strong and expanded in birds which fly with great force and rapidity. In the ostrich and cassowary, it is imperfect, the lateral branches not uniting together.

The ornithorhynchus and ichthyosaurus both possess a peculiar form of sternum, resembling the *furcula* of birds. The echidna is the only known land quadruped that has a similar *furcula* and clavicles. A cartilaginous rudiment of a furcula occurs also in the dasypus.

Fu′scite. An opaque mineral of a greyish or greenish-black colour, found in Norway, in masses of granular quartz.

Fu′siform. Spindle-shaped, swelling in the centre with the ends tapering; intermediate between the conical and the oval.

Fu′sus. A subfusiform univalve, ventricose in its middle or lower part, with a canaliculated base, and no varicose sutures; an elongated spire, a smooth columella, and the lip not slit. The genus comprises many species. The genus *fusus* comprises all shells with a salient and straight canal, which are destitute of varices. Fusi are found at depths varying to eleven fathoms, in mud, sandy mud, and sand.

Fu′sus contra′rius. A species of fusus found in the crag of Suffolk, a sinister shell. The fusus contrarius, or reversed whelk, which is synonymous with the murex contrarius of Linnæus, appears to be a shell characteristic of the crag deposit.

# G

Ga′bbro. A crystalline granular mixture of labradorite, diallage, and euphotida.

Ga′dolinite. A mineral thus named after Gadolin, who first ascertained its composition. Its colour is greenish-black; that of its powder greenish-grey. Occurs massive; in granular and prismatic concretions. Fracture conchoidal and glassy. According to Berzelius its constituent parts are yttria 45·93, silica 24·16, protoxide of cerium 16·90, protoxide of iron 11·34. It was first discovered at Ytterby, in Sweden, by Capt. Arhenius, in white felspar; it is found also in Ceylon, in granite.

Ga′hnite. Thus named from Gahn; another name for automalite.

Gaillone′lla ferrugi′nea. The name assigned to a minute animalcule, the case of which is found in bog-ore.

Ga′lactite. (γαλακτίτης, Gr.) Milkstone.

Ga′lea. (Lat.) A genus of echini, found fossil only. They are distinguished by an oval base, from which the shell rises in a vaulted, helmet-like form.

Gal′eola. A genus of echinites possessing the same characters as the galea, but differing in size. This circumstance induced Klein to divide them into two genera, but Leske deeming a mere difference of size as insufficient to affect the genus, included them both under the genus *echinocorys*.—*Parkinson*.

Ga′leated. (*galeatus*, Lat.) Helmet-shaped; covered as with a helmet. In botany, plants bearing flowers of a helmet shape, as the monk's-hood.

Gale′na. (*galena*, Lat.) A shining metallic ore, composed of sulphur and lead; sulphuret of lead; lead-glance. Its colour is bluish-grey, resembling lead. Occurs regularly crystallized, frequently in cubes and cubo-octahedrons. Before the blow-pipe it decrepitates and melts, emitting a sulphureous smell. It is found in every lead-mine. There are two varieties, common galena and compact galena.

GALERI′TES. A genus of Radiaria: nineteen species have been determined, all in a mineralized state, and they have been distinguished by various names, as scolopendrite, bufonite, cap-stones, &c.

GALLINA′CEÆ. The fourth order of the second class Aves. So called from their affinity with the domestic cock.

GALLINÆ. The fifth order of the class Aves, comprising the fowl, grouse, partridge, &c.

GALLIA′RD. } Provincial names for a
CALLIARD. } trappean sandstone of a hard, smooth, flinty character.

GALT. } A provincial name for a stiff
GAULT. } marl, varying in colour from a light grey to a dark blue. The upper and lower beds of the green-sand are in many places separated by the galt; it has been also called Folkstone Marl. The galt abounds in fossil remains, remarkable for their beauty, the pearly covering of the shells being in many instances preserved. The galt is a member of the cretaceous group, passing, in its lower parts, into calcareous marl. The fossils hitherto found in the galt belong to forty-three species, among which are several species of ammonites and hamites; nautili and belemnites; nuculæ and inocerami; caryophilleæ, &c. The galt rarely exceeds 100 feet in thickness; although in some parts of Sussex it is not less than 250. It is a soil that must rank, says Mr. Young, among the finest in this or any other country, being pure clay and calcareous earth. The occurrence of this mud, says Sir H. De La Beche, marks a modification of the causes that transported and brought detrital matter to rest during the deposit of the sands beneath the chalk, and the area over which it took place was considerable. The gault is frequently composed of clay in the upper, and marls in the lower part, containing disseminated specks of mica. It effervesces strongly with acids.

GAMOPE′TALOUS. Another term for monopetalous. Having the petals united by their edges; a corolla, the petals of which are all united by their edges.

GAMOSE′PALOUS. In botany, a term used for a calyx when the sepals of which it is composed are all united.

GANGUE. (Called also Matrix.) The substance in or on which a mineral is found.

GANISTER. (Called also Calliard.) The provincial name for a fine grained silicious stone, a member of the coal series.

GA′NOID. Belonging to the order Ganoidial.

GANOI′DIA. (from γάνος, Gr. splendour, from the brightness of their enamel.) The second order of fishes, according to the arrangement of M. Agassiz. The families of this order are characterized by angular scales, composed of horny or bony plates, covered with a thick plate of enamel. The bony pike and sturgeons are of this order. It contains more than sixty genera, of which fifty are extinct.—*Prof. Buckland.*

GANOIDIAN. Shining scaled. Belonging to the order Ganoidia. The ganoidian order of fishes with the placoidean prevailed, exclusively, in all formations till the termination of the oolitic series, when they ceased suddenly and were replaced by genera of new orders, the Ctenoidean and Cycloidean, then for the first time introduced.—*Ib.*

GAP. In conchology, an opening, in multivalves and bivalves, when the valves are shut, as in the pholades, myæ, &c.

GA′RNET. *(grenat, sorte de pierre précieuse, d'un rouge foncé, comme le gros vin,* Fr. *granáto,* It. *pietra preziosa.)* Prof. Jameson, in his system of Oryctognosy, constitutes a family of earthy minerals which he denominates the garnet family; this consists of Leucite, Vesuvian, Grossular, Melanite, Allochroite, Garnet, Grenatite, Pyrope, and Cinnamon Stone. The precious garnet is found in rhombic dodecahedrons, in mica-slate, amongst the oldest, or primary, rocks, in many parts of the world. It is of a beautiful red colour, sometimes with shades of yellow or blue. Those from the kingdom of Pegu are most esteemed, and it is supposed that this was the carbuncle of the ancients. It is harder than quartz, and consists of nearly equal parts of silex, alumine, and oxide of iron, with traces of manganese. Common garnets are more opaque, of a duller colour, and less hard than the precious garnet, though harder than quartz. They are abundant in similar localities in all countries, sometimes constituting nearly the whole mass of a rock.

GAS. The name given to all permanently elastic, or aëriform, fluids, except the atmosphere. The term was first used by Van Helmont, who appears to have intended to denote by it every thing which is driven off from bodies in the state of vapour by heat.

GASTE′ROPOD. Belonging to the order Gasteropoda.

GASTERO′PODA. (from γαστήρ, the belly, and πούς, the foot, Gr.) The third order of Mollusca; they have the head free, they crawl upon the belly, or upon a fleshy disk, situated under the belly, which serves them as feet. They are univalvular or multivalvular, but in no case bivalvular. The back is furnished with a mantle which is more or less extended, takes various forms, and in the greater number of genera, produces a shell. The tentacula are very small, situated above the mouth, and do not surround it, varying in number from two to six; sometimes they are wanting altogether. The eyes are very small, and sometimes wanting. Several are entirely naked; others have merely a concealed shell, but most of them are furnished with one that is large enough to receive and shelter them. Most of the aquatic gasteropoda, with a spiral shell, have an operculum, a part sometimes horny, sometimes calcareous, attached to the posterior part of the foot, which closes the shell when its occupant is withdrawn into it and folded up. The limax or slug is an example of the class. Cuvier divides this class of Mollusca into nine orders.

GA′STRIC. (from γαστήρ, Gr. *gastrique,* Fr. *gastrico,* It.) Belonging to the stomach or belly.

GAULT. See *Galt.*

GA′VIAL. A subdivision of the genus crocodile, characterized by the narrow, elongated, almost cylindrical jaws, which form an extremely lengthened muzzle. The muzzle is narrow, cylindrical, extremely elongated, and a little swelled out at the ends. The length of the cranium is scarcely one-fifth of the entire length of the head. The teeth are nearly equal, twenty-five to twenty-seven on each side below. The first two and fourth two of the lower jaw pass into notches of the upper, and not into hollows. The cranium has large foramina behind the eyes, and the hind feet are indented and palmated, like those of the crocodiles proper.

The living species of the crocodile family are thirteen in number. Teeth of the fossil gavial have been found in the Tilgate strata. Dr. Mantell observes, "it appears that the strata of Tilgate Forest contain the remains of at least two, if not four, species of crocodiles: that one of these (that with slender curved teeth) resembles the gavial of Caen, and probably was about twenty-five feet in length."

GÉHLENITE. A mineral allied to Vesuvian, and so named after Gehlen. A combination of silica, alumina, lime, and oxide of iron. It occurs in rectangular four-sided prisms, nearly approaching in their dimensions the form of a cube; sometimes isolated, generally invested by calcareous spar, aggregated irregularly in groups, or massive, including pleonaste. Colours: various shades of grey, green, and brown. Surface: commonly, rough and dull; when sufficiently brilliant for the use of the reflective goniometer, the crystals afford angles of 90° in every direction. Before the blow-pipe, when alone, it is but little affected, with borax it fuses into a glass coloured by iron. Heated in muriatic acid, it gelatinizes.—*Prof. Phillips.*

GE´LATINE. ⎫ (*gelee*, Fr. *gelatina*, It.)
GE´LATIN. ⎭ A concrete animal substance; the principle of jelly. If a piece of the fresh skin of an animal, after every impurity is carefully separated, be put into a quantity of cold water, and boiled for some time, part of it will be dissolved. Let the decoction be slowly evaporated till it is reduced to a small quantity, and then put aside to cool. When cold, it will have assumed a solid form, and precisely resemble that tremulous substance known as jelly. This is, what is called by chemists, *gelatin*. If the evaporation be carried farther, the substance becomes hard, semitransparent, breaks with a glassy fracture, and is, in short, that substance known under the name of glue. *Gelatine* therefore is the same with *glue*, only that it is free from those impurities with which glue is so often contaminated. Gelatine is colourless and transparent; when thrown into water it soon swells, and assumes a gelatinous appearance, and gradually dissolves away. By evaporating the water it may again be obtained unaltered in the form of jelly. Cold water dissolves it slowly, but water at a temperature of 90° rapidly. Gelatin is insoluble in alcohol.

GEM. (*gemma*, Lat.)
1. In mineralogy, any precious stone. Gems may be divided into natural and artificial; the latter are made of what is termed paste, coloured with different metallic oxides.
2. In botany, the bud, a small conoid body, covered with scales, formed during summer on the branches of trees, containing the rudiments of a future plant, or part of a plant: some buds produce flowers and leaves, others leaves only, and some flowers only.

GE´MINATED. In conchology, marked with a double elevated striæ connecting the wreaths.

GE´MINOUS. (*geminus*, Lat.) Double; in pairs.

GE´MMIPAROUS. Producing buds, or gems. The animals forming the class Poriphora are gemmiparous.

GE´MMULE. A little bud.

GENE´RIC. ⎫ (*générique*, Fr. *genérico*,
GENE´RICAL. ⎭ It.) That which comprehends the genus, or distinguishes from another genus, but does not distinguish the species.

GENI´CULATE. ⎫ (*geniculatus*, Lat.)
GENI´CULATED. ⎭ Knotted; jointed;

applied to culms bent like the knee; also to peduncles bent at the joints.

GE'NESSEE SHALE. One of the Devonian rocks of America, it is a black shale, with flagstones and sandstones.

GE'NUS. (Lat.)
1. That which is predicated of many things, as the material or common part of their essence.
2. A subdivision of any class or order of natural beings, whether of the animal, vegetable, or mineral kingdoms, all agreeing in certain common characters.

GE'ODE. (γεωδης, from γέα, Gr.) A roundish piece of mineral matter, sometimes only an incrustation, generally more or less hollow, usually lined with crystals, or in some cases loose earthy matter. The geodes found in the greensand near Sidmouth, says Mr. Bakewell, in his valuable Introduction to Botany, are composed of opaque chert on the outside, and contain within, mammillated concretions of beautiful chalcedony, and occasionally perfect minute rock-crystals.

GEO'GNOSTIC. } Pertaining to a
GEO'GNOSTICAL. } knowledge of the structure of the earth.

GEO'GNOSY. (from γῆ and γνῶσις, Gr.) That branch of natural history which treats of the structure of the earth. Werner and his disciples, as well as some of the French geologists, have substituted geognosy for geology, the former derived from γῆ and γνῶσις, the latter from γῆ and λόγος: for this change no sufficient reason can be asssgned, and it is contrary to the established analogies of language. Nothing can be more unmeaning than the apologies that have been offered for substituting γνῶσις, knowledge, for λόγος, reason. By the same rule we ought to change meteorology, physiology, &c., into meteorognosy, physiognosy, &c. Prof. Jukes advocates the retention of the word geognosy, and would divide geology into three heads:—1. Geognosy, or the study of the structure of rocks independently of their arrangement into a chronological series, and he would again divide geognosy into lithology, and petrology. 2. Palæontology, and 3, Stromatology.

GEO'LOGY. (from γῆ and λόγος, Gr.) Geology may be defined to be that branch of natural history which investigates the successive changes that have taken place in the organic and inorganic kingdoms of nature. It is a science founded in exact observation and careful induction; it may be termed the physical history of our globe; it investigates the structure of the planet on which we live, and explains the character and causes of the various changes in the organic and inorganic kingdoms of nature. It has been emphatically termed the sister science of astronomy, ranking, undoubtedly, in the scale of sciences, next to astronomy, from the sublimity of the objects of which it treats.

Geology is as intimately related to almost all the physical sciences, as is history to the moral. As the historian should, if possible, be at once profoundly acquainted with ethics, politics, jurisprudence, the military art, theology; in short, with all those branches of knowledge, whereby any insight into human affairs, or into the moral and intellectual nature of man can be obtained; so is it desirable that the geologist should be well versed in chemistry, mineralogy, zoology, botany, comparative anatomy; in short, in every branch of science relating to organic and inorganic nature. "It was long," says Sir C. Lyell, "ere the distinct nature and legitimate objects of geology were fully recognised, and it was at first confounded with many other branches of inquiry, just as the limits of history, poetry, and mythology, were ill defined in the infancy of civilization."

Werner appears to have regarded geology as little other than a subordinate department of mineralogy, and Desmarest included it under the head of physical geography. Dr. Hutton, in his treatise, published in 1795, first endeavoured to draw a positive line of demarcation between geology and cosmogony, declaring that geology was in no ways concerned with questions as to the origin of things; and, in fact, geology differs as widely from cosmogony, as hypothesis concerning the mode of man's first creation differs from history. Philosophers for some ages past neglected the examination of the earth, contenting themselves with vain speculations respecting its formation; and to Strabo, who flourished under Augustus, and died under Tiberius, about the year 25, and to the old philosophers, who studied the local phenomena of their countries, would the title of geologists with more propriety be given than to Burnet and Buffon, whose systems of cosmogony have more the air of a system of romance, than of a serious generalization of facts. In tracing the history of geology from the close of the seventeenth to the end of the eighteenth century, we find the science retarded by the wild and visionary speculations of a host of writers; to enter on these would, however, far exceed the limits of a work of this kind, and those desirous of so doing, I would refer to Sir C. Lyell's admirable work, Principles of Geology.

Hutton, following the example of Newton in astronomy, endeavoured to give fixed principles to geology; but, at that time, too little progress had been made in the science, to enable him to realize so noble a project. A brighter period has now dawned, and the following out the only true method, namely, that of keeping within the boundary of inductive philosophy, has led to the most important results.

Happily, facts have multiplied so rapidly that geology is daily emerging from that state when an hypothesis, provided it were brilliant or ingenious, was sure of advocates and temporary success, even when it sinned against the laws of physics and facts themselves.

One of the greatest difficulties with which geology has had to contend, is the false notion entertained by many well-meaning but weak persons, that geology was opposed to Scripture revelation, and that geological researches might prove injurious to religion. Unfortunately prejudice and ignorance have too frequently called in the aid of religious feeling to thwart and oppose the progress of scientific knowledge; and it is too much to be feared that did the same power now exist, the geologists of the present day might suffer the same persecutions that Gallileo Gallilei did, and that the works of Lyell, Buckland, De la Beche, Conybeare, Murchison, Phillips, Sedgwick, Mantell, and a host of others, would swell the catalogue of the forbidden list.

Why, it may be asked, should persons whose opinions are founded on the basis of immutable truth fear the elicitation of truth? or what has religion to fear from the minutest, the most searching, investigation? Let it ever be borne in mind that, on the one hand, truth can never be opposed to truth, and, on the other, that error is only to be effectually confounded by searching deep and tracing it to its source.

Nothing can be more unfounded than the objection which has been taken against the study of natural philosophy, and, indeed, against all science, that it fosters in its cultivators an undue and overweening self-conceit, leads them to doubt the immortality of the soul, and to scoff at revealed religion. Its natural effect on every well regulated mind is, and must be, directly the reverse. Minds which have long been accustomed to date the origin of the universe, as well as that of the human race, from an era of about six thousand years back, receive reluctantly any information, which, if true, demands some new modification of their present ideas of cosmogony, and, as in this respect, geology has shared the fate of other infant sciences, in being for a while considered hostile to revealed religion, so, like them, when fully understood, it will be found a potent and consistent auxiliary to it, exalting our conviction of the power, wisdom, and goodness of the Creator. The consideration of the evidences afforded by geological phenomena may enable us to lay more securely the very foundation of natural theology, inasmuch as they clearly point out to us a period antecedent to the habitable state of the earth, and consequently antecedent to the existence of its inhabitants. When our minds become thus familiarized with the idea of a beginning and first creation of the beings we see around us, the proofs of design, which the structure of those beings afford, carry with them a more forcible conviction of an intelligent Creator, and the hypothesis of an eternal succession of causes, is thus at once removed.

"It may fairly be asked," says Chalmers, "of those persons who consider physical science a fit subject for revelation, what point they can imagine short of a communication of omniscience, at which such a revelation might have stopped, without imperfection or omission, less in degree, but similar in kind, to that which they impute to the existing narrative of Moses. A revelation of so much only of astronomy as was known to Copernicus, would have seemed imperfect after the discoveries of Newton; and a revelation of the science of Newton would have appeared defective to Laplace. And unless human nature had been constituted otherwise than it is, the above supposed communication of omniscience would have been imparted to creatures utterly incapable of receiving it, under any past or present moral or physical condition of the human race. Does Moses even say, that when God created the heavens and the earth he did more, at the time alluded to than transform them out of previously existing materials? Or does he ever say that there was not an interval of many ages between the first act of creation, described in the first book of Genesis, and said to have been performed 'in the beginning,' and those more detailed operations, the account of which commences at the second verse, and which are described as having been performed in so many days?

"Let no one, therefore, be checked in his enquiries into the history of the globe by anything but the good rules of philosophical induction, which are essential to the right use of the intellectual strength which God has conferred upon man, to be exercised on the mighty works of nature; and least of all let him be deterred from the pursuit of truth by the vain and impious dread that he may go too far, and penetrate too deeply into those mysteries, which, among their other uses have this one, namely, that they continually excite to activity the soul of man; and, the more they are studied, lead to deeper delight, and more awful contemplation of their glorious and beneficent Author."

Geology reveals to us the extraordinary fact that as the globe passed from one condition to another, whole races of animals perished, and were succeeded by others with organizations adapted to the altered state of the planet. On this phenomenon is based the fundamental principle of the identification of strata by their imbedded remains; the passage from one deposit to another being marked by a change in the animals which lived and died during the accumulation of each. Thus, although the fossils of any one great series of beds possess a common character, yet those which are found in the lowest and highest strata of a great formation are for the most part dissimilar in species, and often in genera.

Geology, aided not only by the higher branches of physics, but by recent discoveries in mineralogy and chemistry, in botany, zoology, and comparative anatomy, is enabled to extract from the archives of the interior of the earth, intelligible records of former conditions of our planets, and to decipher documents, which were a sealed book to our predecessors. Thus enlarged in its views, and provided with fit means for pursuing them, geology extends its researches into regions more vast and remote, than come within the scope of any other physical science, except astronomy. —*Davy. Buckland. Herschell. Chalmers. Lyell. Phillips. Mantell. Bakewell.*

Geosau′rus. A fossil saurian of the oolite and lias formations, discovered by M. de Sœmmering in the environs of Manheim, and intermediate between the crocodile and monitor. Cuvier considered the geosaurus to constitute a new subgenus of the saurian order. The specimen above mentioned is supposed to have been from twelve to thirteen feet in length.

Germ. (*germe*, Fr. *germe*, It. *germen*, Lat.) 1. In botany, the swollen base of the pistil, forming the rudiment of the fruit and seed.
2. The embryo; so long as the offspring has no independent existence, but participates in that of its parent, it is called a germ. The separation of the germ is called generation.

Gervi′llia. A fossil genus of bivalve shells; they are subequivalve, oblong, oblique, the hinge long, straight, having small irregular, transverse, ligamentary pits, and likewise a series of parallel internal ribs. The general form resembles avicula, the umbo having an auricle on each side. Five species are known as belonging to the carboniferous, seven to the jurassic, and five to the cretaceous strata of the British Isles.—*Lycett.*

Gey′ser. The name given to certain boiling springs or fountains in Iceland. The water of these geysers holds a considerable proportion of silex in solution. "These intermittent hot springs occur in a district situated in the south-western division of Iceland, where nearly one hundred of them are said to break out within a circle of two miles. They rise through a thick current of lava which may, perhaps, have flowed from Mount Hecla, the summit of that volcano being seen from the spot at a distance of more than thirty miles. Few of the geysers play longer than five or six minutes at a time, and the intervals between their eruptions are, for the most part, very irregular. The great geyser rises out of a spacious basin at the summit of a circular mound, composed of siliceous incrustations deposited from the spray of its waters. The diameter of this basin is fifty-six feet in one direction by forty-six in another. In the centre is a pipe seventy-eight feet in perpendicular depth, and from eight to ten feet in diameter, but gradually widening as it rises into the basin. The **circular basin is** sometimes empty, **but is** usually filled with beautifully transparent water in a state of ebullition. During the **rise of the** boiling water in the **pipe,** especially when the ebullition **is** most violent, and when the water is thrown up in jets, subterranean noises are heard, like the distant firing of cannon, and the earth is slightly shaken. The sound then increases, and **the** motion becomes more violent, **till** at length a column of water **is** thrown up, with loud explosions, to the height of one or two hundred feet. After playing for **a time** like an artificial fountain, and giving off clouds of vapour, the pipe or tube is emptied, and a column of steam rushing up with amazing force and a thundering noise, terminates the eruption. If stones are thrown into the crater, they are instantly ejected, and such is **the explosive force,** that very hard rocks are **sometimes** shivered by **it** into small pieces,"—*Lyell.*

Sir George Mackenzie thus describes an eruption of the Great Geyser, "The eruption commenced with a sound resembling the distant discharge of a piece of ordnance; the sound was repeated irregularly and rapidly, and I had just given the alarm to my companions, (Drs. Bright and Holland) who were at a little distance, when the water, after heaving several times, suddenly rose in a large column, accompanied by clouds of steam, from the middle of the basin, to the height of ten or twelve feet. The column seemed as if it burst, and sinking down it produced a wave, which caused the water to overflow the basin in considerable quantity. After the first propulsion, **the water was** thrown up again **to the height of** about fifteen feet. There was now a succession of jets to the number of **eighteen,** none of which appeared **to exceed** fifty feet in height; they lasted **about** five minutes. Though the wind blew strongly, yet the clouds of vapour were so dense, that after the first two jets I could only see the highest part of the spray, and some of it that was occasionally thrown out sideway. After the last jet, which was the most furious, the water suddenly left the basin, **and sunk into** the pipe **in the** centre. **The temperature of the water in the** basin was ascertained to be 209° Fahrenheit." In a subsequent eruption, witnessed by Sir George, there was a succession of magnificent jets, the highest of which was at least ninety feet.

Dr. Black gives the following analysis of the waters of the geysers of Iceland. Soda 5·56, alumina 2·80, silica 31·50, muriate of soda 14·42, sulphate of soda 8·57

The siliceous deposits from the geysers are well known. Sir G. Mackenzie describes the leaves **of** certain **trees** converted into stone, every **fibre** being discernable. Grasses, rushes, and peat are in every state of petrifaction.

Gi′bbous. *(gibbus,* Lat. *gibbeaux,* Fr. *gibbóso,* It.) Bossed; convex; bunched. In botany, **applied to** fleshy leaves having **one or both** sides convex, arising from the great abundance of pulp.

Gi′bbsite. A mineral of a dirty white colour, found in America, and named after Mr. Gibbs. Its analysis gives alumina 64·8, water 34·7. Specific gravity = 2·4.

Gill. The lung, or respiratory organ of the fish. The gills, or branchiæ, lie in openings on each side of the head; their form is semicircular;

they have a vast number of fibrillæ standing out on each side of them like a fringe, and very much resemble the vane of a feather. There are, in most fishes, four gills on each side, resting on an equal number of arched portions of cartilage or bone, connected with the os hyoides. In some cartilaginous fishes there are five gills on each side; in the lamprey there are seven. The larger Crustacea have their branchiæ situated **on the under** side of their body, not **only** in order to obtain protection from the carapace, which is folded over them, but also for the sake of being attached to the haunches of the feet, jaws, and thoracic feet, and thus participating in the movements of those organs. They may be seen in the lobster and in the crab, by raising the lower edge of the carapace.

In the greater number of mollusca these important organs, although external with respect to the viscera, are **within** the shell, and are generally situated near its outer margin. They are composed of parallel filaments, arranged like the teeth of a fine comb; and an opening exists in the mouth for admitting the water which is to act upon them. These filaments appear, in many instances, to have the power of producing **currents of water** in their vicinity by **the action of minute** cilia, similar to those belonging to the tentacula of many polypi, where the same phenomenon is observable. In the Acephala, or bivalve mollusca, the gills are spread out, in the form of laminæ, round the margin of the shell, as is exemplified in the oyster where it is commonly known by the name of *beard*. The aërated water is admitted through a fissure in the **mouth,** and when it has performed **its office** in respiration, is usually **expelled by** a separate opening.

All the sepiæ have their gills **enclosed in two lateral cavities,** which communicate with a funnel-shaped opening in the middle of the neck, alternately receiving and expelling the water by the muscular action of its sides. The forms assumed by the respiratory organs in this class are almost infinitely diversified. In fishes the gills form large organs, and the continuance of their action is more essential to life than it appears to be in any of the inferior classes. When their surfaces are minutely examined, they are found to be covered with innumerable minute processes, crowded together like the pile of velvet; and on these are distributed myriads of blood-vessels, spread like a delicate net-work, over every part of their surface. A large flap, termed the operculum, extends over the whole organ, defending it from injury, and leaving below **a** wide fissure for the **escape of the** water, which has **performed its** office in respiration. **For this** purpose the water is taken in by the mouth, and forced by the muscles of the throat through the apertures which lead to the branchial cavities; in this action the branchial arteries are brought forward and separated to a certain distance from each other, and the rush of water through them unfolds, and separates, each of the thousand minute filaments of **the** branchiæ, so that they all receive **the full** action of that fluid as it passes by them. When a fish is taken out of the water, the animal vainly reiterates its utmost efforts to raise the branchiæ, and relieve the sense of suffocation it experiences in consequence of the general collapse of the filaments of those organs, which adhere together in a mass, and can no longer receive the vivifying influence of oxygen. "It has been generally stated," says Dr. Roget, "by physiologists, even of the highest authority, that the principal reason why fishes cannot maintain life, when surrounded by air instead of water, is that the branchiæ become dry, and lose the power of acting when thus deprived of their natural moisture. The rectification of this error is due to Flourens, who pointed out the true cause."

GI'NGLYMOID. (from γιγγλυμὸς and εἶδος, Gr.) Resembling a hinge; pertaining to that species of hinge-like joint which admits of flexion and extension.

GI'RASOLE. (from *gyro*, to turn, and *sol*, the sun, Lat.) The name given to a variety of opal. The silex girasol of Brongniart, and quartz resinite girasol of Haüy. The *girasole* is of a milk-white colour, but it possesses a remarkable property of reflecting a red colour when turned towards the sun, or any bright light. From this peculiar property it obtained its name. Girasole is sometimes strongly translucent, and the finest specimens resemble translucid jelly.

GLA'CIER. (*glacier*, Fr. *Amas de montagnes de glace, qui se trouvent en quelques endroits de la Suisse de la Savoie et du Dauphiné, au sommet des montagnes.*) Great accumulations or extensive fields of ice, common in mountainous countries. The presence of glaciers depends on the accumulation of a large mass of snow, subject to variations of temperature sufficient partially to thaw and then to reconsolidate the mass in its downward course. They have been compared, aptly enough, to gigantic icicles. At Mont Blanc, the glacier of Montanvert is said to descend 1,200 feet below the summit of the mountain. "I was much struck," says Mr. Darwin, "by observing the great difference between the matter brought down by torrents and by glaciers: in the former case a spit of gravel is formed, but in the latter a pile of boulders." Glaciers are instruments of the degradation of land, inasmuch as they drive before them and transport such substances as they may have power to move. In front of glaciers there is generally a pile of rubbish, composed of pieces of rock, earth, and trees, which they have forced forward, known in Switzerland by the name of *moraine*. If there be a line of moraine some distance from the front of the glacier, it is considered that the glacier has retreated to the amount of that distance; but if there be no other than that which the glacier immediately drives before it, it is considered to be on the increase.

GLA'CIAL DEPOSITS. These consist of clays, sands, and gravels, sometimes stratified, sometimes rudely piled together, and containing great blocks of rock. These deposits are generally known as "drift," and "erratic block groups."

GLANCE. (*glantz*, Germ.)
1. A name given to some minerals which possess a metallic, or pseudo-metallic lustre.
2. An order of minerals, containing eight genera.—1. Copper-glance; 2. Silver-glance; 3. Lead-glance; 4. Tellurium-glance; 5. Molybdena-glance; 6. Bismuth-glance; 7. Antimony-glance; 8. Melane-glance.

GLANCE-COAL. (*glanzkohle*, Germ.) A variety of coal, known also as anthracite. This is the glanzkohle of Werner, the glance-coal of Jameson, the native mineral carbon of Kirwan, and the blind-coal of some authors. There are several varieties of the glance-coal, namely, *conchoidal glance-coal*, or that having a conchoidal fracture and splendent lustre; *slaty glance-coal*, or that with a slaty structure; *columnar glance-coal*, and *fibrous glance-coal*. This combustible, at first view, strongly resembles coal, from which, however, it materially differs. Its colour is black, or rather grayish and iron-black, sometimes tinged with blue or brown. It, perhaps, never possesses the pure deep black of coal. Glance-coal, like the diamond, appears to be essentially composed of pure carbon, but in a very different state of aggregation. The glance-coal of Kilkenny contains about 97 per cent. of carbon; that of Rhode Island about 94 or 95. It occurs in beds in the coal formation, in the secondary class of rocks; it is occasionally found among rocks of the primary and transition series. It is sometimes nearly allied to graphite. It may be distinguished from coal by the difficulty with which it burns, by its greater specific gravity, and by its composition: it differs from graphite in being less heavy; its trace on paper is dull and blackish, whereas that of graphite is a shining metallic gray; and graphite is unctuous to the feel, whereas glance-coal is not.

GLAND. (*glande*, Fr. *glandula*, It.)
1. Bodies employed to form or alter the different liquids in the animal body. There are two distinct sets of glands, the conglobate, and the conglomerate. Great variety is observable both in the form and structure of different glands, and in the mode in which their blood-vessels are distributed. In some glands, the minute arteries suddenly divide into a great number of smaller branches, like the fibres of a camel-hair pencil; this is called the pencillated structure. Sometimes, the minute branches, instead of proceeding parallel to each other, after their division, separate like rays from a centre, presenting a stellated arrangement. In the greater number of instances, the smaller arteries take a tortuous course, and are sometimes coiled into spirals. It is only by means of microscopic aid that these minute structures can be rendered visible.

2. In botany, a small transparent tumour or vesicle, discharging a fluid, either oily or watery, and situated on various parts of plants, as the stalk, calyx, leaves, &c. These glands are composed of closely compacted cells, which perform the functions of secretion, or the conversion of the nutritious juices into particular products required for various purposes in the economy of the plant. The perfume of the flowers and leaves of plants arises from secretions from glands.

GLANDI′FEROUS. (from *glandifer*, Lat.) Bearing acorns, or fruit resembling acorns.

GLA′NDULAR. } (*glanduleux*, Fr. *glan-*
GLA′NDULOUS. } *dulóso*, It. *glandulosus*, Lat.)
1. Pertaining to glands; containing glands; full of glands.
2. In botany, applied to the margins of leaves having glands.

GLA′NDULOUS GNEISS. A variety of gneiss, (in which the mica is sometimes arranged in undulated layers,) presenting numerous small masses of felspar or quartz, of a globular or elliptical form, interspersed like glands through the mass. From this circumstance it has obtained its name.

GLASSY PUMICE. The Glasiger Bimstein of Werner and Harsten. A subspecies of pumice, for a description see *Pumice*.

GLAU′BERITE. An hydrous sulphate of soda and lime. A mineral of a white or yellow colour; crystallised in oblique four-sided prisms; consisting of 51 parts sulphate of soda, and 49 parts sulphate of lime. It is less hard than carbonate of lime, but scratches sulphate of lime. It is found in New Castile, in Spain, disseminated in muriate of soda.

GLA'UCONITE. Called also firestone. The Glauconie crayeuse of the French geologists. This comprises the chalk-marl, without any intermixture of green particles, as well as the limestone, called in Sussex malm-rock, and the sands and arenaceous marls and limestones full of green particles of silicate of iron, and termed the *upper green sand* and *firestone.—Mantell.*

GLAUCON'OME. A genus of fossil corals. Under this name Goldfuss has described five fossils, four of which, according to De Blainville and Milne Edwards, belong to the genus Vincularia, previously established by De France. The fifth species, common at Dudley, possesses, however, characters essentially different from those of Vincularia, and even from those assigned by Goldfuss. Instead of the stem being impressed on *all* sides with rows of cells, it has them only over *half* the surface, the other half being striated longitudinally. It is probable, adds Sir R. Murchison, that the position of the fossil in the matrix prevented that author from detecting the true characters of the coral.

GLAU'COUS. (*glaucus*, Lat. azure, γλαυκὸς, Gr.)
1. Of a sea-green colour; grey or blue; azure.
2. In botany, applied to the leaves or stems of plants, when covered with a fine mealiness of a sea-green colour.

GLI'MMER. (Germ.) The name given by Werner to mica.

GLO'BULUS. A genus of fossil univalve shells, belonging to the family of the Naticæ; as their name implies, they are remarkable for their globose figure and minute spire. Eight species have been procured from the English tertiary formations, and three from the carboniferous limestone and shale.—*Lycett.*

GLOM'ERIS. (from *glomero*, Lat. to gather into a round heap.) A myriapod, resembling a wood-louse, which, like the armadillo, when alarmed, rolls itself up into a spherical ball.

GLOSSOPTE'RIA. The name proposed by Brongniart for a genus of fossil ferns, whose elongated leaves or fronds are covered by fine arched dichotomous, often anastomosing, nervures. Specimens are met with in the carboniferous and oolitic strata.

GLO'TTIS. (from γλῶττα, Gr.) The upper opening of the larynx, at the root of the tongue.

GLUCI'NA. (from γλυκὺς, Gr. sweet.) An earth obtainable from the emerald, beryl, and euclase, of all which it forms a constituent part. Sir H. Davy discovered that glucina consisted of three parts glucinum and one part oxygen. Glucina is soluble in the liquid fixed alkalies, in which respect it agrees with alumina. It is insoluble in ammonia, but soluble in carbonate of ammonia. It combines with all the acids, and forms with them sweetish salts, from which circumstance it obtained its name. It was first discovered by Vauquelin in 1798.

GLUCINE. (from γλυκὺς, sweet, Gr.) An earth composed of oxygen 30, and glucinum 70. When pure it is a white powder, soft and somewhat unctuous; specific gravity nearly 3.

GLUCI'NUM. The metal which is the base of the earth glucina; it has not yet been obtained in a separate state.

GLUME. (*gluma*, Lat.) The husk of corn; the chaff; the outer husk of corn and grasses; the calyx of corn and grasses. In the grasses, and plants resembling them, the floral envelopes are not called calyx and corolla, but bracteæ. The two outer bracteæ are termed *glumes*.

GLUMI'FERÆ. The first order of the fourth class, Endogens, comprising the grasses, &c.

GLUTI'NOUS HAG. Called also the ramper eel, or poisonous ramper. The name assigned to a species of suctorii. "The suctorii, one of which, the glutinous hag, has scarcely any brain." This creature was regarded by Linnæus as belonging to the class vermes.

GLYCE'MERIS. ⎫ A transverse bivalve,
GLYCI'MERIS. ⎭ gaping at both extremities; hinge callous, without tooth. Fossil and recent.

GLYPTOLE'PIS. (from γλυπτὸς, carved, and λεπίς, scale, Gr.) A genus of ichthyolites found in the lower old red sandstone. The characteristic parts of the creature are its scales, these are of great size compared with that of the animal. One specimen, not more than half a foot in length, exhibits scales three-eights of an inch in diameter. In another specimen there are scales a full inch across, and yet the length of the ichthyolite to which they belonged appears not to have exceeded a foot and a half. Each scale consists of a double plate, an inner and an outer. The structure of the inner is not peculiar to the family or the formation; it is formed of a number of minute concentric circles, crossed by still minuter radiating lines. The outer plate presents an appearance less common. It seems relieved into ridges that drop adown it like *sculptured* threads, some of them entire, some broken, some straight, some slightly waved; and hence the name given to this ichthyolite. There are several species.—*Hugh Miller.*

GLYPTO'POMUS. An ichthyolite of the old red sandstone, known also as Platygnathus.

GNEISS. The name given by the German mineralogists to a schistose or slaty granite, abounding in mica. It is a member of the metamorphic rocks. By some geologists, gneiss has been called secondary granite. Granite frequently may be observed passing by scarcely perceptible gradations into gneiss: when granite contains but little felspar, and the proportion of mica is increased, the mica being arranged in layers, it becomes schistose, and we find a true gneiss. Again, when the mica becomes very abundant, gneiss passes into mica-slate. Gneiss occurs in Ireland and Scotland; but it is rarely found in England or Wales. It is most abundant in Sweden. Gneiss is composed of the same ingredients with granite, namely, felspar, quartz, and mica, its texture being equally crystalline. According to the Huttonian theory, the materials composing gneiss were originally deposited from water, but from the influence of subterranean heat, became altered, so as to assume a new texture. The structure of gneiss is always more or less distinctly slaty, when viewed in the mass, although individual layers may possess a granular structure. When this mineral is broken perpendicularly to the direction of its strata, its fracture has commonly a striped aspect. This rock, though slaty in its structure, is rarely perfectly fissile. Gneiss, like granite, never contains any fossil remains; when it occurs with granite, it usually lies immediately over the granite; or, if the strata be highly inclined, it appears rather to rest against the granite than to be incumbent upon it. Gneiss is more or less distinctly stratified, and the strata are often inclined to the horizon at a very great angle; indeed, they are sometimes nearly, or quite, vertical. Mountains composed of gneiss are seldom so steep as those of granite, and their summits, instead of presenting those needle-like points, or aiguilles, which characterize granite mountains, are usually rounded. It has been remarked, with much truth, that abruptly conical hills are characteristic of the formation which Humbolt designates as gneiss-granite. Few of the primary rocks are so metalliferous as gneiss. Its ores occur both in beds and veins: more frequently in the latter.

GOLD. (Sax.) A metal, when pure, of a rich yellow colour: specific gravity 19·3. It does not readily combine with oxygen; hence it does not rust when exposed to the air, and it may be melted and re-melted frequently with scarcely any diminution of its quantity. It is said to have been kept in a state of fusion for nearly eight months without undergoing any perceptible change. In ductility and malleability it surpasses all other metals, and it may be beaten into leaves so exceedingly thin, that one grain of gold shall cover fifty-six square inches, such leaves having the thickness only of one 282,000th part of an inch. Its tenacity is inferior to that of iron, copper, platinum, and silver. Gold is soluble in nitro-muriatic acid, and in a solution of chlorine. The gold coins of this country contain one-twelfth part of copper alloy: jeweller's gold is a mixture of gold and copper, in the proportions of three-fourths of pure gold and one-fourth of copper.

GONI'ATITES. A sub-genus of ammonites, in which the last whorl covers the spire. Seven species have been found in the carboniferous system, and seventeen in the primary strata.

GONIO'METER. (from γωνία, an angle, and μέτρον, a measure, Gr.) An instrument invented by M. Carangeau, for the accurate measurement of crystals. Dr. Wollaston also invented a *goniometer* called the *reflecting goniometer*; this is a very superior instrument. There are two kinds of goniometers, the plain and the reflecting. Of the latter Mr. Phillips says, "in almost every instance in which I have tried it upon the planes produced by good fractures, the success has been complete. The great value of this instrument, which may be used with readiness and ease, demands the attention of every one who has the slightest pretension to crytallographical research. The reflecting goniometer is of great use to the geologist, as well as to the mineralogist. He finds those rocks which are termed primary, and many of those which are called transition, to consist, not of one homogenous mass, but of two or more minerals, so intermixed and associated that a reference to the chemist is of little avail to him: by such means he may indeed become informed whether a particular earth or alkali is to be found in the mass, but the various substances of which it is compounded are often too minute, and therefore too intimately associated with the others, to allow of a determination as to which of the component substances may contain the earth or the alkali so discovered. Hence structure, if it exist, becomes a character of essential importance, for it will be found that fragments far too minute for analysis will often afford brilliant planes, well adapted to the use of the reflecting goniometer.

The surfaces produced by cleavage are sometimes very small, and therefore are not adapted to the common goniometer; while for the reflecting goniometer, it matters not if the surface be small, provided it be perfect and brilliant; a surface of the 100th part of an inch in length and breath will suffice.

GONYLE'PTES. A genus of the second

order of Pseudorachnidans, the posterior legs exhibiting a raptorious character.

GORGO'NIA. A genus of Ceratophyta, of the family Corticati, class Polypi. These animals have a horny skeleton, are carnivorous, feeding upon living animalcules. The polypi of several species have been observed, and they are found to possess eight denticulated arms, a stomach, &c. To a species found fossil in the Wenlock limestone, Mr. Lonsdale has assigned the specific name *assimilis*; he says, "it is impossible to determine if this fossil be a true Gorgonia, but from its great resemblance to the axis of some existing species, I have ventured to place it in that genus." Mr. Lonsdale thus describes G. Assimilis:—axis round, branched, striated longitudinally, branches anastomosed, no projecting papilla, or impressed pores.—*Murchison's Silurian System.*

GO'SSAN. In mining, a technical term for an ochreous substance which, from its appearance, may be termed an argillaceous oxyd of iron. The presence of *gossan* is considered by the miner to be an indication of neighbouring riches, and is denominated *kindly*, or *very kindly*, in proportion to the darkness of its hue, and the loosenesss of its texture.

GRA'LLÆ. } An order of aquatic
GRA'LLATORES. } birds. Waders; frequenting marshes and streams; having long naked legs; long neck; cylindrical bills. In this order are included the crane, stork, heron, bittern, &c. &c.

GRA'MINA. (*gramen*, Lat.) The fourth order in Linnæus's division; the grasses.

GRAMINI'VOROUS. (from *gramen*, grass, and *voro*, to devour, Lat.) Subsisting wholly on grass or vegetable food. Animals which subsist wholly on vegetable food are called graminivorous, while those which live on flesh alone are called carnivorous; those feeding on both are called omnivorous; while those feeding solely on fishes are denominated piscivorous.

GRA'MMATITE. (The name given by Brongniart to Tremolin.) A mineral, a variety of hornblende, confined almost entirely to primary rocks. Colours white and blue. Disposed in fibrous, radiated, and granular coneretions.

GRA'NATINE. A name given by Mr. Kirwan to a granular aggregate containing three ingredients, but those differing from the ingredients of granite. A compound embracing two ingredients only, he termed a *granitell*; when three ingredients are present, but not the three forming granite, he called it a *granatine*; when more than three ingredients form the compound, he termed it a *granilite*.

GRA'NATITE. } The Prismatoidischer
GRE'NATITE. } granat of Mohs, Granatit of Werner, Staurotide of Haüy. A reddish-brown mineral, occurring in primary rocks in the Shetland Isles, and in many parts of Scotland, aud in America. According to Vauquelin, it consists of alumina 45, silica 33, oxide of iron 13, oxide of manganese 4, and lime 4. Its form and infusibility distinguish it from the garnet. See *Staurotide.*

GRANI'FEROUS. (from *granum* and *fero*, Lat.) Pods which bear seeds like grains.

GRA'NILITE. An aggregate containing more than three constituent parts; thus named by Mr. Kirwan.

GRA'NITE. (*granito*, It. *granit, ou granite*, Fr. *Pierre fort dure, qui est composée d'un assemblage d'autres pierres de différentes couleurs.*) An aggregate of felspar, quartz, and mica, whatever may be the size or figure of the several ingredients, or their relative proportions, is denominated granite. Felspar is generally the predominating, as mica is the least considerable ingredient of the rock. In some varieties the quartz is wanting; in others the mica; these varieties have received particular names. There are many varieties of granite; as porphyritic granite, in which large crystals of felspar occur; sienitic granite, in which hornblende supplies the place of mica; **chloritic, or** talcy granite, composed of quartz, felspar, and talc or chlorite, instead of mica; felspathic granite, &c. &c. Granite is a compound plutonic or **igneous rock**, unstratified and crystalline, of a granular structure, whence its name. From its great relative depth, granite is rarely met with but in mountainous situations, where it appears to have been forced through the more superficial covering. "It was at one time supposed that granite was peculiar to the lowest portions of the rocks composing the crust of the earth, and that, in fact, it constituted the fundamental rock upon which all others had been formed, and was not discovered higher in the series. This opinion has given **way** before facts, for we find granitic **rocks** in situations where they must have been ejected subsequently to the period during which the cretaceous group was deposited, as also in other places, into which they must have been thrust at intermediate periods **down to the oldest rocks** inclusive." Granite is said to contain forty-eight per cent. of oxygen. Granite being **an** igneous rock, no organic fossil remains could be expected to be found therein, nor have any ever been discovered; nevertheless granite is occasionally found overlying strata containing fossil organic remains, as in Norway; a mass of granite has been discovered superincumbent on secondary limestone, which contains orthocerata, &c. From these circumstances **there can no** longer exist a doubt but that granite has been formed at different **periods**, and is of various ages. A comparatively modern granite may be observed in the Alps penetrating **secondary** strata, such secondary strata containing fossils, such as belemnites, referrible to the age of the English lias. Although granite has often pierced through other strata, it has seldom, or ever, been found to rest upon them, as if it had overflowed. Granite almost invariably constitutes the central ridges of mountain chains, occupying the highest and the lowest position in their stratification. That it has been **forced up by some violent convulsions of** nature, which **have** shaken our planet to its very centre, is beyond doubt. The indented ridges, the ragged precipices, the bristling peaks, by which these chains are characterised, prove to demonstration the violence exerted in their production. In this respect they present a marked contrast to those more convex mountains, and undulating ridges of hills, whose mass was quietly deposited by the last retiring sea, and has subsequently remained undisturbed by any violent revolution. Felspar is by far the largest constituent of granite, and in some kinds it is found in large whitish crystals of irregular forms, occasionally of one or two inches in length. Granite of this kind, however beautiful it may be **to the** eye, is **not well** adapted for buildings, the felspar being subject to decomposition from the continued action of the atmosphere. Waterloo-bridge is unfortunately built of this perishable kind of granite. It may be considered as a general law, that wherever granite rises to any height above the surface of the earth, the strata of other surrounding rocks rise towards it. The highest point at which granite has been discovered in any part of the world is Mont Blanc, 15,683 feet above the level of the ocean. Saussure, who has published an account of his ascent of Mont Blanc, infers from his observations that the verticals beds of granite were originally horizontal and have been upheaved by some violent convulsions of nature, and he states that what now forms the summit of the mountain must at some former period have been more than two leagues below the surface.

GRA'NITEL.  ⎫  A name given by
GRA'NITELL. ⎭  Kirwan to a binary aggregate composed of any two of the following ingredients: felspar, mica, shorl, quartz, garnet, steatite, hornblende, jade.

GRANI'TIC AGGREGATE. A granular compound, consisting of two, three, or four simple minerals, among which only *one* of the essential ingredients of granite is present. Among the granitic aggregates, which contain only one of the essential ingredients of granite, may be enumerated combinations of quartz and hornblende,—quartz and actinolite,—felspar and schorl, —mica and hornblende,—quartz, hornblende, epidote, &c. &c.

GRANITI'FEROUS. All rocks admit of a twofold division, namely, into the primary or granitiferous, and the secondary or derivative classes. The granitiferous **rocks** comprise various series of massive and schistose rocks, which are not fragmentary, nor fossiliferous, and which are always inferior, in their original position, to those of the secondary class.—*Boase.*

GRA'NITINE. An aggregate of three mineral constituents, one or more differing from those which compose granite. For example, an aggregate of quartz, felspar, and shorl is a granitine, as is one of quartz, mica, and shorl; or quartz, hornblende and garnet; and many others.

GRANI'VOROUS. (from *granum* and *voro*, Lat.) Eating grain; subsisting on grain.

GRA'NULAR LIMESTONE. A subspecies of carbonate of lime, the result of a confused or irregular crystallization. Structure foliated and granular. The grains are of various sizes, from coarse to very fine, sometimes, indeed, so fine that the mass appears almost compact. When these grains are white and of a moderate size, this mineral strongly resembles white sugar in solid masses. Its fracture is foliated, and when the structure is very finely granular, the fracture often becomes a little splintery. It is more or less translucent, but in the dark-coloured varieties, at the edges only. Its colour is most commonly white or gray, often snow-white, and sometimes grayish black. Some varieties are flexible when sawn into thin slabs. Granular limestone is sometimes a pure carbonate of lime. It occurs in very large masses, and is almost exclusively found in primary rocks; sometimes it occurs among secondary, but then its relative age is easily determined by the shells it contains, or the accompanying minerals. In the Pyrenees vertical beds of *granular limestone* alternate with granite and trap, or the limestone is sometimes intermixed with those rocks.

There are few countries in which *granular* limestone is not found. Italy and Greece furnished the ancients with valuable quarries. Both granular and compact limestone furnish numerous varieties of marbles, but those which belong to the former exhibit a more uniform colour, are generally susceptible of a higher polish, and are consequenty most esteemed for statuary and other purposes.

GRANULA'TION. *(granulation,* Fr.)
1. The operation by which metals are reduced into small grains.
2. The act of forming into bodies resembling aggregates of grains.

GRA'NULITE. A crystalline aggregate of quartz and felspar, in which the quartz occurs in thin flakes.

GRA'PHIC GRA'NITE. Called also Pegmatite. A variety of granite, composed of felspar and quartz, so arranged as to produce an imperfect laminar structure. When a section of graphic granite is made at right angles to the alternations of the constituent minerals, broken lines, resembling Hebrew characters, present themselves; hence its derivation.

GRA'PHITE. Another name for black-lead, or plumbago; carburet of iron. Graphite is of a dark steel-gray, or nearly iron-black. It leaves on paper a well defined, shining trace, which has very nearly the colour of the mass, and consists of minute grains. It is perfectly opaque, easily scraped by a knife, and soils the fingers. It is a conductor of electricity, and when rubbed on sealing-wax till a metallic trace appears, communicates no electricity to the wax. Specific gravity from 1·98 to 2·26. Constituent parts, carbon 92, iron 8.

GRA'PTOLITE. *(graptolithus,* Linnæus.) Graptolites form a genus of the family of sea-pens. "These pen-like, serrated, fossils have," says Sir R. Murchison, "a great vertical range in the older or Protozoic rocks, being found from the lower part of the Ludlow formation, down to very ancient beds in the Cambrian system." "Very different opinions," says the Danish naturalist, Dr. Beck, "have been entertained as to the place graptolites hold in the series of living beings, but that of Prof. Nilsson may come nearest to the truth, who conceives the graptolite to be a polyparium of the ceratophydian family. Yet I am more inclined to regard them as belonging to the group Pennatulinæ, the Linnæan Virgularia being the nearest form in the present state of nature to which they may be compared. I am now acquainted with six or seven species of graptolites, all occurring in the oldest fossiliferous strata, where they are associated with Trilobites, Orthoceratites, &c. This genus was established by Linnæus. The graptolite of the most ancient fossiliferous rocks occurs in greatest abundance in a finely levigated mudstone, for it too was a dweller in the mud."— *Hugh Miller.* A fossil zoophyte, found in the Silurian shales.

GRAVITA'TION. *(gravitation,* Fr. *gravitazione,* It.) The difference between gravity and the centrifugal force induced by the velocity of rotation or revolution: the force which causes substances to fall to the surface of the earth, and which retains the celestial bodies in their orbits; its intensity increases as the squares of the distance decrease. "Gravitation," says Mrs Somerville, " not only binds satellites to their planets, and planets to the sun, but it connects sun with sun throughout the wide extent of creation, and is the cause of the disturbances, as well as of the order, of nature; since every tremor it excites in any one planet is immediately transmitted to the farthest limits of our system, in oscillations, which correspond in their periods with the cause producing them."

GRA'VITY. *(gravité,* Fr. *gravitá,* It.) The reciprocal attraction of matter on matter. The force of gravity is everywhere perpendicular to the surface, and in direct proportion to the quantity of matter.

GRA'YWACKE. ⎫ (from *grauwacké,*
GRAU'WACKE. ⎬ Germ. a com-
GRAUWACKE'. ⎪ pound of *grau,*
GRE'YWACKE. ⎭ grey, and *wacké,*

a provincial term used by miners.) The name given to a group of rocks, being the lowest members of the secondary strata. Sir C. Lyell comprises in this group the Ludlow, Wenlock and Dudley, Horderly and May Hill rocks, the Builth and Landeilo flags, and the Longmynd rocks. The French have changed the name *grauwacke* for *traumate*, a word as little euphonic as the one repudiated. Sir R. Murchison says " It would appear that the Silurian loved to dwell amid the relics of the old greywacke of the Scottish region, as well as along the Welsh border; and thus I rejoice in having substituted a pleasing name, full of glorious British recollections, for the foreign term, Grauwacke." The grauwacke group may be regarded as a mass of sandstones, slates, and conglomerates, in which limestones are occasionally developed. Sandstones which mineralogically resemble the old red sandstone of the English, not only occupy the upper part, but frequently also other situations in the series. In the lower portions of the grauwacke group, stratified compounds, resembling some of the unstratified rocks, are by no means unfrequent. In speaking of "grauwacke" Sir R. Murchison says " a name which, until recently, comprehended every rock from the roofing slates to the beds immediately beneath the old red sandstone. It may indeed be said that grauwacke was at that time considered the limit, on reaching which all stratigraphical and geological definition ceased. This word should cease to be used in *geological* nomenclature, and it is also *mineralogically* valueless, because rocks undistinguishable from the so called grauwacke occur both in the old red sandstone and in the coal measures." Sir R. Murchison adds "I might, I believe, assert that rock specimens, which many mineralogists would term greywacke, may be found in every stage of the geological series, even in the tertiary deposits." Mr. Bakewell observes, " Graywacke, in its most common form, may be described as a coarse slate containing particles or fragments of other rocks or minerals, varying in size from two or more inches to the smallest grain." When the imbedded particles become extremely minute, graywacke passes into common slate. When the particles and fragments are numerous, and the slate in which they are cemented can scarcely be perceived, graywacke becomes coarse sandstone, or gritstone. When the fragments are larger and angular, graywacke might be described as a breccia with a paste of slate. When the fragments are rounded it might not improperly be called an ancient conglomerate. The old red sandstone is a graywack, coloured red by the accidental admixture of oxide of iron; it possesses all the mineral characters, and occupies the geological position, of graywacke. The rock, though composed of substances of various colours, usually exhibits some shade of gray or brown; it is sometimes of considerable hardness, and susceptible of a high polish. Graywacke is often distinctly stratified, but the strata are not usually parallel to those of the subjacent rocks. The common and slaty varieties often alternate with each other, and both are traversed by veins of quartz. This rock is remarkably metalliferous; and its ores occur both in beds and veins. Most of the mines of the Hartz are contained in greywacke.

Of the fossils of the greywacke, zoophytes and crinoidea are the most numerous. The trilobite is characteristic of this era, and the orthocera, of which certain species are also found in the carboniferous limestone, but in no deposit more recent. Some of the shells belong to more recent genera, as the terebratula. The only vertebrated remains hitherto found are a few bones of fishes.

The most abundant shells of the grauwacke belong to Orthoceras, Productæ, Spirifer, and Terebratula. Some orthoceratites are found of large size, of a yard or more in length. Productæ are common in the grauwacke as well as in the carboniferous group, they are also found in the zechstein or magnesian limestone, above which they disappear. Spirifers are found as high up as the lias, and consequently continued to exist to a later period than Productæ. Terebratulæ, however, have survived all the great changes which have occurred, and are to be found amongst existing genera.

Graywacke abounds in Germany and in Scotland; indeed, nearly all the mountains of Scotland north of the Frith of Forth are chiefly composed of it. In the neighbourhood of Mont Blanc, and in other parts of the Alps, it occurs at a great elevation, forming large masses in vertical beds.

GRA'YWACKE SLATE. A variety of graywacke, in which the grains are so minute as to be scarcely perceptible by the naked eye.

GREEN-EARTH. The Grün Erde of Werner; the Talc Zographique of Haüy; the Chlorite Baldogée of Brongniart. A variety of talc, occurring in vesicular cavities in amygdaloid. Its colour is a pleasant green, more or less deep, sometimes bluish or grayish-green, and passing to olive and blackish-green. Its fracture is dull, and fine-grained earthy, or slighty conchoidal. It is somewhat unctuous to the touch, and adheres to the tongue. Easily reducible to powder. Specific gravity 2·63.—*Kirwan.*

According to Vauquelin it consists of silex 52, magnesia 6, oxide of iron 23·4, alumina 7, potash 7·4, water 4. It is met with in the mountainous districts of England and Scotland. It is the mountain-green of artists; and, when ground with oil, is employed as a paint.

GREEN-SAND. *(glauconie sableuse,* Fr. *grünsand,* Germ.) A member of the chalk formation, called also Shanklin sand. The beds of sand, sandstone, and limestone, which form the lowermost strata of the chalk formation, have obtained the name of green-sand, from the circumstance of their containing a considerable quantity of chlorite, or green earth, scattered throughout their substance. "The colouring matter," says Dr. Turner, "of green-sand sometimes appears in the rock of its ordinary green tint, and sometimes in grains of so deep a green that they seem black. The former generally occurs in sand, or where the sandstone is porous, and in this state an ochreous appearance is often observed, due to the green particles being partially decomposed and their iron having passed into a higher state of oxidation; whereas the black looking grains are met with in a highly calcareous sandstone, where the texture is too firm to admit of the percolation of water. From either kind of rock the green matter may be obtained by washing with water and subsidence, since the colouring matter subsides less readily than grains of quartz, and more readily than calcareous and argillaceous substances. On reducing the samples obtained to powder, washing away the finer particles with pure water, and separating any adhering carbonates by dilute muriatic acid, the colouring matter is left, mixed only with small grains of quartz. It then always appears in the form of earthy particles of a deep green tint. The green matter, when not previously weathered, is very feebly attacked by concentrated acids. It gives out water when heated, and becomes brown from its iron passing into the state of **peroxide**. As it has been supposed to owe its green colour to the presence of phosphoric acid, it **was** carefully examined, with the **view** of detecting that acid, if **present**. The result proved that lime and phosphoric acid are not essential constituents of the colouring matter of green-sand, and their presence must be regarded as casual. The green particles have been found to consist of silica 48·5, black oxide of iron 22, alumina 17, magnesia 3·8, water 7, and potash some traces: M. Berthier gives of potash 10, as found in some green particles from near Havre." In describing the group of deposits to which the name of green-sand, or Shanklin-sand, is appropriated, geologists state that they admit of a triple division; the first, or uppermost, consists of sand, with irregular concretions of limestone and chert, sometimes disposed in courses oblique to the general direction of the strata. The second consists chiefly of sand, but in some places is so mixed with clay, or with oxide of iron, as to retain water. The third, and lowest group, abounds much more in stone; the concretional beds being closer together and more nearly continuous. The total thickness of the green-sand, where it is fully developed, is more than 400 feet. The animal remains of the green-sand are exclusively marine. The **fossils** principally hitherto discovered in the green-sand group are, Inoceramus, Cucullæa, Pecten, Vermetus, Solarium, Mya, Nucula, Natica, Nautilus, Dentalium, Belemnites, Ammonites, Hamites, Thetis, Siphonia, Corbula, Sphæra, Trigonia, Diceras, Modiola, Avicula, Gervillia, Terebratula, Limo, Litorina, Rostellaria, Fusus, &c. The French have denominated this for**mation** glauconie crayeuse, **and** craie chloritée. It is very common to divide the greensand into the upper green-sand and the lower green-sand, the two being separated by the gault.

GREENSTONE. The Grünstein of Werner; Roche Amphibolique of Haüy; the diabase of the French geologists. A granular rock composed of hornblende and felspar, in the state of grains, or sometimes of small crystals. Greenstone contains a larger quantity of felspar than basalt, and the grains both of hornblende and felspar are less amalgamated. It is a variety of trap rock. The hornblende usually predominates, and frequently gives to this aggregate a greenish hue, from which circumstance it obtains its name. Greenstone occurs in beds of considerable magnitude, and sometimes forms whole mountains. It often appears in conical hills, or presents high, mural precipices, whose fronts are frequently composed of numerous columns of various sizes, resembling basalt. Sometimes it forms only the summits of mountains. Small veins of actinolite, epidote, felspar, prehnite, quartz, &c., &c., frequently are found traversing greenstone. When greenstone is decomposed it produces a reddish brown soil. The presence of green-stone, even when concealed from view, is oftentimes declared by the reddish brown colour of the soil, the tint varying **in** depth according to the proportion of the hornblende in the rock.

GREISS-STONE. A rock composed of quartz and mica in a peculiar state of aggregation. Greiss-stone occurs in Cornwall.

GRE′NATITE. Prismatoidal garnet. See *Granitite* and *Staurotide*.

GRÈS DE VOSGES. "A very extensive deposit," says De la Beche, "occurs in the Vosges, and has thence obtained its name. A difference of opinion seems to exist between M. Élie de Beaumont and M. Voltz respecting the exact member of the red sandstone series to which this rock should be referred; the former considering it the equivalent of the Rothe Todte Liegende, which occurs beneath the zechstein; the latter, that it is the lower portion of the red or variegated sandstone, which rests on the zechstein. It is essentially composed of amorphous grains of quartz, commonly covered by a thin coating of red peroxide of iron; among which are discovered others which appear to be fragments of felspar crystals. The rock contains quartz pebbles, sometimes so abundantly as to present a conglomerate with an arenaceous cement. It is also often marked by cross and diagonal laminæ.

GRÈS BIGARRÉ. (from two French words, signifying a stone formed of grains variously coloured.) The name given by French geologists to that member of the red sandstone group known as the new red sandstone; the bunter sandtein of the Germans.

GREY MARL. (The Craie Tufeau of French geologists.) One of the members of the chalk formation, placed above the firestone or upper green sand and below the chalk without flints. Where, as is sometimes the case, the firestone is absent, the grey marl is found reposing immediately upon the gault. In Sussex, says Dr. Mantell, this deposit constitutes the foundation of the chalk hills, its outcrop forming a fillet or zone round their base, and connecting the detached parts of the range with each other.

The marl is commonly soft and pliable, but indurated blocks occur which possess the hardness of limestone. It is of a light grey colour, inclining to brown, and frequently possesses a ferruginous tinge, derived from oxide of iron. It consists principally of carbonate of lime and alumine, with an intermixture of silica, a very small proportion of iron, and perhaps of oxide of manganese.

The mineralogical productions, says the same talented author, in his Geology of the South-east of England, of the grey marl are few, and offer but little variety: they consist of various modifications of sulphuret of iron, and crystallized carbonate of lime. In organic remains, the grey marl is very rich, and these differ both in their nature, and in the mode of their preservation, from those of the lower chalk above, and of the gault beneath. Ammonites, hamites, nautilites, turrilites, scaphites, pectenites, madreporites, inocerami, rostellariæ, and auriculæ, together with the teeth and vertebræ of sharks, are found in different parts of this deposit. No parts of England are supposed to be so rich in the various species of turrilites as the marl pits in the neighbourhood of Lewes. A specimen of turrilites tuberculata nearly two feet long, the only instance in which traces of a siphunculus are visible, was discovered in the grey marl at Middleham, near Lewes.

GRE′YSTONE. A rock of greyish or greenish colour, composed of felspar and augite, the former being more than 75 per cent. of the whole.

GREY-WEATHERS. The name given to large boulders of siliceous sandstone. There is a singular assemblage of these erratic blocks in a field on the borders of Wiltshire, not far from Marlborough. The immense blocks forming, as is supposed, the

Druidical temple at Stonehenge, are composed of this siliceous sandstone. "Boulders of druid sandstone," observes Dr. Mantell, "also occur in the shingle bed, and calcareous deposit at Brighton, and may be observed lying on the sea-shore in considerable numbers, after a recent fall of the cliff. Upon comparing the sandstone of Stonehenge with that of Sussex, no perceptible difference can be detected.

GRIT. The provincial term for a coarse siliceous sandstone. Some of the strata of this description have been worked for mill-stones, from which circumstance they have been called mill-stone grit. The mill-stone grit is an important deposit in the north of England, from the Coquet to the Tyne, and on the hills between the dales of Durham and York, from the Tyne to the Ribble.

GRO'SSULAR. (from *groseille*, Fr. a gooseberry, thus named from its gooseberry green colour.) A very rare mineral. The asparagus green variety of dodecahedral garnet. It is found in Siberia. Its constituents are, silica 40·50, alumina 20·10, lime 33·80, oxide of iron 5·00, oxide of manganese 0·50.

GRO'WAN. 1. A Cornish name for a soil formed of disintegrated granite. The growan occupies a very considerable area of the peninsula of Cornwall, constituting no less than three hundred thousand acres. The fertility of the growan soil varies greatly, and is in proportion to the amount of felspar in the subjacent granite. The hard growan varies extremely in its hardness, but in respect to its composition, it is perfectly like the soft growan. Some of the hard growan after being exposed for a few months to the weather, has dissolved completely into a soft growan. The soft growan is nowhere hard enough to support itself. The hard growan is jointed in all directions.

2. Those veins that are called lodes by the miners are divided into two classes, those formed of granite being called *growan*, and those of porphyry *elvan*.

GRUM. The name given by the miners to a dull dark red and green impure concretionary limestone, in parts ferruginous, each geode being enveloped in red shale. It occurs in the coal measures, and sometimes attains a thickness of twenty feet. The productus hemisphæricus and another species are most abundant in it, together with the corals Lithostrotion floriforme, Syringopora reticulata, and Lithodendron irregulare.—*Sir R. Murchison.*

GRYPHÆ'A. (from *gryps*, Lat. a griffin.) An inequivalved bivalve; the lower valve concave, terminated by a beak, and curving upwards and inwards; the upper valve much smaller, like an operculum; the hinge toothless, the pit oblong and arched: one impression in each valve. From the curved beak of the shell, Linnæus placed it among the anomiæ; but Lamarck placed it under a distinct genus.

GRY'PHITE. (*gryphites*, Lat.) A fossil bivalve belonging to the genus gryphæa. This deeply-incurved bivalve is so abundant in some of the beds of lias in France, as to have occasioned them to be called Calcaire à gryphites. These shells are known in this country by the provincial name of "miller's thumbs."

GRY'PHITE GRIT. A local term assigned by Sir R. Murchison to a stratum of the inferior oolite of Gloucestershire: it is thus named from the prevalence in it of the Cryphæa Cymbium.

GUAYACAN'ITE. A newly discovered mineral from the Cordilleras of Chili, consisting of copper, arsenic, and sulphur. Hardness 3·5 to 4. Specific gravity 4·39.

GYMNOSPE′RMIA. (from γύμνος, *nudus*, naked, and σπέρμα, *semen*, seed.) The first order in the fourteenth class, Didynamia, in Linnæus's artificial system; having four naked seeds in the bottom of the calyx, with the exception of one genus, Phryma, which is monospermous.

GYNA′NDRIA. (from γυνή, a woman, and ανήρ, Gr. a man.) The name of the 20th class in Linnæus's sexual system, consisting of plants with hermaphrodite flowers, having the stamens growing upon the style, or having the stamina united with, or growing out of, the pistil, and either proceeding from the germen or the style. The orders of this class are taken from the number of the pistils, but botanists are not agreed as to the admission of some of them into the class.

GYNA′NDROUS. A term applied to a particular class of plants, in which the stamens and pistils are united. The orders of this class depend on the number of the anthers.

GY′PSUM. (γύψος, Gr. *gypsum*, Lat. *gypse*, Fr. *gesso*, It.) The chaux sulfatée of Brongniart and Haüy. Sulphate of lime; it is composed of sulphuric acid 46, lime 33, water 21. It is insoluble in acids, and does not effervesce like chalk and dolomite, the lime being already combined with sulphuric acid, for which it has a stronger affinity than for the other acids. It possesses double refraction. There is one variety known by the name of anhydrite, or anhydrous gypsum, which contains no water. Sulphate of lime is an abundant mineral salt, presenting itself under various forms, crystallized or otherwise. From gypsum is obtained plaster of Paris, the gypsum being burnt in a kiln, and the water thereby driven off. The varieties of gypsum are numerous; the crystallized transparent varieties are known as selenite; the fibrous and earthy as gypsum; and the granular and massive as alabaster. The salt mines of this country afford examples of nearly all the varieties. The white powder obtained by exposing gypsum to a strong heat has obtained the name of plaster of Paris, from the extreme abundance of this mineral in the neighbourhood of that capital. Its inferior hardness, together with its chemical characters, will serve to distinguish it from the carbonate, fluate, and phosphate of lime. "The gypsum formation consists of alternating beds of gypsum and argillaceous and calcareous marl, which are regularly arranged, and preserve the same order of succession wherever they have been examined. The gypsum forms three distinct masses. The lowest consists of thin strata of gypsum containing crystals of selenite, which alternate with strata of solid calcareous marl, and with argillaceous shale. The middle is like the lowest mass, except that the strata of gypsum are thicker, and the beds of marl are not so numerous; it is chiefly in this mass that fossil fish are found. The uppermost mass is the most remarkable and important of all; it is in some parts more than seventy feet thick; there are but few beds of marl in it; the lower strata of gypsum in this mass have a columnar structure. In this upper mass of gypsum the skeletons and scattered bones of birds and unknown quadrupeds are discovered. Remains of turtles and crocodiles have been also found in the same strata." For a further and more interesting detail, see Bakewell's Introduction to Geology.

GYROCA′RPUS. In botany, a genus of plants containing few species, but these widely distributed.

GYRO′DUS. (from γύρος, round, and οδούς, a tooth, Gr.) A genus of fossil fishes, established by Agassiz, of the family of Pycnodonts, or thick-toothed fishes, found in the oolite of Durrheim, in Baden.

GYRO'GONITE. (from γύρος, *curvus*, and γένος, genus.) Petrified seed-vessels of the Chara. These bodies are found in fresh-water deposits, and were, not very long since, supposed to be microscopic shells, indeed they are thus described by Parkinson, who concludes his notice of them by stating, "Lamarck observes, that it has the form of a very small seed of some species of lucern; and hesitating at determining it to be really a multilocular shell, only assumes it as such for the present." See *Chara*.

# H

HACKL'Y. A term used in mineralogy to designate a fracture with a peculiarly uneven surface, similar for instance to that of pure copper. This term, says Mr. Allan, relates to a fracture which is peculiar to the malleable metals.

HADE. A term used by miners; to dip.

HADE. ⎫ The dip from the perpendicular line of descent;
HA'DING. ⎭ the angle of inclination of a vein.

HÆ'MATITE. ⎫ (αἱματίτην, Gr. *hæma-*
HE'MATITE. ⎭ *tites lapis*, Lat.) Blood-stone, an iron ore; it occurs in masses of various shapes, both globular and stalactitic.

HALF-MOON SHAPED. A figure formed by the portion of a circle cut off by the segment of a larger circle.

HALINI'DA. An order, the lowest, of the class Poriphera, and thus denominated from the composition of the skeleton, which consists of minute *silicious* crystalline spicula.

HALIO'TIS. (from ἅλα, *mare*, and ὠτὸς, *auris*, Gr.) The sea-ear. A shell, both fossil and recent, obtaining its name from the excessive amplitude of its aperture, and the flatness of its spire, whence it has been likened to an ear. The shells of this genus are said to be among the rarest fossils. The recent shells are littoral, and found adhering to rocks; they are very beautiful, and are remarkable for the pearly irridescence of the inner surface, and for the shell being perforated along the side of the columella by a series of holes; they are amongst the most highly ornamented of all the gasteropoda. The sea-ears protect their open side by fixing themselves to the rocks, and preserve a communication with the atmosphere, or water, without elevating their shells, by means of a line of apertures, under the thickest margin, near the apex: these apertures begin, when the animal is young, near the spire, and as it grows it stops up one and opens another, as its occasions require. "I have," says the Rev. W. Kirby, "a very large specimen, in which there are traces of eighteen apertures, and all but six are stopped up." The soft parts of the inhabitant of this shell are eaten in some places, and are esteemed delicious.

HA'LITHERIUM. The name assigned to an extinct genus of mammalia of the Eocene period.

HA'LOIDE. (from ἅλς, salt, and εἶδος, Gr. form or appearance.) An order of earthy and metalliferous minerals; tasteless; specific gravity from 2.2 to 3.3.

HA'LLOSITE. An earthy mineral of a white colour with a slightly bluish tinge, occurring with ores of lead, zinc, and iron. Thus named by Berthier in honour of M. Omalius d'Halloy.

HALT'ERES. (ἁλτῆρες, Gr. *halteres*, Lat.) The poisers, so named from their supposed use in balancing the body, or adjusting with exactness the centre of gravity when the insect is flying. In those insects which compose the order *Diptera*, we meet with two organs, consisting of cylindrical filaments, terminating in a clubbed extremity; one arising from each side of the thorax, in the situation in which the second pair of wings originate in those insects that have four wings; these are called *halteres*. Whatever may be their real utility, they may still be regarded as rudiments of a second pair of wings.

HA'MITE. (from *hamus*, Lat. a hook.) A genus of fossil multilocular hook-formed shells. Parkinson states that the hamite has no evident siphunculus, but this is a mistake; the siphuncle of the hamite, like that of the ammonite, is placed on the back, or outer margin of the shell, and in some species this marginal siphuncle has a keel-shaped pipe raised over it. The external shell is fortified by transverse folds or ribs, which serve to strengthen both the outer and the air chambers. The hamite is sometimes found of large size, more especially that species known as Hamites grandis; some of them are of the diameter of a man's wrist.

HAPLACA'NTHUS. The name assigned to a fossil fish of the old red sandstone, one species of which has been described by Agassiz, namely, H. Marginalis.

HARDNESS. In mineralogy, one of the physical or external characters used in determining minerals. To Prof. Mohs we are indebted for a scale easily formed, and at the same time distinct and accurate. It is as follows:—

1. Talc, of a white or greenish colour.
2. Rock-salt, a pure cleavable variety; or gypsum uncrystallized, and only semi-translucent.
3. Calcareous spar, any cleavable variety.
4. Fluor spar, presenting good cleavage.
5. Apatite, the asparagus stone from Saltzburgh.
6. Adularia, any perfectly cleavable variety.
7. Rock crystal, limpid and transparent.
8. Topaz, any simple variety.
9. Corundum stone from Bengal, which affords a smooth surface when fractured.
10. The diamond.

In employing this scale, we endeavour to find the degree of hardness of a given mineral by trying which number of the series is scratched by it; or, still better, by passing with the least possible force the specimens under comparison over a very fine file. From the resistance these bodies afford to the file, from the noise occasioned by their passing over it, and from the quantity of powder left on its surface, their mutual relations in respect to hardness are deducible with great correctness. When, after repeated trials, we are satisfied which member of the scale the mineral is most closely allied to, we say its hardness is equal to seven, (suppose it to be rock crystal) and write after it H. $= 7.0$. If, however, the mineral under examination do not exactly correspond with any one member of the scale, but is found to be between two of them, we say H. $= 7.5$, or $7.75$ if it approximate to the next higher number. The file should be cut fine, and of the hardest steel.

HAR'MOTOME. (from ἁρμός, a joint, and τέμνω, to divide.) The Kreutzstein of Werner; Cross-stone of Jameson; Pierre cruciform of Brochant; Staurolite of Kirwan. See *Cross-stone.*

HA'RPA. A genus of shells placed by Cuvier in the family Buccinoida, order Pectinibranchiata, class Gasteropoda. A beautiful genus of shells, distinguishable from all others by the regular longitudinal ribs that mark the external surface, in some degree resembling a stringed instrument, from which the name is derived. The genus is both fossil and recent; the shells are marine, and are inhabitants of warm climates.

HA'RPAX. (Parkinson.) A proposed fossil genus of bivalve shells founded upon the well known Plicatula spinosa of the Lower Lias. Harpax has long since by universal consent been merged in Plicatula; recently however, (1858) M. Eudes Deslongchamps has proposed to re-establish Harpax, in an elaborate memoir, entitled, "Essai sur les Plicatules fossiles des terrains du Calvados," wherein he has figured and described many species of Harpax from the Jurassic rocks of Calvados; compared with Plicatula the distinctions resides chiefly in the hinge, and in the figure of the borders of the fossa which receive the cardinal teeth.—*Lycett.*

HASTINGS BEDS. } The name assigned by Dr.
HASTINGS SANDS. } Fitton to the central group of the Wealden formation, from their great development in the neighbourhood of Hastings. The Hastings beds comprise the Horsted sand; the Tilgate Forest strata; and the Worth sandstone. These consist of numerous strata of sands and sandstones, often ferruginous, and occasionally intermingled with shales. The organic remains found in these beds are not numerous as regards species; they are, however, characterized by containing bones of the Iguanodon. The Hastings sands attain a thickness in some places of 400 feet and upwards. These sandy clays, says Dr. Fitton, have, in general, great variety of composition and colour; being in some places almost totally composed of sand,—in others of clay, or Fuller's earth, frequently mottled with various shades of tea-green, and dark purplish red. The Hastings sands consist throughout of beds of sand, calciferous grit, clay and shale, with argillaceous iron ore, and limestone abounding in shells: and of these, clay intermixed with sand forms so very large a proportion, that the name of the stratum, might, perhaps, with equal propriety, have been taken from the former substance. Dr. Fitton gives the following as the order of the beds of the Hastings sands in Sussex:—

1. Ferruginous and fawn-coloured sands, and sand-rock, including small linear portions of lignite, with stiff grey loam.
2. Sand-rock, with concretional courses of calciferous grit.
3. Dark coloured shale, ten to twelve feet thick.
4. The white sand-rock of the Hastings cliffs; about one hundred feet thick.
5. Clay, shale, thin beds of sandstone; lignite, and silicified wood.
6. Sand-rock, without concretions; dividing naturally into rhomboidal masses; numerous veins of argillaceous iron ore and of clay, approaching to pipe-clay at the lower part.
7. Dark coloured shale, with roundish masses of sand-rock, and several layers of rich ironstone, thin layers of lignite, and innumerable fragments of carbonized vegetables.

HA'TCHETINE. A variety of bitumen, known also as mineral adipocere, found in the iron ore of Merthyr Tydvil in Glamorganshire. It is insoluble in water, but soluble in alcohol and ether. It fuses at 160°. It is of the hardness of soft tallow.

HAUSMA'NNITE. } Pyramidal manga-
HAUSSMA'NNITE. } nese-ore. The Manganese oxydé hydraté of Haüy. It occurs in porphyry, in veins, in America and Germany. It is of a brownish-black colour. It consists, according to Turner, of 98·10 of red oxide of maganese, silica 0·34, oxygen 0·21, baryta 0·11, water 0·43.

HAUYNE. Dodecahedral Zeolite, or Lapis Lazuli.

HEA'VY SPAR. Sulphate of barytes, barosclenite, or prismatic heavy spar. The Baryte sulphatée of Haüy; the Schwer spath of Werner; and Prismatischer halbaryte of Mohs. There are several varieties of this genus, namely, the compact heavy spar, having a splintery and uneven fracture; the fibrous heavy spar; the straight and curved lamellar heavy spar; the radiated heavy spar; the fetid heavy spar, giving out, on friction a hepatic odour, whence it is also called hepatite; the earthy heavy spar; the prismatic heavy spar.

Heavy spar consists of baryta 66 parts and sulphuric acid 34 parts. It frequently contains a trace of silex, alumine, oxide of iron, and sometimes of sulphate of strontian. It occurs in veins, both massive and crystallized, in many parts of England, Ireland, Scotland, and Germany, being found in primary, transition, and secondary rocks. It is of different colours. It strongly decrepitates when heated, and fuses into a white enamel, which in the course of some hours falls into powder. One of the most striking characters of this mineral is its great specific gravity, which varies from 4·29 to 4·50. It is from this circumstance it has obtained its name. It is harder than crystallized carbonate of lime, but may be scratched by fluate of lime. Heavy spar may be confounded with sulphate of strontian, but its specific gravity is greater. After fusion, the enamel produced from heavy spar, if applied to the tongue, produces a taste similar to that of rotten eggs, this does not occur in the enamel of sulphate of strontian.

HEDENBER'GITE. An earthy mineral so named after Hedenberg, who first analysed it; it is a variety of hornblende, and is found near Tunaberg, in Sweden.

HE'LICAL. (*hélice*, Fr. ἕλιξ, Gr.) Spiral; winding.

HE'LIOLITES. A genus of fossils found in the Silurian rocks.

HE'LIANTHOIDA. Just as we may see the Helianthoida and Ascidioida of our seas fixed to their boulders and rocky skerries.—*Hugh Miller.*

HE'LIOTROPE. (*héliotrope*, Fr *eliotropia*, It. *heliotropium*, Lat. ἡλιοτρόπιον, Gr. from ἥλιος and τρέπω.)

1. A plant, the turnsole.

2. Heliotrop of Werner. Quartz-agathe vert obscur et ponctué of Haüy. Called also Blood-stone. A variety of rhombohedral quartz, of a deep green colour, with disseminated spots of yellow and red jasper. It is more or less translucent, in which respect it differs from jasper. It occurs massive. By some mineralogists it is supposed to be calcedony, coloured by chlorite or green earth. Fracture imperfectly conchoidal. Specific gravity about 2·63. It consists of silica with about seven per cent. of alumina and five per cent. of iron. It is infusible before the blow-pipe. The finest specimens are brought from Siberia and Bucharia. Like agate, it is employed in forming

ornamental articles, and is much esteemed.

HE´LIX. (ἕλιξ, Gr.) Pl. Helices.
1. The outer bar, or margin, of the external ear.
2. The snail. A globular or orbicular shell; spire short, convex or conoidal, last whorl ventricose; opening entire, being wider than long; no operculum. The helix aspera, or common snail, is a well-known illustration. Parkinson observes of the fossil helix, "Shells of this genus are rarely found in a state of petrifaction. The circumstances of conservation in which they are found are, generally, such as are explicable on the supposition of their having become involved in the gradually accreting tufaceous matter, which is deposited by certain streams and rivers; or in the stalactitic concretions forming in the cavities of limestone rocks, of comparatively modern formation."
3. A screw, a spiral.

HELMI´NTHOLITE. The name given to what have been considered fossil impressions of earth-worms.

HELMINTHO´IDA. The name given to the second sub-kingdom of the animal kingdom, called also Diploneura. This sub-kingdom comprehends the classes Polygastrica, Rotifera, Suctoria, Cirrhopoda, and Annelida.

HE´LVINE. Tetrahedral garnet.

HEMI´PTERA. (from ἥμι and πτέρον, Gr. So called, because their wing-covers at the base are of a substance resembling horn or leather, and at the tip are membranous.) The eighth order of the class Insecta. These insects have four wings, either stretched straight out, or resting across each other; the superior are coriaceous at their base, with a membranous apex. The mouth of hemipterous insects is adapted for extracting fluids by suction only.

HE´MISPHERE. (hémisphère, Fr. emis-foro, It. hemisphærium, Lat. ἡμισφαίριον, Gr.) The half of a globe when it is supposed to be cut through its centre, in the plane of one of its greatest circles; one-half of the globe, or sphere, when divided into two by a plane passing through its centre. The equator divides the sphere into two equal parts, called the northern and southern hemispheres. The horizon also divides the sphere into two parts, the upper and lower hemispheres.

HENDE´CAGON. (from ἕνδεκα, and γωνία, Gr.) A figure of eleven faces or angles.

HEPA´TIC.      } (hepaticus, Lat. hépati-
HEPA´TICAL.  } que, Fr. epático, It.) Belonging to the liver; pertaining to the liver; resembling liver in form or colour.

HEPA´TIC CINNABAR. A dark-coloured, steel-grey variety of the mercuro sulphuret of Haüy, or cinnabar.

HEPA´TICÆ. The first order of plants in the class Anogens, comprising the liver-worts.

HEPA´TIC PYRI´TES. Hepatic sulphuret of iron. A variety of prismatic iron-pyrites, of a yellow colour, which, on exposure to the atmosphere, acquires a brown tarnish. This embraces those varieties of sulphuret of iron, which are susceptible of a peculiar decomposition, by which the sulphur is more or less disengaged. During this process the pyrites is converted, wholly or in part, into a compact oxide of iron of a liver-brown colour, from which circumstance it obtains its name. The decomposition begins at the surface, and gradually passes into the centre.

HE´PATITE. A mineral; a variety of heavy spar, or sulphate of barytes. This variety is distinguished by its emitting a fetid smell when rubbed, resembling that of sulphuretted hydrogen, arising from its containing a portion of sulphur.

HE´PATULE. A name given by Kirwan

to certain combinations, called by others hydrosulphurets.

**HEPTACA'PSULAR.** (from ἑπτά, Gr. and *capsula*, Lat.) Having seven cells.

**HE'PTAGON.** (*heptagone et eptagone*, Fr. *ettagono*, It. ἑπτά and γωνία, Gr.) A figure having seven sides and as many angles.

**HERBA'CEOUS.** (*herbaceus*, Lat.) Plants that perish annually down to the roots, having succulent stems or stalks. Of herbaceous plants, some are annual, these perish, stem and root, every year; some are biennial, the roots subsisting two years; others are perennial, being perpetuated for many years by their roots, a new stem springing up every year.

**HERMA'PHRODITE.** (from Ἑρμῆς, Mercury, and Ἀφροδίτη, Venus, *hermaphrodite*, Fr. *ermaphrodito*, It. A fabulous name, compounded of those of Mercury and Venus, Hermes and Aphrodité.)
1. Having both the male and female parts of generation.
2. In botany, plants are so called which contain both stamens and ovarium, in contra-distinction to monœcious and diœcious plants.

**HERMETIC SEAL.** (from Hermes, the reputed inventor of chemistry.) When the neck of a glass vessel or tube is heated to the melting point and then twisted with pincers till it become air-tight, the vessel or tube is said to be hermetically sealed, or to have received the seal of Hermes.

**HERPETO'LOGY.** (ἑρπετόν and λόγος, Gr.) That branch of natural history which treats of reptiles, their habits, &c.

**HE'RSCHELITE.** A mineral of a white colour, found by Herschel in olivine, and named after its discoverer.

**HETEROCERCAL.** Professor Jukes says, Fish have two forms of tail, the one *homocercal*, that is in which the caudal fin is equally spread round the termination of the vertebral column, as in the cod, perch, &c.; and the other *heterocercal*, where the vertebral column is, as it were, continued into the upper lobe of the caudal fin, the extremity of the body in reality being slightly bent up, as in the dog-fish, shark, and sturgeon. It is a very remarkable fact, that in *all* fish, whether osseous or cartilaginous, found in the rocks more ancient than the lias, the tails are *heterocercal*, while in the lias and more recent rocks, the majority of fish have *homocercal* tails.

**HETERO'CLITAL.** In conchology, reversed; a term applied to shells whose spires turn in a contrary direction to the usual way; sinistral. In shells, the axis of revolution is termed the *columella*, and the turns of the spiral are denominated *whorls*. In consequence of the situation of the heart and great blood-vessels relatively to the shell, the left side, or the mantle, is more active than the right side, so that the lateral turns are made in the contrary direction. Sometimes, in consequence of the heart being placed on the right side the turns of the spiral are made to the left: this left-handed convolution seldom occurs among the shells of land or fresh-water mollusca.

**HETERO'PORA.** The name given by De Blainville to a genus of corals. A species established by Mr. Lonsdale, and by him named H. crassa, found in the Wenlock limestone, is thus described in Sir R. Murchison's Silurian System; "branched, the branches thick, tubes small, radiating from a centre; transverse fracture; concentric layers formed of two systems of tubes, one visible to the naked eye, the other microscopic, numerous and close together.

**HETERO'STROPHE.** (ἑτερόστροφα, Gr. ex ἕτερον et στροφή.) A term applied to reversed shells, or shells whose spires turn in a contrary direction to the usual way; sinistral shells. Synonymous with Heterocrital.

HEU'LANDITE. Hemi-prismatic zeolite, or foliated zeolite. A mineral thus named after Heuland. It occurs in drusy cavities, in secondary trap rocks, in the Hartz mountains, in Ireland, and in Scotland. It is of different colours, white, grey, brown, and red. Its constituent parts are silica 59, alumina 16·87, potash 8, (or, according to others, lime 9,) water 16·5. Specific gravity 2·20. Hardness 3·5 to 4. It commonly is found crystallised in right oblique angled prisms (two of its opposed lateral planes being longer than the other two) generally modified on the angles and one lateral edge. It also occurs massive, frequently in a globular form, and is easily frangible.

HE'XAGON. (*héxagone*, Fr. *esàgono*, It. *ἑξάγωνος*, Gr. from *ἕξ* six, and *γωνία*, angle.) A figure of six sides or angles; the most capacious of all the figures that can be added to each other without any interstice. The cells of the honeycomb are hexagons.

HE'XAGYN. (from *ἕξ*, six, and *γυνή* Gr. a woman.) A plant having six pistils.

HEXAGY'NIAN. Having six pistils.

HEXAHE'DRAL. Having six sides equal to one another; cubal.

HEXAHE'DRON. (from *ἕξ* and *ἕδρα*, Gr. *héxaèdre*, Fr. *esaedra*, It.) A cube; a solid body having six equal faces.

HEXA'NDER. (from *ἕξ*, six, and *ἀνήρ*, Gr. a man.) As plants which have six pistils are denominated hexagyns, so those with six stamens are termed hexanders.

HEX'ANDRIAN. Having six stamens; belonging to the sixth class in Linnæus's sexual method.

HEXA'NGULAR. from *ἕξ* and *angulus*.) Having six angles.

HE'XAPED. } *ἑξοπόδης*, (from *ἕξ* and
HE'XAPOD. } *πούς*, Gr.) An animal with six feet.

HEXAPE'TALOUS. (from *ἕξ*, six, and *πέταλον*, Gr. a leaf.) A plant having six petals; a corolla consisting of six petals.

HEXAPHY'LLOUS. (from *ἕξ* and *φύλλον*, Gr.) Six-leaved.

HIG'HTEA. The name assigned by Mr. Bowerbank to a genus of fossil fruits found in the London clay, and thus named in honour of an able botanist, John Hight, Esq. The following is Mr. Bowerbank's description, "pericarp one celled, valveless. Placenta central, usually five, rarely four, or six angled, obconical and very large, with one or more seeds attached to each angle. Seeds and placenta enveloped in a mass of downy or filamentous structure, which fills up the whole of the remaining space within the pericarp. Seed about two or three times its own diameter in length, angular and somewhat curved. Testa reticulated. Ten species have been distinguished by Mr. Bowerbank, and are beautifully figured in his "History of the Fossil Fruits of the London Clay."

HI'LUM. (*hilum*, Lat.) In botany, the small mark on seeds showing the spot where they were joined to the fruit.

HINGE. In conchology, the point by which bivalve shells are united: it is formed by the teeth of the one valve inserting themselves between those of the other, or by the teeth of one valve fitting into the cavities or sockets of the opposite valve.

It is on the peculiar construction of the *hinge* that the generic character of bivalve shells is mainly founded, in connection with the general form of the shell. Some *hinges* have no visible teeth, and are termed *inarticulate*; *hinges* with few teeth are termed *articulate*; those having many teeth are called *multiarticulate*.

HINNI'TES. A genus of eared irregular bivalves, separated from the pectens by the irregularity of surface. Several species are known in the Jurassic, cretaceous, and tertiary deposits.—*Lycett.*

HIPPOPO'TAMUS. (ἱπποπόταμος, Gr. *hippopotame*, Fr. *ippopótamo*, It.) Bochart conceived this animal to be the Behemoth of the book of Job. One existing species only is at present known, but Cuvier considered he had discovered four in the fossil state. The first so nearly resembles the species now existing that the Baron could at first scarcely distinguish them. The second is about the size of the wild boar; the third was intermediate between the two, and the fourth about the size of a guinea-pig. The hippopotamus, or river-horse, belongs to the order Pachydermata, or thick-skinned animals, class Mammalia. This genus of quadrupeds has four fore-teeth in the upper jaw, disposed in pairs at a distance from each other; and four prominent fore-teeth in the under jaw, the intermediate ones being longest. There are two tusks in each jaw, those of the under being long, and obliquely truncated; in both they stand solitary, and are recurvated. No other animal has tusks formed like the hippopotamus. Those of the elephant are larger, but neither angular nor striated; those of the morse are very much striated towards the root, but not angular; the tusk of the narwhal is straight, but twisted spirally by the striæ of the surface. In the hippopotamus there are five striæ, concentrical to the contour of the tooth. The feet are hoofed on the edges. The head is of an enormous size, and the mouth vastly wide. The ears are small and pointed. The eyes and nostrils are small. The body is naked. The tail is about a foot long, taper, compressed and naked. The legs are short and thick; the belly reaching to the ground. One species only is known, the hippopotamus amphibius, confined to the rivers of Africa.

Fossil bones and teeth of the hippopotamus are abundantly found in England, France, Germany, and Italy; several extinct species have been determined by Cuvier. The head of a hippopotamus was found in Lancashire, under the peat; bones and teeth are found in large quantities in the Val d'Arno, in Italy, and in alluvial deposites in the neighbourhood of Rome.

HI'PPOPUS. (from ἵππος, a horse, and πούς, a foot.) A subtransverse, equivalve, inequilateral shell; lunule closed with crenulated edges; the hinge formed of two long compressed entering teeth in one valve, and three in the other; the crescent closed. Parkinson states that he does not know of any fossil shells of this genus having been found, but Sowerby mentions fossil species.

HIPPOTHE'RIUM. (from ἵππος, and Θηρίον, Gr.) An extinct animal, allied to the horse, belonging to the miocene period.

HIPPOPO'DIUM. A fossil genus of thick, inequilateral, equivalve, bivalve shells; the hinge is much incrassated, with one rugged oblique tooth. One species H. ponderosum, is found in the lower lias. —*Lycett.*

HI'PPURITE. A fossil conical shell, having an under shell of great depth, with a flat lid or operculum. This genus is believed to be wholly extinct. The operculum is sometimes convex, but generally it is concave. The particular economy of the inhabitant of this shell is not known. Lamarck has placed the hippurite among his chambered cephalopoda.

HISI'NGERITE. A mineral of a dark colour, occurring massive, found in the cavities of calcareous spar.

HOG-TOOTH SPAR. A dodecahedral

variety of calcareous spar. The Matastatique of Haüy.

HOLOPTY'CHIUS. A genus of fossil fishes, discovered in the old red sandstone formation. The name has been assigned to ichthyolites of this genus by M. Agassiz from the folds of the scales. A species has been named H. Nobilissimus, in honour of the Rev. James Noble, who discovered a splendid specimen, measuring two feet four inches by twelve inches, in the old red sandstone at Clashbinnie, in Scotland.

"The generic characters," says M. Agassiz, "consist in the peculiar structure of the scales, the enamelled surface of which is marked by large undulating furrows. Another characteristic feature is in the distant position of the ventral fins, far removed towards the tail, and much nearer the arms and anal fins than in any other genus of the family of ganoids."—*Murchison's Silurian System*.

The Holoptychius is the most characteristic and most abundant ichthyolite of the upper old red sandstone. Agassiz states that the Holoptychius of the coal measures is the largest of all osseous fishes. —*Hugh Miller*.

HOLOTHU'RIA. A zoophyte belonging to the order Pedicellata, class Echinodermata. The holothuria is covered with a thick coriaceous skin, which, by means of longitudinal and circular bands of muscular fibres, the animal can shorten or lengthen at pleasure. The body is oblong, and open at each end; numerous tentacula surround the mouth. There are many species.

HO'LMITE. A mineral, a variety of carbonate of lime, named after Mr. Holme, who analysed it. Its constituents are lime, carbonic acid, oxide of iron, silica, alumina, and water.

HOMACA'NTHUS. A genus of ichthyolites of the old red sandstone, one species of which, H. Arcuatus, has been described by Agassiz.

HO'MALONOTUS. The name assigned by Mr. Konig to a genus of trilobites, in which the longitudinal division into lobes is scarcely perceptible. Four species have been described, namely, H. Knightii, H. Ludensis, H. Delphinocephalus, and H. Herschelii. These are all found in the Upper Silurian rocks.

HO'MOCERCAL. See *Heterocercal*.

HOMOGENE'ITY. (*homogénéité*, Fr. *omogeneità*, It.) Of the same nature; having the same nature throughout.

HOMOGE'NEAL. } (ὁμογενής, Gr. *homogène*, Fr. *omogeneo*, It.) Similitude of kind; sameness of nature.
HOMOGE'NEOUS. }

HOMO'LOGOUS. (*homologue*, Fr. *omologo*, It. ὁμόλογος, Gr.) Having the same manner or proportions.

HOMOTHO'RAX. A genus of fossil fishes of the old red sandstone.

HONE. (*hœn*, Sax.) Whetstone slate. A variety of talcy slate, containing particles of quartz: when these particles are exceedingly minute, and the slate possesses a certain degree of hardness with a uniform consistence, it yields hones of the best quality. Kirwan gave to this mineral the name novaculite, from *novacula*, the Latin for a razor.

HO'NEY-STONE. Pyramidal mellite. The Honigstein of Werner; Mellite of Haüy and Brongniart; and Mellilite of Kirwan. It is pyramidal, its primitive form being a pyramid of $118°\ 4'$ and $93°\ 22'$. It was first discovered in Thuringia, between the layers of wood-coal. It is of a honey-yellow colour, sometimes a little tinged with brown. Transparency considerable. When heated it whitens, and in the open air burns without being sensibly charred. It yields neither flame, smoke, nor odour. It is composed of alumina, water, and an acid to which Klaproth gave the name of mellitic acid: the latter constitutes 46 per cent. of the whole. It differs from amber in its weak electricity, double refraction, and chemical character.

Ho'rnblende. The amphibole of Haüy. A mineral of a black or dark green colour, often intermixed: heavier than either quartz or felspar, but less hard, its specific gravity being between 3·15 to 3·38. It enters largely into the composition, and forms a constituent part of several of the trap rocks, and appears to connect the primary with the volcanic. When breathed on, it yields a peculiarly bitter smell. Before the blow-pipe it melts easily into a black or grayish-black, glass. There are many varieties of hornblende, known as carinthine, tremolite, actynolite, calamite, amianthus, &c. &c. The constituents of hornblende are, silica 45·60, magnesia 18·50, lime 14, alumina 1·18, protoxide of iron 7·50, fluoric acid 1·50. Massive hornblende is generally coarsely granular and lamellar; when intermixed with felspar in large lamellar grains, it forms sienite. This very common mineral may, generally, be easily recognised. Though sometimes in regular and distinct crystals, it is more commonly the result of a confused crystallization, and appears in masses composed of laminæ, acicular crystals, or fibres variously aggregated. Though inferior to schorl in hardness, hornblende usually will scratch glass, and, though with difficulty, will yield sparks with steel. Its laminated structure, its inferior hardness, and its inability of becoming electric by heat, distinguish it from schorl. It is less hard and more easily fusible than augite. It differs from epidote in hardness, and the results of fusion. Its powder is not soft to the touch like that of asbestus.

Ho'rnblende schist. One of the metamorphic rocks, composed principally of hornblende, with an uncertain proportion of felspar, and sometimes grains of quartz: its colour is usually black. Dr. M'Culloch observes, "hornblende schist may at first have been mere clay, for clay or shale is found changed by trap into Lydian stone, a substance differing from hornblende schist almost solely in its compactness and in the uniformity of its texture. Argillaceous schist, when in contact with granite, is sometimes converted into hornblende schist, the schist becoming first siliceous, and, at the contact, hornblende schist."

Horne'ra. The name given by Lamouroux to a genus of fossil corals. One species, found in the Wenlock limestone, has been named by Mr. Lonsdale, H. crassa, and is by that gentleman thus described, "Branches short, *thick*, flat, dichotomosed; opening of the cells large, elevated, and irregularly disposed on one side; opposite side striated; internal structure not ascertained."

H'ornstone. The silex corné of Brongniart; the quartz agathe grossier of Haüy; Hornstein of Werner. Prof. Jameson divides Hornstone into three sub-species, namely, Splintery Hornstone, Conchoidal Hornstone, and Woodstone. A siliceous mineral; a sub-species of quartz. It has usually a dull and splintery fracture, but sometimes it is conchoidal. In appearance it closely resembles compact felspar, but it differs from felspar in being infusible without the addition of an alkali; it occurs massive, and in extraneous external shapes; lustre dull or glimmering, opaque or translucent on the edges; sometimes the whole mass, if thin, has the strong translucency of certain horns. According to Kirwan, its constituents are, silica 72, alumina 22, carbonate of lime 6. "A specimen yielded to Faraday," says Phillips, "silica 71·3, alumina 15·3, protoxide of iron 9·3, and a trace of lime." Its colours are numerous, and commonly dull. Its infusibility by the blow-pipe distinguishes it from petrosilex and jade. Its translucency serves to render it distinct from jasper. It is generally more dull than flint, and emits sparks more feebly with steel. It is deficient in lustre in comparison with quartz.

HO'RNSTONE PORPHYRY. The Hornstein porphir of Werner. A variety of porphyry of a red, brown, purple or blackish colour. Fracture splintery and conchoidal. Emits sparks when struck with steel. Is susceptible of a fine polish.

HO'RSTED SAND. (So called from Horsted in Sussex.) The Horsted sand forms one of the members of the Hastings beds, or central group of the Wealden formation. It consists of grey, white, fawn coloured, and ferruginous sands and friable sandstone, with numerous small portions of lignite. It has been found principally in Sussex and Kent The Horsted sand is characterized by the dissemination throughout of small portions of lignite, which appear to have originated from the carbonization of plants of the fern tribe.—*Mantell.*

HO'RTUS SICCUS. (Lat.) A collection of dried plants.

HU'MBOLDITE. A rare mineral, thus named after Humboldt.

HU'MITE. A reddish-brown mineral, found at Somma, thus named after Sir A. Hume.

HY'ACINTH. (ὑάκινθος, Gr. *hyacinthus*, Lat. *hyacinthe*, Fr.) The Hiazinth of Werner; Zircon hyacinthe of Brongniart. A variety of pyramidal zircon, of a hyacinth-red colour; it occurs also, but more rarely, of a yellow, grey, and green colour, and sometimes, though very rarely, it is found white. It is crystallized, and when in distinct crystals its ordinary form is a four-sided rectangular prism, terminated by four rhombic planes. Each plane angle at the summit is 73° 44'. It is found in beds of streams and rivers along with rubies, sapphires, &c., but sometimes it occurs in the primary rocks. It consists of, zirconia 70, silica 25, oxide of iron 0·5. It is considered a gem, but is little used as such. The finest specimens are brought from Ceylon, but it is found in Europe in many places. It is cut by the jewellers, like the diamond, into rose and table forms, and possessing, as it does, though in an inferior degree, the play of the diamond it is frequently sold for that more precious gem.

HYACI'NTHINE. A mineral of a brown or greenish colour, usually crystallized in rectangular eight-sided prisms. Fracture imperfectly conchoidal. Transparent. Causes double refraction.

HYALÆ'A. (So named from its semi-transparent shell.) The *Hyalæa* has the appearance of a bivalve with soldered valves, the upper one being the largest; this difference of size of the seeming valves causes an aperture through which the animal sends forth two large yellow and violet wings, or sails, rounded and divided at their summit into three lobes. When its wings, or sails, are unfolded, it moves with great velocity on the surface of the sea. The Pteropods, both from the beautiful colouring of their filmy wings, and from their number and symmetry, are better entitled to the appellation of the butterflies of the ocean, than the escalop shells, which have sometimes been so called.—*Rev. W. Kirby.*

HY′ALINE. (from *hyalus*, Lat. ύαλος, Gr.) Transparent.

HY′ALITE. (ύαλος, Gr.) The Hyalith of Werner; quartz hyalin concretionné of Haüy. A yellowish or greyish variety of uncleavable quartz or opal. It exhibits the usual appearance of a concretion, and differs but little from calcedony, except in possessing a vitreous lustre, and sometimes a loose texture. It is found in secondary trap rocks; it occurs in grains, filaments, and botryoidal masses. It is infusible before the blow-pipe. It is nearly all silica, its component parts being silica 92·00, water 6·33. Specific gravity from 2·4 to 2·11.

HYBER′NACLE. ⎫ (*hibernacula*, Lat.) 1.
HIBER′NACLE. ⎭ The winter dwelling or residence of animals.
2. In botany that part of the plant which defends the embryo from injuries arising from frost.

HY′BODONT. (from ύβός, Gr. and *dens*, Lat.) A sub-family of sharks, according to the arrangement of M. Agassiz. They seem to have begun with the coal formation, to have continued throughout the oolitic deposition, and to have ceased at the beginning of the chalk formation. The teeth of this sub-family possess characters intermediate between the blunt crushing teeth of cestracionts, and the sharp cutting teeth of squaloids.

HYB′ODUS. A genus of fishes that prevailed throughout the oolitic period. Not a single genus of all that are found in the oolitic series exist at the present time.

HY′BRID. (ύβρις, Gr. *hybrida*, Lat. *hybride*, Fr.) Mongrel; a term applied both to plants and animals, when of a cross-breed.

HYDA′TID. (ύδατις, Gr. *hydatide*, Fr.) An order of internal worms. In hydatids there has not been discovered any vascular system. Hydatids, so frequently found in the liver and other parts of the body in mammalia, have been considered by some as animals, consisting merely of a stomach; by others, as a matrix, or womb, from something like young hydatids being frequently found adhering to their inner side. Hydatids immersed in warm water, immediately after being obtained from a living animal, are observed to have a contractile power, but they have no external opening; they are pellucid spherical bodies, of different sizes and kinds. Each consists of two coats, the inner of which is extremely delicate. They do not possess any visible blood-vessels, though the sac containing them has abundance of vessels, nerves, &c., derived from those of the organ with which it is connected. Hydatids ought not to be confounded with watery vesicles, connected occasionally with the kidney, &c., which are not enclosed in cysts, have no small hydatids adhering to their inner surface, and want contractility. There are many genera.

HYDNOPHO′RA. The name given by M. Fischer to a genus of stony polypifers, fixed, and incrusting other bodies; either forming a sub-globose, gibbons, or located mass, or spread in subfoliaceous lobes; the upper surface is set with small stars raised in pyramids. The stars project and are conical; the central axis is solid, simple, or dilated, surrounded by radiating adhering lamellæ. Many species, both recent and fossil, have been described. Lamarck formed the genus *monticularia* for their reception.

HY′DRATE. (from ύδωρ, water, Gr.) A chemical compound in definite proportions of a solid body with water, still retaining a solid form. It must, however, be borne in mind that when water in combination with other bodies contributes (as in crystalized bodies) to their regular form and transparency, it is then termed water of crystallization. The hydrates are numerous, as hydrates of alumine, baryta, cobalt, copper, lime, magnesia, potassa, soda, &c., &c. 2. Any uncrystallized substance which **contains** water in a fixed **definite** proportion.

HYDRAU'LIC ⎫ (ὑδραυλίς, Gr. *hy-*
HYDRAU'LICAL. ⎬ *draulicus*, Lat. *hy-*
HYDRAU'LICK. ⎭ *draulique*, Fr. *idraulico*, It.) Relating to the conveyance of water through pipes. The siphuncle of the nautilus, ammonite, belemnite, &c., forms a very beautiful and complete hydraulic apparatus.

HY'DRAULICS. The science of the motion of fluids, and the construction of all kinds of instruments and machines relating thereto.

HY'DRO. As a prefix, denotes that hydrogen enters into the composition of the substance which it signifies.

HY'DROGEN. (from ὕδωρ and γεννάω, Gr.) One of the simple, or elementary bodies. Inflammable air, proved by Cavendish to be the basis of water, from which circumstance it has obtained its name. It is obtained by the action of iron or zinc on dilute sulphuric acid. Hydrogen is colourless, and has commonly a slight odour of garlic; it is not absorbable by water; it is devoid of taste, and is destructive of life when respired for any time. It is the lightest body known, 100 cubic inches weighing only 2.25 grains, or being nearly thirteen times lighter than atmospheric air. It is combustible, and, when pure, burns with a yellowish-white flame. Hydrogen enters into the composition of all waters, and is evolved in a compound state from volcanos, from certain fissures in the earth, and in districts where coal is found. Two volumes of hydrogen unite with one of oxygen in the production of water. As far as the superficies of our planet is concerned, water so predominates, that, at first sight, hydrogen might be considered as constituting a substance of more relative abundance than it really does. The quantity of hydrogen locked up in coal is considerable. According to Dr. Thomson, cannal coal contains 21 per cent. of it, although in the Newcastle caking coal the proportion is but a trifle more then 4 per cent. Hydrogen may be considered **as the most important** substance of its class **next to** oxygen, which enters into the composition of the earth's crust. Symbol H.

HYDRO'GENATED. Combined with hydrogen.

HYDRO'GENOUS. (from ὕδωρ. water, and γεννάω, to produce, Gr.) Produced by the **action of** water, as hydrogenous rocks or strata. Among the older rocks, as among the newer, two entirely distinct classes occur, namely, pyrogenous and *hydrogenous* deposits.

HYDRO'GRAPHY. (from ὕδωρ and γράφω, Gr. *idrografia*, It. *hydrographie*, Fr.) Description of the watery part of the globe; the art of measuring and describing the sea, rivers, canals, lakes, &c. With regard to the sea, it gives an account of its tides, counter-tides, soundings, bays, gulfs, creeks, &c., as also of the rocks, shelves, sands, shallows, promontories, harbours; the distance and bearing of one point from another; with every thing that is remarkable either at sea or on the coast.

HYDRO'IDA. The second order of the class Hydrozoa, comprising the medusa, hydra, &c.

HYDRO'METER. (from ὕδωρ and μέτρον, Gr. *hydromètre*, Fr. *idrometro*, It.) An instrument for measuring the extent or depth, gravity, density, and velocity of liquids.

HY'DROPHANE. (from ὕδωρ and φαίνω, Gr.) Called also changeable opal. Silex Hydrophane of Brongniart.

Quartz resinite hydrophane of Haüy. A variety of opal which is opaque and white when dry, but by immersion in water becomes transparent. Hydrophanes, or changeable opals, are highly prized by collectors. In order, however, to preserve their beauty, care must be taken never to immerse them in any but pure water, and to remove them from the fluid as soon as they have acquired their transparency; otherwise, their pores will soon become filled with earthy particles, and their hydrophanous properties destroyed. "Hydrophane has a porous structure, and when immersed in water bubbles of air escape from its pores, while the water enters, and its weight is increased. When the pores are thus filled with water, a less portion of the light is reflected during its passage through the mineral, than if the same pores were filled with air; consequently more light is transmitted, and the transparency increased."—*Cleaveland*.

HYDRO'PHANOUS. Those substances are termed hydrophanous which, when dry, are opaque, but become transparent on being wetted. This property especially occurs in some varieties of common and semi-opal.

HY'DROPHYTE. (from ὕδωρ and φυτὸν, Gr.) A plant which lives and grows in water. Sir C. Lyell observes, "the number of hydrophytes is very considerable, and their stations more varied than could have been anticipated; for while some plants are covered and uncovered daily with the tide, others live in abysses of the ocean, at the extraordinary depth of one thousand feet: and although in such situations there must reign darkness more profound than night, at least to our organs, many of these vegetables are highly coloured."—*Principles of Geology*.

HYDROST'ATICS. (from ὕδωρ and στατικὸν, Gr. *hydrostatique*, Fr. *idrostatica*, It.) The science which treats of the nature, gravity, pressure, and equilibrium of fluids, and of the weighing of solids in them.

HY'DRURET. A compound of hydrogen with a metal.

HYDROZ'OA. The second class of the sub-kingdom Cœlenterata, comprising two orders, namely, Lucennaroida and Hydroida.

HYGRO'METER. (from ὑγρὸς and μέτρον, Gr. *hygromètre*, Fr. *igrometro*, It.) An instrument for measuring the degree of moisture of the atmosphere. There are various kinds of hygrometers; for whatever either swells by moisture, or shrinks by dryness, is capable of being formed into an hygrometer.

HYLÆOSAU'RUS. (from ὕλη, wood, weald, or forest, and σαῦρος, a lizard, Gr., the Forest Lizard.) A fossil lizard discovered in the wealden formation of Tilgate forest by Dr. Mantell, in 1832. Its probable length was about twenty-five feet. It is characterised by a series of long, flat, and pointed bones, which appear to have formed a large dermal fringe, resembling the bones on the back of the modern iguana. In this reptile the osteology of the lizard seems blended with that of the crocodile.

HYMENO'PTERA. (from ὑμενόπτερος, Gr. *alas membranaceas habens*: membrane-winged.) The first order of the class Insecta, comprising the Ichneumon, Saw-fly, &c. It is divided into six families. They have four membranous wings, and the tail of the female is usually armed with a sting. Though the insects of this order are included in the mandibulate section, for their mouth is furnished with mandibles and maxillæ, yet they do not generally use them to masticate their food, but for purposes usually connected with their sequence of instincts, as the bees in building their cells; the wasps in scraping particles of wood from posts and rails for a similar purpose, and likewise to seize their prey; but the great instrument by which they collect their food is their tongue; this the bees particularly have the power of inflating, and can wipe with it both convex and concave surfaces; and with it they *lick*, but not *suck*, the honey from the blossoms, for Reamur has proved that this organ acts as a tongue, and not as a pump.

HYPANTHOCRINI'TES. (from $\dot{\upsilon}\pi\dot{o}$, under, $\overset{\text{"}}{\alpha}\nu\theta o \varsigma$, a flower, and $\kappa\rho\iota\nu o\nu$, a lily.) A genus of encrinites belonging to the Silurian system, and established by Phillips. The name of this genus is taken from the floriform aspect of the basal portion of the body. In this new genus, says Prof. Phillips, the lowest plates clearly seen appear to correspond to the first costals of the genus Actinocrinites. One species only is described, namely, H. decorus, which is described and figured in Sir. R. Murchison's Silurian System.

HY'PERSTHENE. } (from $\dot{\upsilon}\pi\dot{\epsilon}\rho$. above,
HY'PERSTENE. } and $\sigma\theta\dot{\epsilon}\nu o\varsigma$, strength, Gr., in allusion to its difficult refrangibility.) Prismatoidal Schiller-spar. Labrador Schiller-spar. A mineral of a greenish-black colour, but on the cleavage of a copper-red. Occurs in **granular** and lamellar concretions, and **massive**. It is found in Labrador, in the Isle of Skye, in Banffshire, and in the Shetland Isles. It is composed of silica, magnesia, alumina, lime, and oxide of iron, the last of which is said to form one-fourth of the whole.

HYPOCRATE'RIFORM. Salver-shaped; a term applied to a monopetalous corolla, the limb of which being placed on a tube, spreads out horizontally.

HYPOGA'STRIC. (from $\dot{\upsilon}\pi\dot{o}$ and $\gamma\alpha\sigma\tau\dot{\eta}\rho$, Gr.) Belonging to that region of the abdomen which is called the hypogastrium.

HYPOGA'STRIUM. The lower anterior region of the abdomen, from a little below the umbilicus, or navel, to the pubes.

HY'POGENE. (from $\dot{\upsilon}\pi\dot{o}$, under, and $\gamma\iota\nu o\mu\alpha\iota$, to be born, Gr.) A term applied to rocks, expressing that they have assumed their form or structure, at a depth from the surface.

Sir C. Lyell, who proposes to give this term to certain rocks, observes, "It will appear that the popular nomenclature of geology, in reference to the so called 'primary' rocks, is not only imperfect, but in a great degree founded on a false theory; inasmuch as some granites and granitic schists are of origin posterior to many secondary rocks. In other words, some primary formations can already be shown to be newer than many secondary groups, a manifest contradiction in terms." To obviate this difficulty, Sir C. Lyell prefers the term hypogene, as one not of chronological import, but implying the theory that such rocks are netherformed, and have not assumed their form and structure at the surface.

HY'RACOTHERE. } An extinct genus
HYRACOTHE'RIUM. } of pachydermal mammalians, remains of which have been found in the London clay. Two species, H. Cuniculus, and H. Leporinum have been described by Prof. Owen. The nearest existing ally to the Hyracothere is now peculiar to America.

HY'STRIX. ($\overset{\text{"}}{\upsilon}\sigma\tau\rho\iota\xi$, Gr.) The porcupine.

# I J

**Ja′de.** The Nephrit of Werner; Nephrite of Jameson; called also nephritic stone, nephrite, and axe-stone. It was formerly much celebrated for its supposed medicinal properties in nephritic affections, or diseases of the kidneys. It is found in Hungary and Siberia, America, Egypt, and China. The inhabitants of New Zealand form it into axes, and other cutting instruments, from which circumstance it has obtained the name of axe-stone. Its surface is smooth; fracture splintery. It has a greasy feel. Colour, dark leek green. In hardness, jade is, at least, equal to quartz; it possesses a peculiar tenacity, which renders it difficult either to break, cut, or polish. Brochant states its fresh fracture to present a paler green than that of its surface. Before the blowpipe it fuses easily, and with a slight ebullition, into a bead of white semi-transparent glass. Its analysis is very variously given by different authors; its constituents are, according to some, silica, carbonate of magnesia, alumina, and carbonate of lime; others add chrome, oxide of manganese, soda, and potassa. In consequence of its tenacity, it has been wrought into chains and other delicate works.

**Jamb.** A term used by miners for a thick bed of stone which prevents their pursuing a vein.

**Ja′mesonite.** A mineral, thus named after Prof. Jameson, by Haidinger; axotomous antimony-glance. It was first discovered in Cornwall, in clay-slate, and it has since been found in Germany and Siberia. Its colour is steel-grey. It consists of lead, antimony, sulphur, and iron.

**Janthine.** ⎫ (from *ianthum*, Lat. a
**Ja′nthina.** ⎭ violet.) The violet snail. A purple-coloured univalve shell, nearly resembling a snail in its form: it is recent, and commonly found in the Mediterranean. The inhabitant of this shell is said, when irritated, to discharge a purple secretion. The Janthina was considered a helix until the time of Lamarck, who placed it in a distinct genus. He ascertained that the inhabitant of this shell differed essentially from the snail. The organ which, in the snail, would constitute a foot, is, in the janthina, formed for swimming, not for crawling.

**Ja′nthina fragilis.** A species of janthina. Sir C. Lyell remarks, "janthina fragilis has wandered into almost every sea, both tropical and temperate. This common oceanic shell derives its buoyancy from an admirably contrived float, which has enabled it not only to disperse itself so universally, but to become an active agent in disseminating other species, which attach themselves, or their ova, to its shell.—*Principles of Geology.*

According to the account given by Bosc, the *janthina* exhibits many remarkable peculiarities. When the sea is calm, these animals may be seen collected often in large bands, swimming over the surface by means of a floating apparatus consisting of aërial vesicles produced by their foot. During this action their head is very prominent, and the foot or line of vesicles forms an angle with the middle of the shell. When the sea is rough, the animal absorbs the air from its vesicles, changes the direction of its foot, contracts its body, and lets itself sink. It does the same when in danger from any enemy, and like the cuttle-fish, has the power of emitting a coloured fluid, which, by darkening the surrounding water, serves to conceal it from view. If the floating apparatus be injured or destroyed, there exists a reproductive power in the foot, by which it can be restored.

JA′RGON. (*jargon*, Fr.) The zircon jargon of Brongniart. A mineral, a variety of zircon. This variety or sub-species of zircon is of a grey, brown, or yellow colour, and is met with in small transparent or translucent prismatic crystals. It is obtained principally from Ceylon: according to the analysis of Vauquelin, it consists of zirconia 66·0, silica 31·0, oxide of iron 2·0.

JA′SPER. (*jaspe*, Fr. *pierre dure et opaque, de la nature de l'agate; jaspide*, It.) "Mineralogists have not been able," says Prof. Jameson, "to ascertain the origin of the word Jasper. We only know that it is of high authority, because it occurs in the Hebrew and Greek languages. We are also ignorant of the particular stone denominated Jasper by the ancients." A species or variety of rhombohedral quartz. It is an ingredient in the composition of many mountains. It occurs usually in large amorphous masses, and sometimes also crystallized in six-sided prisms. Fracture conchoidal. It is said to compose the substance of entire ranges of the Asiatic mountains. When quartz is combined with a considerable proportion of iron and alumine, it loses its translucency and becomes jasper. There are many varieties of jasper, distinguished principally by their different colours, and the arrangement of those colours.

Chemically, jasper differs from agate solely in containing a larger portion of iron; mineralogically, it may be distinguished by its opacity. Some mineralogists divide jasper into five sub-species, namely, Egyptian jasper, striped jasper, porcelain jasper, common jasper, and agate jasper. Jasper is commonly somewhat less hard than flint, or even common quartz, it will, however, give fire with steel. It is entirely opaque, or, sometimes, feebly translucent at the edges; it presents almost every variety of colour. Specific gravity from 2·30 to 2·70. It is infusible before the blow-pipe, but loses its colours. With the compound blowpipe, Professor Silliman says, it fuses into a greyish-black slag with white spots. Jasper is often traversed by metallic veins, or by veins of quartz. It sometimes contains fossil shells and marine plants. It is never, says Professor Cleaveland, porphyritic: the base of that, which has been called jasper porphyry, is fusible, and is either petrosilex or compact felspar. By some it has been supposed that jasper has been formed by the filtration of silex into beds of ferruginous clay.

JA′SPER OPAL. The Opal Jaspis of Werner. A sub-species of opal of different shades of red, brown, and yellow, the colour being often distributed in spotted, veined, or clouded delineations. It occurs massive only, and is found in Saxony, Hungary, Turkey, and Siberia. Before the blow-pipe it is infusible. Lustre shining, between vitreous and resinous. Fracture conchoidal. Specific gravity from 1·89 to 2·08. According to Klaproth it consists of silica 43·5, oxide of iron 47·0, water 7·5.

ICE′BERG. (from *ice* and *berg*, Germ.) A large mass of ice, met with in cold regions, floating upon the sea, sometimes of an enormous magnitude and great height. Icebergs have been seen of the great height of 300 feet, and as it has been ascertained that for every foot above the surface of the sea-water there are **eight feet below**, the whole thickness must be immense. In a geological point of view, icebergs are to be viewed as very important and powerful agents, inasmuch as they are the means of transporting to great distances, animals, plants, and rocks.

ICE-SPAR. So named from its resemblance to ice. A mineral occurring both massive and in flat crystals. It is found at Mount Somma, near Naples.

ICHTHY'ODON. The **name** used by some authors to designate a tooth of any fossil fish.

ICHTHYODO'RULITE. The fossil dorsal spine of certain fishes, armed with tooth-like hooks, or prickles. These were long supposed, says Professor Buckland, to be jaws, and true teeth; more recently they have been ascertained to be dorsal spines of fishes, and, from their supposed defensive office, have been named *ichthyodorulites*, from the Greek words $i\chi\theta\dot{v}\varsigma$, a fish, $\dot{\epsilon}\dot{o}\rho v$, a spear, and $\lambda\iota\theta o\varsigma$, **a stone.** The icthyodorulites of the **old red sandstone,** observes M. Agassiz, belong to distinct species of the genera *onchus* and *ctenacanthus*. These bony spines are more or less arched, and grooved by longitudinal furrows, separated by round ridges forming ribs. The great distinction between this genus of Ichthyodorulites, and the large species of Hybodus **of the** lias and new sandstone, to which they have some resemblance in the arrangement of their longitudinal furrows, is, that their posterior edge has no sharp points or teeth, while in the genus *hybodus* there are on that side strong points which are arched downwards.—*Murchison*.

That highly talented, and much to be lamented author, the late Hugh Miller, says, "there occasionally turn up, in the sandstones of Perthshire, *ichthyodorulites* that in bulk and appearance resemble the teeth of a harrow, rounded at the edges **by a few months wear.**"

I'CHTHYOLITE. (from $i\chi\theta\dot{v}\varsigma$, and $\lambda\iota\theta o\varsigma$, Gr. *ichtyolite*, Fr.) The name given to any fossil fish; a petrified fish. Fossil fishes occur in all the English formations, from the old red sandstone to the tertiary deposits inclusive.

ICHTHYOLI'TIC. Pertaining to fossil fishes, as the ichthyolitic formation. "In the ichthyolitic formation, immediately over the Silurian, that of the old red sandstone, the ganoids first appear."—*Hugh Miller*.

ICHTHYO'LOGY. (from $i\chi\theta\dot{v}\varsigma$, a fish, and $\lambda\dot{o}\gamma o\varsigma$, Gr. discourse; *ichtyologie*, Fr. *ictologia*, It.) That branch of zoology which treats of the structure, classification, habits, and history of fishes. The study of fossil ichthyology, says Prof. Buckland, is of peculiar importance to the geologist, as it enables him to follow an entire class of animals, of so high a division as the vertebrate, through the whole series of geological formations; and to institute comparisons between their various conditions during successive periods of the earth's formation. Professor Agassiz has already extended the number of fossil fishes to two hundred genera, and eight hundred and fifty species. No existing genus is found among the fossil fishes of any stratum older than the chalk formation.

ICHTHYO'PHAGOUS. (from $i\chi\theta\dot{v}\varsigma$, a fish, and $\phi\dot{a}\gamma\omega$, Gr. to eat; *ichthyophage*, Fr. *colui che non si ciba d'altro fuorchè di pesci*, It.) Feeding on fish.

ICHTHYOSAUR. } (from $i\chi\theta\dot{v}\varsigma$, a fish,
ICHTHYOSAU'RUS. } and $\sigma a\hat{v}\rho o\varsigma$, Gr. a lizard.) A fish-like lizard; an immense fossil marine-saurian or reptile, having an intermediate organization between that of a

lizard and a fish. The name appears to have been given to it by M. König. The genus comprises many species; some of these attain a magnitude not inferior to that of young whales. The head of the ichthyosaurus resembled that of a dolphin, its teeth were conical, sharp, and striated, and exceedingly numerous, in some cases amounting to nearly two hundred, not enclosed in separate sockets, but, as in the crocodile, ranged in one continuous groove, or furrow, of the maxillary bone; as also in the crocodile, abundant provision was made for replacing the old teeth, as they were lost,-by a supply of new ones. The eye was of enormous magnitude, the orbit in some instances measuring fourteen inches in its longer diameter, and Prof. Buckland states, "We have evidence that it possessed both microscopic and telescopic properties. The bony sclerotic of the ichthyosaurus approaches to the form of the bony circle in the eye of the Golden Eagle; one of its uses being to vary the sphere of distinct vision, in order to descry their prey at long or short distances. The soft parts of the eyes of the ichthyosaurus have, of course, entirely perished; but the preservation of this curiously constructed hoop of bony plates, shows that the enormous eye, of which they formed the front, was an optical instrument of varied and prodigious power, making this marine saurian to descry its prey at great or little distances, in the darkness of night and in the ocean's depths." The beak was that of the porpoise; the teeth, as before mentioned, those of the crocodile. In order to demonstrate the close relation between the ichthyosaurus and the lacerta family, it is desirable to state, that the lower jaw in that family, instead of having, like other quadrupeds, a single bone on either side, exhibits not fewer than six of these; one called the dental, which carries the teeth, forms the whole anterior extremity of the jaw, and continues to cover the upper portion, and to lap over the exterior; the rest of the outer face is formed, in the posterior part by a second bone, the coronoid; the bottom by a third, the angular; and the inner face by a fourth, the opercular: in addition to these four bones, the articular, which is placed at the posterior extremity for the purpose which its name denotes, and a small crescent-shaped bone, which sometimes forms the coronoid process, complete the stated number. Of these bones, the five first have been proved to have been possessed by the Ichthyosaurus, occupying situations closely corresponding to those which they possess in the crocodile. The crescent-shaped bone is the only one of the series not yet detected. The vertebræ nearly resembled those of the shark, being hour-glass shaped, and having both faces of their body deeply concave as in fishes; the vertebral column was composed of more than one hundred pieces; the ribs were slender, and the majority of them bifurcated, or forked, at the top; the bones of the sternum were strong and largely developed, and combined nearly in the same manner as in the ornithorynchus or platypus. The ichthyosaurus had four paddles, the form of its extremities deviating from the saurians and approaching the mammalians, being converted from feet into fins; these fins, or paddles, were composed of numerous bones enclosed in one fold of integument; the fore-paddle was composed of nearly one hundred bones, and like the mammalians it possessed a humerus, or shoulder bone, a radius and ulna, or the bones of the fore arm, and phalanges; the bones of the phalanges were polygonal and exceedingly numerous, as before stated. The hind-paddles were very much smaller, containing only from thirty to forty bones. The general conformation of the ichthyosaurus must have greatly resembled that of the porpoise or grampus. Its teeth would have sufficiently proved it to have been carnivorous, but the subsequent discovery of its fæcal remains, now called coprolites, and the finding within the intestinal canal the half-digested remains of fishes and reptiles, render this point quite certain; like the crocodile, it must have gorged its prey entire; its stomach was exceedingly capacious, forming a sort of pouch, or sac, and extending through nearly the whole body. The fossil remains of the ichthyosaurus have been most abundantly discovered in the lias **formation**, and it appears to have become extinct at the termination of the secondary series of geological formations: the debris of ichthyosauri have been found more abundantly in England than in any other country, and they are to be met with in every formation from the new red sandstone up to the green sand inclusive. It is however, the opinion of Bakewell that the ichthyosaurus, or some species of a similar genus, is still existing in the present seas, and with his remarks the description of the fish-like lizard will be concluded. "About sixteen years since, a large animal was seen for several summers in the Atlantic, near the coast of the United States, and was called the great sea-serpent. I am informed by Professor Silliman, that many persons who attested the existence of the sea serpent from their own observations, were so highly respectable, both for intelligence and veracity, that their evidence could not be disputed. I remember **one of the** most particular descriptions of the sea-serpent was given by an American captain, who saw the animal raise a large portion of its body from the water: he represented it as of great length, and about the bulk of a large water cask; it had paddles somewhat like a turtle, and enormous jaws like the crocodile. This description certainly approaches to, or may be said to correspond with, the ichthyosaurus, of which animal the captain had probably never heard." — *Bakewell, Conybeare, De la Beche.*

Ichthyophtha'lmite. (from ἰχθὺς, and ὀφθαλμὸς, Gr.) Fish-eye stone; apophyllite; pyramidal zeolite; the fischaugenstein of Werner; mesotype epointée of Haüy. It is of a white colour, and semi-transparent, or translucent. Occurs both crystallised and massive. The primitive form of its crystals is a four-sided prism, with rectangular bases. It is easily divisible by percussion into laminæ, whose broader surfaces are splendent and somewhat pearly. It scarcely scratches glass, and does not yield sparks when struck with steel. Specific gravity 2·46. Before the blow-pipe it exfoliates, froths, and eventually melts into an opaque bead. It is composed of silica, 50, lime 23, potash 4, water 18, with a trace of fluoric acid. It is found in secondary trap-rocks in the Hebrides and other parts of Scotland, in Sweden, and Iceland.

Ichthyospon'dyle. The name given by many writers on oryctology to signify a vertebra of different species of fossil fishes.

I'cius. The termination of adjectives in *icius* and *aceous* express a resemblance to a material; those in *eus* and *ous* indicates the material itself: thus, membranaceous, resembling skin; membranous, skin itself; coriaceous, leathery; latericious, resembling bricks.

ICOSAHE'DRAL. (from *icosahedron*.) Having twenty equal sides or faces.

ICOSAHE'DRON. (εἰκοσάεδρος, Gr. *icosedre*, Fr. *isosaedro*, It.) A regular solid, consisting of twenty triangular pyramids, whose vertices meet in the centre of a sphere supposed to circumscribe it; and therefore have their height and bases equal: wherefore the solidity of one of these pyramids multiplied by twenty, the number of bases, gives the solid contents of the icosahedron.

ICOSA'NDRIAN. (from εἴκοσι and ἀνήρ, Gr.) The twelfth class in Linnæus's sexual method, consisting of plants with hermaphrodite flowers, furnished with twenty or more stamens, inserted into the calyx. The first order of this class consists of trees bearing for the most part stone fruits, surrounded by a pulp, as the plum, peach, cherry, &c.; in the second order we find the apple, pear, &c.; in the third order, the genus rosæ. In this class the stamens grow out of the sides of the calyx, as in the strawberry, and it is important to observe, that such a mode of insertion indicates the wholesomeness of the fruit; we are not aware that there is a single exception to this rule, so that a traveller, who might meet with an unknown fruit, need not scruple to eat it if he find the stamens thus inserted. This character of the insertion of the stamens into the calyx holds good in other classes, as well as in the class Icosandria; thus, in the genus Ribes, including the gooseberry and currant, which belong to the class Pentandria, the stamens grow out of the calyx, and these fruits are well known to be wholesome, while many of the berries of the same class, whose stamens have not a like insertion, are often very deleterious.

I'DOCRASE. (from ἰδέα, form, and κρᾶσις, mixture, Gr.) The term idocrase was given to this mineral by Haüy in reference to its form, which is a mixed figure. Idocrase is a silicate of lime, combined with a silicate of alumina. A mineral found in lava, and formerly mistaken for the hyacinth; it is the Vesuvian of Werner. See *Vesuvian*.

JE'FFERSONITE. A mineral found in New Jersey; colour olive-green, passing into brown. It is named after Mr. President Jefferson.

JET. (from *Gaga*, a river of Asia; *jayet*, Fr.) The Jayet of Haüy; Lignite Jayet of Brongniart; Pech Kohle of Werner. A mineral substance, found in detached kidney-formed masses in many countries. It is of a firm and very even structure, harder than asphaltum, and susceptible of a good polish. It becomes electrical by rubbing, attracting light bodies, like amber. In many respects it resembles cannel-coal, its colour is full-black, and it does not soil the fingers. It is, however, easily distinguished from cannel-coal, in being specifically lighter than water, which cannel-coal is not, and in possessing electrical properties which cannel-coal does not. Some persons have supposed that jet is a true amber, differing only in the mere circumstance of colour. During combustion it emits a bituminous smell. It is never found in strata or continued masses, but always in separate and unconnected heaps.

It is formed into various trinkets, and is particularly used for making mourning ornaments, such as earrings, brooches, bracelets, buttons &c.

JEWS-STONE.

1. An extraneous fossil, being the elevated spine of a very large egg-shaped sea-urchin, or echinus.

2. A local term for basalt. In Salop and Worcestershire this name is applied to any hard trap-rock.

I'GNEOUS ROCKS. Those rocks are termed igneous which are considered to have once been in a fluid state from the action of heat upon them, and in that state to have overflowed, to have been injected among, or to have been propelled through, other rocks.

IGUA'NA. A species of lizard, a native of many parts of America and the West Indies, rarely met with any where north or south of the tropics. It is from three to five feet long, from the end of the snout to the tip of the tail. It inhabits rocky and woody places, and feeds on insects and vegetables. Cuvier states that the iguana subsists upon fruit, grain, and leaves: Bosc, that it lives principally upon insects. It nestles in hollow rocks and trees. The female lays its eggs, which have a thin skin like those of the turtle, and are about the size of those of a pigeon, in the sand. Though not amphibious, they are said to be able to remain under water an hour. When they swim, they do not use their feet, but place them close to their body, and guide themselves with their tails. Capt. Belcher found, in the Island of Isabella, swarms of iguanas, that appeared omnivorous. This statement proves both Cuvier and Bosc to be correct. The teeth of the iguana are not fitted for comminuting its food, and it is said to swallow it whole.

IGUA'NODON. An extinct fossil colossal lizard, discovered in the strata of Tilgate Forest by that indefatigable historian of the chalk and Wealden formation, the late Dr. Mantell. He observes, "the discovery of the teeth and other remains of a nondescript herbivorous reptile in the strata of Tilgate Forest, a reptile pronounced by Cuvier to be 'encore plus extraordinaire que tous ceux dont nous avons connoissance,' is one of the most gratifying results of my labours." The remains of one of these immense animals have lately been found in the Kentish rag, near Maidstone. The Kentish rag is a grey arenaceous limestone, belonging to the Shanklin sands. From the great resemblance in the dentature, as well as in many other extraordinary characteristics, of this immense reptile to that of the iguana, Dr. Mantell determined [on naming it the Iguanodon, signifying an animal having teeth like the iguana. In the perfect teeth, and in those which have been but little worn, the crown is somewhat of a prismatic form; widest, and most depressed, in front; convex posteriorly, and rather flattened at the sides. As soon as the tooth emerges from the gum it gradually enlarges, and its edges approach each other and terminate in a point, making the upper part of the crown angular; the edges forming the side of this angle are deeply serrated, or dentated; and the teeth exhibit two kinds of provisions to maintain sharp edges along the cutting surface: the first the serrated edge already described; the second, a provision of compensation for the gradual destruction of this edge, by substituting a plate of thin enamel, to maintain a cutting power in the anterior portion of the tooth, until its entire substance was consumed. These teeth were sometimes two inches and a half in length. While the crown of the tooth was diminishing above, an absorption of the fang was proceeding below, caused by the pressure of a new tooth rising to replace the old one, until by continual consumption, both above and below, the middle portion of the older tooth was reduced to a hollow stump, which fell from the jaw to make room for its more efficient successor. The size attained by the iguanodon appears to have been enormous, the average length from the snout to the tip of the tail being estimated by Dr. Mantell at seventy feet, while, he considers, some may have been one hundred feet in length. This last calculation Prof. Buckland deems improbable, but he gives a length of seventy feet to the iguanodon. A thigh-bone in the possession of Dr. Mantell is three feet eight inches long, and thirty-five inches round, at its largest extremity. The length of the hind foot is supposed to have been six feet and a-half; the circumference of the body, fourteen feet and a-half. A most remarkable appendage possessed by the iguanodon was a horn of bone, placed upon the nose, equal in size, and resembling in form, the lesser horn of the rhinoceros; here was a further analogy between the extinct fossil iguanodon and the recent iguana. The base of this nasal horn was of an irregular oval form, and slightly concave. It possessed an osseous structure, and appears to have had no internal cavity. It is evident that it was not attached to the skull by a bony union, as are the horns of the mammalia.

ILLÆ'NUS. A genus of trilobites thus named by Dalman. One species, Illænus perovalis, so named by Sir R. Murchison, is described by him as found in the Lower Silurian rocks. It is of an elongated oval form; the central lobe of the body slightly prolonged into the caudal portion.

I'MBRICATED. (*imbricatus*, Lat.) Laid one over the other at the edges, like the tiles of a house. In botany, applied to leaves when so placed.

IMPE'RMEABLE. Not admitting the passage of fluids through its pores or interstices, as clay or marl, which are impermeable to water.

I'NCIDENCE. (from *in*, upon, and *cado*, to fall, Lat.) The direction in which one body falls on or strikes another: the angle which the moving body makes with the plane of the body struck is called the angle of incidence.

INCI'SOR. (from *incisores*, Lat.) A fore or cutting tooth.

INCRUSTA'TION. (*incrustatio*, Lat. *incrustation*, Fr. *incrostatúra*, It.) An adherent covering; something superinduced; a coating of siliceous matter.

INDECI'DUOUS. (from *in* and *deciduus*, Lat.) Not falling off; not shed, as the leaves of trees, but evergreen.

I'NDIANITE. A whitish or grey mineral, brought from the Carnatic, found in masses, of a foliated structure, and having a shining lustre.

I'NDICOLITE. (from *indigo*, and λίθος, Gr.) An indigo-coloured mineral found in Sweden. It occurs crystallised, and is considered a variety of shorl.

INDI'GENOUS. (*indigena*, Lat. *indigène*, Fr.) Native to a country; originally born, or produced, in a particular country. The term is more usually applied to plants than animals; thus plants, the natural produce of any particular country, are said to be indigenous to that country.

INDU'CTION. (*inductio*, Lat. *induction*, Fr. *induzione*, It.) A consequence drawn from several propositions or principles first laid down; reasoning from particulars to generals, as when from several particular propositions we infer one general. The process by which a new principle is collected from an assemblage of facts, has been termed *induction*.

INDU'CTIVE REASONING. That kind of philosophic reasoning which ascends from particular facts to general principles, and then descends again from these general principles to particular applications and exemplifications.

INDU'SIA. (*indusia*, Lat.) The case

or covering of certain larvæ; generally used plurally, *indusiæ*.

INDU'SIAL LIMESTONE. A fresh-water limestone to which the name *indusial* has been given, from its containing the indusiæ, or cases, of the larvæ of Phrygania.

INÆQUILA'TERAL. ⎫ Having unequal
INEQUILA'TERAL. ⎭ sides; in conchology, when the anterior and posterior sides make different angles with the hinge.

INE'QUIVALVE. ⎫ Where one valve
INEQUIVA'LVULAR. ⎭ is more convex than the other, or dissimilar in any respect, as in the common oyster.

INFLORE'SCENCE. (*inflorescentia*, Lat.) A word used to express the particular manner in which flowers are placed upon a plant; this by older writers was denominated the modus florendi, or manner of flowering. Botanists distinguish many kinds of inflorescence, under the names whorl, cluster or raceme, spike, corymb, fascicle, tuft, umbel, cyme, panicle, bunch, &c.

INFUNDIBU'LIFORM. (from *infundibulum* and *forma*, Lat.) Funnel-shaped: in botany, applied to a monopetalous corolla, having a conical border placed upon a tube.

INFUSO'RIA. ⎫ Beings so
IMFU'SORY ANIMA'LULES. ⎭ extremely minute as to be invisible to the naked eye, and which have only been discovered since the invention of the microscope. The infusoria have been divided into two orders, the Rotifera and Homogenea. The order Rotifera comprises many genera, Brachionus, Furcularia, Tubicolaria, and Vaginicola: the Homogenea comprises Ureolaria, Trichoda, Leucophra, Kerona, Himantopes, &c., &c. The most extraordinary genus of all is the Proteus. It is not possible to assign to them any determinate form; their figure changes momentarily; sometimes rounded, sometimes divided. The bodies of the infusoria, are, for the most part, gelatinous.

When we place a drop of any infusion of animal or vegetable matter under a powerful microscope, and throw a light through that drop, and through the microscope to the eye, we discover in the drop of water various forms of living beings, some of a rounded, some of a lengthened form, and some exhibiting ramifications shooting in all directions, but all apparently of a soft, transparent, gelatinous, and almost homogeneous texture. These beings constitute the lowest form of animals with which we are at present acquainted, and they were at first considered astomatous, that is, without any mouth, and agastric, or possessing no stomach, and were called infusoria, a denomination explanatory merely of their habitat, but not of their structure. Upon further examination, it was discovered that there existed animalculæ of a higher denomination; these exist in every stagnant pool of water, in every river, and in the ocean. Upon examining with great care many years since the effects of coloured infusions upon these minute animalculæ, it was found that they devoured great quantities of the coloured matter in which they were placed, and that they conveyed it into internal cavities or stomachs, which are sometimes extremely numerous in them. Those cavities exist in almost every known genus. Sometimes there are nearly 200 stomachs in a single animalcule. Animalcules are found so exceedingly minute that nearly five hundred millions are contained in a single drop of water, that is, as many as there are individuals of our own race on the face of the earth. In those minute beings which constitute the simplest forms of animals, there are numerous stomachs, the lowest class is therefore called Polygastrica. They are the food of higher classes, particularly of zoophytes. There is no proper skeleton in the entire class of animalcules called Polygastrica. Some of the polygastrica exude on their surface a secretion which agglutinates, and lays hold of, foreign particles floating in the waters which surround them, and thus form for themselves a partial covering. The earthy matter, however, is not their own produce. —*Prof. Grant.*

Prof. Buckland observes, "We are more perplexed in attempting to comprehend the organization of the minutest *infusoria*, than that of a whale; and one of the last conclusions at which we arrive, is a conviction that the greatest and most important operations of nature are conducted by the agency of atoms too minute to be either perceptible by the human eye, or comprehensible by the human understanding."

Ehrenberg has ascertained that the *infusoria*, which have heretofore been considered as scarcely organized, have an internal structure resembling that of the higher animals. He has discovered in them muscles, intestines, teeth, different kinds of glands, eyes, nerves, and male and female organs of reproduction. He finds that some are born alive, others produced by eggs, and some multiplied by spontaneous divisions of their bodies into two or more distinct animals. Their powers of reproduction are so great, that from one individual a million were produced in ten days; on the eleventh day four millions, and on the twelfth sixteen millions. Ehrenberg has described and figured more than 500 species of animalcules; he has found them in fog, in rain, and in snow.

I'NGUINAL. (from *inguen*, Lat.) Pertaining to the groin.

INK-BAG. A bladder-shaped sac found in some species of cephalopods, containing a black and viscid fluid, resembling ink, by ejecting which, in case of danger from enemies, they are enabled to render the surrounding water opaque, and thus to conceal themselves. Examples of this contrivance may be seen in the Sepia vulgaris and Loligo of our seas. To the late Miss Mary Anning we owe the discovery of numerous fossil ink-bags, found in the lias of Lyme Regis, still distended, as when they formed parts of the living animals. The contents of the ink-bags of cephalopods is used in drawing, the sort preferred is from an oriental species of sepia; some of that extracted from a fossil ink-bag found in the lias was used by Sir Francis Chrantrey, on the request of Dean Buckland, and was by a celebrated painter, who was ignorant of the particulars, pronounced to be sepia of excellent quality. This extreme indestructibleness of sepia arises from its being chiefly composed of carbon.

INOCERA'MUS. A genus of fossil bivalvular shells of great range, according to some authors, extending from the Silurian to the chalk, inclusive. Pictet refers the Inoceramus of the Silurian and Devonian periods to Posidonomya.

INOSCULA'TION. Union by junction of the extremities; the union, or junction, of the mouths of vessels, as arteries with veins.

I'NSECT. (*insecta*, Lat. *insecte*, Fr. *insétto*, It.) The third class of articulated animals provided with articulated legs; they possess a dorsal vessel analogous to the vestige of a heart, but are wholly destitute of any branch for the circulation. Insects breathe atmospheric air by means of tracheæ, which are most freely ramified through all parts of the body; they possess compound eyes. All insects, which possess wings, metamorphose, or pass through certain changes, before they arrive at their perfect form. In their first state, after leaving the egg, they form larvæ, or caterpillars. The bodies of insects are divided into head, corslet, pectus, abdomen, and members. The head is joined to the body, in some, by ball and socket; in others, by plain surfaces; in others, after the manner of a hinge. In some, the connection is entirely ligamentous, the different motions corresponding with the nature of the joint. The corslet or thorax, is situated between the pectus and head. The first pair of feet are joined to this, and it contains the muscles for moving those and the head. To the upper and lateral part of the pectus, the wings, when present, are fixed, and the four posterior feet to its under part. To the upper part a horny process is frequently fixed, termed scutellum, or escutcheon. The pectus contains the muscles which move the wings and four pair of the feet. —*Fyfe.*

Cuvier divided insects into twelve orders, but modern classification places Insecta as the first class of the sub-kingdom Annulosa, dividing the class into nine orders.

INSECTI'VORA. (from *insect* and *voro*, Lat.) In Cuvier's arrangement, a family of animals which lead a subterraneous life, and having grinders studded with conical points. They live principally on insects, and many of them, in cold climates, pass the winter season in a state of torpidity. The hedgehog and mole are examples.

IN SITU. A mineral is said to be *in situ*, when in its natural place or position.

I'NTEGRAL. (*intégral*, Fr. *integrále*, It. *integer*, Lat.) A portion of a whole, being similar to the whole and not an elementary portion. Thus the smallest portion of carbonate of lime is still carbonate of lime, but if by any means we separate the carbonic acid from the lime, we no longer have in these, separately, integral portions but the elementary parts.

INTERCO'STAL. (from *inter* and *costa*, Lat. *intercostal*, Fr.) Anything between the ribs, as the intercostal muscles, intercostal arteries, nerves, or veins.

INTERNO'DAL. (from *inter* and *nodus*, Latin.) Applied to flower-stalks proceeding from the intermediate space of a branch between two leaves.

I'NTERNODE. The space between one knot or joint and another; a term used both in conchology and botany.

INTERO'SSEAL. } (from *inter* and *os*,
INTERO'SSEOUS. } Latin.) Placed between bones, as interosseous muscles, arteries, veins, &c.

INTERRU'PTEDLY. In botany, applied to compound leaves when the principal leaflets are divided by intervals of smaller ones; applied also to spikes of flowers, when the larger spikes are divided by a series of smaller ones.

INTESTI'NA. Linnæus divided the class Vermes, or worms, into five orders, the first of which he named *intestina*; these mostly inhabit the bodies of other animals; they are denominated the most simple animals, being perfectly naked, and without limbs of any kind. Cuvier has divided them into cavitaria, or nematoidea, and parenchymata. The cavitaria or nematoidea are worms having cavities or stomachs, or an intestinal canal floating in a distinct abdominal cavity, such canal extending from the mouth to the anus. The parenchymata comprises those species in which the body is filled with a cellular substance, or with a continuous parenchyma; the only alimentary organ it contains being ramified canals.

INTO'RSION.  ⎫ A twisting or turning
INTO'RTION. ⎭ in any particular direction. A term used in botany and conchology.

I'NULIN. (from *inula*.) A vegetable product, resembling starch, obtained from the roots of the Inula Hellenium, or elacampane, by boiling them in water. It was thus named by Mr. Rose.

INVE'RSION. (*inversio*, Lat.) Change of order or position so that the upper may be lower, or the lower upper; the first last, or the last first. In the order of superposition of the different stratified rocks, some strata may be wanting altogether, but there will not be found an *inversion* of the regular order of superposition.

INVE'RTEBRAL. (from *in* and *vertebral*.) Not possessing any vertebral column, or hard bony tube for the spinal cord, or medulla spinalis; not having a back-bone.

INVE'RTEBRATE. ⎫ All those animals
INVE'RTEBRATED. ⎭ are *invertebrated* which are included in the three great divisions, mollusca, or cyclogangliata; articulata, or diploneura; and radiata, or cyclo-neura. The other great division includes the *vertebrata*, or spini-cerebrata. In the cephalopodes, the *invertebrate* form of the lower divisions is beginning to be lost, and the *vertebrate* form of that division, to which man belongs, to appear. The first of the true *vertebrated* animals, is the class of fishes; from this class upwards, including pisces, amphibia, reptilia, aves, and mammalia, all are *vertebrated*. From the class Pisces downwards, including cephalopoda, pteropoda, gasteropoda, conchiphera, tunicata, of the division mollusca; crustacea, arachnida, insecta, myriapoda, annelida, cirrhopoda, rotifera, entozoa, of the division articulata; echinoderma, acalepha, polypiphera, poriphera, polygastrica, of the division radiata, all are *invertebrated*. There is one remarkable distinction which separates the vertebrated from the invertebrated animals, namely, that in the former, the muscles have no *external* points of attachment; and in the latter, with a few partial exceptions, no *internal* ones.

INVE'RTED. (from *inverto, inversus*, Lat.) Turned upside down; turned inwards; placed in contrary order to that which was before, or which is usual.

INVO'LUCEL. A small or partial involucre.

INVOLU'CRE. ⎫ (*involucrum*, Lat. *cui*
INVOLU'CRUM. ⎭ *aliquid involvitur*.)
1. Any membranous covering.
2. In botany, a species of calyx, remote from the flower, and bearing a great resemblance to bracteæ: the involucre is composed of many small leaves placed at the foot of the general umbel; in umbelliferous plants, the involucre, accompanying the partial umbels, is called the involucella.

INVOLU'CRET. A small, imperfect, or partial involucre, an involucel.

I'NVOLUTE. ⎫ (from *involvo*, Lat.
I'NVOLUTED. ⎭ 1. In botany, applied to leaves, when the margins are rolled inwards upon each other.
2. In conchology, where the exterior lip is turned inwards at the margin, as in all the cypreæ.

I'ODATE. A compound salt formed by the combination of iodine, oxygen, and a salifiable base; as the iodates of ammonia, soda, &c.

I'ODIDE. A compound of iodine and some metallic substance; as iodide of iron, iodide of lead, &c. Also a compound of iodine with a simple non-metallic substance. When iodine combines with metals in more than one proportion, it forms a protiodide, or a periodide.

I′ODIN. } (from ἰοειδής, ex ἴον, violet,
I′ODINE. } and εἶδος, appearance, Gr.)
This substance, which was discovered by Courtois, a manufacturer of salt-petre, at Paris, in 1812, obtained its name from the colour of its vapour, which is a beautiful violet. Iodine is procured from sea-water and from marine vegetables. It is of a greyish-black colour and shining metallic lustre. It is crystallizable; the primitive form of the crystals being a rhombic octahedron. Iodine possesses an extensive range of combination, forming acids both with oxygen, hydrogen, and chlorine. Iodine forms one of the simple or elementary bodies, included in those that are non-metallic.

I′OLITE. (from ἴον, violet, and λίθος, a stone, Gr.) A stone of a violet colour. The Prismatischer Quartz of Mohs, the Iolith of Werner, the Iolithe of Haüy, the Dichroite of Cordier, and the Cordierite of Leonhard. It is found massive and disseminated, and crystallized, in Finland, Norway, Greenland, Switzerland, and Spain; in gneiss and granite. It occurs in regular six and twelve-sided prisms. Its fracture is conchoidal and uneven. It is of a deep blue colour when seen along the axis, and of a brownish yellow when seen in a direction perpendicular to the axis of the prism. When we look along the resultant axes, which are inclined 62° 50′ to one another, we see a system of rings which are pretty distinct when the plate is thin; but when it is thick, and when the plane passing through the axis is in the plane of primitive polarisation, branches of blue and white light are seen to diverge in the form of a cross from the centre of the system of rings. It consists of silica, nearly 50 per cent., alumina, magnesia, oxide of iron, and oxide of maganese. It was first brought to France by Launoy, from Cape de Gatte, in Spain. Specific gravity 2·560. It scratches glass easily, quartz with difficulty.

IRIDE′SCENCE. (from *iris*, Lat. the rainbow.) The quality of shining with many colours, resembling those of the rainbow.

IRIDE′SCENT. Shining with the colours of the rainbow. Many membranous shells exhibit on several parts of their internal surface, a glistening, silvery, or iridescent appearance. This appearance is caused by the peculiar thinness, transparency, and regularity of arrangement, of the outer layers of the membrane, which, in conjunction with the particles of carbonate of lime, enter into the formation of that part of the surface of the shell. The surface, which has thus acquired a pearly lustre, was formerly believed to be a peculiar substance, and was termed mother-of-pearl; Sir David Brewster has, however, satisfactorily proved in the Philosophical Transactions, that the iridescent colours exhibited by these surfaces are wholly the effect of the parallel grooves, consequent upon the regularity of arrangement in the successive deposites of shells.

IRI′DIUM. (from *iris*, Lat.) An excessively infusible metal to which this name has been given from some of its salts having varied tints like those of the rainbow, and from the variety of colours exhibited by its solution. It was discovered by Mr. Tennant, in 1803, who, in examining the black powder left after dissolving platina, found that it contained two distinct metals, which he named iridium and osmium. It is of a pale steel-grey colour. It occurs in grains, in alluvium, in South America. From the researches of Berzelius, who estimates the equivalent of Iridium at 98·8, it appears to have three degrees of oxidation; and it is the rapid transition of these oxides into each other, that occasions the variable tints of iridium.

I'ris. (*iris*, Lat. *iris*, Fr. *iride*, It. *ίρις*, Gr.)
1. The rainbow.
2. The membrane round the pupil of the eye, deriving its name from its various colours. The colour of the iris corresponds in general with that of the hair, being blue or grey where the hair is light, and brown or black where the hair and complexion are of a dark colour. It floats in the aqueous humour, and serves to regulate the quantity of light sent to the bottom of the eye.
3. A genus of plants; order Monogynia, class Triandria; the flag-flower.

I'risated. A mineral is described as *irisated* which exhibits the prismatic colours either externally or internally; the latter is generally the consequence of some injury sustained by the mineral.

Iron. One of the most generally diffused of all solid minerals. Of all the metals, the oxides of which are neither alkalies nor earths, iron, geologically considered, is the most important. "Calculating the mean," says De La Beche, "of thirty kinds of rocks, and neglecting iron ores, properly so called, of every kind, iron constitutes, as an oxide, 5·5 of the lowest stratified rocks, amounting to 14·72 per cent. in mica slate with garnets, and 15·31 per cent. in chlorite slate. It forms 12·62 per cent. in hypersthene rock, and about 20 per cent. in basalts. Oxide of iron constitutes about two or three per cent. of the mass of granites and gneiss, and between three and four per cent. of the mass of greenstone and the more common trappean rocks. When we consider the large amount of iron which exists either in the state of an oxide, a carbonate, a carburet, a silicate, or a sulphuret, therein including all iron ores of importance, we shall probably not errr greatly if we estimate iron as constituting about 2 per cent. of the whole mineral crust of our globe. There is scarcely a rock without iron.—*Geological Researches.*

It is to the presence of iron that rocks and stones most frequently owe their colour, earths when pure being white. The specific gravity of all stones or earthy minerals if it much exceed 2·5 may be attributed to the presence of iron.

In its natural state iron is very unlike what we are hourly accustomed to see it. It presents itself everywhere only as an earthy mass, a dirty impure rust; and even when found in the mine with a metallic lustre, it is still far from possessing those qualities which are necessary to fit it for the endless uses to which it is applied. Man has only to purify gold, silver, &c. but he has, as it were, to create iron. It does not appear to have been known so early, or wrought so easily, as gold, silver, and copper. For its discovery we must have recourse to the nations of the East. The writings of Moses furnish us with the most ample proof at how early a period it was known in Egypt and Phœnicia. He mentions furnaces for working iron, "and brought you out of the iron furnace;" the ores from which iron was extracted, "a land whose stones are iron;" and he states that swords, knives, axes, and tools for cutting stones, were at that time made of iron, "and if he smite him with an instrument of iron, so that he die, he is a murderer," "and his hand fetcheth a stroke with the axe to cut down the tree," "thou shalt not lift up any iron tool upon them." The knowledge of iron was brought over from Phrygia to Greece by the Dactyli, according to Hesiod, as quoted by Pliny, who settled in Crete during the reign of Minos I. about 1431 years before Christ. It would appear that a knowledge of iron obtained even before the deluge, for in Genesis we read "And Zillah, she also bare Tubal-Cain, an instructer of every artificer in brass and iron."

Iron forms a constituent part of many animal and vegetable substances; it enters into the composition of the blood; and the various shades of hue of some of the most delicate flowers are more or less owing to its presence.

Iron is of a bluish-white colour, and, when polished, has a considerable degree of brilliancy. It has a styptic taste, and emits a smell when rubbed. Its specific gravity is 7·77.

Iron is placed the eighth in order, as regards its malleability, possessing this quality in a less degree than gold, silver, copper, tin, platinum, lead, and zinc. In ductility it ranks fourth, being inferior only to gold, silver, and platinum, and it may be drawn out into wire as fine as a human hair. In tenacity it ranks first, iron one-twelfth of an inch in diameter being capable of supporting 995 pounds without breaking. Iron is fusible at a temperature of 1797 Fahr.

Iron is found native, and is then generally considered to be of meteoric origin, being alloyed with nickel and other metals; these masses are called meteoric iron, and it certainly appears that they have fallen from the atmosphere. A mass was discovered in Siberia by Prof. Pallas, weighing 1680 lbs. A mass discovered in Bahia, in Brazil, is estimated to weigh 14,000 lbs.

A singular structure is frequently observed in the argillaceous iron ores of coal districts. The substance of the iron ore is formed into conical sheaths, involving one another, and marked by concentric undulations and radiating striæ. Large spheroidal masses of iron ore, weighing at least a ton, are thus found, in connexion with the coal, at Ingleton, in Yorkshire; and in the coal fields of Staffordshire and South Wales, it is a well known form of aggregation.

The quantity of iron manufactured in Great Britain is enormous; in the year 1827 it was calculated at 690,000 tons; nearly one-half of which, or 296,000 tons, was manufactured in Wales, and upwards of 200,000 in Staffordshire. For the manufacturing of this immense quantity, three millions seven hundred and ninety-five thousand tons of coals would be required.

In a supplementary note to Professor Buckland's Bridgewater Treatise, it is stated, "Ehrenberg has ascertained that a soft yellow ochreous substance, Raseneisen, which is found in large quantities every spring in marshes about Berlin, covering the bottom of ditches, and in the footsteps of animals, is composed of iron secreted by infusorial animalcules of the genus Gaillonella. This iron may be separated from the siliceous shields of these animals, which retain their form after the extraction of the iron.

I'RONSTONE. A heavy mineral, possessing sometimes a specific gravity of 3·6, and composed chiefly of iron combined with oxygen, carbonic acid, silex, and water, with, in some instances, calcareous earth. When of a superior quality, it will yield upwards of 36 per cent. of iron. Mr. Bakewell observes "We know nothing certain, respecting the formation of ironstone; but it appears to have been deposited in fresh water, as it occurs in fresh-water strata in the regular coal formation, and in the coal strata of the oolites of Yorkshire, and among the clay and sandstone strata in the wealds of Kent. Few geologists have attempted to explain the formation of ironstone. The manufacture of iron was formerly carried on to a considerable extent in the county of Sussex; Fuller in his Worthies observes, "it is almost incredible how many great guns were made of the iron of this county."

IRON FLINT. A mineral, thus named by Jameson.

I′RON-GLANCE. Rhombohedral iron-ore. A peroxide of iron, of a dark steel-gray colour. There are several varieties; the red varieties are called red iron ore, and the fibrous, hematite.

I′RON SAND. A variety of octohedral iron-ore, in grains.

ISCHIA′TIC. (*ischiadicus*, Lat. ἰσχιαδικὸς, Gr.) Pertaining to the ischium, as the ischiatic notch, &c.

I′SCHIUM. (*ischium*, Lat. ἰσχίον, Gr.) One of the bones of the pelvis, situated in the lowest part thereof, and being that bone upon which we sit. It forms the under, and largest portion of the acetabulum or cup which receives the head of the thigh bone.

I′SERIN. } (from *eisen*, Germ.) A
I′SERINE. } mineral of an iron-black colour, from which it derives its name. It consists of 48 per cent. of oxide of titanium, an equal proportion of oxide of iron and four per cent. of uranium. It occurs in small obtuse angular grains, being a kind of metallic sand. It appears to differ but little from menachinite.

ISOCA′RDIA. A heart-shaped shell, with separated involuted and diverging beaks. The hinge formed by two flattened cardinal inserted teeth, and an isolated lateral tooth under the cartilage slope.

ISOCHEI′MAL. (from ἴσος, equal, and χεῖμα, winter, Gr.) Of the same winter temperature: lines drawn through places having the same winter temperature are denominated *isocheimal lines.*

ISO′CHRONAL } (*isochrone*, Fr. from
ISO′CHRONOUS.} ἴσος, and χρόνος, Gr.) Having equal times; uniform in time. The isochronal vibrations of a pendulum are such as are performed in the same space of time; as all the swings or vibrations of the same pendulum are, whether the arches it describes are longer or shorter.

I′SOGEOTHE′RMAL LINES. (from ἴσος, γῆ, and θερμὸς, Gr.) Certain lines or divisions in the earth's crust possessing an equal degree of mean annual temperature. If we draw lines through all the points which have the same terrestrial temperature, these *isogeothermal* lines resemble the isothermal, as they are parallel to the equator, but diverge from it in several points.

ISOME′RIC. (from ἴσος, equal, and μέρος, a part, Gr.) A term applied to substances which consist of the same ingredients in the same proportion, and yet differ essentially in their properties.

ISOMO′RPHISM. (from ἴσος and μορφὴ, Gr.) That quality which a substance possesses of replacing some other substance in a compound body, without any alteration of its primitive form.

ISOMO′RPHOUS. That has the property of retaining its primitive form when united with other substances in a compound body.

ISOPERIME′TRICAL. (from ἴσος, πέρι, and μέτρον, Gr.) Such figures as have equal perimeters or circumferences, of which the circle is the greatest.

ISO′PODA. (from ἴσος and πούς, Gr.) An order of crustaceans, thus named from the formation of their feet which are fourteen in number.

This order embraces the genus Oniscus.

ISO'SCELES. (ἰσοσκελής, Gr.) That which hath only two sides equal.

I'SOPYRE. A mineral of a greyish or black colour. Occurs massive. Found in Cornwall, imbedded in granite.

ISO'THERAL. (from ἴσος and θέρος, summer, Gr.) Of the same summer temperature: lines drawn through places having the same summer temperature are denominated *isotheral lines*.

ISOTHE'RMAL. (from ἴσος and θέρμη, Gr.) Possessing equal temperature. Lines drawn upon a map through a series of places having the same annual mean temperature are termed *isothermal* lines, or lines of equal temperature. Sir C. Lyell observes, "it is now well ascertained that zones of equal warmth, both in the atmosphere and in the waters of the ocean, are neither parallel to the equator nor to each other. It is also discovered that the same mean annual temperature may exist in two places which enjoy very different climates, for the seasons may be nearly equalized or violently contrasted. Thus the lines of equal winter temperature do not coincide with the lines of equal annual heat, or *isothermal lines*. If lines be drawn round the globe through all those places which have the same winter temperature, they are found to deviate from the terrestrial parallels much farther than the lines of equal mean annual heat. The lines, for instance, of equal winter in Europe, are often curved so as to reach parallels of latitude 9° or 10° distant from each other, whereas the *isothermal lines* only differ from 4° to 5°."

ISOTHE'RMAL ZONES. As the isothermal lines are as numerous as the places, and as diversified as numerous, geographers have grouped them into bands or zones. Humboldt has divided the northern hemisphere into six isothermal zones or bands.

I'SOTOME. Isotomes are those bodies which have the same crystalline form, and similar formulæ, and equal atomic volumes.

ISTIU'RUS. A genus of that family of saurians called iguanida, and thus named by Cuvier. The distinguishing character of the genus Istiurus is an elevated and trenchant crest, extending along a portion of the tail, and supported by spinous apophyses of the vertebræ.

IU'LUS.  }
JU'LUS.  } (ἴουλος, Gr.)

1. In botany, a catkin; a species of inflorescence consisting of chaffy scales arranged along a stalk; they are worm-like tufts, which at the beginning of the year grow out, and hang pendular down from the hazel, walnut, filberd, &c.

2. In zoology, a genus of insects of the order Aptera. The feet are very numerous, being on each side twice as many as the segments of the body; the antennæ are moniliform; there are two articulated palpi; and the body is of a semi-cylindrical form. There are many species.

I'VORY. (*ebur*, Lat. *ivoire*, Fr. *avório*, It.) A hard, solid, and firm substance, of a white colour, and capable of a very good polish. It is the tusk of the elephant. The ivory from Ceylon is more valuable than any other, from its not becoming yellow in the wearing, as nearly all other ivory does.

JU'RA LIMESTONE. (*calcaire de Jura, Jura kalk.*) The name given by some continental geologists to that group of rocks comprised in the oolite. The Jura limestone group is composed of limestones of various qualities, clays, sands, and sandstone, and contains the same fossils as those found in the oolitic group of England. In the range of the

**Jura** and the outer ranges of the Alps, the calcareous formations are of such immense magnitude, and the beds are often so higly indurated and crystalline, that it is only from their relative position and imbedded fossils, that we can trace their analogy to the English strata.

**Jurassic.** (from Jura.) The name given to certain strata composing the mountain-chain of the Jura, or the fossil remains therein contained; it is synonymous with oolitic; the Jurassic system of rocks consist of three great divisions, the lower of which is the Lias formation; the middle includes the Inferior oolite, the Great oolite, the Cornbrash, the Kelloway rock, and the Oxford clay; the upper is the Coralline oolite formation, with its subordinate calcareous grits, and the Portlandian formation, with its subordinate members of Kimmeridge clay, Portland oolite, and Purbeck limestone.—*Lycett.*

**Juxta-posi'tion.** (*juxta-position*, Fr. *juxta* and *positio*, Lat.) The state of being placed in nearness or contiguity; apposition.

# K

**Kainoz'oic.** A term used by Palæontologists as applicable to strata containing recent fossil remains; it is synonymous with tertiary, and has been divided into the Human, Historical, or Recent period; the Pleistocene period; the Pliocene period; the Miocene period; and the Eocene period.

**Kamm.** A provincial term, used in Cornwall, for that portion which lies over the bed or principal division of a mineral deposit. The mineral deposit may be divided into two parts, the bed or floor, and the *Kamm* or overlyer.

**Kaneelstein.** } Names given by Werner, Haüy, and others to Cinnamon Stone.—See *Cinnamon Stone.*
**Kannelstein.** }

**Kangaroo'.** An animal of the genus didelphys, the Didelphys gigantea of Linnæus. It is a native of New Holland. When of full growth, it attains the size of a large sheep. The fore-legs are short, the hind legs of considerable length, so that it advances by leaping rather than walking or running.

**Ka'olin.** (The Porzellan Erde of Werner; the Argile Kaolin of Brongniart; and Feldspath décomposé of Haüy.) Porcelain clay. The name of an earth which is used as one of the ingredients in the manufacture of oriental porcelain. Mr. Bakewell observes, "I believe it is the soft earthy granite from the mountains of Auvergne which supplies the *kaolin* used in the porcelain manufacture at Sevres. Mons. Brongniart shewed me a specimen of their best kaolin: it contained crystals of pinite." M. Bromare says that by analysing some Chinese kaolin, he found it was a compound earth, consisting of clay, to which it owed its tenacity; of calcareous earth; of sparkling crystals of mica; and of small quartz crystals. He says that he has found a similar earth upon a stratum of granite, and conjectures that it may be a decomposed granite. The *kaolin* used in most countries for the manufacture of fine porcelain or china, is generally produced from the felspar of decomposing granite, in which the cause of decay is the dissolution and separation of the alkaline ingredients. Cleveland says, "Kaolin is essentially composed of silex and alumine; the proportions are variable, but the silex usually predominates. When pure kaolin is employed in the manufacture of porcelain, some ingredient must be added as a flux, as, when pure, it is infusible. There is satisfactory evidence that kaolin has, in most cases, if not in all, originated from the decomposition of rocks abounding in felspar, more particularly from graphic granite, which consists almost entirely of quartz and felspar. According to Werner, it is the carbonic acid which has changed the felspar in granite and gneiss into kaolin. The quantity of kaolin, derived from the felspar of decomposing granite, shipped from Cornwall to Worcestershire for the china manufactories amounted, in 1816, to 1775 tons.

Prof. Phillips gives the following as the analysis of kaolin: "the kaolin of China consists of silex 71·15, alumine 15·86, lime 1·92, water 6·73; the kaolin of Cornwall is composed of alumine 60·00, silex 40·00

KA'RPHOLITE. } (from κάρφος, straw,
CA'RPHOLITE. } and λίθος, a stone, Gr.) A straw-coloured mineral, occurring in thin prismatic concretions, and of a fibrous structure. According to Stromeyer, it consists of silica 36·15, alumina 28·60, protoxide of iron 2·29, protoxide of manganese 19·16, lime 0·27, fluoric acid 0·47, water 10·78. It is found in the tin mines of Schlackenwald in Bohemia.

KEEL. (*Kiel*, Germ, *quille*, Fr.
1. In conchology, the longitudinal prominence in the Argonautæ.
2. In botany, the term keel is applied to two of the petals in papilionaceous flowers: the keel is composed of two petals, separate or united, and encloses the internal organs of fructification.
3. In entomology, a sharp, longitudinal, gradually rising elevation upon the inferior surface.

KEE'LED. Applied to leaves when the back is very prominent longitudinally.

KELVE. In the south of Ireland, carbonaceous shale is called Kelve.

KENTISH RAG. The name given to a siliciferous limestone with disseminated dark green particles: it is found in the lower green sand.

KERA'TOPHYTE. A name given to the horny zoophyte.

KERATO'SA. One of the three orders into which the class Poriphera has been divided. The axis of the animal is entirely composed of *horny* anastomosing filaments, from which circumtance the name has been applied.

KE'ESANTON. A greenstone rock, composed essentially of hornblende and mica, in which some felspar is often mixed.

KE'ROLITE. An earthy mineral occurring in Silesia and in Saxony, associated with serpentine. According to Pfaff, it contains silica 37·95, alumina 12·18, magnesia 16·02, water 31·00. It is of a white, yellow, or green colour, and is found in kidney shaped masses, which have a lamellar or compound structure. Feels greasy, but does not adhere to the tongue.

KE'UPER. The name given by the German geologists to one division of rocks of the Triassic period. The Marnes Irisées of the French. It consists principally of red and green marl.

KI'LLAS. A provincial name for a coarse argillaceous schist; a variety of slate. Mr. Bakewell, in mentioning the *killas* of Cornwall, says, "perhaps the best designation of the *killas* rock in this situation is, that of a minutely grained and highly indurated gneiss that had lost its schistoze character." Mr. Hawkins considers the common *killas*, or slate of the mining district, is an intimate mixture of quartz, with mica, talc, chlorite, and, perhaps, in some instances, with felspar. Mohs says that *killas* is an intermediate substance between mica-slate and clay-slate. Dean Conybeare states that the common *killas* has been at last admitted on all hands to be a genuine clay-slate. Dr. Boase proposes to apply to the *killas* the name of Cornubianite. See *Cornubianite*. Kirwan gives the following analysis of killas: 100 grains contained silica 60, alumina 25, magnesia 9, iron 6.

KI′LLINITE. A mineral of a pale green colour, occuring in veins of granite at Killiney, near Dublin.

KI′MMERIDGE CLAY. A blue and greyish-yellow slaty clay of the upper oolite formation; a member of the oolite group, thus called from its being found abundantly at Kimmeridge, in the Isle of Purbeck. It contains gypsum and bituminous shale. It is a marine deposit. Kimmeridge clay forms the base of the Isle of Portland. The bituminous shale found in the Kimmeridge clay on the coast of the Isle of Purbeck, has obtained the name of Kimmeridge coal, and is used as fuel. The most interesting remains contained in the Kimmeridge clay are those of the extinct genera allied to the order Lacerta, evidently calculated for a marine abode; the vertebræ, paddles, &c., of a species of Ichthyosaurus differing from those in the lias; the vertebræ, phalanges, and head of another saurian, perhaps a variety of plesiosaurus, have been found at Kimmeridge and Headington; bones, apparently of cetacea, likewise appear.

KING-CRAB. (an entomostracan, or shelled insect.) The Limulus polyphemus, known also as the horse-shoe. It is very common on the coast of New Jersey. The king-crab is placed by Cuvier amongst the pœcilopods.

KNOB. (*knöbel, knopf,* Germ.) A hard protuberance. In conchology, any part of a shell bluntly rising above the rest.

KOA′LA. An extraordinary quadruped inhabiting the continent of Australia. Cuvier placed the *koala* in marsupialia, or the fourth order of Mammalia. This animal has a short stout body, short legs, and no tail: it has the five toes, or fingers, of the fore-feet divided into two groups, the thumb and index forming one group, and the three remaining toes or fingers the other. On the hind-feet the thumb is altogether wanting. Carrying its young for a long period on its back, this separation of the toes of the fore-feet enables it to take firmer hold of the branches of the trees, on which it passes a portion of its time.

KOTH. A name given by the Spaniards to an earthy slimy substance ejected from the volcanoes of South America. It is of a blackish brown colour, an earthy texture, and is but slightly coherent. The natives call it Moya.

KOU′PHOLITH. (from κούφος, light, and λίθον, a stone, Gr.) The prehnite koupholite of Brongniart: prehnite lamelliforme rhomboidal, of Haüy; a variety of prehnite. It occurs in minute, rhomboidal plates, is of a greenish or pale yellow colour, glistening, and slightly pearly. It has been found in the Pyrenees.

KY′ANITE. (from κύανος, blue colour, Gr. This is frequently written Cyanite.) The Cyanit of Werner; the Disthene of Haüy, and the Sappare of Kirwan. Kyanite occurs both massive and crystallized, the crystals, often very long, are frequently grouped. Its colour, as its name imports, is blue, varying from an intense to a light sky-blue. It is infusible, except under the compound blow-pipe. It consists of alumina 64, silica 34, with a small quantity of oxide of iron and a trace of lime. It is found in the primary rocks in Scotland and in America.

# L

**LABE′LLUM.** (*labellum*, Lat. a little lip.) A term applied, in botany, to one of the three pieces forming the corolla in orchideous plants. The calyx and corolla consist of three pieces each, and one of those forming the latter, differs very much in size and form from the other two; it is called the *labellum*, and is often spurred.

**LA′BIATE.** ⎫ In botany, plants are so
**LA′BIATED.** ⎭ called which have the segments or divisions of their corollas resembling the form of lips.

**LABIATÆ.** There is a large class of plants called *labiatæ*, which have irregular monopetalous corollas, and these, generally, bibaliate and ringent; the mint, nettle, &c. are examples.

**LA′BIUM.** (*labium*, Lat.)
1. In entomology, the lower lip of insects is called the *labium*; the upper, the *labrum*. The lower pair of jaws are behind the mandibles, and between them is situated the *labium*, or lower lip, which closes the mouth below, as the labrum does above. The labium of insects consists of two chief parts, each of which may be considered as a separate organ; namely, the chin and the tongue.
2. In conchology, the inner lip of the shell.

**LA′BRUM.** (*labrum*, Lat.)
1. In entomology, the upper lip of insects. The labrum is situated above, or rather in front of, the mandibles, it is generally of the form of the segment of a circle, or a triangular, or quadrangular, somewhat convex, corneous plate, which is united posteriorly by a membranous hinge with the clypeus.
2. In conchology, the outer lip; that edge of the aperture which is placed at the greatest distance from the axis of the shell.

**LA′BRADOR FELSPAR.** ⎫ So named from
**LA′BRADOR STONE.** ⎭ having been found on the coast of Labrador. This mineral was at one time called Labrador Hornblende, but its present name has been substituted for what was incorrect. Labrador felspar has been found massive and disseminated only. Its laminæ are slightly curved; lustre nearly metallic, and pearly on the perfect cleavage faces. It is distinguished by its splendid changeability of colour, reflecting very beautiful colours when the light falls upon it in certain directions. This mineral has been also found in different parts of Europe.

**LA′BYRINTH.** (*labyrinthus*, Lat. $\lambda\alpha\beta\acute{\nu}\rho\iota\nu\theta\circ\varsigma$, Gr. *labyrinthe*, Fr. *laberinto*, It.) The name given to several cavities of the ear, from their flexuous position. The internal parts of the ear compose what is designated, from the intricacy of its winding passages, the *labyrinth*.

**LABRYRI′NTHODON.** Called also Mastodonsaurus. A name proposed to be given by Prof. Owen to a genus of reptiles, the remains of which have been discovered in the Warwick sandstone. The Labryrinthodon belongs to the extinct order Labyrinthodonta.

**LABYRI′NTHODONTA.** An extinct order of the class Amphibia, comprising Labryrinthodon; Capitosaurus; Metopias; Trematosaurus; Zygosaurus; Odontosaurus; Archegosaurus; Rhinosaurus; and Telerpeton.

**LACERT′ILIA.** The third order of the class Reptilia.

**LAC′ERTA.** (The lizard.) A genus of reptiles of the order Lacertilia.

LACI′NIATE. } (*laciniatus*, Lat. *lacinié*,
LACI′NIATED. } Fr.) Ragged at the edges; jagged. In botany, applied to leaves cut into numerous irregular portions.

LA′CRYMAL. (from *lachryma*, *vel lacryma*, Lat. *lacrymale*, Fr.) Certain parts about the eye, connected with the secretion and passage of the tears, as the lacrymal glands, the lacrymal ducts, &c. This word is also written lachrymal.

LA′CTEAL. (from *lac*, Lat.) The lacteals are numerous minute tubes commencing, by open and very minute orifices, from the inner surface of the intestines, and uniting successively into larger vessels, till they form trunks of considerable magnitude. The office of the lacteals is to take up the chyle and transmit it to the heart. It is only among the vertebrata that lacteals are met with; in invertebrated animals, the absorption of the chyle is performed by veins instead of lacteal vessels. The chyle of the higher orders of animals often contains a multitude of globules, which give to it a milky appearance, from which circumstance the vessels containing it have obtained their name.

LACU′STRINE DEPOSITS. Purely lacustrine deposits are almost unknown among any of the stratified rocks of a date earlier than the tertiary period, and it was not until the publications of Cuvier and Brongniart, on the environs of Paris, that the attention of geologists was much directed to the study of those numerous fresh-water deposits from which we may obtain a knowledge of the ancient condition of the land. "If we drain a lake," says Sir C. Lyell, "we frequently find at the bottom a series of deposits, disposed with great regularity one above the other; the uppermost, perhaps, may be a stratum of peat, next below a more compact variety of the same, still lower a bed of laminated marl, alternating with peat, and then other beds of marl, alternating with clay. Now if we sink a second pit through the same continuous *lacustrine deposit*, at some distance from the first, we commonly meet with nearly the same series of beds, yet with slight variations; some, for example, of the layers of sand, clay, or marl may be wanting, one or more of them having thinned out, and given place to others, or sometimes one of the masses, first examined, is observed to increase in thickness to the exclusion of other beds. At length we arrive at a point where the whole assemblage of lacustrine strata terminate, as, for example, when we arrive at the borders of the original lake-basin."

LAGA′NUM. A fossil echinite thus named by Klein, called also pancake.

LAG′OMYS. (from λάγος, a hare, and μῦς, a rat, Gr.) A rat hare. A genus of animals forming a link between the hare and the rat. The *lagomys* is placed by Cuvier in the order *rodentia*. They have been found in Siberia, and are described by Pallas. There are several species. The *lagomys* has ears of a moderate size; legs nearly alike; clavicles almost perfect; and no tail. The Rev. W. Kirby observes of the *lagomys*, "it ought rather to have been called the *hay-maker*, since man may, or might, have learned that part of the business of the agriculturist, which consists in providing a store of winter provender for his cattle, from this industrious animal. The Tungusians, who inhabit the country beyond the lake of Baikal, call it *Pika*, which has been adopted as its trivial name. These animals make their abode between the rocks, and during the summer employ themselves in making hay for a winter store. About the middle of August these little animals collect with admirable precaution their winter's provender, which is formed of the choicest grasses and the sweetest herbs, which they bring near their habitations and spread out to dry like hay. In September, they form heaps or stacks of the fodder they have collected, under the rocks, or in other places sheltered from the rain or snow. Where many of them have laboured together, their stacks are sometimes as high as a man, and more than eight feet in diameter. A subterranean gallery leads from the burrow, below the mass of hay, so that neither frost nor snow can intercept their communication with it.—*Bridgewater Treatise.*

LAGOO'N. ⎫ (*laguna*, It.) A salt-water
LAGU'NE. ⎭ lake. This term is more particularly applied to those pools of water which are found in the centre of coral reefs.

LAMA'NTIN. ⎫ The manatus of Cuvier.
LAMA'NTINE. ⎭ A species of herbivorous cetacea, living upon the plants which grow at the bottom of the sea. The *lamantin* appears to have existed during the miocene and pliocene periods. Fossil remains have been discovered in France. The fossil remains of the lamantins differ sufficiently from the analogous parts of existing species to justify the inference of specific distinction. It would also appear that they belonged to a lost species. The existing species of the *lamantine* are found near the mouths of rivers in the hottest parts of the Atlantic ocean and in the torrid zone, and the discovery of their fossil remains in Europe adds another link to the long chain of evidence of a diminished temperature of the climate of Europe.

LAMBDOI'DAL. (from the Greek letter λαμβδα, and εἶδος, form.) The name given to one of the sutures of the cranium, from its supposed resemblance in form to the Greek letter Λ.

LA'MELLATED. ⎫ Composed of thin
LAME'LLAR. ⎭ plates, layers, or scales. In conchology, when a shell is divided into thin and distinct plates or layers, overlying each other with the edges produced.

LAME'LLI BRANCHIA'TA. In De Blainville's conchological arrangement, the third order of Acephalopora, containing ten families of bivalves. Lamelli branchiata or Acephala comprises the cockle, mussel, oyster, Venus, and all ordinary bivalve shells.

LAMELLICO'RNES. In Cuvier's arrangement, the sixth family of pentamerous coleoptera; they have foliated horns, from which circumstance they obtain their name.

LAMELLI'FEROUS. (from *lamella*, a small plate, and *fero*, to bear, Lat.) Having a structure composed of thin layers; having a foliated structure.

LAMELLI'FORM. Consisting of lamellæ or regular and parallel plates.

LAMELLIRO'STRES. In Cuvier's arrangement, the fourth family of the order of Palmipedes. The lamellirostres have a thick bill, the edges of which are furnished with laminæ, from which circumstance they have obtained their name.

LA'MINA. (*lamina*, Lat.) A thin plate or scale; a thin layer of a stratum.

LA'MINATED. (*laminé*, Fr.) Disposed in layers, placed one over another.

LAMINA'TION. Arrangement in layers. Lamination prevails amongst all the varieties of gneiss, mica schist, chlorite schist, hornblende schist, &c. It is often observable in primary limestone, and sometimes in quartz rock. All the members of the carboniferous series display lamination, though in unequal degrees. By laminæ and lamination are always meant layers of deposition.

LAMN'ODUS. A genus of ichthyolites of the old red sandstone formation, called also Dendrodus, (but which latter has been supplanted,) of which three species have been described by Agassiz.

LA'NATE. } (from *lanatus*, Lat.)
LA'NATED. } Woolly; covered with a sort of pubescence resembling short woolly hairs.

LA'NCEOLATE. Lance-shaped; narrow and tapering.
1. In conchology, applied to a shell of an oblong shape, and gradually tapering at each end.
2. In entomology, in describing the figure of the superficies, when the base is not so broad as the centre, and the lateral margins slightly, but equally swollen, gradually tapering towards the apex, where it terminates in a point, and the longitudinal diameter more than three times the length of the transverse.—*Burmeister*.
3. In botany, applied to leaves of a narrow oblong form, gradually tapering towards each end.

LA'NCIFORM. (from *lancea* and *forma*, Lat.) Spear-shaped; lance-shaped.

LA'NDSLIP. A portion of land that has separated from the main body, in consequence of long-continued rains, or the expansive powers of severe frosts, and has fallen to a lower situation. Landslips must necessarily be often attended by fatal consequences, as in the falls of avalanches. We are informed that when the mountain of Piz fell, in 1772, three villages, with the entire population, were covered; and that when part of Mount Grenier, in Savoy, fell, in 1248, five parishes were buried, the ruins ocupying an extent of nine square miles.

LA'NGOUSTE. (Fr. *sorte d'écrevisse de mer*.) The name given by the French to the Palinurus vulgaris of Leach; the cray-fish or thorny lobster.

LA'NTANE. } (from λανθάνω, to con-
LA'NTHANUM. } ceal, Gr.) One of the elementary bodies, its symbol being La.

LANU'GINOSE. } (from *lanuginosus*,
LANU'GINOUS. } Lat.) In entomology, when longish curled hair is spread over the surface; covered with soft hair resembling wool.

LA'PIDES JUDA'ICI. A name given to certain fossil spines of echinites, formerly supposed to be petrified olives.

LAPIDIFICA'TION. (*lapidification*, Fr. from *lapis*, a stone, and *fio*, to make or become, Lat.) The conversion into stone of some other substance; the act of forming stone.

LAPI'LLI. (*lapillus*, Lat.) Volcanic cinders, abounding in minute globular concretions.

LA'PIS LA'ZULI. The Lazurstein of Werner; Azure-stone of Jameson; Lazulite of Haüy; Dodecaedrischer Kuphon-spath of Mohs. When lapis lazuli is pure, it is a mineral of a fine azure-blue colour; it occurs in rhombohedral dodecahedrons, massive, and disseminated. Structure finely granular, almost compact; fracture uneven or conchoidal; lustre feeble; a little translucent at its edges. It scratches glass, but gives sparks with steel with difficulty. Specific gravity about 2·30. Its analysis is very differently given by different authors. It contains silica, nearly fifty per cent. carbonate of lime, alumina, potash, soda, oxide of iron, and sulphuric acid. It occurs associated with primary rocks, especially granite. It is accompanied by garnets, quartz, felspar, &c., with some of which it is often intermixed. It is found chiefly in China, Persia, and Russia. It is capable of a high polish, and is much esteemed. Its chief use, however, is to furnish the ultramarine blue, used by painters, a pigment remarkable for the durability of its colour.

LA'PIS ŒTI'TES. Eagle-stone. A mineral which derives its name from the ancient belief that it was found in the nests of the eagle. It is a variety of iron ore, commonly met with in the argillaceous mines of this country. Its supposed virtues are described by Dioscorides, Œtius, and Pliny, who assert that if tied to the arm it will prevent abortion; if fixed to the thigh it will facilitate delivery.

LA'RYNX. (λάρυγξ, Gr.) The upper part of the wind-pipe or trachea; that cartilaginous projection in the throat known as the pomum Adami, which strictly is formed by one of the cartilages of the larynx only, namely, the thyroid. The larynx consists of five cartilages, the cricoid, thyroid, two arytænoid, and the epiglottis.

LA'TERITE. The name given to a ferruginous clay, mottled red and yellow. When first dug from its bed, it is soft and easily fashioned into the form of bricks or large square masses for building; it rapidly indurates on exposure to the atmosphere, and, as far as is yet known, is destitute of fossils.

LATIRO'STROUS. (from *latus*, broad, and *rostrum*, a beak, Lat.) Broad beaked.

LA'TITUDE. (*latitudo*, Lat. *latitude*, Fr. *latitudine*, It.) The latitude of a place on the earth's surface is its angular distance from the equator, measured on its own terrestrial meridian: it is reckoned in degrees, minutes, and seconds, from 0 up to 90°, and northwards or southwards according to the hemisphere the plane lies in. Thus the observatory at Greenwich is situated in 51° 28′ 40″ north latitude. Latitude may also be thus defined, the angular distance between the direction of a plumb-line at any place and the plane of the equator.

LA'TROBITE. A mineral, thus named after Latrobe, having been found by him on the coast of Labrador. Colour, pale pink; specific gravity 2·8. Occurs massive and crystallized.

LA'TTICED. In conchology, shells having longitudinal lines or furrows decussated by transverse ones, resembling lattice-work.

LA'RVA. (*larva*, a mask, Lat.) An insect in its caterpillar state, before it has attained its winged or perfect state. Some insects, as the butterfly, moth of the silkworm, &c., pass through four distinct states, namely, the egg; the *larva*, or caterpillar; the pupa, or chrysalis; and the imago, or perfect insect. The egg, which is deposited by the perfect insect, gives birth to a caterpillar, or *larva*; an animal, which, in outward shape, bears not the slightest resemblance to its parent, or to the form it is itself afterwards to assume. It has, in fact, both the external resemblance, and the mechanical structure, of a worm. The same elongated cylindric shape, the same annular structure of the denser parts of the integument, the same arrangements of longitudinal and oblique muscles connecting these rings, the same apparatus of short feet, with claws, or bristles, or tufts of hair, for facilitating progression; in short, all the circumstances most characteristic of the vermiform type are equally exemplified in the different tribes of caterpillars, as in the annelida. These external investments, which hide the real form of the future animal, have been compared to a mask; so that the insect, while wearing this disguise, has been termed larva, the Latin name for a mask.—*Roget*.

LA'VA. (This word, according to Kirwan, is derived from the Gothic, *lopa* or *lauffen*, to run, and is applied to the melted and liquified matter, discharged from the mouths of volcanoes.) The matter which flows in a fused, or melted, state from a volcano.

Lava, whatever be its chemical composition, puts on very different appearances, according to the circumstances which accompany its consolidation, hence by some authors it has been divided into compact lava, cellular lava, and cavernous lava. The mineral called felspar forms, in general, more than half of the mass of modern lavas. When this is in great excess, lavas are called trachytic; when, on the other hand, augite prevails, they are called basaltic. Lavas occur of an intermediate composition, and these, from their colour, have been called gray-stones. When lava is observed as near as possible to the point whence it issues, it is found to be, for the most part, a semi-fluid mass of the consistence of honey, but occasionally so liquid as to penetrate the fibre of wood. It soon cools externally, and consequently exhibits a rough uneven surface; but, from its being a bad conductor of heat, the internal mass remains liquid long after that portion which is exposed to the air has become solid. That of 1822, some days after it had been ejected, raised the thermometer from 59° to 95°, at a distance of twelve feet; at a distance of three feet, the temperature greatly exceeded that of boiling water. The temperature at which lava continues in a state of fluidity is sufficiently great to melt glass and silver; even stones are said to have been fused when thrown into the lava of Etna and Vesuvius. The length of time during which streams of lava retain their heat is quite astonishing: the current of lava which flowed from Etna in 1669 retains a portion of it to the present time. That which was poured from Jorullo, in Mexico, in the year 1759, was found to retain a high temperature half a century afterwards. Sir W. Hamilton lighted small pieces of wood in the fissures of a current of Vesuvian lava four years after it had been ejected. The streams of lava often become solid externally, even while yet in motion, and their sides may be compared to two rocky walls, which are sometimes inclined at an angle of 45°. Of the immense bodies of lava thrown out during volcanic eruptions, few persons entertain a just idea. Etna, which rises upwards of 10,000 feet in height, and embraces a circumference of 180 miles, is composed entirely of lava. "In the structure of this mountain," says Dr. Daubeny, "every thing wears alike the character of vastness." The products of the eruptions of Vesuvius may be said almost to sink into insignificance, when compared with these *coulées*, some of which are four or five miles in breadth, fifteen in length, and from fifty to one hundred feet in thickness. Still the eruptions of Etna are nothing when compared with that of Skaptár Jokul. "On the 11th of June Skaptár Jokul threw out a torrent of lava which flowed down into the river Skaptá, and completely dried it up. The channel of the river was between high rocks, in many places from four hundred to six hundred feet in depth, and near two hundred in breadth. Not only did the lava fill up these great defiles to the brink, but it overflowed the adjacent fields to a considerable extent. The burning flood, on issuing from the confined rocky gorge, was then arrested for some time by a deep lake, which it entirely filled. The current then again proceeded, and reaching some ancient lava, full of subterraneous caverns, **penetrated and melted down part of it.** On the 18th of June, another ejection of liquid lava rushed **from the volcano,** which flowed down **with amazing** velocity over the surface **of the first** stream. After flowing for several days, it was precipitated down a tremendous cataract called Stapafoss, where it filled a profound abyss, which that great waterfall had been hollowing out for ages, and again the fiery current pursued its onward course. On the 3rd of August, fresh floods of lava still pouring from the volcano, a new branch was **sent** off in a different direction. When the fiery lake which filled up the lower portion of the valley of the Skaptá had been augmented with new supplies, the lava flowed up the course of the river to the foot **of** the hills whence the Skaptá takes its rise. This eruption did not entirely cease till the end of two years, and although the population of Iceland did not exceed fifty thousand, not fewer than twenty villages were overwhelmed, besides those inundated by water, **and** more than nine thousand human beings perished, together with an immense number of cattle. Of the two branches of liquid lava, which flowed in nearly opposite directions, *the greater was fifty, and the lesser forty miles in length.* The extreme breadth which the Skaptá branch attained in the low countries was from twelve to fifteen miles, that of the other about seven. The ordinary height of both currents was one hundred feet, but in narrow defiles it sometimes amounted to six hundred." The sources from which lava ascends are deeply seated beneath the granite; but it is not yet decided whether the immediate cause of an eruption be the access of water to local accumulations of the metalloid bases of the earth and alkalies; or whether lava be derived directly from that general **mass** of incandescent elements, which **may** probably exist at a depth of about one hundred miles beneath the surface of our planet.—*Lyell.*

LAU'MONITE. Diatomous zeolite. Lomonit of Werner: Mesotype Laumonite of Brongniart: Zeolite efflorescente of Brochant: Diatomous Kouphone Spar of Mohs. A mineral long known under the name of efflorescent zeolite, and to which its present name has been assigned by Werner, in honour of M. Gillet de Laumont, to whom we are indebted for our first knowledge of it. It is of a white, or grayish-white colour, sometimes tinged with red. It occurs regularly crystallized, and in distinct granular concretions. Its crystals are four-sided prisms, slightly oblique, sometimes terminated by dihedral summits, sometimes truncated on their lateral edges. The height of the prism is to the edges of the terminal faces, in the ratio of eight to seven. This prism divides in a direction parallel to all its planes, but much more easily longitudinally than on its terminal surfaces; this division takes place also with greater facility on two of the opposite sides than on the two others. When Laumonite has not been altered by exposure to the atmosphere, it cuts glass easily. By exposure to the air, Laumonite disintegrates, and is at length reduced to a white powder. If, however, recent specimens be immersed for two or three **hours in** a strong mucilage of gum, the action of the atmosphere upon them, and their efflorescence, will be prevented. By the action of acids, Laumonite is reduced to a state of jelly. Under the blow-pipe it fuses, with a slight degree of ebullition, and affords a perfectly opaque and beautifully white enamel. Specific gravity 2·23. Laumonite consists of silica 52, alumina 21·20, lime 10·50, and water about 14. It occurs in secondary traprocks in France, Scotland, Iceland, and America.

LA'ZULITE. A mineral of a light blue colour, supposed by some mineralogists to be a sub-species of lapis lazuli.

LEAD. *(læd*, Sax.) Lead is of a bluish-grey colour, with considerable lustre, but soon tarnishes by exposure to the atmosphere. By friction, this metal exhales a peculiar, and somewhat disagreeable odour. Its specific gravity is 11·35, or nearly eleven and a-half times heavier than water. It is soft and easily melted, being fusible at about 600° Fahrenheit. It is the softest of all the durable metals; it can be scratched by the nail, and is easily cut by a knife. Its elasticity, ductility, and tenacity are comparatively low; it cannot be drawn into wire thinner than a line in diameter. Lead is very malleable, and can be beaten into thin leaves; but these, from its imperfect tenacity, are easily torn. All the salts of this metal are highly poisonous; they are, however, most shamefully employed by unprincipled persons to correct or conceal the acidity of cider and wines. The presence of lead in these liquors may be detected by the following means. Dissolve 120 grains of sulphuret of lime, and 180 grains of supertartrate of potash in 16 ounces of distilled water, by repeatedly shaking the mixture; when perfectly dissolved, leave the mixture to settle, and pour off the clear liquid into clean phials, adding about twenty drops of hydrochloric acid to each. A small quantity of this poured into a wine-glass of the suspected wine, will detect the smallest quantity of lead, if any be present, by producing a black precipitate.

Several instances of the occurrence of native lead have been mentioned, though in few of them does the fact appear to be well established. In the island of Madeira, it is found in small masses, in lava, and has undoubtedly been reduced to its present state by volcanic fire. Next to iron, lead may be considered the most abundantly diffused of all the metals; it has been known from the earliest ages. The lead of our mines is in a state of combination with sulphur, forming a sulphuret of lead; this is called galena, or lead-glance. By exposure to a strong heat, the sulphur is driven off and pure lead is obtained: the average produce of metal from the Derbyshire ore is about 66 per cent. The present annual produce of lead in the United Kingdom may be computed at from 47,000 to 50,000 tons.

LEE'LITE. (thus named after Dr. Lee, of Cambridge.) An earthy mineral, first noticed by Dr. Clarke, is found at Gryphitta, in Sweden. It is compact and massive, of a deep flesh-red colour. Texture wax-like, with the lustre and transparency of horn. Fracture resembling that of flint. Specific gravity 2·71. Its constituents are silica 75, alumina 22, manganese 3, or manganese 2·50, and water 0·50.—*Phillips.*

LEGU'ME. *(legumen,* Lat. *légume,* Fr. *legúme,* It.) A species of fruit; a pod; a seed vessel peculiar to leguminous plants, formed of two oblong valves having no longitudinal partition; the seeds are attached to one of its margins only; the bean, pea, vetch, and all the natural order of leguminosæ furnish examples.

LEGUMINO'SÆ. An order of plants, calyx five-toothed, inferior, the odd segment anterior, or farthest from the axis; corolla papilionaceous, rarely regular; stamens definite or indefinite, perigynous, either distinct, monadelphous or diadelphous; ovarium superior, one-celled, many seeded, style and stigma simple; fruit a legume, or, rarely, a drupe; seeds occasionally with an arillus; embryo, exalbuminous; cotyledons, either remaining under ground, or appearing above, in germination; leaves compound, stipulate, alternate; leaflets stipulate; inflorescence usually axillary, but various.

One genus of this order, Detarium, has a drupe for its fruit, and Mimosa has a perfectly regular corolla.

The order Leguminosæ is most important to man both for its beauty and utility. The pea, bean, harico, vetch, liquorice, clover, sainfoin, lucerne, tamarind, indigo, gum arabic, &c., &c., belong to it. Generally, the order is innocent, if not wholesome; but some few genera are poisonous.

LEGUMINOSI'TES. The name given by Mr. Bowerbank to certain fossil seeds found in the London clay. He describes them as seeds of true *Leguminosæ*, the pericarps of which are not known. These seeds present, both in form and structure, all the characteristic features observable in numerous recent genera of true leguminous fruits. Mr. Bowerbank has described eighteen species, and says it would, however, be extremely difficult, if not impossible, to identify these seeds with those of existing genera, supposing them to belong to such, without the assistance of their pericarps.

LEGU'MINOUS. Belonging to the order Leguminosæ; bearing pods; having a legume for a pericarp.

LEHM. A German word, being another name for the deposit now commonly known as loëss. See *Loëss*.

LE'MMING. The Lapland marmot. The lemming has short ears and a tail, with the toes of its fore-feet peculiarly adapted for digging.

Cuvier places the lemming in the order Rodentia, class Mammalia. Bones of the lemming have been found fossil in a breccia at Cette.

LE'MNIAN EARTH. A mineral found in the island of Lemnos, in the Egean Sea, whence its name. It is also called Sphragide, from σφραγὶς, sigillum, a seal. It is of a reddish colour and has a soapy feel. It is dug once a year, with much ceremony, in the Isle of Lemnos. It was formerly used in medicine; when impressed with the seal of the Grand Seignior, it was sold under the name of *terra sigillata*.

LENS. (*lens*, a lentil, Lat. *lentille*, Fr. *lénte*, It.) So named from its resemblance to a seed of the lentil. A transparent substance having its two surfaces so formed that the rays of light, in passing through it, have their direction changed. Of lenses there are various sorts; a *spherical lens*, is a sphere, all the points in its surface being equally distant from the centre. A *double convex lens*, is a solid formed by two convex spherical surfaces, having their surfaces on opposite sides of the lens. When the radii of its two surfaces are equal, it is said to be *equally convex;* when the radii are unequal, it is said to be *unequally convex*. A *plano-convex lens*, is a lens having one of its surfaces convex and the other plane. A *double concave lens* is a solid, bounded by two concave spherical surfaces, and may be either equally or unequally concave. A *plano-concave lens*, is a lens one of whose surfaces is concave and the other plane. A *meniscus*, is a lens one of whose surfaces is convex and the other concave, and in which the two surfaces meet if continued. A *concavo-convex lens*, is a lens one of whose surfaces is concave and the other convex, and in which the two surfaces will not meet if continued.

LENTICE'LLÆ. The name given by

De Candolle to certain points, which appear as dark spots, on the surface of the bark of plants.

LENTI′CULAR. (*lenticularis*, resembling a lentil, Lat. *lenticulaire*, Fr.) Having the form of a lens.

In entomology, a round body, with its opposite sides convex, meeting in a sharp edge. In conchology, doubly convex shells.

In mineralogy, crystals nearly flat, and convex above and beneath.

LENTI′CULAR ORE. The name given by Jameson to obtuse octahedral arseniate of copper; called also lenticular arseniate of copper.

LENTICULI′NA. A sublenticular, multilocular, spiral univalve; a genus of microscopic foraminifera. Distinguished from Nautilus by having no syphon.

LENTI′CULITE. A fossil shell of a lenticular form.

LE′NZINITE. A mineral found in Germany, and thus named after Lenzius, a German mineralogist. There are two kinds of lenzinite, the opaline and the argillaceous; the former of a milk-white, the latter of a snow-white colour.

LE′PADITES. The goose-barnacle. An order of Cirripedes, the species of which are distinguished by a tendinous, contractile, and often long tube, fixed by its base to some solid marine substance, supporting a compressed shell, consisting of valves united to each other by a membrane; and by having six pairs of tentaculated arms. They are usually found in places exposed to the fluctuations of the waves.--*Kirby.*

LE′PAS. (λεπὰς, Gr. *lepas*, Lat. *lepas*, Fr.) Linnæus included under the name lepas all the cirripedes or multivalves. These animals are known in this country by the name of Barnacles. The lepas, or barnacle, constitutes a connecting link between molluscous and articulated animals; the gills are attached to the bases of the cirrhi, or jointed tentacula. In the Linnæan system, the lepas constitutes the second genus of multivalve shells. The animal a triton; shell affixed at the base, and consisting of many unequal erect valves. They are without eyes, or any distinct head have no powers of locomotion, but are fixed to various bodies. Their body, which has no articulations, is enveloped in a mantle: their mouth is armed with transverse toothed jaws in pairs, and furnished with a feeler. This genus consists of two families, or divisions, very different in their form, the first of which is the *Balanites*, or Acorn-barnacles, having a shelly instead of a tendinous tube, with an operculum or lid, consisting generally of four, but sometimes of six valves, and being of a sub-conic form. The second family consists of the Lepadites or Goose-barnacles, the species of which are distinguished by a tendinous, contractile, and often long tube or pedicle, which, being of a flexible nature, allows the animal, fixed by its base to some solid marine substance, to writhe about in quest of food. The animals of this genus have only been found in the ocean.

LEPIDODE′NDRON. (from λεπὶς, a scale, and δένδρον, wood, Gr.) An extinct genus of fossil plants, of very frequent occurrence in the coal formation. Count Sternberg has thus divided the Lepidodendra:—

Tribus 1. (Lepidotæ) squamis convexis.
*a.* Scutatæ.
*b.* Escutatæ.
Tribus 2. (Alveolariæ) squamis subconcavis.

It is stated by Lindley and Hutton that plants of this genus are, next to the calamites, the most abundant of the fossils in the coal formation of the north of England. Lepidodendra are sometimes found of enormous size, fragments of stems occurring upwards of forty feet in length. Their internal structure has been ascertained to be intermediate between coniferæ and lycopodiaceæ. In some points of their structure they resemble coniferæ, but in other respects, setting aside their great magnitude, they may be compared to lycopodiaceæ. To botanists, this discovery is of very high interest, as it proves that those systematists are right, who contend for the possibility of certain chasms now existing between the gradations of organization, being caused by the extinction of genera, or even of whole orders, the existence of which was necessary to complete the harmony which it is believed originally existed in the structure of all parts of the vegetable kingdom. By means of *Lepidodendron*, a better passage is established from flowering to flowerless plants, than by either equisetum or cycas, or any other known genus.—*Lindley and Hutton.*

LE'PIDOIDS. A family of extinct fossil fishes, found in the oolitic series; they were remarkable for their large rhomboidal bony scales, which were of great thickness, and covered with enamel. The scales of lepidoids had a remarkable structure in being furnished on their upper margin with a hook-like process, placed like the hook or peg near the upper margin of a roofing tile; this hook fitted into a depression on the lower margin of the scale placed immediately above it.

LEPI'DOLITE. (from λεπίς, a scale, and λίθος, a stone, Gr.) The Lepidolith of Werner; Hemiprismatischer Talk-glimmer of Mohs. A mineral of a peach-blossom, red, and sometimes grey colour, occurring massive and in small concretions. This mineral, at first view, appears to be composed of small grains, sometimes extremely minute; but these grains, among which little pearly scales are often interspersed, are themselves composed of a great number of minute foliæ or spangles, like those of mica, from which circumstance it has obtained its name. Its constituents have been variously stated. According to some authors, it contains silica 50·36, alumina 28·32, potash 9·0, oxide of manganese 1·25, fluoric acid and water 5·40, lithia 5·50. It exhibits two axes of double refraction, from which circumstance it has been called Di-axial mica. From the beauty of its colour it has been cut into snuff-boxes.

LEPIDO'PTERA. (from λεπίς, a scale, and πτερόν, a wing, Gr.) Scaly-winged insects. Lepidoptera form the tenth order of insects in Cuvier's arrangement; they have four wings, both sides of which are covered with small coloured scales, resembling farinaceous dust. This order comprises butterflies, moths, and sphinxes. The scales are attached so slightly to the membrane of the wing as to come off when touched with the fingers, to which they adhere like fine dust. When examined with the microscope, their construction and arrangement appear to be exceedingly beautiful, being marked with parallel and equidistant striæ, often crossed by still finer lines. The former of these scales are exceedingly diversified, not only in different species, but also in different parts of the same insect. The proboscis of the Lepidoptera is a double tube. In the classification of the most modern naturalists, Lepidoptera constitutes the fifth order of the class Insecta; sub-kingdom Annulosa.

LEPIDO'PTERAL. } Belonging to the
LEPIDO'PTEROUS. } order Lepidoptera; having wings covered with scales.

LEPIDO'STEUS. } (The Lepisosteus of
LEPISO'STEUS. } Lacépède. A genus

of fishes inhabiting the rivers of North America, one of the two living representative genera of Sauroid fishes. Teeth of a fish related to Lepidosteus, or Lepisosteus, have been **found in the Tilgate beds and in those of Stonesfield**.

LEPIDO'STROBUS. } A genus of extinct
LEPIDO'TUS. } fresh water fishes, with very **thick**, enamelled, rhomboidal **scales, and** obtuse hemispherical teeth; the latter are called fishes' eyes by the common people who collect them.

LEPTÆ'NA. (from λεπτὸς, tenuis, Gr.) A sub-division of the family of Terebratula, established by Dalman. The name Leptæna may be deemed synonymous with Productus or Producta, but the latter being objected to, it has been thought desirable to adopt that of Leptæna instead. Some authors separate Leptæna from Productus from the character of the hinge, which is compressed and rectilinear, frequently exceeding the width of the shell.—See *Producta*.

LEU'CIN. } (from λευκὸς, white, Gr.)
LEU'CINE. } The name given by M. Braconnot to a white substance obtained from muscular fibre, by treating it with sulphuric acid, and subjecting it to a peculiar process.

LEU'CITE. (from λευκὸς, white, Gr.) A mineral of a white colour, found in volcanic rocks. Notwithstanding its name, leucite is not always white, its common colours are yellowish and greyish white, but it is occasionally of a reddish white, and translucent. Before the blowpipe it is infusible, a circumstance which serves to distinguish leucite from garnet and analcime; when mixed with borax it fuses into a brownish diaphanous glass. Its constituents are silex 53·75, alumine 24·62, potash 21·35; specific gravity from 2·45 to 2·49; hardness = 5·5. It scratches glass with difficulty. It occurs **regularly** crystallized; **in** granular concretions, and in roundish grains. It is often embedded in lava and in basalt. All lavas do not contain crystals of leucite. In the lava **of** Vesuvius they are abundant, but **in** that of Etna they are rarely found.

LEUCO'NIDA. An order of the class poriphera, thus named from the body being of a white colour, the spicula of the skeleton being also calcareous.

LEU'TTRITE. A mineral found in Leuttra, in Saxony, and thus named from that circumstance. Colour, grayish-white, tinged in places with an ochreous brown.

LI'AS. A provincial name, now become conventional amongst geologists, for a kind of limestone, which, with its associated beds, form a particular group of the secondary series. The name of *lias* has been very generally adopted for a formation of argillaceous limestone, marl, and clay, usually found in comfortable stratification to the **rocks** of the oolite group. Mr. Bakewell considers that the name *lias* was probably given to this formation by the provincial pronunciation of the word *layers*, as the strata of lias limestone are generally very regular and flat, and can easily be raised in slabs from the quarry. "The great bed of dark argillaceous limestone, divided into thin strata, called *lias*, is the best characterized of all the secondary strata, (except chalk) both by its mineral characters and the fossil remains imbedded in it. The *lias* cannot be mistaken for any of the lower strata; it serves as a key to the geology of the secondary formations in England; and the first enquiry which the student should make, when he is in doubt respecting the position of any of the secondary beds, should be, *does it occur above or below the lias?*"

The upper portion of the deposits constituting the lias, including about two thirds of their total depth, consists of beds of a deep blue marl, containing only a few irregular and rubbly limestone beds. In the lower portion, the limestone beds increase in frequency, and assume the peculiar aspect which characterises the lias, presenting a series of thin stony beds, separated by narrow argillaceous partings; so that quarries of this rock at a distance assume a striped and riband-like appearance: in the lower beds of this limestone, the argillaceous partings often become very slight and almost disappear, as may be seen in the lias tract of South Wales: beds of blue marl, with irregular calcareous masses, generally separate these strata from the red marl belonging to the adjacent new red-sandstone formation. The blue lias, which contains much iron, affords a strong lime, distinguished by its property of setting under water; the white lias takes a high polish, and may be employed for the purposes of lithography.

The lias, which may be regarded as the base of the oolitic system, may be traced on the south-west to Lyme Regis, in Dorsetshire, and on the north-east to Whitby, in Yorkshire. Near the latter place, it is more fully developed than in any other part of the kingdom, and has been divided by Prof. Phillips into three parts.

The building stone obtained from the lias is, in general, when first taken from the quarry, dressed with great ease, but on exposure to the atmosphere it parts with the moisture which it possessed when underground, and becomes much harder.

The lias group is placed below the oolite, and above the variegated sandstone, in this country; in France and Germany, below the oolite, and above the Muschelkalk. It is in the lias that the petrified ink-bags of Loligo have been found. Proofs are not wanting of intervals between the depositions of the component strata of the lias. Gryphites are so abundant in it, that in France it has obtained the name of Calcaire à Gryphites; and, indeed, the Gryphite appears peculiar to, and characteristic of, the lias formation.

Lɪᴀ'ssɪᴄ. Belonging to the lias. The remains of reptiles, those of saurians in particular, are very common in the *liassic* rocks in certain parts of England.

Lɪ'ʙᴇʀ. In botany, a layer on the inner surface, or that which is contiguous to the wood, or the bark of trees; the innermost layer of the bark. The liber appears to be formed from the cambium.

Lɪ'ᴄʜᴇɴs. The third order of the class Thalogens, in the vegetable kingdom.

Lɪᴇ'ᴠʀɪᴛᴇ. An earth mineral, called also Yenite; Lievrite after Le Lieore, and Yenite, in commemoration of the battle of Jena. Lievrite is of a brown, or brownish-black colour, sometimes dull externally, but the crystals are often brilliant, and opaque; it scratches glass, but is scratched by adularia. It occurs amorphous, acicular, and also crystallized. Analysis, silex 29·278, lime 13·779, alumine 0·614, black oxide of manganese 1·587, water 1·268. Specific gravity 3·8. It is found in Elba, and also in Siberia and Norway.

Lɪ'ɢᴀᴍᴇɴᴛ. (*ligamentum*, Lat. *ligament*, Fr. *ligaménto*, It.) A strong, flexible, tough, compact, membrane, serving to keep together certain parts.

"Nothing," says Dr. Roget, "can be more artificially contrived than the interweaving of the fibres of ligaments; for they are not only disposed, as in a rope, in bundles placed side by side, and apparently parallel, to each other; but, on careful examination, they are found to be tied together by oblique fibres curiously interlaced in a way that no art can imitate. It is only after long maceration in water, that this complicated and beautiful structure can be unravelled."

In conchology, the membranaceous substance which connects the valves together; the true ligament is always external.

Li'GNIN. } Lignin is deposited, during the growth of the plant, with the intention of forming a permanent part of the vegetable structure, constituting the basis of the woody fibre, and giving mechanical support and strength to the whole fabric of the plant.
Li'GNINE. }

Li'GNITE. (from *lignum*, Lat.) Wood-coal. Lignite is brown or black. Some lignite has the appearance of jet, is of a velvet-black, does not soil the fingers, is very brittle, and burns with a bright flame. Lignite is a much more recent formation than that of common coal. By some, lignite is considered to be an imperfect coal, as wood not yet mineralized, or passed into a state of coal; while others doubt whether lignite ever becomes true coal. Lignite, like coal, is of vegetable origin, but it differs in many respects from common coal. There are several varieties of lignite; these mostly burn with flame, but they neither swell nor cake like coal.

"Lignites," says Dr. Ure, "which are manifestly bituminized wood, hold an intermediate place in the gradation between vegetable matter and pit coal. They have the fibre of the former, with the jetty lustre and fracture of the latter. Some lignites closely resemble peats in their chemical characters; others seem to graduate into perfect coal. **Lignite** has generally a woody aspect; coal always that of a rock."

"I may remark," says Hugh Miller, " that independently of their well-marked organisms, there is a simple test by which the lignites of the newer formations may be distinguished from the true coal of the carboniferous system. Coal, though ground into an impalpable powder, retains its deep black color, and may be used as a black pigment; lignite, on the contrary, when fully levigated, assumes a reddish, or rather umbrey hue."

Lignite consists of carbon 69·3; hydrogen 6·6; and oxygen and nitrogen 25·3.

Li'GULATE. } (from *ligula*, a strap, Lat.) Strap-shaped. A term applied to the radical florets of compound flowers, when shaped like a strap or ribbon. The projecting parts of the limb of an irregular corolla are called *lips*; when one lip is very long and narrow, compared to the length of the tube, the corolla is called *ligulate*, or strap-shaped.
Li'GULATED. }

Li'GURITE. (from *liguria*, Lat.) A mineral of an apple-green colour, occasionally speckled. It ranks as a gem.

Li'LALITE. Another name for the mineral *lepidolite*.

LILIA'CEOUS. (*liliaceus*, Lat.) Resembling a lily; lily-like. A corolla having six regular petals is termed a *liliaceous* corolla.

LI'LY E'NCRINITE. (The encrinites moniliformis.) So called, because the arms, when folded, resemble the head of the lily. This is one of the most beautiful of the fossil crinnoïdea, hitherto found only in the muschelkalk of the new red sandstone group. Mr. Parkinson states that, independently of the number of pieces which may be contained in the vertebral column, and which, from its probable length, may be very numerous, the fossil skeleton of the superior part of the *lily encrinite* consists of at least twenty-six thousand six hundred pieces. The body is supported by a long vertebral column attached to the ground by an enlargement of its base. It is composed of many cylindrical thick joints, articulating firmly with each other, and having a central aperture, like the spinal canal in the vertebræ of a quadruped, through which a small alimentary cavity descends from the stomach to the base of the column. From one extremity of the vetebral column to the other, and throughout the hands and fingers, the surface of each bone articulates with that adjacent to it, with the **most perfect** regularity and nicety **of adjustment.** So exact and methodical is this arrangement, even to the extremity of its minutest tentacula, that it is just as improbable, that the metals which compose the wheels of a chronometer should **for** themselves have calculated and arranged the form and number of the teeth of each respective wheel, and that these wheels should have placed themselves in the precise position, fitted to attain the end resulting from the combined action of them all, as for the successive **hundreds** and thousands of little bones that compose an encrinite, to have arranged themselves, in a position subordinate to the end produced by the combined effect of their united mechanism; each acting its peculiar part in harmonious subordination to the rest, and all conjointly producing a result which no single series of them acting separately, could possibly have effected. The pelvis of the *lily encrinite* resembles in shape a depressed vase, and, by some, it is supposed that its upper part was closed by an integument, in the centre of which was placed the mouth. The encrinite differs from the pentacrinite in having its plates, or vertebræ, rounded, whereas in the pentacrinites they are pentagonal.

LI'MA. (*lima*, Lat. a file.) A genus of bivalve shells placed by Lamark in the family Pectinides. The limæ are nearly equivalve, inequilateral, compressed, oblique, and auriculated, the anterior auricle being smaller than the other; the **hinge** consists of a triangular disk separating the umbones, divided in the centre by a triangular ligamentary pit without teeth; the anterior border has an hiatus more or less wide, a feature which, together with the triangular disk and oblique form, distinguishes them from the pectens. The Limæ are found in the shallows among corals and at small depths, they swim freely and rapidly, or rather flutter through the water, working the valves of their shell like fins or paddles. M. Deshayes in his tables records 13 species fossil in the tertiary formations, other species occur in the cretaceous and oolitic systems of rocks; in the latter more especially, the number of species exceeds that recorded from the tertiary series. Some species spin a byssus, others are free.—*Lycett.*

LIMAX. (*limax*, Lat. a snail.) The cochlea terrestris, or snail, so called from its sliminess.

LIMAX. The Latin name for those air-breathing naked gasteropodous mollusks, vernacularly known by the name of slugs. Linnæus employed the term *limax* as a generic appellation for the naked slugs, and thus defines **them**; body oblong, repent, with a fleshy shield above and a longitudinal flat disk below. A dextral lateral foramen for the genitals and excrements. Four tentacles above the mouth.

LIMB. (from *limbus*, Lat.)
1. An edge or border, as the sun's *limb*, the moon's *limb*, &c.
2. An extremity of the body, as the arm or leg.

3. In botany, the outer spreading portion of a monopetalous corolla.

LI'MBILITE. (from Limbourg, in Swabia.) A compact mineral of a honey-yellow colour, supposed to be a decomposed olivine. On exposure to the action of the blowpipe, it fuses into a compact, shining, black enamel.

LIME. The protoxide of calcium, one hundred parts consisting of 72 of calcium, its metallic basis, and 28 of oxygen. Lime does not exist in a pure state in nature, it has so strong an affinity for carbonic acid as to absorb it from the atmosphere, when it becomes converted into carbonate of lime, constituting the different kinds of marble, chalk, and limestone, and forming extensive strata, and the largest mountain ranges. Lime is a white or light grey earth, fusible only by the heat of a galvanic battery, or of a gas blow-pipe; it is exceedingly caustic, and if water be sprinkled upon it, great heat is produced, the water unites with the lime, forming a hydrate of lime. Lime is partially soluble in water, and there is a singular circumstance connected with this, namely, that cold water dissolves a larger proportion than hot water. Specific gravity 2·3.

LI'MESTONE. A genus of minerals comprising many species. However various in external appearance limestone may be, it is, if pure, essentially composed of 57 parts of lime and 43 carbonic acid; but in some rocks the limestone is intermixed with magnesia, alumine, silex, or iron. The specific gravity of limestone varies from 2·50 to 2.80. All limestones may be scraped with a knife. They are infusible; but when impure, by an intermixture with a portion of other earths, they vitrify in burning. All limestones effervesce when a drop of strong acid is applied on the surface; and they dissolve entirely in nitric or muriatic acid. The specific gravity, hardness, and effervescence with acids, taken collectively, distinguish limestone from all other minerals.

The whole of the limestone deposits have been arranged into the following suite. In the inferior or primary order, crystalline marbles; in the transition and carboniferous orders, compact and sub-crystalline limestones; in the secondary, less compact limestone, calcareous freestone, and chalk; in the tertiary or supracretaceous order, loose earthy limestones.

Primary limestone has always a granular structure; but the size of the grains is variable, and seems, in some degree, to correspond with the relative age of the mineral. Thus the limestone, which occurs in beds in gneiss, and which is supposed to belong to the older formations, has usually a coarse texture, and large granular concretions. But when its beds exist in mica slate, or argillite, its texture becomes more finely grained, and its colour less uniform. Transition limestone has a texture more or less compact; its colours are much variegated; and it often contains petrifactions. Secondary limestone has a compact texture, a dull fracture, and usually contains shells, and sometimes other organic remains. It is always stratified; but the strata are sometimes inclined, sometimes horizontal.

Sir H. De La Beche states the quantity of lime in granite composed of two-fifths quartz, two-fifths felspar, and one-fifth mica, to be 0·37; and in greenstone, composed of equal parts of felspar and hornblende, to be 7·29.

LIMNÆ'A. } (from λιμνας, a marsh, or
LIMNE'US. } pool, Gr.) A genus of fresh water univalves, placed by Cuvier in the order Pulmonea, class

Gasteropoda; and by Lamarck in the family Limnacea. The Limnea is an ovato-conical, or turretted univalve, it has an oblong spire, and the aperture higher than it is wide; it may be distinguished from the *bulini* by the very oblique fold on the columella. The Limnea has been found fossil in the neighbourhood of Paris. The recent Limnea inhabits our lakes and pools; its shell is of a light amber colour.

LIM'NITE. A fossil limnea.

LIM'OPSIS. The name assigned by Sassi to a genus of bivalve conchifera, belonging to the family Arcacea.

LI'MULUS. The Molucca crab. A genus of crustaceans, or entomostracans, having a distinct carapace or buckler, with two eyes in front of the shield. The limulus appears to approximate towards the trilobite, and Buckland says, "the history of this genus is important, on account of its relations both to the existing and extinct forms of crustaceans; it has been found fossil in the coal formations of Staffordshire and Derbyshire, and in the Jurassic limestone of Aichstadt, near Pappenheim; a small fossil species is found in the ironstone nodules of Coalbrook Dale. Of the tail of the recent limulus, savages form a point to their arrows; and, when thus armed, they are much dreaded. The eggs of the limulus are eaten by the Chinese.

LINE OF BEARING. } Strata almost always decline,
LINE OF DIP. } or dip down, to some point of the horizon, and, of course, rise towards the opposite point. A line drawn through these points is called the *line of their dip*. If a book be raised in an inclined position, with the back resting lengthwise upon the table, the leaves may be supposed to represent different strata, then a line descending from the upper edges to the table, will be the *line of dip*, and their direction lengthwise will be their *line of bearing*.

LINES OF GROWTH. In conchology, those concentric lines or markings in a shell, formed by successive layers of shelly matter, which mark its growth. The external layer is always the most recent.

LI'NEAR. (*linearis*, Lat.)
1. In entomology, a figure having the lateral margins very close together, and parallel throughout.
2. In conchology, composed of lines; being marked with lines.
3. In botany, a term applied to narrow leaves, when they are of equal breadth throughout, the two edges being straight, and equi-distant from each other.

LI'NEATE, (*lineatus*, Lat.) Marked with lines; marked with longitudinal depressions.

LI'NGULA. A genus of bivalve shells composed of two valves, nearly equal, truncated anteriorly; the hinge having no teeth: the beak of the valves pointed, and united to a tendinous tube, serving for a ligament of attachment. This is the only bivalve shell which is pedunculated. The recent lingula inhabits the Indian ocean; it has thin, horny, and greenish valves. It is also found fossil in the Ludlow limestone, and in the oolitic rocks of Yorkshire.

LI'NGULATE. (*lingulatus*, Lat.) Tongue-shaped; an epithet for leaves shaped like a tongue.

LIQUEFA'CTION. (*liquefactio*, a melting, Lat. *liquéfaction*, Fr. *liquefazióne*, It.) The act of melting; the state of being melted. This word is sometimes used synonymously with fusion, sometimes with deliquescence, and at others with solution.

LI'RICONITE. (from λειρὸς, pale, and κονία, sand or dust, Gr.) A name given to arseniate of copper. It is of a blue or green colour, and occurs in copper mines.

LI′THIA. A new alkaline substance, discovered by M. Arfwedson, a Swedish chemist, in 1818, in the mineral called Petalite: in its obvious properties, lithia approaches to potassa and soda, but it possesses a greater neutralizing power. It has subsequently been obtained from lepidolite, spodamene, and some kinds of mica. It is composed of lithium 56·50, and oxygen 43·50. It received its name from having been discovered in a stony mineral.

L′ITHIUM. The metallic base of Lithia, discovered by Sir H. Davy. This metal is obtained from lithia by de-oxidising processes; but if placed in contact with the atmosphere, it returns to the state of lithia too rapidly to admit of an accurate examination.

LI′THOCARP. (from λίθος, a stone, and καρπὸς, fruit, Gr.) Petrified, or fossil fruit.

LITHODE′NDRON. (from λίθος, a stone, and δένδρον, a tree, Gr.) A name given to coral, from its likeness to petrified wood.

LITHO′DOMUS. (from λίθος, a stone, and δέμω, to build, Gr.) A transverse, elongated, cylindrical, marine equivalve. Affixed at first by byssus to rocks, which it subsequently penetrates, and remains ever after in the cavity. It is a littoral shell, found at depths varying to ten fathoms.

LITHO′GENOUS. (from λίθος, stone, and γεννάω, to produce, Gr.) Belonging to the class of animals which form coral.

LITHOI′DAL. (from λίθος, a stone, and εἶδος, resemblance, Gr.) Resembling stone; of a stony structure.

LITHOLO′GICAL. (from λίθος, and λογικὸς, Gr.) Relating to the science of stones; in geology, a term used to express the stony character or structure of a mineral mass.

LITHO′LOGY. (from λίθος, a stone, and λόγος, discourse, Gr. lithologie, Fr.) That branch of natural history which treats of stones. "By lithology," says Prof. Jukes, "I would mean the study of the internal structure, the mineralogical composition, the texture, and other characters of rocks, such as could be determined in the closet by the aid of hand specimens."

LI′THOMARGE. (The steinmark of Werner: argile lithomarge of Haüy.) Called also stone-marrow; a variety of talc. It has commonly a fine grain, of a white, gray, yellow, red, or brown colour, these colours being sometimes disposed in spots, clouds, veins, or stripes; unctuous or greasy to the touch, and adheres to the tongue. In water it falls to powder, and does not form a paste. Specific gravity 2·4. It is infusible before the blow-pipe. It occurs massive, disseminated, globular, and in irregular lumps, in gneiss, porphyry, serpentine, &c. Jameson divides lithomarge into two sub-species, friable lithomarge and indurated lithomarge: the friable is characterised by its scaly particles, soiling, and low degree of coherence; the indurated, by fracture, streak, softness, and sectility. The Chinese are said to use it, when mixed with the root of veratrum album, instead of snuff.

LITHO′PHAGI. (from λίθος and φαγεῖν, to eat, Gr. lithophage, Fr.) Molluscs which eat holes in stones and rocks. While the lithodomi penetrates rocks by chemical action, dissolving the stony matter, lithophagi mechanically perforate, or bore into them. They belong to Lamarck's family of Lithophagidæ.

LITHOPHA′GIDÆ. A family of terebrating bivalves.

LI'THOPHYTE. (from λίθος, a stone, and φυτὸν, a plant, Gr. *lithophyte*, Fr)
1. A stony plant; a coral.
2. The animal which secretes coral.

LITHO'RNIS. (from λίθος, a stone, and ὄρνις, a bird, Gr.) The name assigned by Prof. Owen to a subgenus of fossil birds of the vulture kind, one species of which, Lithornis Vulturinus, he has described.

LITHO'XYLE. (from λίθος, stone, and ξύλον, wood, Gr.) Silicified wood.

LITTO'RNIA. A genus of turbinated univalve shells, they are thick, have few whorls with a short spire, the aperture is large, entire, rounded anteriorly, the outer lip thickened and the columella rather excavated, the operculum is horny and spiral; this latter character separates it from the Turbo. The species are numerous, both recent and fossil.—*Lycett*.

LI'TUITE. A fossil shell found in the transition limestone together with the Orthoceratite. The *lituite* is a chambered shell, partially coiled up into a spiral form at its smaller extremity, its larger end being continued into a straight tube of considerable length, separated by transverse plates, outwardly concave, and separated by a siphuncle.

LITU'OLA. A multilocular univalve; a genus of microscopic foraminifera, partly spiral, the last turn being straight at the end: found both recent and fossil.

LI'ZARD. (*lézard*, Fr. *lacertus*. Lat.) In Cuvier's arrangement the lizards form the second genus of Lacertinida. They are distinguished by the tongue, which is thin, extensible, and terminating in two threads. The extremity of the palate is armed with two rows of teeth, which are generally either recurvated or conical. Lizards have usually a single perforated eye-lid, which, when closed by its orbicular muscle, exhibits merely a horizontal slit. The body is naked, with four feet and a tail; and they possess the property of reproducing the tail should it be lost.

LLANDEI'LO FLAGS. The name given to the lower member of the lower Silurian rocks, forming the base of the Silurian system. "They have been named," says Sir R. Murchison, "after the town of Llandeilo, in Caermarthenshire, where they are largely developed. They consist of hard, dark-coloured flags, sometimes slightly micaceous, frequently calcareous, with veins of white crystallised carbonate of lime, and are especially distinguished by containing the large trilobites, Asaphus Buchii and Asaphus tyrannus; Agnostus M'Coyii, Trinucleus fimbriatus, &c."

LOAM. (*lehm*, Germ.) An earthy mixture, in which sand and clay form large proportions: when the compound contains much calcareous matter it is usually called marl. Any soil which does not cohere so strongly as clay, but more strongly than chalk, is designated loam. Loam may be defined, a soft and friable mixture of clay and sand, enough of the latter being present for the mass to be permeable by water, and to have no plasticity.

LO'BATE. ⎫
LO'BATED. ⎬ 1. In entomology, when the margin is divided by deep undulating, and successive incisions.
LO'BED. ⎭

2. In botany, applied to leaves, when the margins of the segments are rounded; according to the number of lobes, the leaf is termed bilobate, trilobate, &c.

LOBE. (*lobus*, Lat. *lobe*, Fr. *lobo*, It.)
1. A rounded portion of certain bodies, as the lobes of the brain, the lobe of the ear, the lobes of the lungs, liver, &c.
2. In botany, the cotyledon of the seed is also called the lobe.

LO'BULE. The diminutive of lobe; a little lobe.

LOCOMO'TION. (from *locus* and *motio*, Lat.) The power of moving at will from one place to another; of transferring the whole body from one place to another. The power of locomotion constitutes the most general and palpable feature of distinction between animals and vegetables. Excepting a few among the lower order of creation, such as molluscs and zoophytes, all animals are gifted with the power of spontaneously changing their situation.

LOCULI'CIDAL. In botany, a particular kind of dehiscence. Some fruits open by the dividing of each carpellum at its midrib, so that the dissepiments stick together, and to two halves of contiguous carpella; this is called *loculicidal dehiscence*.

LODE. (a mining term.) A word used to signify a regular vein or course, whether metallic or not; but most commonly it signifies a metallic vein. When the substances forming the lodes are reducible to metal, the lodes are said to be alive; otherwise, they are termed dead lobes.

LOËSS. (Germ.) A provincial German term for an alluvial tertiary deposit of calcareous loam, occurring in detached patches throughout the valley of the Rhine. In Alsace the loëss is termed *lehm*. It encloses freshwater and land shells, as well as some mammiferous remains. Sir C. Lyell observes, "the loëss is found reposing on every rock, from the granite near Heidelberg to the gravel of the plains of the Rhine. It overlies almost all the volcanic products, even those which have the most modern aspect; and it has filled up, in part, the crater of the Rodenberg, at the bottom of which a well was sunk in 1833, through seventy feet of loëss."

LOLI'GO. (*loligo*, Lat.) A genus of the family of Sepiæ. In the loligo is found that peculiar provision for defence, the ink-bag, a bladder-shaped sac, containing a black and viscid ink, the ejection of which, by rendering the surrounding water dark and opaque, defends the animal from the attacks of its enemies. In the lias of Lyme Regis, ink-bags of the fossil loligo are preserved, still distended, as when they formed parts of the organization of living bodies, and retaining the same juxta-position to an internal rudimentary shell resembling a horny pen, which the ink-bag of the existing loligo bears to the pen within the body of that animal.

LO'MONITE. Diatomous Geolite. So named after its discoverer, Gillet Laumont. See *Laumonite*.

LONCHO'PTERIS MANTE'LLI. A species of fossil fern found in the shales and clays of Tilgate Forest, and thus named after Dr. Mantell. It is characterised, by the distribution of the nervures of the leaves. This fern probably did not exceed a few feet in height.

LONDON CLAY. (The name London Clay has been assigned to this great argillaceous formation from the circumstance of its forming the general substratum of London and its vicinity, occurring immediately beneath the vegetable soil, excepting when occasional deposits of alluvial, or diluvial, gravel, sand, &c., intervene.) This formation consists of a bluish or blackish clay, including, in some localities, beds of grey limestone and sandstone, lying immediately over the plastic clay and sand, and is an upper member of the arenaceous and argillaceous formation that covers the chalk. Its thickness is very considerable, sometimes exceeding 700 feet. It contains layers of ovate, or flattish masses of argillaceous limestone. These masses, called septaria, are sometimes continued through a thickness of two hundred feet; of these, Parker's cement is made. The septaria lie horizontally, and are disposed at unequal distances from each other in seemingly regular layers. They frequently include portions of wood pierced by teredines, nautili, and other shells; and it is a curious fact that septa of calcareous spar frequently intersect the substances contained in the septaria. These septaria were at one time deemed characteristic of the London Clay, but they have been found in other formations, more particularly in the upper parts of the Wealden. Sulphuret of iron, phosphate of iron, and selenite are found interspersed throughout the London Clay; on which account the water issuing from it is not fit for domestic purposes. Amber and fossil copal or resin have been found in this deposit. From the London Clay three or four hundred species of testacea have been procured, but the only bones of vertebrated animals are those of reptiles and fish. Remains of turtles have been dug out of this deposit at Highgate and Islington, and some bones of a crocodile were discovered by Mr. Parkinson; nautilites also are found in it. The shells of the London clay mostly belong to genera inhabiting our present seas. The following is principally extracted from Conybeare and Phillips's Geology of England and Wales.

An idea of the nature and composition of the London Clay formation may be arrived at by the following section afforded by a well sunk at Tottenham, in Middlesex.

1. Immediately below the surface was found brick earth and coarse yellow sand, stiff clay, and marl. 20
2. Blue clay of various intensity of colour and degrees of stiffness, adapted for tile-making. It effervesced slightly, and enclosed hard and irregular masses of a lighter colour, full of minute appearances of charred vegetable matter, and septaria, which also effervesced. 60
3. Blue clay of a greasy aspect and somewhat greasy to the touch; it did not effervesce. 20
4. Purple, blue, red and brown clay mixed, having greatly the appearance of some varieties of lithomarge; it did not effervesce. 10
5. Blue, white, and brown clay mixed, much heavier than the preceding: it contained very compact, hard, and nearly cylindrical masses, six to twelve inches long, and of a yellowish-white colour: it effervesced strongly. 10
6. Yellowish white clay, frequently compact and hard, and equally heavy with the preceding: it effervesced strongly. 3
7. Rock bored through. 2

*Feet*, 125

There are very few genera of recent shells which have not some representative imbedded in this formation, but the specific character is usually different; on the other hand but few of the extinct genera, so common in the older formations, occur in this, so that it seems to hold a middle character in this respect between the earlier and more recent beds. Thus though nautilites resembling those of the Indian seas are common, specimens of the cornu ammonis and the belemnite are so rare, that it is very doubtful whether they have ever been found. Echinites, so common in the chalk, are very rare in this formation.

The most interesting facts connected with the vegetable remains found in the London clay, are those which have been observed in the Isle of Sheppey: the quantity of fruit or ligneous seed vessels is extraordinary. Mr. Crowe, of Faversham, has procured from this productive situation a very large collection, from which he has selected 700 specimens, none of which are duplicates, and very few of which agree with any known seed-vessels. Among these are many which appear to belong to tropical climates. Mr. Bowerbank observes, "among the numerous and highly interesting fossils found in the London clay, none are more abundant than the remains of fruits and seeds." He also states that during a few years 120,000 fruits and seeds have passed through his hands. In these beautiful remains of an extinct flora, the minute and delicately formed vegetable tissues are preserved in the most perfect manner. As regards the extent of the London clay deposit, it forms the superior stratum of the chalk basin of London, except where it is partially covered by the sands of the upper marine formation, or by alluvial sands, gravel, and loam. It extends uninterruptedly and in a south-westerly direction from Orford, on the coast of Suffolk, about 20 miles north-east of Harwich, and a little to the north of Ipswich, in that county; to the south of Coggeshall, and thence to Roydon, in Essex; from this place it turns nearly south, extending to a little on the west of Edmonton, in Middlesex, and thence in a north-westerly direction by Chipping Barnet and South Mims to the north of Ridge Hill; here it turns suddenly southward, and afterwards **south**-west by Harfield and Ux**bridge** to the eastward of Colebrook; **it then** passes away nearly west, **crosses** the Thames by Windsor and Twyford, and passes to about three miles south-west of Reading, which is its most western point. It then turns to the south-east, in an irregular line, to within a very short distance of Farnham and Guildford, in Surrey, and on by Epsom, and a little to the north of Croydon to Deptford, in Kent. The London clay, therefore, constitutes a very large part of the soil of Suffolk, nearly the whole of Essex, quite to the sea, the whole of Middlesex, and portions of Berkshire, Surrey, and Kent. In this last county it shows itself on the northern side of the Medway; it constitutes the whole of the Isle of Sheppy, rapidly disappearing under the force of the waves, the cliff from Whitstable to Reculver, and extends nearly to Canterbury, and thence to Boughton Hill. In the chalk basin of Hampshire this deposit is also extensively developed, forming the whole line of coast from Worthing in Sussex to Christchurch in Hampshire, extending thence inland by Ringwood, Romsey, Fareham, and to the southward of Chichester to Worthing. It is found also in the Isle of Wight, thrown with the subjacent chalk into a nearly vertical position. With this exception, the beds of the London clay are nearly horizontal.

It has been remarked, that if the description of the Paris rocks had not preceded that of the country round London and of the Isle of Wight, it never would have been considered that the Plastic clay was separated from the London clay, but rather that they constituted different terms of the same series. The London clay belongs to the eocene period.

LONGICO'RNES. (from *longus* and *cornu*, Lat. long horned.) A family of insects in Cuvier's arrangement, and so named from the length of their antennæ, which are filiform and cetaceous, and usually as long, often longer, than the body of the insect.

LONGIP'ENNES. (from *longus* and *penna*, Lat. long wings.) A family of birds in Cuvier's arrangement, including those birds, which, from the great strength of their wings, whence they derive their name, are to be met with in all latitudes: the bill, in some genera is hooked at the end, in others simply pointed.

LONGIRO'STRES. (from *longus* and *rostrum*, a beak, or bill, Lat.) A family of birds comprising the waders, or birds with long bills.

LO'NGITUDE. (*longitudo*, Lat. *longitude*, Fr. *longitúdine*, It.) The distance of any part of the earth to the east or west of any place. The meridian passing through the observatory at Greenwich is assumed by the British as a fixed origin, whence terrestrial longitudes are measured. As each point on the surface of the earth passes through 360°, or a complete circle, in twenty-four hours, at the rate of 15° in a hour, time becomes a representative of angular motion. Hence, if the eclipse of a satellite happens at any place at eight o'clock in the evening, and the nautical almanack shows that the same phenomenon will take place at Greenwich at nine, the place of observation will be 15° of west longitude. In the case of stations differing only in *latitude*, the same star comes to the meridian at the same *time*, but at different *altitudes*. In that of stations differing only in *longitude*, it comes to the meridian at the same *altitude*, but at different *times*. Supposing, then, that an observer is in possession of any means by which he can certainly ascertain the time of a known star's transit across his meridian, he knows his *longitude*; or if he knows the difference between its time of transit across his meridian and across that of any other station, he knows the difference of longitudes between those two places.

LOPHI'ODON. (from λοφὶς and οδοὺς, Gr.) A fossil genus of mammalia, now entirely extinct, allied to the tapir, rhinoceros, and hippopotamus, and connected with the Anoplotherium, and Palæotherium; so named from certain points, or eminences, on the teeth. Although this genus is not widely removed from the tapirs, Cuvier thought it desirable to separate it. Like the tapirs the Lophiodon has six incisors and two canines in each jaw.

LOWER CHALK. The chalk formation or series is generally divided into six distinct members, namely, the lower green-sand; the gault; the upper green-sand; *the chalk without flints, or the lower chalk;* the chalk with flints, or the upper chalk, and the Maestricht beds. This arrangement is, however, altered by some writers, inasmuch as a more minute subdivision of some of the members is concerned, depending on local appearances. Generally speaking, the lower chalk may be distinguished from the upper by the absence of flints, and by the superior hardness of the chalk, which is sometimes used for building-stone. The lower is regularly stratified. In the north of England, Professor Phillips states, "The *lower chalk* is of a red colour, and flints are found in it. The only mineral found in the lower chalk is sulphuret of iron. The fossil remains are very numerous and all of them marine."

LOWER SILURIAN ROCKS. In this division of the Silurian rocks, Sir R. Murchison places the Caradoc sandstones and the Llandeilo flags; these are described under their particular names.

LOWER NEW RED SANDSTONE. (The Rothe-todte-liegende of the Germans; the Grès des Vosges—couches inférieures, of the French. The lowest subdivision of the New Red System. For a description see *Rothe-todte-liegende*. It being now fully proved by Sir R. Murchison that the lower new red sandstone graduates into the coal-measures, a practical acquaintance with this division of the new red system becomes a matter of great national importance. The maximum thickness of the lower new red sandstone hitherto observed is 1000 feet; this formation is interpolated between the magnesian limestone and the coal-measures.

- LOWER OLD RED SANDSTONE. This formation found in Scotland is the representative of the Tilestone of the Old Red System of England, but in extent of vertical development it far exceeds it. The tilestones compose the least of the three divisions in England; their representative in Scotland forms by much the greatest of the three.—*Hugh Miller*.
- LOWER LUDLOW ROCK. The name given to the third or lowest subdivision of the Ludlow formation. The strata of the lower differ from those of the upper Ludlow rock in being more argillaceous, less sandy and calcareous, with rarely a trace of mica. They constitute, in fact, says Sir R. Murchison, a great argillaceous mass strictly entitled to the provincial name of *mudstone*. The organic remains of the lower Ludlow rock are upon the whole very different from those of the upper, as well as the Aymestry limestone: for although some species of shell are common to the whole formation, the lower Ludlow rock is characterized by many peculiar remains, including two new genera which have not been observed in any overlying stratum; namely, the conchifer Cardiola, and the chambered shell Phragmoceras. These, with the Orthoceras filosum

and Orthoceras pyriforme; the Lituites giganteus, and the Graptolites Ludensis, are peculiar and distinguishing fossils.

LO'ZENGED. In entomology, of a quadrangular shape, with two opposite angles acute, and two obtuse.

LUCERNARO'IDA. The first order of the class Hydrozoa, comprising sertularia, and similar zoophytes.

LUCI'NA. A genus of equivalve bivalve shells belonging to the Nymphacea of Lamark. They are orbicular, flattened, and radiately striated; the hinge has usually two small cardinal teeth, and two lateral teeth on each side of the umbo in one valve, and one in the other; the ligament is external, which distinguishes it from Amphidesma. The species are numerous, both living and fossil, more especially the latter.

LUCU'LLITE. (from *Lucius Lucullus*, a celebrated Roman, who is said greatly to have admired it.) A black variety of transition limestone, a black marble.

LUDLOW FORMA'TION. } The term has
LUDLOW ROCKS. } been selected because the town of Ludlow is built upon the upper beds. The Ludlow rocks form one of the four divisions of the Silurian system, and have been sub-divided into the Upper Ludlow rocks, the Aymestry limestone, and the Lower Ludlow rocks. They consist of beds of sandstone, shale and limestone. The upper Ludlow rocks of slightly micaceous, gray-coloured, thin-bedded sandstone: the Aymestry limestone of sub-crystalline grey or blue argillaceous limestone; the lower Ludlow rocks of sandy, liver and dark coloured shale and flag, with concretions of earthy limestone. These rocks are developed principally in Shropshire, Herefordshire, Worcestershire, Staffordshire, and Gloucestershire, in England, and in Radnorshire, Brecknockshire, Montgomeryshire, Glamorganshire, and

Caermarthenshire, in Wales. Organic remains are found in all three sub-divisions. The Ludlow formation, says Sir R. Murchison, is the key which accurately reveals to us the relations of the inferior masses to the overlying strata.

Lu′nated. } (*lunatus*, Lat.) Crescent-shaped; formed like a half-moon.
Lu′nulated. }

Lu′nule. In conchology, a crescent-like mark or spot, situated near the anterior and posterior slopes in bivalve shells.

Lu′nulet. In entomology, a half-moon shaped spot in insects, of a different colour from the rest of the body.

Lunuli′tes. A genus of foraminated polypifers. A free, stony, circular polypifer, with one side convex, the other concave. The convex side striated in rays, with interstitial pores; the concave side radiated with diverging rugæ and grooves.

Lu′stre. In mineralogy, one of the physical characters by which minerals may be recognized. It is of several kinds, as metallic and pseudo-metallic, adamantine, pearly, silky, resinous, vitreous, waxy, &c. In the absence of lustre, a mineral is said to be dull.

Lutra′ria. A genus of bivalves, placed by Lamarck in the family Mactracea. A thin, transverse, inequilateral shell, gaping at the extremities; two oblique and diverging hinge-teeth accompanying a large pit for the cartilage. No lateral teeth, in which feature it differs from Mactra.

Lycopodia′ceæ. The first order of the class Anogens. The club-mosses; or club-moss tribe. Plants of an inferior degree of organization to coniferæ, some of which they greatly resemble in their foliage. This tribe, at the present day, contains no species more than three feet high, while many of the fossil species are as large as recent coniferæ, having attained to the size of forest trees. The affinities of existing *lycopodiaceæ* are intermediate between ferns and coniferæ on the one hand, and ferns and mosses on the other. They are related to ferns in the want of sexual apparatus, and in the abundance of annular ducts contained in their axis; to coniferæ in the aspect of their stems; and to mosses in their general appearance. The leaves of existing lycopodiacea are simple, and arranged in spiral lines around the stem, and impress on the surface scars of rhomboidal, or lanceolate form, marked with prints of the insertion of vessels.

Lycopodi′tes. 1. Fossil plants of the genus Lycopodium.
2. A genus of plants of the club-moss tribe.

Ly′dian stone. The Lydischer-stein of Werner; La pierre de Lydie of Brochant; Basanite of Kirwan; Lydian stone of Jameson. A variety of siliceous slate; a black siliceous flint-slate, called by some black jasper. It differs but little from the common variety of siliceous slate. Colour grayish or bluish-black, sometimes quite black. Prof. Jameson constitutes Lydian stone a sub-species of flinty-slate. Specific gravity from 2·58 to 2·62. It is employed as a test or touchstone to determine the purity of gold and silver, whence its name Basanite, given to it by Kirwan, from βάσανος, Gr. the trier. It obtained its present name from having been first noticed in Lydia. It is described by Pliny and Theophrastes.

Lymph. (*lympha*, Lat. water.) A colourless liquid, found in the lymphatics.

Lymph of plants. During the vegetation of plants, there is a juice continually ascending from their roots. This is called the sap, or *lymph* of plants. From experiments made by Vauquelin, it was ascertained that the lymph of the common elm consisted as follows:—Of 1039 parts, 1027·904 water and volatile matter, 9·240 acetite of potash, 1·060 vegetable matter, 0·796 carbonate of lime.

LYMPHA'TICS. Minute vessels pervading every part of the body, absorbing and conveying the absorbed matter into the thoracic duct, to be afterwards conveyed into the blood. The lymphatics are supplied within with valves, and without with glands.

LY'RATE. (from *lyra*, a harp, Lat.) Lyre-shaped; a term applied to leaves divided transversely into several segments, the segments gradually increasing in size as they approach the extremity of the leaf.

# M

MAC'ACUS. The fossil monkey, of which two species are described by Professor Owen, namely, M. Eocœnus, and M. Rhesus. "Cuvier says, our great comparative anatomist had met with no evidence of any species more highly organized than a bear or a bat, in the fossiliferous strata which formed the theatre of animal life anterior to the record of the human race. I have been so fortunate, in my researches of the fossil mammalia of Great Britain, as to determine not only the remains of extinct pachydermal animals (Lophiodon and Hyracotherium) in the Eocene beds called the London clay, but, likewise of a Quadrumane, or monkey, in a sandy stratum of the same formation."

MACHAI'RODUS. The name assigned by Dr. Kaup to an extinct genus of carnivorous mammalia, distinguished by their falciform teeth, and found in the newer tertiary deposits of Italy, Germany, France, and Great Britain. Prof. Owen, in describing one species, the M. Latidens, says, it was a feline animal as large as the tiger, and, to judge by its instruments of destruction, of great ferocity. The canines curved backwards, in form like a pruning knife, having the greater part of the compressed crown provided with a double-cutting edge of serrated enamel; that on the concave margin being continued to the base; the convex margin becoming thicker there, like the back of a knife, to give strength. Thus each movement of the jaw with a tooth thus formed, combined the power of the knife and saw.

MACHAI'RODUS. An extinct animal of the order Mammalia, referrible to the Miocene period, and allied to the bear.

MACI'GNO. The Italian word for a kind of stone, a siliceous sandstone sometimes containing calcareous grains, &c. Macígna piétra, a very hard stone; an eocene rock of Italy.

MA'CLE. The Hohl spath of Werner; Hollow spar of Jameson. Macle occurs only in crystals, the form of which is a four-sided prism. But each crystal, when viewed at its extremities, or on a transverse section, is obviously composed of two very different substances; and its general appearance is that of a black prism, passing longitudinally through the axis of another prism, which is whitish. These crystals, often long, are sometimes very minute; in some instances their edges are rounded. The crystals of Macle present a considerable number of natural joints, which lead to an octohedron for their primitive form. Macle scratches glass; its powder is soft and unctuous. It is opaque, or sometimes translucent. Colour, white or gray, often shaded with yellow, green, or red. Specific gravity 2·94. It is found, generally, imbedded in black argillaceous slate.

MACLU'REA. A genus of fossils of the Silurian formation.

MACLU'REITE. Called also Brucite and Chondrodite. A mineral occurring in imbedded grains in small massive pieces, and in longish granular concretions. Colours yellow, straw-colour, orange, red, and brown; translucent; scratches glass; fracture imperfectly conchoidal. Specific gravity 3·15 to 3·50. It consists of magnesia 54·0, silica 36·60, fluoric acid 4·0, oxide of iron 2·30, potash 2·0, manganese a trace.

MACROPO'MA. The name given by M. Agassiz to a genus of sauroid fishes, the fossil remains of which have been discovered in the chalk formation. The scales of the Macropoma are studded with hollow tubes, through which, it is stated, there flowed a fluid which served to lubricate the surface of the body.

MACRODA'CTYLUS. (from μακρὸς, long, and δάκτυλος, finger, Gr.) The name given to a family of birds in Cuvier's arrangement, having very long toes. The coot, rail, &c., are examples.

MA'CRODON. (*Lycett*.) Figure resembling Byosoarca, with a similar aperture in the lower border; several teeth placed at the anterior extremity of the hinge plate are parallel and directed obliquely downwards and backwards; one or two elongated teeth or plaits, occupy the length of the hinge plates. Macrodon Hirsonensis occurs in the great and inferior oolite of the Cotteswolds.

MACROSPO'NDYLUS. A fossil saurian found in the oolite and lias formations.

MACROSTO'MATA. (from μακρὸς, long, and στόμα, mouth, Gr.) A family of univalves, belonging to the order Trachellipoda, comprising the genera Stomata, Stomatella, and Haliotis.

MACROU'RA. (from μακὸς, long, and οὐρά, a tail.) A family of crustaceans, including the lobster, prawn, shrimp, &c. They are so named from their having a long tail, which is, at least, as long as the body, and provided at its termination with appendages which most frequently form a fin on each side. This tail is always composed of seven distinct segments. Fossil genera of the family Macroura have been found in the Muschelkalk and in the lias.

MA'CTRA. (μάκτρα, Gr. *mactra*, Lat. a kneading-trough.) A genus of equivalve, inequilateral, transverse bivalves, slightly gaping at the extremities; the hinge, or middle tooth, complicated; lateral teeth rather remote, compressed, and inserted. Shells of this genus have only been found to inhabit the ocean, at depths varying from ten to twelve fathoms, in sands and sandy mud. The French naturalists divide *Mactra* into two genera, Mactra and Lutraria. In Turton's Linné twenty-seven species are described; twelve are inhabitants of our seas. The fossil species belong to the tertiary formations.

MACTROMY'A. A genus of fossil bivalve shells found in the Jurassic rocks, they are equivalve, rather globose and rugose; the hinge is without teeth, on the posterior side of the umbo is an hiatus surrounded by a thickened laminal plate to support the ligament. One species occurs in the Lias, two in the Inferior and Great Oolite in England.—*Lycett*.

**MA′DREPORE.** (*madrépore*, Fr. *corps marin pierreux, qui resemble à des rameaux, à une végétation.*) "A stony polypifer, fixed, subdendroidal, ramified; the surface furnished on every part with projecting, muricated cells: the interstices porous. The cells scattered, distinct, cylindrical, tubular, and prominent, hardly any stella; the lamellæ of the internal parietes very narrow."—*Parkinson.* In a living state, the stony matter is covered with a skin of living gelatinous matter, fringed with little bunches of tentacula; these are the polypi: the skin and the polypi contract on the slightest touch. Madrepores are sometimes united and sometimes detached; where the laminæ take a serpentine direction they are called meandrina, or brain-stone.

The term Madrepore is generally applied to all those corals which have superficial star-shaped cavities. Madrepores raise up walls and reefs of coral rocks with astonishing rapidity, in tropical climates.

**MA′DRE′PORITE.**
1. Fossil madrepore.
2. A variety of limestone, found in large rounded fragments, composed of numerous, small prisms, nearly cylindrical. Opaque; surface dark brown; fracture conchoidal and black. Constituent parts, carbonate of lime 63, silex 13, alumine 10, oxide of iron 11.

**MAE′STRICHT BEDS.** The name given to the uppermost member of the cretaceous group, from Maestricht, a town of the Netherlands. The Maestricht beds are marine, and composed of a soft yellowish-white limestone, resembling chalk, and containing siliceous masses, ammonites, hamites, hippurites, baculites, &c. The siliceous masses found in these beds are not composed of black flint, but of chert and calcedony. The Maestricht beds repose on the upper chalk with flints. Similar beds occur also at Faxoe, in Denmark. Some of the fossils are cretaceous, but none are tertiary. Deshayes has been unable to identify any of the shells of the Maestricht beds with those of the tertiary deposites.

**M′AGAS.** A genus of Brachiopodous shells. They are equilateral, inequivalve, one valve convex with a triangular area, divided by an angular sinus in the centre. The other valve flat, with a straight hinge line and two small projections, a partial longitudinal septum with appendages attached to the hinge within. Two species are recorded from the chalk of Norfolk.—*Lycett.*

**MA′GILUS.** A genus of univalve shells belonging to the family Cricostomata, according to the arrangement of De Blainville, and to the order Tubulibranchiata of Cuvier. The shell is thick, tubular, and irregularly contorted, having a longitudinally carinated tube, at first regularly spiral, and then extending itself in a line more or less straight. The young of the genus Magilus has a very thin shell of a crystalline texture, but, when it has attained its full size, and has formed for itself a lodgment in a coral, it fills up the cavity of the shell with a glassy deposite, leaving only a small conical space for its body; it continues to accumulate layers of this material, so as to maintain its body at a level with the top of the coral to which it is attached, until the original shell is quite buried in this vitreous substance.—*Roget. Cuvier. Sowerby.*

**MAGNESIA.** (*magnesis*, Fr.) An earth with a metallic basis called magnesium. Magnesia consists of magnesium 61·4, oxygen, 38·6. Magnesia is rarely found pure in a native state. It enters into the composition of some of the primary rocks, to which it usually imparts a saponaceous feel, producing also a striated texture, and frequently a greenish shade. Magnesia first became known about the beginning of the eighteenth century. Little, however, was known concerning its nature, till Dr. Black published his celebrated experiments on it 1755. Magnesia may be thus prepared: sulphate of magnesia, a salt, composed of magnesia and sulphuric acid, is to be dissolved in water, and half its weight of potass added. The potass having a stronger affinity for the sulphuric acid than the magnesia has, seizes the sulphuric acid, and the magnesia is precipitated. Magnesia is often present in chalk; some of the strata in France are said to contain ten per cent. of magnesia. Magnesia is present in all the inferior stratified rocks, with the exception of quartz rock, (without mica,) and certain eurites, or compact felspars. In the detrital rocks it is also common, particularly when mica forms any considerable portion of them. There are few limestones which do not contain magnesia. It is an essential ingredient of dolomite, carbonate of magnesia constitutes more than 40 per cent. of that rock. Magnesia is also disseminated through the waters of the ocean, muriate of magnesia forming from ·004 to ·005 of their mass.

MAGNE'SIA MICA. Called also Biotite. See *Biotite*.

MAGNE'SIAN LI'MESTONE. (The Zechstein of the German, the Calcaire Alpin of the French geologists.) The Magnesian limestone, though somewhat extensively deposited in England and Germany, appears but little known in France. A marine deposit, belonging to the new red sandstone group. It lies above the red conglomerate and below the variegated sandstone. It is composed of carbonate of magnesia, the proportion of the latter amounting to nearly one half in some instances. It effervesces much more slowly and feebly with sulphuric, nitric, or muriatic acids than does the common limestone. The magnesian limestone of this country is a dolomite of a yellow or yellowish-brown colour; it is distinctly stratified, the strata varying from a few inches to several feet in thickness. This deposit is fossiliferous, and certain shells, *productæ*, appear for the first time in the magnesian limestone. Magnesian limestone forms the most durable building-stone, and it is of this that the two new houses of parliament are built. It is to be lamented that Waterloo Bridge was not built of magnesian limestone instead of felspathic granite, a very perishable kind of stone. Where the magnesia is in excess the land is sterile, but when it is not in excess, the soil is fruitful, and, as a subsoil, healthful.

MA'GNESITE. A mineral of a white or yellowish-white colour. It occurs massive, tuberose, reniform, and versicular: fracture is conchoidal: opaque: sp. gr. 2·881: infusible, and before the blow-pipe becomes so hard as to scratch glass. It is composed of one equivalent of carbonic acid and one of magnesia.

MAGNE'SIUM. The metallic basis of magnesia.

MAGNE'TIC IRON ORE. The Fer oxydulé of Haüy; Magnet eisenstein of Werner. A black ore, possessing a slight metallic lustre. Occurs regularly crystallized; in granular concretions; massive and disseminated. It is magnetic, sometimes sufficiently so to take up a needle. It occurs in beds in primary and transition rocks. This ore is very common in Sweden.

MA'GNETISM. (*magnétisme*, Fr.) The tendency of the iron towards the magnet, and the power of the magnet to produce that tendency; the power of attraction. Very delicate experiments have shewn that all bodies are more or less susceptible of magnetism. Many of the gems give signs of it; titanium and nickel always possess the properties of attraction and repulsion. But the magnetic agency is most powerfully developed in iron, and in that particular ore of iron called the loadstone, which consists of the protoxide and peroxide of iron, together with small portions of alumina and silica. A metal is often susceptible of magnetism if it contain only the 130,000th part of its weight of iron, a quantity too small to be detected by any chemical test. One of the most distinguishing tests of magnetism is polarity, or the property a magnet possesses when freely suspended, of spontaneously pointing nearly north and south, and always returning to that position when disturbed. Induction is the power which a magnet possesses of exciting temporary or permanent magnetism in such bodies in its vicinity as are capable of receiving it. By this property the mere approach of a magnet renders iron or steel magnetic, the more powerfully the less the distance. Iron acquires magnetism more rapidly than steel, yet it loses it as quickly on the removal of the magnet, whereas the steel is impressed with a lasting polarity.

MAJO'LICA. A local name for a variety of white compact limestone.

MA'LACHITE. (*malachite*, Fr *malachite*, It.) Green carbonate of copper.

M'ALACOLITE. A variety of augite, of a darkish green colour.

MALACOPTE'RYGII. (from μαλακὸς, soft, and πτερὸν, a wing or fin, Gr.) One of the two great orders into which Cuvier divided all bony fishes. The rays of the fins are thin, flexible, articulated, and branched; each ray somewhat resembles a jointed bamboo, with this difference, that what veins a single ray at bottom, branches out into three or four rays atop. The gold-fish serves as a familiar illustration of this order.

MALACOPTE'RYGIONS. Soft finned; belonging to the order Malacopterygii.

MALACO'STRACAN. An order of Crustaceans, distinguished by having sessile eyes, imbedded in the substance of the head.

MALACTI'NIA. The name of a class of animals belonging to the order Cyclo-neura or radiata. These are soft, free, aquatic animals, of a simple structure, entirely marine, generally of a transparent gelatinous texture, and radiated structure or form, luminous, and emitting an acrid secretion from their surface, which is capable of irritating and inflaming the human skin like the sting of a nettle.

MALLEABI'LITY. (*malleabilité*, Fr.) The property or capability of being hammered into different forms without breaking.

MA'LLEABLE. (*malleable*, Fr. *malleábile*, It.) That may be spread out by hammering. Of all the metals, the most malleable is gold, five grains of which may be hammered out so as to cover a surface of 273 square inches, the thickness of the leaf not exceeding $\frac{1}{282020}$th of an inch.

MA'LLEUS. (*malleus*, Lat. a hammer.) 1. One of the bones of the ear, thus named from its supposed resemblance to a hammer.
2. A bivalve shell of the family Malleacea.

MALM ROCK. The name given to a variety of firestone, a member of the chalk series.

MA'LTHA. A variety of bitumen;

N N

called also mineral pitch, from its resemblance to pitch. Colour black or dark brown. Specific gravity from 1·45 to 2·07.

MA'MMA. (*mamma*, Lat.) The breast.

MA'MMAL. An animal belonging to the class mammalia.

MAMMA'LIA. (from *mamma*, Lat. the breast.) The highest class of **ani**mals is that which comprehends man, **and** animals, which, **like** man, possess a viviparous mode of generation. The class mammalia is divided by some authors on natural history into two principal sections, namely, Placentalia and Marsupialia, or Implacentalia. These animals have a heart consisting of four cavities, two auricles and two ventricles; they have hot and red blood. Their most essential character is that of their being viviparous, and suckling their young on milk, secreted in mammary glands, which open by ducts or teats, and they are thence called *Mammalia*. The *mammalia* are placed at the head of the animal kingdom, not only because it is the class to which man himself belongs, but also because it is that which enjoys the most numerous faculties, the most delicate sensations, the **most** varied powers of motion, and in which all the different qualities seem combined in order to produce a more perfect degree of intelligence —the one most fertile in resources, most susceptible of perfection, and least the slave of instinct. The muscular system and the living movements of mammalia are more varied than in any other vertebrated class, for some are organized to plough the deep, or to clamber on the rocky coasts; some to burrow in the earth, or to bound over the plains; some to gambol on lofty trees or cliffs, or to wing their way through the air, and these different conditions affect more especially the organs of motion. The bones of all the mammalia are nearly of the same colour and general appearance as those of man; they are covered with a periosteum, and contain marrow, which in the whale tribe is fluid. The skeleton of mammalia is divided into head, trunk, and extremities. Next to man, the **ape** tribe is found to have the largest skull in proportion to the face, but, even in the ape, it is found to be small when compared with that of man. The facial angle, which in the adult European is about $85°$, is in the ourang outang $67°$, gradually descending in some of the monkeys as low as $30°$, till among the other genera of quadrupeds it does not sometimes exceed $20°$. It may be mentioned, that the cerebral system not only exhibits an ascending series of advances, from the lowest fish up to the highest mammal, but that, in the highest mammal, a parallel series of advances obtains, from the fish-like condition of the brain, at an early fœtal age, up to its complete development. Among the vertebrate classes, the brain is the most completely developed in the mammalia, and, among the mammalia, in man. The characters which serve to distinguish the brain of mammalia from that of other vertebrated animals consists, according to Cuvier, in the existence of a corpus callosum, a great comissure, fornix, cornu ammonis, and tuber annulare; in the situation of the tubercula quadrigemina; in the absence of ventricles, or cavities, in the thalami nervorum opticorum; in the position of the above mentioned thalami within the hemispheres, and in the alternate lines of **grey** and white within the corpora **striata**. The young of mammalia are nourished for some time after birth by milk, a fluid peculiar to animals of this class, which is produced by the mammæ at the time of parturition, and remains as long as it is necessary. Mammalia constitutes the first class of Spini-Cerebrata or Vertebrata, and this class has been divided by some into ten orders.

1. Bimana, or two-handed; the thumbs separate on the superior extremities only. Example, man.

2. Quadrumana, or four-handed; the thumb, or great toe, capable of being opposed to the other fingers or toes on each of the four extremities. The ape is an example.

3. Bradypoda, or slow moving animals, with their bodies generally covered by a hard crust. The armadillo is an example.

4. Cheiroptera, wing-handed, or animals having their fingers elongated for the expansion of membranes, which serve as wings. This membrane commences at the side of the neck, extends between the feet and toes, serves to support them in the air, and enables such of them to fly as have their hands sufficiently developed for that purpose. The bat is an example.

5. Glires, or Rodentia. Gnawing animals, having large incisors in each jaw, separated from the molars by an empty space, by which they divide hard substances. They have no canine teeth; they cannot seize living prey nor tear flesh; they cannot, even with their teeth, cut their food, but they gnaw or file it, hence their name. The squirrel, mouse, hare, &c., are familiar examples.

6. Feræ, or predaceous and carnivorous animals. They have large canine teeth, the molares forming pointed prominences for tearing and cutting the food. The bear, hedge-hog, &c., are examples.

7. Solidungula, or Solipeda. Animals having a single toe, or hoof, on each foot. These have six incisor teeth in each jaw, and are all of them herbivorous. The horse is an example.

8. Ruminantia, or Pecora. The term ruminantia indicates the peculiar property possessed by these animals of chewing the cud, that is, of masticating their food a second time, by bringing it back to the mouth after having swallowed it. This property depends upon the structure of their stomachs, of which they have four, the three first being so disposed that the food may enter into either of them, the œsophagus, or gullet, terminating at the point of communication. They have two toes on each foot, and no incisors in the upper jaw. The sheep, goat, ox, &c., are familiar examples.

9. Pachydermata, or Belluæ. Thick-skinned animals. Animals of unshapely form and a thick tough hide. They have more than two toes on each foot, some having three, four, or five; some of them have large tusks, and a proboscis. The elephant, rhinoceros, hippopotamus, &c., are placed in this order.

10. Cetacea. These are mammiferous animals, destitute of hind feet; their trunk terminates in a thick horizontal tail with a cartilaginous fin. They live in the sea, and their external form is that of fishes, the fin of the tail excepted, which in cetacea is *horizontal*, while in fishes it is always *vertical*. Their respiring by lungs, instead of gills; their possessing warm blood; their viviparous production; and their having mammæ with which they suckle their young, all entitle them to be placed in the class to which they belong. The arrangement of mammalia by Cuvier somewhat differs from the above, and is as follows:—1. Bimana; 2. Quadrumana; 3. Carnaria; 4. Marsupialia; 5. Rodentia; 6. Edentata; 7. Pachydermata; 8. Ruminantia; 9. Cetacea.

Mammalia have been by a more recent author divided into two

sub-classes, Placentalia and Implacentalia; Placentalia comprising 13 orders, namely, Bimana, Quadrumana, Carnivora, Artiodactyla, Perissodactyla, Proboscidea, Toxodontia, Sirenia, Cetacea, Cheiroptera, Insectivora, Bruta, Edentata, and Rodentia; and Implacentalia comprising 2 orders, namely, Marsapialia and Monotremata.

MAMMA'LIAN. Belonging to the class Mammalia.

MAMMALI'FEROUS. (from *mammalia* and *fero*, to produce, Lat.) A term applied to strata containing mammiferous remains. As the *mammaliferous crag* of Norfolk, &c.

MA'MMARY. Pertaining to the mammæ, as the *mammary glands*, the *mammary arteries*, &c.

MA'MMIFER. (from *mamma*, a breast, and *fero*, to bear.) All animals having breasts and suckling their young are included amongst the mammifers. To these Linnæus assigned the name Mammalia. Cuvier, however, called them *Mammifera*; but, as has been observed, there appears no good reason for altering the original term.

MA'MMILLARY. Having small rounded prominences, or projections something resembling teats or nipples; studded with rounded projections.

MA'MMILLATED. A term, like the one immediately preceding it, applied to certain minerals, which have the appearance of small bubbles, or rounded protuberances. Flint containing calcedony, is generally mammillated.

In conchology, the apex of a shell when rounded like a teat, is termed *mammillated*.

MA'MMOTH. (The etymology of this word does not appear quite agreed on; some state it to be from a Russian word, *mamant;* some, that it is of Tartar origin; others, that it is derived from Behemoth, an Arabic word, signifying elephant. *Mammut*, Germ.) The mammoth appears to be quite extinct; from the fossil remains of it which have been discovered, it appears to have had the feet, tusks, trunk, and many other particulars of conformation in common with the elephant; but it differed from the elephant in its grinders. Five species have been distinguished. The bones of the mammoth are found in great abundance in Siberia, and not only the bones, but portions of the flesh and the skin, and even whole animals have been found in icebergs and in frozen gravel. Towards the close of the last century, the entire carcase of a mammoth was exposed, and at length fell to the ground from a cliff of ice and gravel, on the banks of the river Lena. This animal was nine feet high, and about sixteen feet in length; the tusks were nine feet long. The skin was covered with hair, and it had a mane upon the neck.

The mammoth appears to have survived in England when the temperature of our latitudes could not have been very different from what it now is; for remains of this animal have been found in a lacustrine formation at North Cliff, in Yorkshire, in which all the land and freshwater shells can be identified with species now existing in that county. We have no great certainty at what period the mammoth ceased to exist; it is commonly supposed that it became extinct previous to the commencement of the modern group, but of this there is no good proof. It is supposed, from the prodigious number of bones found in certain places, that the mammoth must have existed in herds of hundreds, or even thousands. According to Pallas, there is scarcely a river, from the Don, or the Tanais, to the extremity of the promontory Tchuskoinosa, in the banks of which the bones of the mammoth are not most abundant. There are two large islands near the mouth of the river Indigerska, which are said to be entirely composed of bones of the mammoth, intermixed with ice and sand. Remains of the mammoth have been found from Spain to the coasts of Siberia, and throughout all North America.

The grinders of the mammoth are formed of two substances only; an internal bony substance, and a thick covering of enamel. The form of their crown is generally rectangular, the crown being divided by spreading grooves into a certain number of transverse risings, each of which is divided, in the contrary direction, into two large obtuse, and somewhat quadrangular and pyramidical points; the whole crown, when not worn, being beset with large points, arranged in pairs. In consequence of several of these teeth being much worn down, not only to the base of the pyramids, but even so low as only to leave one square surface edged with enamel, it has been concluded that they were used in the trituration of vegetable food. M. Cuvier particularizes three sorts of these grinders; nearly square, with three pairs of points, generally much worn; rectangular, with eight points, less worn; and others with five pairs of points, and a single smaller one, scarcely the least worn. Cuvier considered he had distinguished five different species.

MANA'TUS. A genus of herbivorous cetacea, placed by Cuvier in the family Cetacea herbivora. The manatus appears to have inhabited the seas of our latitude during the Miocene and Pliocene periods. The recent species are now found near the coasts and mouths of rivers, in the torrid zone. They have an oblong body, terminated by an elongated oval fin: the grinders have a square crown, marked with two transverse elevations; they are eight in number throughout. Lamantin, a name often given to manatus, is said to be merely a corruption of Manatus.

MA'NDIBLE. (*mandibulum*, Lat. a jaw.) In insects, the upper jaws are called mandibles; the under jaws, maxillæ. The mandibles of insects are two strong, corneous, somewhat bent hooks, the inner margin being more or less dentate; they articulate with the cheeks at their broad basis, move by ginglymus, and are opposed to each other like the blades of scissors.

MANDI'BULAR. Pertaining to the jaw.

MANGANE'SE. (*manganèse*, Fr. *manganese*, It.) A metal but little known in its pure or metallic state, to which it is reduced with much difficulty, in consequence of its great affinity for oxygen. It was first obtained in a metallic form by Gahn, from the black oxide of manganese, a substance first investigated by Scheele. When pure, it has a grayish-white colour, with some lustre. Its texture is granular; hardness, nearly that of iron. Specific gravity from 7·0 to 8·0. It has little or no malleability, and is difficult of fusion. It absorbs oxygen by exposure to the atmosphere, and its melting point is about 160° W.

In its metallic state, manganese is not applied to any use. It is obtainable in small quantities from the black oxide by heating it in an intense furnace, with charcoal and a little oil.

The common ore of manganese is the black, or peroxide, a valuable substance to chemists, as that from which oxygen is most easily obtained; it consists of 27·7 manganese, and 16 oxygen. When added in small quantities to glass, it removes the greenish or yellowish tinge which arises from iron or other impurities; but if added in considerable quantity, it communicates to glass or enamel a violet or purple colour. The ores of manganese present much diversity in their external characters. All minerals containing any considerable quantity of this metal, when melted with borax and a little nitre, yield a violet glass. One of the ores of manganese, known by the name of *Black Wadd*, is remarkable for its spontaneous inflammation when mixed with oil.

MA'NON. A genus of zoophytes, found in the cretaceous group and determined by Goldfuss.

MANGANE'SIAN GARNET. Grenat manganésie of Brongniart; Spessartine of Beudant. A species of garnet occurring massive and in dodecahedral crystals variously modified. Colour, a deep hyacinth or brownish red; edges slightly translucent. Fusible before the blow-pipe, and when fused with borax and nitre the borax is violet. According to Klaproth it contains protoxide of mangane 35·00, silica 35·00 alumine 14·25, protoxide of iron 14·00. It occurs in granite.

MANTE'LLIA. A genus of fossil animals belonging to the chalk deposit. This genus is thus described, "an animal with a fusiform or ramose, root-like pedicle, a stem and body formed of tubuli, anastomosing in a basket-like, texture, with openings on the internal surface." The name Mantellia has also been given by Adolphe Brongniart, to a genus of cycadeous fossil plants found in the petrified forest of the Isle of Portland.

MA'RBLE. (*marbre*, Fr. *marmo*, It. *marmor*, Germ. *marmor*, Lat.) Any limestone possessing sufficient hardness to take a polish may be called marble. Many of these are fossiliferous; but statuary marble, which is also called saccharine limestone, from its possessing a texture resembling that of loaf-sugar, is devoid of fossils, and belongs to the metamorphic series of rocks.

MA'RGARATE. A compound of margaric acid with potash, soda, or some other base, and so named from its pearly lustre.

MARGA'RIC ACID. An oleaginous acid, formed from different animal and vegetable fatty substances.

MARGARITA'CEOUS. Reflecting the prismatic colours like mother-of-pearl. —*Shuckard.*

MA'RGARITE. A mineral, of a grayish-white colour, occurring massive and in thin crystalline laminæ, intersecting each other in all directions. It bears some resemblance to silvery mica.

MA'RGARODITE. A mineral which is described as forming the matrix of black tourmaline.

MA'RGIN. (*marge*, Fr. *màrgine*, It.) The border or edge. In conchology, the whole circumference or outline of the shell in bivalves.

MA'RGINATE. } (*marginatus*, Lat.)
MA'RGINATED. }

1. In conchology, having a prominent margin or border.
2. In entomology, when the sharp edge is margined, and surrounds the surface with a narrow border.

MARGINE'LLA. An ovato-oblong, smooth univalve, with a short spire. The lip thickly marginated on the outside. The base of the aperture slightly notched; the columella plicated. Marginella differs from Voluta in the reflection of its outer lip. Recent Marginellæ are found in sand and sandy mud. Several fossil species have been discovered in the calcaire grossier.

MA'RINE ALLU'VIUM. Shingle thrown up by the sea; materials cast upon the land by a wave of the sea, or those which a submarine current has left in its track.

MARI'NE VEGETA'TION. The marine vegetation, says Sir C. Lyell, is less known; but we learn from

Lomouroux, that it is divisible into different systems, apparently as distinct as those on the land, notwithstanding that the uniformity of temperature is so much greater in the ocean. The number of hydrophytes, or plants growing in water, is very considerable, and their stations are found to be infinitely more varied than could have been anticipated; for while some plants are covered and uncovered daily by the tide, others live in abysses of the ocean, at the extraordinary depth of one thousand feet; and although in such situations there must reign darkness more profound than night, at least to our organs, many of these vegetables are highly coloured.

MARL. (*mergel, märgel*, Germ.) A combination of common clay and calcareous earth; a mixture of clay and lime.

MARL SLATE. A brown indurated fossil shale, with occasional beds of thin compact limestone: a rock of the Permian period.

MARLSTONE. The name given to a member of the lias which (the lias) Sir R. Murchison states to be capable of a four-fold division. This member of the lias, he says, is well exposed in the hill on which the church of Prees is built, both in quarries and by the sides of the rocks, dipping to the N.N.E. at low angles. The upper beds are composed of yellowish and greenish thin-bedded sandstone, slightly micaceous, and in parts calcareous; the middle, of other yellowish sandstones, some of which are more calcareous; and the lowest beds, of sandy, dark-coloured slaty marl, and shale with flattened spheroids of impure lias limestone, which are undistinguishable from the cement-stone of the Yorkshire coast. The marlstone is a well marked division of the lias, being more arenaceous, though still fine-grained, and often bound by calcareous or ferruginous cement into a hard stone. In Gloucestershire it is divisible into the "hard rock bed" above, and the sands below.—*Jukes.*

MA'RLY. Composed of marl; containing marl; resembling marl.

MA'RNES IRISÉES. The French geologists designate by this term the upper party-coloured marls or clays of the new red formation. They are the same as the keuper-marls of the Germans, and the gypseous and saliferous marls of Cheshire, Worcestershire, &c., of England.

MA'RSIPOBRANCHII. The fifth order of fishes, comprising the lampreys, &c.

MARSU'PIAL. (from *marsupium*, a pouch, Lat.) Having a pouch; belonging to the order Marsupialia. New Holland is known to contain a most singular assemblage of mammiferous animals, consisting of more than forty species of the *marsupial* family.

MARSUPIA'LIA. (The term Marsupialia is derived from the presence of a large *marsupium*, or pouch, fixed on the abdomen, in which the fœtus is placed after a very short period of uterine gestation. Fœtus ad úterum maternum haud intermedio placentæ veræ annexus.) Animals possessing abdominal pouches. The marsupialia form the fourth order of Mammalia, in Cuvier's arrangement. The economy of marsupialia is in many respects most singular. One most striking peculiarity is the premature production of the fœtus, whose state of development at birth is extremely small. Immediately on their birth they pass into a sort of second matrix. Incapable of motion, and scarcely displaying any germs of limbs or external organs, these diminutive beings attach themselves to the mammæ of the mother, where they remain fixed by the mouth, until they have acquired a growth and development, resembling that of other newly-born animals. The skin of the animal is so arranged round the mammæ as to form a pouch in which not only the imperfect fœtus, attached to the nipple by its mouth, remains till fully developed, but into which, long after it is able to run about, it leaps when alarmed, or when wishing to conceal itself. The order Marsupiala holds an intermediate place between viviparous and oviparous animals, forming a link, as it were, between Mammalia and Reptiles. The order Marsupialia contains many genera, both herbivorous and carnivorous. The kangaroo and opossum are familiar examples. Another peculiarity in these animals consists in this; that the members of two litters are sometimes sucking at the same time. The New Holland opossums are very voracious, and devour carcasses as well as insects; they enter into the houses, where their voracity is very troublesome. That most common, the Didelphys Virginiana, attacks poultry in the night, and sucks their eggs. It is said to produce sixteen young ones in one litter, which, when first born, do not weigh more than one grain each; though blind and almost shapeless, when placed in the pouch they instinctively find the nipple, and adhere to it till they attain the size of a mouse, which does not take place till they are are fifty days old, at which period they begin to see. The discovery of marsupials, both in the secondary and tertiary formations, shows that, so far from being of more recent introduction than other orders of Mammalia, this order is in reality the first and most ancient condition, under which animals of this class appeared upon our planet, and as far as our present discoveries go, it was their only form during the secondary period.

MARSUPIOCRINI'TES. (from μάρσυπος, Gr. *marsupium*, Lat. a purse, and κρίνον, Gr. a lily.) The name given by Mr. Phillips to a genus of encrinites belonging to the Silurian rocks.

MA'RSUPITE. (from *marsupium*, a purse, Lat.) The name given by Dr. Mantell, from their resemblance to a purse, to a genus of Crinoïdea found in the chalk. The marsupite was a molluscous animal, of a sub-ovate form, having the mouth in the centre, and surrounded by arms, or tentacula. The skeleton was composed of crustaceous, hexagonal plates: the arms, subdivided into numerous branches of ossicula, or little bones: the whole invested with a muscular tissue, or membrane. When floating, the creature could spread out its tentacula like a net, and by closing them, seize its prey, and convey it to its mouth. Fossil remains of this zoophyte have been found in the upper chalk of Sussex, Wiltshire, and Yorkshire. The name of '*cluster-stones*,' given to them by the quarry-men of Sussex, conveys an idea of their general appearance. They may, however, so far as their body is concerned, be compared to the fruit of the pine. The body is orbicular, contained in a pelvis composed of sixteen convex, radiated, angular, crustaceous plates. The marsupite was once placed among the encrinites, from which it differs most essentially, inasmuch as it possesses neither a vertebral column nor any processes of attachment: from which circumstances it is clearly manifest that it was not attached to one place or point, but floated about in the surrounding sea. The marsupite may be considered as forming a link between the Crinoïdea and the Stellaridæ.

MARSU'PIUM. The name given to a dark-coloured membrane situated in the vitreous humour of the eye of birds. The use of the marsupium is not ascertained, but it is present in almost every bird having extensive powers of vision.

MA'SCAGNINE. A native sulphate of ammonia, found, by M. Mascagni, near the warm spring of Sasso, in Tuscany, and named after its discoverer. It has also been called Sassolin, from the place near which it was found.

MA'SSETER. (from μασσάομαι, Gr. to chew.) A muscle connected with the under jaw of insects, and which assists in masticating.

MA'SSIVE. A term used in mineralogy to describe a substance of no determinate form, whatever its internal structure may be : it is, however, generally applied to minerals possessing regular internal structure, but having no characteristic external form.

MA'STODON. (from μαστός, a breast or udder, and ὀδούς, a tooth, Gr.) A name given by Cuvier to a genus of fossil mammalia. The remains of the mastodon were first discovered at Albany, near Hudson river, about the year 1711. The first specimens brought to Europe were obtained from the neighbourhood of the Ohio, from which circumstance the French called the mastodon the animal of the Ohio. Several species have been distinguished. "The Great Mastodon," says Griffith, "is one of the most remarkable and apparently the most enormous of all the fossil species. Bones of this species are found in abundance over all North America, from the 43rd degree of north latitude, as far south as Charlestown, in Carolina, in thirty-three degrees, and, as far as we at present know, do not exist in any other country of the globe. They are found at moderate depths, exhibiting few marks of decomposition, and none of detrition."

The great mastodon was very similar to the elephant in the tusks and entire osteology, the cheek-teeth excepted. It is supposed to have possessed a trunk ; its height, about which very false and exaggerated notions once prevailed, does not appear to have exceeded ten or twelve feet, but its body was longer than that of the recent elephant, its limbs were thicker, and its belly less bulky. The structure of its molars, is sufficient to constitute it a different genus.

"Two dental characters exist which distinguish, in a well-marked and unequivocal manner, the genus Mastodon from the genus Elephas. The first is the presence of two tusks in the lower jaw of both sexes of the mastodon, one or both of which are retained in the male, while both are early shed in the female. The second character is equally decisive ; it is the displacement of the first and second molars, in the vertical direction, by a tooth of simpler form than the second, a true *dent de remplacement*, developed above the deciduous teeth in the upper, and below them in the lower jaw."—*Prof Owen*.

MA'STOID. (from μαστός, the breast, and εἶδος, likeness, Gr.) Shaped like the breast, or like a nipple. Applied to some prominences of bones ; to a foramen ; to a muscle ; and to cells in the ear.

MA'TRIX. (*matrix*, Lat. *matrices*, pl.) In mineralogy, the earthy or stony matter in which a fossil is imbedded. Called also gangue.

MAXI'LLA. (*maxilla*, Lat.) The jaw. The lower jaws of insects are called maxillæ ; they are placed behind the mandibles, and between is situated the labium, or lower lip. The maxillæ are employed principally for holding the substances on which the grinding apparatus of the mandible is exerted.

MEAN QUANTITIES. Such as are intermediate between others that ar greater and less. The mean of any number of unequal quantities is equal to their sum divided by their number. For instance, the mean between two unequal quantities is equal to half their sum.—*Mrs. Somerville.*

MEAG'RE. A term used in mineralogy relating to the touch or feel of a mineral. Chalk is said to be meagre.

MEANDRI'NA. Brain-stone; brain-coral. Madrepores in which the laminæ assume a meandering direction are called meandrinæ. Meandrinæ are large hemispherical corals, having their surface covered with serpentine ridges and depressions, resembling the convolutions of the cerebrum, or brain, from which circumstance they have been called brain-stone. "On the convex side," says Mr. Parkinson, "are excavated, open, winding, ambulacræ, lamellated on each side. The lamellæ are transverse and parallel, adhering on each side of hillock-shaped ridges. The lamellated ridges occupy the interstices of the tortuous vallies which hold the polypes and thus separate them. Fossil meandrinæ are found both in the cretaceous and oolite series."

MEA'TUS (Lat.) A passage, as that leading to the ear, called the meatus auditorius, &c.

MEDIAL ORDER. A term proposed by the Rev. J. Conybeare for that assemblage of rocks which contains not only the great coal deposit itself, but also the limestone and sandstone on which it reposes. "This series of rocks," says Dean Conybeare, "is by some geologists referred to the flœtz, and by others to the transition class of the Wernerians. We have preferred instituting a particular order for its reception, a proceeding justified by its proportional importance in the geological scale, its peculiar characters, and the many inconveniences arising from following either of the above conflicting examples. For this order we have proposed the name of *Medial*, wishing to adopt an appellation entirely free from theory; and indicating only the central position of this groupe in the five-fold division of the geological series which results from assigning to it a separate class."

MEDU'LLA. (*medulla*, the marrow, Lat.)
1. In botany, the pith of plants.
2. The marrow in the cavities of bones.

MEDULLARY. (*medullaris*, Lat.)
1. Relating to the brain, or to the marrow. The medullary substance composes the greater part of the brain, spinal marrow, and nerves.
2. In botany, relating to the pith of plants.

MEDU'LLIN. A name given by Dr. John to the porous pith of the sunflower.

MEDU'SA. A genus of marine molluscous animals belonging to the class Acalepha. The *medusæ* approach nearly to the fluid state, appearing like a soft and transparent jelly, which by spontaneous decomposition after death, or by the application of heat, is resolved almost into a limpid watery fluid. The usual form of a medusa is that of a hemisphere, with a marginal membrane, like the fold of a mantle, extending loosely downwards from the circumference.

*Medusæ* are met with of very various sizes; the larger abound in the seas around our coasts, but immense numbers of the more minute, and often microscopic, species occur in every part of the ocean. In some parts of the Greenland seas the number of Medusæ is so great, that in a cubic inch, taken up at random, there are not fewer than 64. In a cubic foot this will amount to 110,592; and in a cubic mile, the number is such, that allowing one person to count a million in a week, it would have required 80,000 persons from the creation of the world, to complete the enumeration.

MEGA'CEROS. A sub-genus of the genus cervus, and so named from the immense size of its antlers; the Megaceros Hibernicus, or Irish fossil elk, is a species of this sub-genus. The first tolerably perfect specimen of the Irish elk was found in the Isle of Man. A very complete and well articulated skeleton is in the British Museum, another in the Woodwardian Museum, at Cambridge, and a third in the Hunterian Museum, at the College of Surgeons.

MEGALO'NYX. (from μέγας, great, and ὄνυξ, a claw, Gr.) A huge fossil mammalian, of the order Edentata, and thus named from the great size of its unguical, or claw, bones. The remains of the Megalonyx were discovered in the floor of a cavern, in the limestone of Virginia, in America. Cuvier approximated the Megalonyx to the Megatherium, considering that these two animals must have constituted an intermediate genus between the bradypi and the ant-eaters. He concludes that they were both herbivorous. The Megalonyx has hitherto been found fossil only.

MEGALI'CHTHYS. (from μέγας, great, and ἰχθύς, a fish, Gr.) The name given to a fossil sauroid fish, first discovered, by Dr. Hibbert, in the limestone near the bottom of the coal formation, near Edinburgh.

MEGALO'DON. A genus of fossil bivalve shells, found hitherto only in the Devonian system of rocks; they are equivalve, longitudinal, thick, the hinge forming an incrassated septum across the cavity of the shell, with a large compressed bifid tooth in the right valve, and one irregular and pointed tooth in the left. Two species are described.— *Lycett.*

MEGALO'SAUR. } (from μέγας, great,
MEGALOSAU'RUS. } and σαῦρος, a lizard, Gr.) A genus of fossil amphibious animals, of great size, belonging to the saurian tribe; holding an intermediate place between the crocodiles and monitors. Cuvier concludes, from a comparison of the fossil bones with those of existing lizards, the megalosaurus to have been an enormous reptile measuring from forty to fifty, or even seventy, feet in length, and partaking of the structure of the crocodile and monitor, but more nearly approximating the latter. Remains of the megalosaurus have been found in the Oolite and in the Wealden. This huge creature appears to have been carnivorous, from the form of its teeth, and its head terminated in a straight and narrow snout.

MEGAPHY'TON. (from μέγας, great, and φυτὸν, a plant.) An extinct genus of plants belonging to the order Conifera. In the genus Megaphyton the stem is not furrowed, and the leaf scars are very large, resembling the shape of horse-shoes, and arranged on each side of the stem in two vertical rows. It is found in the coal strata.

MEGATHE'RIUM. (from μέγας, great, and θηρίον, a beast.) An extinct animal, of great size, of the family tardigrada, belonging to the order Edentata, nearly allied to the sloth, and, like the sloth, presenting an apparent monstrosity of external form, accompanied by many singular peculiarities of internal structure. Fossil remains of the megatherium have been discovered in South America, in the alluvial deposites of the Pampas. The megatherium was about eight feet high, its haunches five feet wide, its feet a yard in length, and its body twelve feet long: it united part of the structure of the armadillo with that of the sloth. The relative proportions of the extremities of the megatherium differ greatly from those of the sloth, and indeed from those of any known animal. Its teeth prove that it lived on vegetables, and its forefeet, robust, and armed with sharp claws, show that roots were its chief objects of search. Its hide appears to have been covered with a bony coat of armour of considerable thickness, the use of which was probably defensive, not only against the sharp claws of beast of prey, but also against the myriads of insects that surrounded it. Its tail was long, and composed of vertebræ of enormous magnitude, the body of the largest being seven inches in diameter, and the horizontal distance between the extremities of the two transverse processes, being twenty inches. If to this be added the thickness of the muscles and tendons, and of the shelly integument, the diameter of the tail, at its largest end must have been at least two feet; and its circumference, supposing it to be nearly circular, about six feet, secure within the panoply of his defensive armour, where was the enemy that would dare encounter this behemoth of the Pampas? a creature whose giant carcase was encased in an impenetrable cuirass, and who, by a single pat of his paw, or lash of his tail, could in an instant have annihilated the couguar or the crocodile. The genus megatherium comprehends two species, the megatherium, properly so called, and the megalonyx.

MEI'ONITE. (from μείων, less, Gr.) The Meionit of Werner. A mineral, thus named from its terminating pyramids being lower than those of similar forms in other minerals. Meionite is a prismato-pyramidal felspar. It occurs in grains, or small crystals, whose more common form is an eight-sided prism, truncated on its lateral edges, and terminated by four low-sided pyramids. It is of a greyish-white colour; translucent, and sometimes transparent. It scratches glass, and before the blow-pipe readily melts into a white spongy glass. It is found at Mount Somma, near Vesuvius.

MELA'NIA. (from μέλας, black, Gr.) A genus of univalve fresh-water shells belonging to the order Pectinibranchiata, class Mollusca. The melania is a turreted univalve; the aperture entire, ovate, or oblong, and spread out at the base of the columella, which is smooth. Recent melaniæ are found in rivers and estuaries. Fossil melaniæ are found in the environs of Paris.

ME'LANITE. (from μέλας, black, Gr.) The Melanit of Werner; Grenat noir of Haüy; Grenat melanit of Brongniart. A velvet black, opaque, dodecahedral variety of garnet. It occurs in crystals, which are dodecaëdrons, with truncated edges. Fracture conchoidal. Specific gravity 3·73. Its constituents are silex 35·5, lime 32·5, oxide of iron 25·25, alumine 6, oxide of manganese ·04. It is found at Frascati, near Mount Vesuvius, in Bohemia, and in North America.

MELANO'PSIS. A genus of oval or oblong, fusiform, univalves, belonging to the family Melaniana, in Lamarck's arrangement. Melanopsides are found both recent and fossil; they are distinguished from Melaniæ by a notch in the aperture. Fossil melanopsides are found in the shale of the Wealden, at Pounceford.

ME'LAPHYR. } A compact, fine-
ME'LAPHYRE. } grained, dark

grey, brown, or black rock, consisting apparently, of a feldspathic mineral intimately mixed with augite, hornblende, magnetic iron, &c. It is sometimes vesicular, amygdaloidal, or porphyritic, and and is said to be sometimes slaty. —*Jukes*.

MELA'STOMA. (from μέλας, black, and στόμα, mouth, Gr.) A name given to a **genus of** plants, belonging to the order Melastomacea, from the fruit staining the lips of a black colour.

MELEAGRI'NA. A genus of bivalve molluscans, known as the *pearl-oyster*. Meleagrinæ inhabit the Persian Gulf, the coasts of Ceylon, the sea of New Holland, the Gulf of Mexico, and the coasts of Japan. It attains perfection nowhere but in the equatorial seas; in the pearl fishery of the island of Ceylon it is the most celebrated and productive. The pearls are situated in the fleshy part of the oyster, near the hinge. For one pearl that is found perfectly round and detached between the membranes of the mantle, hundreds of irregular ones occur attached to the interior of the shells, like so many warts: they are sometimes so numerous, that the animal cannot shut its shell, and so perishes.

ME'LILITE. (from μέλι, honey, and λίθος, a stone, Gr.) The name given to a rare mineral, from its honey colour. It occurs only in very minute crystals, perfectly regular and well defined, but not larger than a grain of millet-seed. These grains are of a cubic or prismatic form; their surface is often coated with an oxide of iron. They are glistening, semitransparent, and will scratch glass.

MELI'TA. (from *mel*, honey, Lat.) Honey-cake. A genus of echinites, belonging to Catocysti.

ME'LLATE. The name given to a salt, in which mellitic acid is combined with any salifiable base.

ME'LLITE. (from *mel*, honey, and λίθος, a stone.) Honey-stone. The Honigstein of Werner; La Pierre de miel of Brochant; Pyramidales Melichron-Hartz of Mohs. This mineral was first observed in Thuringia, where it occurs associated with brown-coal. It is of a honey-yellow colour, whence its name, and is usually crystallized in small octahedrons, whose angles are often truncated. Fracture conchoidal. Lustre shining or splendent. By friction the crystals acquire a weak negative electricity. They are more or less translucent, or even transparent, and exhibit double refraction. Mellite may be distinguished from amber by its weak electricity, and double refraction. It consists of mellitic acid 41·0, alumina 14·10, water 44·8.

MELOCRINI'TES. A genus of encrinites established by Goldfuss. It is found in the mountain limestone.

MEMBRANA'CEOUS. (*membranaceus*, Lat.) Resembling membrane. In botany, a membranaceous leaf has no distinguishable pulp between the two surfaces.

MEMBRA'NEOUS. (*membraneus*, Lat.) Consisting of membrane. In this and the preceding word may be observed the difference between words **ending in** *aceous* and *eous*: those **ending in** *aceous* express a resemblance to a material, those ending in *eous* indicate the material itself.

ME'MBRANE. (*membrana*, Lat.) The membranes of animals are thin semitransparent bodies, which envelope certain parts of the body, to which they furnish a covering for their support and protection. Membranes are modifications of cellular texture, the surfaces of the plates cohering so as to obliterate all the cellular interstices, and being impervious to fluids. Membranes also line the interior of all the large cavities of the body: these membranes, after lining the sides of their respective cavities, are reflected back upon the organs which are enclosed in those cavities, so as to furnish them with an external covering. Thus the bowels are covered by the peritoneum, the lungs by the pleuræ; nevertheless, in consequence of these membranes being reflected, the lungs and bowels may be said to be external to their investing membranes.

ME′NACHINE. (from *Menachan*, a valley in Cornwall.) The name given to a metal, to which the name titanium is now more generally applied. The *menachine* of Gregor and the *titanium* of Klaproth are the same substance, and to Gregor is owing the merit of the discovery.

ME′NACHANITE. (from *Menachan*, or *Menaccan*, in Cornwall.) An oxide of titanium, or menachine, combined with iron. It is of a greyish-black colour, and occurs in small grains resembling gunpowder, of no determinate shape, and mixed with a fine grey sand. Specific gravity 4·4. Before the blow-pipe it neither decrepitates nor melts. According to the analysis of Klaproth, it consists of oxide of iron 51·00, oxide of titanium 42·45, silica 3·50, oxide of manganese 0·25.

ME′NILITE. (from *Ménil-montant*, near Paris, where it is found.) The Menilit of Werner; Silex menilite of Brongniart; Quartz résinite subluisant of Haüy. A brown or yellowish-grey tuberose variety of uncleavable quartz. By some authors this mineral is placed as a variety of semi-opal. From its only having been found at Menilmontant, near Paris, and from the resemblance of some of its darker varieties to pitch, it is sometimes called Pitch-Stone of Menil-montant. Hoffmann divided Menilite into two subspecies, namely, Brown Menilite and Grey Menilite. Menilite occurs in small irregular or roundish masses, often tuberose, or marked with little ridges on its surface. It is translucent, often only at its edges. Structure rather slaty; fracture conchoidal or splintery. It scratches glass. Specific gravity 2·18. Infusible before the blow-pipe. Constituents, silica 85·5, alumine 1, lime 0·5, oxide of iron 0·5, water and carbonaceous matter 11.

MENI′NGES. (from μῆνιγξ, Gr., a membrane.) A name given to the membranes which cover the brain.

MENI′SCUS. (from μηνίσκος, Gr. A lens, one of whose surfaces is convex and the other concave, and in which the two surfaces meet if continued. As the convexity exceeds the concavity, a *meniscus* may be regarded as a convex lens.

MEPHI′TIC A′CID. Another name for carbonic acid.

MEPHI′TIC AIR. Another name for nitrogen gas.

MERCURY. (*mercure*, Fr. *mercúrio*, It.) One of the sixty simple or elementary bodies. This metal is of the same colour as burnished silver; when pure and fluid, it is still opaque, and nearly silver-white, with a strong lustre. Its specific gravity is 13·56, or thirteen times and a-half heavier than water, its density being next to those of platinum and gold. Mercury freezes at a temperature of 39° or 40° below the zero of Fahrenheit, that is, at a temperature of 71° below the freezing point of water; under common circumstances we always find it fluid, and in this respect it remarkably differs from all the other metals. It has obtained its name from its fluidity and colour. The boiling point of mercury is somewhere about 680°, at which temperature it is converted into vapour of a highly expansive power; this vapour may be again condensed into the fluid metal, by being received into cold vessels.

Mercury has less affinity for oxygen than most other metals; it may be distilled over five hundred times, without loss of quantity. It combines, however, with oxygen in two proportions, forming a red and a black oxide. By merely heating these in a retort the oxygen may be driven off, and the metal once more obtained in its pure state.

The existence of mercury, even in small quantities, in any of its ores, may be ascertained by mingling the ore with iron filings, and heating this mixture to redness under any cold body, as a plate of polished copper; the mercury is volatilized, and condensed in minute globules on the plate. In consequence of the volatility of mercury, it is usually purified by distillation.

Two of the combinations of mercury with chlorine form most valuable and important medicines; the one called chloride of mercury, or calomel, the other bichloride of mercury, or corrosive sublimate. From the fluid state in which mercury exists, it readily combines with most of the metals, to which, if in sufficient quantity, it imparts a degree of fusibility or softness: these compounds are termed *amalgams*. An amalgam of mercury and tin is employed for silvering the backs of looking glasses, and an amalgam of four parts of mercury, two of bismuth, one of lead, and one of tin, is used for silvering the inside of glass globes, the amalgam fusing on the globe being placed in hot water. The ready combination of mercury with gold or silver, and the facility with which it may be again separated from them by heat, renders it of great value in the obtaining those metals from their ores and alloys in the operations of mining.

Mercury is also most useful in the construction of barometers and thermometers. It was known in the remotest ages, and seems to have been employed by the ancients in gilding, and in separating gold from other bodies, as in the present day. It possesses neither taste nor smell. Native mercury occurs in small globules, disseminated in other metals. These globules are but feebly united to their gangue, and may be liberated by striking or heating the substance which embraces them. It is from the sulphuret of mercury that the metal is principally obtained. Sulphuret of mercury occurs in beds, or large irregular masses, and sometimes in veins. The mines which furnish the ore, sulphuret of mercury, are by no means common; Spain, Germany, and Peru possess the most important. In Spain, at Almaden, these mines are in a mountain of argillaceous slate or shale. The most celebrated, however, are at Idria; these are situated partly in gray compact limestone, and partly in shale. The working these mines is exceedingly injurious to the health and life of those employed. Criminals, and those convicted of political offences, are sent hither to eke out a miserable existence. They soon lose their teeth, and are subject to paralysis, convulsions, and premature old age. It is said that the surrounding district is so affected by the noxious vapours, that cattle cannot be reared there, and that fruit and grain do not come to maturity.

MERE. A deep pool of fresh water.

ME'ROE. In malacology, the name given by Schumacher to certain cowry-shells.

MERSTHAM BEDS. See *Firestone*.

ME'SENTERY. (μεσεντέριον, Gr. from μέσος, middle, and ἔντερον, bowel.) A fatty membrane formed of folds of the peritoneum. This is a fine and delicate membrane which connects the intestines to the spine, and to each other, and which appears to be interposed in order to allow to the intestines that freedom of motion which is so necessary to the proper performance of their functions.

MESENTE′RIC. Pertaining to the mesentery, as the mesenteric glands, &c.

MESOTHO′RAX. (from μέσος, middle, and θώραξ, chest.) In entomology, the mesothorax gives origin to the second pair of legs, and also the first pair of wings, or to the elytra, of insects.

ME′SOTYPE. (from μέσος, middle, and τύπος, form.) Prismatic zeolite; a simple mineral of the zeolite family, occurring in drusy cavities, or in veins in secondary trap rocks. Mesotype is of a white, red, yellow, or yellowish-brown colour. It occurs regularly crystallized. It consists of silica 54·40, alumina 19·75, soda 15·05, lime 1·60, water 9·80. Specific gravity 2·3.

MESOZ′OIC. A palæontological term, corresponding to, or synonymous with, secondary. The Mesozoic, or secondary, epoch comprises the cretaceous, the oolitic, and the triassic periods.

METACA′RPAL. Belonging to the wrist; as the *metacarpal* bones, &c.

METACA′RPUS. (from μετὰ, with, and καρπὸς, the wrist.) That part of the superior extremity which connects the wrist with the fingers; commonly known as the hand, but not including either the wrist or the fingers.

METABO′LIANS. (from μεταβάλλω, Gr. to change.) That sub-class of insects which undergo a metamorphosis, and are usually fitted with wings in their final state.

ME′TAL. (μέταλλον, Gr. *metallum*, Lat. *métal*, Fr. *metállo*, It. *metall*, Germ.) "A metal may be described as a shining opaque body, a good conductor of heat and electricity, insoluble in water; capable, when in a state of oxide, of uniting with acids, and of forming with them metallic salts."—*Phillips*. Metals, as presented by nature, are sometimes pure, or combined with each other only, and are said to exist in a metallic state. But more frequently they are combined with oxygen, sulphur, &c., by which their peculiar metallic properties are more or less disguised; in this case the metal is said to be mineralized, and the oxygen, or sulphur, is called the mineralizer. All the individuals of the class of metals, with the exception perhaps of iron, are perfectly inert and harmless; even arsenic, lead, copper, and mercury, which in certain states of combination constitute some of the most virulent of known substances, exert no action upon the living system, unless they be in union with some other body; but when so united, how valuable do they become, and what various medicinal effects may they not be made to produce! The metals at present known are forty-four in number. Of these, seven were known in the earliest ages, and, in consequence of a superstitious belief in the influence of the stars over human affairs, were first distinguished by the names and signs of the planets; and as the latter were supposed to hold dominion over time, so were astrologers led to believe that some, more than others, had an influence on certain days of the week; and, moreover, that they could impart to the corresponding metals considerable efficacy upon the particular days which were devoted to them. As regards the ages of metals, tin, molybdena, tungsten, and wolfram, are ranked as the most ancient; uranium and bismuth succeed. Gold and copper are deemed relatively and comparatively as new metals. Iron is of all ages. The specific gravity of metals, if we exclude those recently discovered by Sir H. Davy, is always greater than that of minerals; tellurium, the lightest metal, being above 6·0, while the heaviest earthy body is less than 5·0. Metals are opaque; they possess a **peculiar** lustre, which has been **termed** metallic lustre, retaining **it even** when reduced **to** powder. They are mostly malleable, or capable of being hammered into various orders and thin leaves; and ductile, or capable of being drawn into wires of greater or less fineness. They are not soluble in water. They all unite with oxygen, and, probably, all with chlorine. They are fusible, or capable of being rendered fluid by a heat attainable by artificial means; becoming again solid on cooling. They are elastic, hard, heavy, and, generally, sonorous. Some of the metals possess a degree of taste and smell. All the metals are expansible by heat, and their degree of expansibility appears to bear a relation to their fusibility. Two or three of the metals occur in small quantities in the masses of some of the earlier rocks, but in general the metals are found in veins; some in veins traversing the **older** rocks, and rarely or never in those of a **more recent** description; others **most abundantly, or only in** those of newer **formation.**

META′LLIC LU′STRE. One of the most conspicuous properties **of** metals **is** a particular brilliancy which they possess, and which has been called the *metallic lustre.* There are other bodies which apparently possess this peculiar lustre, as, for example, mica, but in them it is confined to **the** surface, and accordingly disappears when they are scratched, whereas **it** pervades every part of metals.

META′LLIC ORE. Metals existing in the state of an oxide, or a salt, or united with a combustible, are called *ores;* and this term is, by analogy, extended to the native metals and alloys. They appear to be the production of every period, but more frequently exist in primary and transition, than in secondary rocks, or than in alluvial earths.

META′LLIC O′XIDE. A metal combined with any proportion of oxygen, is called a *metallic oxide,* provided it does not possess the properties of an acid.

META′LLIC SALTS. Those salts which have a metallic oxide for their base; carbonate of lead is an example.

METALLIC VEINS. "Perhaps," says Mr. Bakewell, "the reader may obtain a clearer notion of a *metallic vein,* by first imagining a crack or fissure in the earth, a foot or more in width, and extending east and west on the surface, many hundred yards. Suppose the crack, or fissure, to descend to an unknown depth, not in a perpendicular direction, but sloping a little to **the** north or south. Now, let us suppose each side of the fissure **to** become coated with mineral matter, of a different kind from the rocks in which the fissure is made, and then the whole fissure to be filled **by** successive layers **of** various metallic and mineral substances; **we shall thus** have a type of a **metallic vein.** Its course from east to west is called its *direction,* and the dip from the perpendicular line of descent, is called its *hading*."

"There is a remarkable circumstance," says Prof. Phillips, "in the distribution of metallic veins in the same class of stratified rocks, —a peculiarity depending on local influences; such, that while the slates of Cornwall near the granitic eruptions, yield tin and copper, and the Snowdonian slates, and those of Coniston Water Head yield copper; those of Loweswater, Borrowdale, Patterdale, and Caldbeck fells yield lead, or lead and copper."

**METALLI′FEROUS.** (from *metallum*, metal, and *fero*, to produce.) Yielding metal; as metalliferous deposits, metalliferous districts, metalliferous veins, metalliferous dykes, &c.

**META′LLURGY.** (from μέταλλον, a metal, and ἔργον, a work, Gr. *métallurgie*, Fr.) Some authors comprehend under this term, the whole art of working metals, from the glebe, or ore, to the utensil; in which sense assaying, smelting, refining, parting, smithery, gilding, &c., are only branches of metallurgy. Others restrain *metallurgy* to those operations required to separate metals from their ores.

**METAMO′RPHIC.** (from μετὰ, trans, and μορφὴ, forma, Gr.) A term proposed for such hypogene rocks as are stratified, or altered by stratification; any stratified primary rock may be termed *metamorphic*. By some authors the metamorphic rocks have been divided into two groups; namely, those which present traces of stratification, and, secondly, those which present no appearance of regular arrangement, but occur in amorphous or shapeless masses. The term metamorphic has been assigned by Sir C. Lyell to a certain division of rocks comprising the crystalline strata or schists, called gneiss, mica schists, clay slate, chlorite schist, marble, and the like, the origin of which is more doubtful than that of the other divisions. The metamorphic rocks are either those in which the original structure and composition are still obvious, or those in which those characters are altogether obscured and replaced by others, produced either by heat, or pressure, or both conjoined. The metamorphic rocks may be divided into two sub-groups, those in which the original mineral structure is still recognizable, the particles, however altered, not having entered into new combinations, and those where such new combinations have been effected.—*Jukes.*

**METAMO′RPHOSIS.** (μεταμόρφωσις, Gr. change into another form, *métamorphose*, Fr. *metamórfosi*, It.) Transformations which insects undergo previously to their arriving at their state of perfection. The progress of *metamorphosis* of insects is most strikingly displayed in the history of the Lepidopterous, or butterfly and moth tribe. The egg, which is deposited by the butterfly, gives birth to a caterpillar; an animal which, in outward shape, bears not the slightest resemblance to its parent, or to the form it is itself afterwards to assume. It has, in fact, both the external appearance, and the mechanical structure, of a worm. But these vermiform insects contain in their interior the rudiments of all the organs of the perfect insect. These organs are, however, concealed from view by a great number of membraneous coverings, which successively invest one another, like the coats of an onion, and are thrown off, one after another, as the internal parts are gradually developed. These successive peelings of the skin are but so many steps in preparation for a more important change. A time comes when the whole of the coverings of the body are at once cast off, and the insect assumes the form of a *pupa* or *chrysalis;* being wrapt as in a shroud, presenting no appearance of external members, and retaining but feeble indications of life. In this condition it remains for a certain period, its internal system continuing in secret the farther consolidation of the organs, until the period arrives when it is qualified to emerge into the world, by bursting asunder the fetters which had confined it, and to commence a new career of existence. The worm, which so lately crawled with a slow and tedious pace along the surface of the ground, now ranks among the sportive inhabitants of the air; and expanding its newly acquired wings, launches forward into the element on which its powers can be freely exerted, and which is to waft it to the object of its gratification, and to new scenes of pleasure and delight.—*Dr. Roget.*

METATA'RSAL. Belonging to the foot, as the metatarsal bones, &c.

METATA'RSUS. (from μετὰ and ταρσὸς, Gr. *métatarse*, Fr.) That part of the foot which lies between the ankle and the toes, corresponding to the metacarpus of the superior extremities. The bones of the metatarsus in the most complete forms of development are always five in number, in each limb.

METATHO'RAX. (from μετὰ, beyond, and θώραξ, the chest.) In entomology, the third and last segment of the thorax, resembling the second in being of a more united structure than the first. The second and third segments are closely united together, but the original distinction into two portions is marked by a transverse line. To the second and third segments are attached both wings and legs, whereas the first segment has legs alone. The third segment consists of seven pieces, which are similar to those of the second. The posterior wings are placed at the anterior angles, and often occupy the whole sides of the *metathorax*. A pergamenteous partition at the posterior margin, which descends in a perpendicular direction, bowing in its middle towards the abdomen, separates the *metathorax* from the abdomen.

METEO'RIC IRON. Colour pale steel-grey; occurs ramose, and disseminated in meteoric stones. Native, or meteoric iron, is composed of iron and nickel, the proportion of nickel varying from one to nearly ten per cent. In some specimens a trace of cobalt has been discovered.

Pallas found a mass of native iron 1680lbs. in weight, in Siberia, which tradition stated to have fallen from the air. Meteoric iron is assuredly unlike any iron of earthly origin, but it has been imitated by fusing iron with nickel.

METEO'RIC STONE. } See *Aerolite.*
METE'ORITE.

METEORO'LOGY. (from μετέωρα, meteors, and λογος, a description, Gr.) The study of the phenomena of the atmosphere. It was not till the 17th century that any considerable progress was made in investigating the laws of meteorology. Previously to that period, the want of proper instruments precluded the cultivation of this science; but the discovery of the barometer and thermometer in the 17th, and the invention of accurate hygrometers in the 18th century, supplied the pre-existing defects, and enabled philosophers to enter on meteorological observations with accuracy and facility.

MI'CA. (from *mico*, to glisten.) This mineral appears to be always the result of crystallization, but is rarely found in regular, well-defined crystals. Most commonly it appears in thin, flexible, elastic laminæ, which exhibit a high polish and strong lustre. These laminæ have sometimes an extent of many square inches, and, from this, gradually diminish till they become mere spangles, discoverable, indeed, by their lustre, but whose area is scarcely perceptible by the naked eye. Mica is said to contain forty-four per cent. of oxygen. The laminæ of mica are easily separated, and may be reduced to a thickness not much exceeding the millionth part of an inch. Mica is easily scratched with a knife, and, commonly, even by the finger-nail. Its surface is smooth to the touch; its powder is dull, usually grayish, and feels soft. Its colours are silver-white, gray, green, brown, reddish, and black, or nearly black. Specific gravity from 2·50 to 2·90. When rubbed on sealing-wax, it communicates to the wax negative electricity. Before the blow-pipe, it fuses into a grey or black enamel. Its constituents parts are, according to Klaproth, silex 48·0, alumine 34·20, potash 8·75, oxide of iron 4·5, oxide of manganese 0·5. According to others, it is a compound of silicium, potassium, magnesium, calcium, &c., combined with oxygen. Mica is one of the component parts of granite, gneiss, and mica-slate; it occurs also in **syenite**, porphyry, and other primary rocks. To quartz and limestone it frequently communicates a slaty texture. It may always be distinguished from talc by the elasticity of its plates, talc being only flexible and not elastic; in its want of unctuosity, and by its communicating negative electricity to sealing-wax. There are several varieties, or sub-species; Jameson enumerates ten. Mica has been employed, instead of glass, in the windows of dwelling-houses. In lanterns it is superior to horn, being more transparent, and not so easily injured by the flame. Mica is a doubly refracting substance, with two optic axes, along which, light is refracted in one pencil.

MICA'CEOUS IRON ORE. A variety of oxide of iron. This occurs generally in amorphous masses, composed of thin six-sided laminæ. Colour iron-black, or steel-grey. Lustre metallic. Opaque. Feel greasy. Hardness 5 to 7. Specific gravity from 4·5 to 5·7. It is said to yield nearly 70 per cent. of iron.

MY'CA SCHIST. } A metamorphic rock,
MY'CA SLATE. } composed of mica and quartz; it passes by insensible gradations into clay-slate, and its texture is slaty. Sometimes the mica and quartz alternate, though commonly they are more or less intimately mingled, the mica usually predominating.

MY'CARELLE. The Pinite of Kirwan. See *Pinite*.

MICROCHO'NCHAS CARBONA'RIUS. A microscopic spiral shell with but few volutions, which when young touch one another, but when old are extended into a free tube, resembling vermetus or vermillia. The shell is sinistral. The lines of growth are strong, somewhat irregular, deficient in parallelism, and oblique to the axis of the tube. Spiral striæ may be perceived, though but faintly. The Microchonchas Carbonarius is a fossil shell of the coal measures.

MI'CROPYLE. (from $\mu\iota\kappa\rho\grave{o}\varsigma$, small, and $\pi\acute{v}\lambda o\varsigma$, gate, Gr.) A botanical term for the foramen in the perfect seed; this foramen is often visible, as in the pea and bean.

MI'CROSCOPE. (from $\mu\iota\kappa\rho\grave{o}\varsigma$, small, and $\sigma\kappa o\pi\acute{\epsilon}\omega$, to behold, Gr. *microscope*, Fr. *microscopio*, It.) A microscope is an optical instrument for examining and magnifying minute objects. Jansen and Drebell are supposed to have separately invented the single microscope, and Fontana and Galileo seem to have been the first who constructed the instrument in its compound form. The single microscope is nothing more than a lens or sphere of any transparent substance, in the focus of which minute objects are placed. The best single microscopes are minute lenses ground and polished on a concave tool; but as the perfect execution of these requires considerable skill, small spheres have often been constructed as a substitute. The most perfect single microscopes ever executed, of solid substances, are those made of the gems, such as garnet, ruby, diamond, &c. Garnet is the best material, as it has no double refraction, and may be procured perfectly pure and homogeneous. When a single microscope is used for opaque objects, the lens is placed within a concave silver speculum, which concentrates parallel or converging rays upon the face of the object next the eye.

When a microscope consists of two or more lenses, or specula, one of which forms an enlarged image of objects, while the rest magnify that image, it is called a *compound microscope*. The ingenuity of philosophers and of artists has been nearly exhausted in devising the best forms of object-glasses and of eye-glasses for the compound microscope.—*Dr. Brewster.*

MICROSCO'PIC. That may be seen only by the aid of a microscope.

MICROSCO'PIC SHELLS. These are found in prodigious abundance and of such extreme minuteness, that in their examination, and in the consideration of them, our mental, like our visual faculties, begin rapidly to fail us when we attempt to comprehend the infinity of littleness towards which we are thus conducted. Of several species of microscopic shells, five hundred scarcely weigh a single grain, and some are so exceedingly minute that one thousand would scarcely weigh a grain. In one ounce and a half of stone from the hills of Casciana, in Tuscany, Soldani obtained 10,454 microscopic shells, the remainder being principally composed of comminuted fragments of shells and spines of echini.

MICROTHE'RIUM. The name assigned to a fossil genus of small anoplotherioid herbivores. Entire crania of the Microtherium, from the lacustrine calcareous marls of the Puy-de-Dome are in the British Museum.

MI'EMITE. A mineral, thus named from having been found at Miemo, in Tuscany. A green variety of Dolomite, occurring in crystals, and in masses with a radiated structure.

MILK QUARTZ. The Milch Quartz of Werner; quartz hyalin laiteux of Haüy. A milk-white sub-species of quartz, and distinguishable only by its colour. It occurs massive. Milk-quartz and rose-quartz are considered by some mineralogists as one and the same sub-species, distinguished merely by their colour, the one presenting a milky appearance, while the other, coloured by a minute quantity of manganese, possesses a rose-red colour, sometimes passing into a flesh-red or crimson-red.

MI'LLEPEDE. ⎫ (*millepeda*, Lat. from
MI'LLIPEDE. ⎭ *mille*, a thousand, and *pes*, a foot.) Insects whose body is generally cylindrical; segments half membraneous and half crustaceous, each half bearing a pair of legs; antennæ seven-jointed, filiform, often a little thicker towards the end. These are called *millipedes*. The *millipedes* belong to the necrophagous tribe, or those which devour dead animals, or any other putrescent substances.—*Kirby.*

MILLE'PORA. ⎫ (from *mille*, a thousand,
MI'LLEPORE. ⎭ and *porus*, a pore.) A genus of lithophytes or zoophytes of various forms, having the surface perforated with numerous small pores or holes. A stony, internally solid, polymorphous, ramose or frondescent polypifer, pierced by simple, not lamellated pores. The pores cylindrical, and perpendicular to the axis, or to the expansions of the polypifer; for the most part small, and sometimes not apparent. —*Parkinson.* In *millepores* the cells are more minute and close

than in *madrepores*, and do not exhibit any star-like radiations.

MILLE'PORITE. A fossil millepore.

MILI'OLA. ⎫ A genus of microscopic
MILLI'OLA. ⎭ multilocular univalves, not larger than a millet seed, with transverse chambers, involving the axis alternately, and in three directions; the opening small and circular, or oblong, at the base of the last chamber. Several species are found to exist on our shores.

MI'LIOLITE. ⎫ The fossil Miliola. So
MI'LLIOLITE. ⎭ numerous are these minute fossils in the neighbourhood of Paris, that some species of them form the principal part of the masses of stone in some of the quarries. The remains of such minute animals as the milliola, have added much more to the mass of materials forming the earth's crust than the bones of the mammoths, whales, and hippopotami.

MI'LLSTONE. Called also Burrhstone. The Quartz agathe molaire of Haüy; Silex meuliere of Brongniart. The exterior aspect of this mineral is somewhat peculiar, being full of pores and cavities, which give it a corroded and cellular appearance. It occurs in amorphous masses, above the marine sand and sandstone. Sometimes the mass is comparatively compact, and the cavities small and not numerous; but in all specimens these cavities or cells are to be found. Millstone is of a white or greyish colour; sometimes with a tinge of blue or yellow; when unmixed it is pure silex. It contains no organic remains, and in the order of superposition of the formations in the neighbourhood of Paris, it constitutes the ninth horizontal bed, counting from the chalk upwards. It is of great use for making into millstones, from which circumstance it has obtained its name.

MI'LLSTONE GRIT. The name given to a silicious conglomerate, composed of the detritus of primary rocks. It has been thus named from some of the strata having been worked for millstones. It constitutes one of the members of the carboniferous, or mountain limestone group. The millstone grit forms a bed of considerable thickness in some situations, amounting to three or four hundred feet; in others, it is of very limited extent; and sometimes it is wholly wanting. The millstone grit is most commonly seen under the form of a coarse-grained sandstone, consisting of quartzose particles of various sizes, (often sufficiently large to give the rock the character of a pudding-stone,) agglutinated by an argillaceous cement. This sandstone differs from those which accompany the coal measures, principally by its greater induration. It sometimes assumes a finer texture, in which the mechanical structure becomes less evident, and even passes into a hard and solid cherty rock.

MI'LOSCHINE. The name given to a hydrated silicate of alumina.

MIMOSI'TES. The name given to fossil fruits belonging to the natural order Mimoseæ.

MINERAL ADIPOCI'RE. A fatty bituminous substance occurring in the argillaceous iron ore of Merthyr, in Wales. It is insoluble in water, and fuses at a temperature of 160°. When cold, it is inodorous; but on being heated, gives out a bituminous odour.

MI'NERAL CAOU'TCHOUC. A variety of bitumen, intermediate between the harder and softer kinds. It sometimes much resembles India rubber in its softness and elasticity, from whence it derives its name, and, like that, removes the traces of the pencil, but, at the same time, it soils slightly the paper. Colour brown, reddish-brown, or hyacinth-red. Specific gravity from 0.90 to

1·23. It burns with a bright flame, emitting, during its combustion, a bituminous odour. It occurs near Castleton, in Derbyshire.

MI′NERAL CHA′RCOAL. A fibrous variety of non-bituminous mineral coal.

MINERALIZA′TION. The process of converting into a mineral some body not previously such.

MI′NERALIZER. That which converts a substance into a mineral. Metals are combined with oxygen, sulphur, &c., by which their peculiar metallic properties are more or less disguised; in this case, the metal is said to be *mineralized*, and the oxygen or sulphur is termed the *mineralizer*.

MI′NERALS. *(minera*, Lat. *minéral*, Fr. *minerále*, It.) Those bodies which are *destitute of organization*, and which *naturally* exist within the earth or at its surface. The term *fossil* is usually appropriated to those *organic* substances which have become penetrated by earthy or metallic particles.

Minerals have been divided into two kinds; simple, or homogeneous, and compound, or heterogeneous. Simple minerals appear uniform and homogeneous in all their parts. They do, in fact, usually contain several different elementary systems; but these are so intimately combined, and similarly blended, in every part, as to exhibit a uniformity of appearance.

Compound minerals more or less evidently discover to the eye, that they are composed of two or more simple minerals, which either merely adhere to each other, or, as is sometimes the case, appear imbedded one in the other. Compound minerals are frequently aggregates or rocks.

The description of minerals, and their arrangement in systematic order, must result from an investigation of their properties. These properties consist in certain relations which minerals bear to our senses, or to other objects. Some of them are discoverable by mere inspection, or, at most, require some simple experiment to be made upon the mineral to ascertain its hardness, structure, gravity, &c., while others cannot be observed without a decomposition of the mineral. All these properties are usually called characters. We hence have a twofold division of the properties or characters of minerals into chemical and physical. —*Cleaveland.*

MI′NERALOGY. That science, says Cleaveland, which has for its object a knowledge of the properties and relations of minerals, and enables us to distinguish, arrange, and describe them.

Jameson defines mineralogy to be that part of natural history which makes us acquainted with the properties and relations of minerals. It is divided, according to that professor, into two grand branches, namely, mineralogy, properly so called, and geology. Mineralogy treats of the properties and relations of *simple minerals;* while geology considers the various properties and relations of the atmosphere, the waters of the globe, the mountain rocks, or those mineral masses of which the earth is principally composed, and the form, density, heat, electricity, and magnetism of the earth.

The history of the materials of the crust of the globe, their properties as objects of philosophical enquiry, and their application to the useful arts and the embellishments of life, with the characters by which they can be certainly distinguished one from another, form the object of mineralogy, taken in its most extended sense.

Mineralogy is a science of such interest, that it would be much to be regretted if its real objects and tendency were misunderstood, or suffered to degenerate into an avidity merely for the collecting of what is brilliant or rare. To the attainment of the science of geology, which is intimately connected with agriculture and the arts of life, that of mineralogy is essentially requisite. The study of mineralogy, therefore, does not include only a knowledge of the more rare and curious minerals: there is nothing in the mineral kingdom too elevated or too low for the attention of the mineralogist, from the substances composing the summits of the loftiest mountains, to the sand or gravel on which he treads.

By the study of what, in opposition to the term aggregated rocks, may be termed simple minerals, the mineralogist becomes enabled to detect the substance with which he holds acquaintance by itself, when aggregated with others in a mass; and thus he becomes qualified for the more difficult and more important study of the science of geology, which embraces a knowledge of the nature and respective positions of the masses and beds composing mountains, and, indeed, of every description of country, whether mountainous or otherwise.

There is no branch of science which presents so many points of contact with other departments of physical research, and serves as a connecting link between so many distant points of philosophical speculation as this. Nor, with the exception of chemistry, is there any which has undergone more revolutions, or been exhibited in a greater variety of forms. To the ancients it could scarcely be said to be known at all, and up to a comparatively recent period, nothing could be more imperfect than its descriptions, or more inartificial or unnatural than its classification. The Arabian writers, however, in the middle ages appear to have cultivated mineralogy with some success; the first foundation of a rational arrangement of minerals was laid by Avicenna at the close of the tenth century. It was only when chemical analysis had acquired a certain degree of precision and universal applicability, that the importance of *mineralogy* as a science began to be recognized, and the connection between a stone and its ingredient constituents brought into distinct notice. The arrangement of simple minerals has always been a subject of controversy with mineralogists; and the discussions to which it has given rise have materially contributed to the advancement of our knowledge of the natural and chemical history of minerals. Berzelius contends for the chemical arrangement, according to which the species are grouped in conformity with their chemical composition and characters. Werner rejects the pure chemical, and adopts the mixed method, in which the species are arranged and determined according to the conjoined external and chemical characters. The writers of the Wernerian school usually divide mineralogy into the five following branches; namely, oryctognosy, chemical mineralogy, geognosy, geographical mineralogy, and economical mineralogy. Of late years, the arrangement according to external characters alone (named the natural history system) has been advocated by Mohs. Among the external characters of a stone, none were, however, found to possess that eminent distinctness which the crystalline form offers; a character in the highest degree geometrical, and affording the strongest evidence of its necessary connection with the intimate constitution of the substance. The full importance of this character was, however, not felt until its connection with the texture or cleavage of a mineral was pointed out, and, even then, it required numerous and striking instances of the critical discernment of Haüy, and other eminent mineralogists, in predicting from the measurements of the angles of crystals which had been confounded together, that differences would be found to exist in their chemical composition, all which proved fully justified in their result before the essential value of this character was acknowledged. The invention of the goniometer by Carangeau, and subsequently the reflecting goniometer by Dr. Wollaston, which last has been improved by Sang, gave a fresh impulse to that view of mineralogy which makes the crystalline form the essential or leading character, by putting it in the power of every one, by the examination of even the smallest portions of a broken crystal, to ascertain the character on which the identity of a mineral in the system of Haüy was made to depend. Mineralogy, however, as a branch of natural history, remains still distinct from either optics or crystallography. But whatever progress may have hitherto been made in mineralogical pursuits, every new advance has opened a wider and more interesting prospect. Mineralogy is in reality essential to the geologist; it is the very alphabet to the older rocks, and it is probably to be attributed in great measure to the want of due preparation for the study of these rocks, by an intimate acquaintance with minerals in the *simple* state, that the primary and transition tracts have been investigated in a far less degree than those of a newer origin. The science is still in its infancy, and in many of its paths can proceed only with a faultering and uncertain step. — *Herschell.*

*Jameson. Cleaveland. Phillips.*

MI′NIUM. (*minium*, Lat.) A red oxide of lead. Minium is of a bright scarlet; it occurs in a loose state, or in masses, composed of flakes with a crystalline texture. It is found in the lead mines of Westphalia. It is used in glass-making, enamelling, and some other arts.

MI′OCENE. ) (from μείων, less, and
MEI′OCENE. ) καινός, recent, Gr.) The name given by Sir C. Lyell to a subdivision of the tertiary strata. He says, "the European tertiary strata may be referred to four successive periods, each characterized by containing a very different proportion of fossil shells of *recent* species." These four periods he names, Newer Pliocene, Older Pliocene, Miocene, and Eocene. The *Miocene* period has been found to yield from eighteen to twenty-five per cent. of *recent* fossils. This was the result of an examination of 1021 fossil species by M. Deshayes. Many shells belong exclusively to the Miocene period. The Miocene strata are largely developed in Touraine, and in the South of France near Bordeaux, in the basin of Vienna, and other localities. The miocene strata contains an admixture of the extinct genera of lacustrine mammalia of the Eocene series, with the earliest forms of genera which exist at the present time. In regard to the relative position of the strata, they underlie the older Pliocene, and overlie the Eocene formations, when any of these are present.

MI′TRA. A genus of shells belonging to the Columellaria in Lamarck's arrangement. It is a subfusiform univalve, with a long, pointed, turretted apex, a notched base, and no canal. Covered with an epidermis of a light brown colour. The columella is plicated; the inferior plicæ being the smallest.

Mitres are found both fossil and recent.

Mo′cha stone. (from *Mocha*, in Arabia.) The quartz agathe arborisé of Haüy; called also dendrite agate. A variety of agate, containing in its interior very beautiful delineations of leafless shrubs, trees, &c., of a brown or dark colour. These dendritic appearances are supposed to be produced by the filtration of the oxides of iron and manganese into the fissures of the agate. Mocha stones resemble those agates which are found on the Sussex coast called *dendrachates*.

Modi′ola. (from *modiolus*, Lat. a little measure.) A genus of shells belonging to the family Mytilacea. A transverse inequilateral bivalve. The mediola is a littoral shell, moored to rocks, stones, and shells. One species, *modiola discors*, floats free, enveloped in its own silky byssus. Fossil species have been found in the neighbourhood of Paris, and in this country.

Mo′lar. (from *mola*, a mill, Lat. *molaire*, Fr.) A grinder-tooth. The large double teeth are called molar teeth, or grinders; these are, however, subdivided according to their different forms; thus, those with two fangs are called bicuspid, or false molar teeth. The posterior molar teeth are differently shaped in carnivorous animals, being raised into sharp, and often serrated, edges, having many of the properties of cutting teeth. In insectivorous and frugivorous animals, their surface presents prominent tubercles, either pointed or round, for pounding the food; while in graminivorous quadrupeds they are flat and rough, for the purpose simply of grinding.

Mola′sse. (from *mollis*, soft, Lat.) The name given to a soft green sandstone found in Switzerland; one of the most recent of the tertiary deposites. In the *Molasse* of Switzerland there are many deposites affording sometimes coal of considerable purity.

Until the place of the *molasse* in the chronological series of tertiary formations be more rigorously determined, the application of this provincial name to the tertiary groups of other countries must be very uncertain, and it will be desirable to confine it to the tertiary beds of Switzerland.—*Lycett*.

Mo′lecule. (*molécule*, Fr. *petite partie d'un corps*.) A minute particle of a mass or body, differing from atom, inasmuch as it is always a portion of some aggregate. The ingredients of granite, and of all other kinds of crystalline rocks, are composed of *molecules* which are invisibly minute, and each of these *molecules* is made up of still smaller and more minute *molecules*, every one of them combined in fixed and definite proportions, and affording, at all the successive stages of their analysis, presumptive proof that they possess determinate geometrical figures.

Mollu′sca. (*mollusca*, a nut with a soft shell, Lat.) According to the arrangement of Cuvier, the second great division of the animal kingdom. This he subdivided into six classes, namely, Cephalopoda, Pteropoda, Gasteropoda, Acephala, Brachiopoda, and Cirrhopoda. A vast multitude of species, possessing in common many remarkable physiological characters, are comprehended in this great division. In all, as their name imports, the body is of soft consistence; and it is enclosed, more or less completely, in a muscular envelope, called the mantle, composed of a layer of contractile fibres, which are interwoven with the soft and elastic integument. Openings are left in this mantle for the admission of the external fluid to the mouth and to the respiratory organs, as well as for the protrusion of the head and the foot, when these organs exist. But a large proportion of animals comprised in this class are *acephalous*, that is, destitute of a head, and the mantle is then often elongated to form tubes, occasionally of considerable length, for the purpose of conducting water into the interior of the body.

The general form of the body, and the kind of motions it performs, vary more in the molluscous than in the articulated classes of animals, and we observe a corresponding diversity in their active organs of motion. The whole skeleton, the solid frame-work of the body, destined to give strength, form, and support to the entire machine, disappears in the class of mulluscous animals. In the molluscous classes there appear much greater variety, diversity, and want of symmetry in the whole muscular system. Many of the lower molluscs are fixed by long peduncles at the bottom of sea; some, as the limaces, creep on the surface of the dry land; the pteropods swim at the surface of the ocean, where the janthinæ hang suspended by floats; the naked cephalopods bound from the surface, and the pholades are fixed deep in cavities of rocks at the bottom; the oyster is fixed to the rock, while the clam skips to and fro by the flapping of its shells; the pinna is anchored to the bottom by its strong byssus, while the cardium swims along the still surface, suspended by its concave expanded foot. So that altthough none of these animals have wings to fly through the air, or jointed legs to creep upon the earth, or spines to oar them through the sea, they possess the means of almost every kind of motion, from the vibratile cilia of the fixed corals to the hands and feet of the finny tribe. The circu-

lation of the mollusca is always double; that is, their pulmonary circulation describes a separate and distinct circle. Their alimentary canal hardly ever passes straight through their body; nor is the anus terminal, as in most of the articulata. Their digestive cavities are more numerous and capacious, the intestine is more lengthened and convoluted, and all the assistant glandular organs are developed on a higher plan, and are more constant throughout the classes. The lowest of the molluscous classes, the tunicated animals, shut up in the interior of a cartilaginous, more or less elastic, and biforate tunic, have no prehensible or masticating organs connected with their mouth. The mouth, in fact, is placed at the bottom of the respiratory sac, and appears to be destitute even of those tentacula, appendices, or lips, which are so much developed, and so various in their forms, in the conchiferous animals.

In the mollusca we have the only instance in creation, of a *unipede* structure, but this one foot answers every purpose of a hand or leg; it spins for the bivalves their byssus, is used by others as a trowel, by others as an augur, and by others for other manipulations, and is generally their sole organ of locomotion; from its soft and flexible substance, it can adapt itself to the surfaces on which it moves, and by the slime that it copiously secretes lubricates them to facilitate its progress. It is probable that the foot may be also employed by these animals as an organ of touch. In the nervous system of mollusca, the ganglia have a circular arrangement. The transition series afford examples of several families and many genera of mollusca, which appear at that period to have been universally diffused over all parts of the world. Some of these, as the orthoceratite, spirifer, and producta, became extinct at an early period in the history of stratified rocks, whilst others as the terebratula and nautilus, have continued through all formations to the present time.—*Cuvier. Grant. Kirby. Roget.*

MOLLUSCOID'EA. The third class of the sub-kingdom Mollusca: this class comprises three orders, namely, Brachiopoda, Polyzoa, and Ascidioida, or Tunicata.

MOLLU'SCOUS. Animals belonging to the division mollusca, or soft, invertebral, inarticulate animals; often protected by a shell.

MOLLU'SKITE. The soft bodies of the testaceous mollusca often occur in a fossil state, changed into a brown carbonaceous substance, to which Dr. Mantell assigned the above name.

MOLY'BDATE OF LEAD. The plomb molybdaté of Haüy; pyramidaler blaibaryt of Mohs. Yellow lead ore. It is of a yellow colour, varying from lemon yellow to yellowish brown. Occurs crystallized and massive. Its specific gravity from 6·5 to 6·9. Fracture uneven, or imperfectly conchoidal. Slightly translucent, especially at the edges. Before the blow-pipe it decrepitates, and fuses into a dark coloured mass. It consists of oxide of lead 58, molybdic acid 38, oxide of iron, 2. It is found at Bleyberg, in Carinthia, and in Mexico, in compact limestone.

MOLYBDE'NA. A mineral of a lead-grey colour, occurring in thin flexible leaves.

MOLYBDE'NUM. (from $\mu o \lambda \acute{v} \beta \delta a \iota v a$, Gr.) A metal discovered by Hielm in 1782; externally of a whitish yellow colour; fracture whitish grey: sp. gr. about 8·6. It is nearly infusible. It is obtained from the mineral molybdena in small grains, agglutinated together in brittle masses.

MO'NAD. (from $\mu o v \grave{a} s$, Gr. an atom, *monade*, Fr.) The recent observations of Professor Ehrenberg have brought to light the existence of *monads*, which are not larger than the 24,000th of an inch, and which are so thickly crowded in the fluid as to leave intervals not greater than their own diameter. Hence he has made the computation that each cubic line, which is nearly the bulk of a single drop, contains 500,000,000 of these monads; a number which equals that of all the human beings on the surface of the globe in one drop of fluid. *Monads*, which are the smallest of all visible animalcules, have been spoken of as constituting "the ultimate term of animality."

MONA'NDRIA. (from $\mu \acute{o} v o s$, one, and, $\grave{a} v \grave{\eta} \rho$, a man, Gr.) The first class of plants in Linnæus's artificial system. The plants of this class have only one stamen; it is a small class, and contains only two orders.

MONI'LIFORM. (from *monile*, a necklace, and *forma*, form, Lat.) Resembling a necklace.

MO'NITOR. (*monitor*, Lat.) A genus of lizards or saurians, species of which are found both fossil and recent; the recent inhabit the tropics. Cuvier places this genus in the family Lacertinida. The monitors frequent marshes, and the banks of rivers, in hot climates; they have received their name from a common but silly notion that they give warning of the approach of crocodiles and caymans by a whistling noise. One species, the Lacerta nilotica, devours the eggs of crocodiles. Fossil remains of the monitor have been discovered in the strata of Tilgate Forest, in Sussex.

MONOCOTYLE'DON. (from $\mu \acute{o} v o s$, one, and $\kappa o \tau v \lambda \eta \delta \grave{\omega} v$, a seed lobe, Gr.)

A plant which has only one cotyledon or seed-lobe.

MONOCOTYLE'DONOUS. Plants, the seeds of which have either only one cotyledon, or if more, those alternate on the embryo, are called monocotyledonous; grasses, lilies, aloes, and palms, are examples. Monocotyledonous plants may be at all times recognised, from the circumstance of the veins of their leaves being parallel, while those of dicotyledonous plants are reticulated.

MONŒ'CIA. (from μόνος, one, and οἰκία, a house, Gr.) The twenty-first class of plants in the artificial system of Linnæus. In this class of plants the stamina and pistils are in separate flowers, but growing on the same individual plant. The orders in this class depend upon the circumstances of their male flowers, and are nine or ten in number.

MONŒ'CIOUS. Plants belonging to the class Monœcia.

MO'NODON. (from μονόδους, Gr. having one tooth.) The sea unicorn, or narwhal; distinguished by its long tusk, or tusks, for there are sometimes two, extended in a horizontal direction. The Monodon belongs to the order Cetacea, class Mammalia.

MONODO'NTA. A genus of univalve shells separated from Trochus on account of a tooth-like process which it forms at its base. Several fossil species are known in the oolitic formations of England.— *Lycett.*

MONOPE'TALOUS. (from μόνος, one, and πέταλον, a petal, Gr.) Flowers are so called which consist of only one leaf or petal; or when the leaves which compose the corolla are united by their edges; the convolvulus, honeysuckle, &c., are examples.

MONOPHY'LLOUS. (from μόνος, sole, and φύλλον, a leaf, Gr.) Having one leaf only, or formed of one leaf; applied to calices consisting of not more than one leaf.

MONOSE'PALOUS. A term applied to the calyx of a flower, when the sepals which compose it are united by their edges: the pink, convolvulus, henbane, &c., are examples.

MONOTHA'LAMOUS. (from μόνος, single, and θάλαμος, a chamber, Gr.) Shells whose chamber is undivided by partitions; these are termed unilocular, or monothalamous: the argonaut is an example.

MO'NOTREME. The *Monotremes* form Cuvier's third tribe of Edentata, comprising two genera, namely, Echidna and Ornithorhynchus. They are found only in New Holland. The *Monotremes* seem connected with the birds; one genus, the ornithorhynchus, having a mouth resembling the bill of a duck, and being almost web-footed; it has also been stated to be oviparous. The *Monotremes* have no marsupial pouch. They suckle their young from a mammary orifice. In the classification of some authors, Monotremata constitutes the second order of the sub-class Implacentalia.

MOO'NSTONE. A variety of felspar, called also *adularia*, possessing a silvery or pearly opalescence. Moonstone is transparent and translucent: colour white, with sometimes a tinge of yellow, green, or red. When held in certain positions, its surface is iridescent. It occurs massive, and in crystals. It is found in the fissures and cavities of granite, gneiss, &c.

MORA'INE. An accumulation of sand, stones, or debris, found upon icebergs, glaciers, &c. In front of glaciers there is usually a pile of rubbish, composed of pieces of rock, earth, and trees, which they have forced forward, known in Switzerland by the name of *moraine*. If there be a line of *moraine* some distance from the front of the glacier, it is considered that the glacier has retreated to the amount of that distance; but if there be no other than that which the glacier immediately drives before it, it is considered to be on the increase.

Moro′xite. A sub-species of apatite, occurring in crystals, of a brownish or greenish-blue colour: found in Norway, in primary rocks.

Mosasau′rus.
Mosæsau′rus.
Mososau′rus.
} "The Mososaurus," says Buckland, "has been long known by the name of the Great Animal of Maestricht, occurring near that city, in the calcareous freestone, which forms the most recent deposit of the cretaceous formation. A nearly perfect head of this animal was discovered in 1780, and is now in the museum at Paris. This celebrated head, during many years, puzzled the most skilful naturalists; some considered it to be that of a whale, others of a crocodile; but its true place in the animal kingdom was first suggested by Adrian Camper, and, at length, confirmed by Cuvier. By their investigations, it is proved to have been a gigantic marine reptile, most nearly allied to the monitor. Some vertebræ of the mososaurus have been discovered in the upper chalk near Lewes, in Sussex; these have the body convex posteriorly, and concave anteriorly, and were one hundred and thirty-three in number. It had four paddles instead of legs. Teeth of the mososaurus have been discovered in the green-sand of Virginia. Portions of jaws, with teeth of the mososaurus, may be seen in the British Museum. The mososaurus was a reptile, holding an intermediate place between the monitor and iguana, about twenty-five feet long, and furnished with a tail of such construction as must have rendered it a powerful oar, enabling the animal to stem the waves of the ocean, of which Cuvier supposes it to have been an inhabitant."

"From the lias upwards, to the commencement of the chalk formation, the ichthyosauri and plesiosauri were the tyrants of the ocean; and just at the point of time when their existence terminated, during the deposition of the chalk, the new genus mososaurus appears to have been introduced, to supply for a while their place and offices, being itself destined, in its turn, to give place to the cetacea of the tertiary periods."

Moss a′gate. A kind of agate which on being cut and polished, presents delicate vegetable ramifications of different shades, resembling small filaments of moss, or fibres of roots, irregularly interwoven. It has been suggested by some authors that these filaments may be really mosses enveloped in the agate.

Moss fir. The name given to a certain kind of wood frequently found in peat mosses or bogs. It much resembles in its colour and general external appearance, ordinary decayed fire-wood; but on examination it appears that the fibre of the wood is strongly imbued with resin, and that all its interstices are filled with resinous matter. It is so highly inflammable as to be employed not only as fuel but as torches.

Mother of coal. In many coals little flakes of mineral charcoal occur, retaining that part of the vegetable structure called the vascular tissue. They are called by the colliers "mother of coal."

Mould. (*muld*, Goth. *mold*, Scel. *mold*, Sax. *mul*, Dan. *mull*, Germ. Dr. Webster says the orthography of this word is incorrect, and that it should be written mold: so far as the etymology of the word is concerned, perhaps Webster is correct, but, assuredly, custom warrants our writing it mould.) The name given to that superficial accumulation of various substances which lies upon the surface of the dry land, and covers the rocks below.

"The process," says Dr. Buckland, "is obvious whereby even solid rocks are converted into soil fit for the maintenance of vegetation, by simple exposure to atmospheric agency; the disintegration produced by the vicissitudes of heat and cold, moisture and dryness, reduces the surface of almost all strata to a comminuted strata of soil, or *mould*, the fertility of which is usually in proportion to the compound nature of its ingredients."

MOU′NTAIN CORK. (The Berg kork of Werner; Suber montanum of Kirwan; Asbeste suberiforme of Brongniart.) A white or grey variety of asbestos, to which the name of mountain cork has been given from its extreme lightness; sp. gr. from 0·68 to 0·99, consequently so light as to swim in water. Its structure is fibrous; the fibres promiscuous and interwoven. Its constituents are silex 56·2, magnesia 26·1, lime 12·7, iron 3·0, alumine 2·0. It occurs in France and Saxony.

MOU′NTAIN BLUE. A species of blue malachite or blue copper. The Cuivre carbonaté bleu of Haüy; Kupfer lazur of Werner. Carbonate of copper. The characteristic colour of mountain blue is azure-blue, often exceedingly beautiful and splendent. Occurs regularly crystallized in scopiform and stellular concretions, radiated, and also curved lamellar. When rubbed on paper, it leaves a light blue streak. Sp. gr. from 3·20 to 3·60. It dissolves with effervescence in nitric acid. It is scarcely fusible alone, but with borax, to which it communicates a fine green, it yields a globule of copper. Its constituents are copper 66, **carbonic acid 18,** oxygen 8, water 2.

MOU′NTAIN LI′MESTONE. (The Calcaire carbonifere, Calcaire anthraxifere, and Calcaire de transition of the French; the Kohlenkalk, and Ueberganskálk of the Germans. Conybeare proposed to designate this rock "Carboniferous Limstone." By some authors it has been termed metalliferous limestone, from its mineral riches; by others, entrochal or encrinal limestone, from its organic remains: it has also been proposed to designate it by the Wernerian name, "first flœtz formation.") A series of marine limestone strata, whose geological position is immediately below the coal measures and above the old red sandstone. To this formation the French have given the name of *Calcaire de transition*. Mountain limestone is one of the most important calcareous rocks in England and Wales, both from its extent, the thickness and number of its beds, **the quantity and variety of its organic remains, and its richness** in metallic ores, particularly of lead. In Derbyshire, where the different beds of limestone have been pierced through by the miners, the average thickness of the three uppermost is about 160 yards; the series is said to exceed, in some instances, 1000 feet: the beds are **separated by beds** of trap or basalt, resembling ancient lavas. The limestone is generally sufficiently hard to bear a polish, and forms what is denominated marble, of considerable beauty. The mountain limestone formation occupies an immense tract in Northumberland, Durham and Yorkshire, from which country it runs out into a curve to encircle on the north, and partially on the south, the group of Cumbrian slate mountains. It also appears in great force in Derbyshire, ranges through Flint and

Denbigh, to St. Orme's head and Anglesea; shows slightly round the Clee hills in Shropshire; and presents picturesque cliffs on the Wye, near Monmouth. The mountain or carboniferous limestone may, according to Mr. Weaver, be considered as the prevalent rock in Ireland; all its counties, with the exception of Antrim, Derry, and Wicklow, being more or less composed of it, and in some instances it attains a thickness of 1700 feet and upwards. The prevailing characteristic organic fossils are madrepores and encrinites; of the latter, some of the upper beds appear to be almost entirely composed. Mountain limestone is generally almost a pure carbonate of lime; its purest beds appearing to contain about 96 per cent. of calcareous matter; but by the admixture of other ingredients, it often passes into **magnesian** limestone, **ferruginous** limestone, **bituminous** limestone, and **fetid** limestone. It is a prevailing character of the mountain limestone to be full of caverns and fissures.

MOU'NTAIN LEATHER. (Bergleder; corium montanum; cuir de montagne.) A variety or sub-species of asbestus, the same as Mountain Cork, *which see*.

MOU'NTAIN MEAL. (The Bergmehl of Fabbroni.) This singular mineral, says Phillips, was found in the form of a bed by Fabbroni, at Santa Fiora, between Tuscany and the Papal dominions; it is manufactured into bricks, so light as to swim in water. Klaproth gives as its analysis, silica 79, alumina 5, oxide of iron 3, water 12.

MOU'NTAIN SOAP. A mineral, a variety of bole, of a black or blackish-brown colour. It is massive, dull, smooth and soapy to the touch, and adheres strongly to the tongue. It writes on paper. Its constituents are silex 44, alumine 26·2, oxide of iron 8, lime 0·5, water 20·10. It occurs in secondary rocks of the trap formation in the Isle of Skye, and in Poland.

MOU'NTAIN WOOD. The berg-kolz of Werner; asbeste ligniform of Haüy; le bois de montagne of Brochant.) A subspecies of asbetus of a woodbrown colour, occurring massive and in plates: it has somewhat the aspect of wood, and is occasionally so hard and compact as to resemble petrified wood. It is infusible before the blow-pipe, according to Jameson, but Phillips states that it fuses into a black slag, and is about twice the weight of water.

MU'CRONATE. (*mucronatus*, Lat. pointed.)
1. In entomology, when from an obtuse end a fine point suddenly proceeds.
2. In botany, when a small point terminates an entire leaf, as in the vetch, house-leek, &c.
3. In conchology, when a shell terminates in a sharp rigid point.

MO'YA. The name given by the natives of South America to the mud and slime ejected from volcanos during the eruptions.

MULA'TTOE. An arenaceous stone, with a calcareous cement, deriving its name from its speckled appearance, which is caused by numerous disseminated spots of green earth. It occurs in Ireland, and agrees altogether in its characters and fossils with the green sandstone.

MULTISPI'RAL. (from *multus*, many, and *spira*, a spire, Lat.) In conchology, a term for a shell whose spire consists of many whorls; also an operculum of many volutions.

MULTILO'CULAR. (from *multus*, many, and *loculus*, a chamber or shell, Lat.) A term applied to shells containing partitions, which divide them into several chambers. Orthoceratites, baculites, hamites, scaphites, belemnites, &c., are all *multilocular* shells; the argonaut, or paper natilus, is a *unilocular* shell.

MU'LTIVALVE. (from *multus*, many, and *valvæ*, valves, Lat.) Some of the mollusca have, in addition to the two principal valves, small supplementary pieces of shell; these have been comprised in the order of *multivalves*.

MU'NDIC. Iron or arsenical pyrites, which strike fire with amazing facility. The term, says Dr. Paris, seems to be derived from this quality.

MU'RCHISONIA. A genus of fossils of the Silurian formation.

MU'RCHISONITE. A mineral, thus named by Mr. Levi, in honour of Sir R. Murchison. Its constituents are silica 68·10, alumina 16·6, potash 14·8. It occurs near Dawlish; it is a variety of felspar.

MU'REX. (*murex*, Lat. *murex*, Fr.) A genus of shells. Animal a limax: shell univalve, spiral, rough, with membraneous sutures; aperture oval, ending in an entire straight, or slightly ascending canal. The murex is an inhabitant of the ocean, found at depths varying from five to twenty-five fathoms, on different bottoms. These shells, besides their long channelled beaks, are remarkable for the beauty and variety of their spines. Murices, or rock-shells, were in high esteem from the earliest ages, on account of the dye that some of them yielded; cloths dyed with it bearing a higher price than others. More than one species yielded a dye; one, according to Bochart, a glaucous or azure colour; the other, a purple. Different species of fossil *murex* are found in the London clay and in the Bognor sandstone, and Lamarck describes many species found in the neighbourhood of Paris.

MU'REX CONTRA'RIUS. The reversed whelk, now more commonly known as **Fusus** contrarius, is a sinistral shell, and is found most abundantly in the crag formation, of which it appears characteristic.

MU'RIACITE. A name given to anhydrite, by Klaproth. See *Anhydrite*.

MU'RICATED. (*muricatus*, Lat.) Clothed with sharp rigid points; beset with short erect spines.

MU'RICITE. The fossil murex.

MUSA'CEA. A family of tropical monocotyledonous plants, including the banana and plaintains.

MU'SCLE BAND. The name given to a bed of ironstone, found about the middle of the coal series in Derbyshire, and thus named from its containing a very large number of different species of mytili.

MU'SCHEL-KALK. (from *muschel*, shell, and *kalk*, lime or chalk, Germ.) A compact hard limestone, of a greyish colour, found in Germany. It belongs to the red sandstone group. The muschel-kalk of Germany cannot be considered to have any precise equivalent among the English strata, and indeed would appear to be unknown in England and the north of Germany. It occasionally is met with of sufficient hardness to be employed as marble. In Bavaria and Wurtemburg the muschel-kalk is interposed between the red sandstone, on which it rests, and the variegated marls which lie over it, and with which, at the junction, it alternates. The muschel-kalk abounds in organic remains.

MU'SCLE-BIND. The name given to a stratum of imperfect ironstone and indurated shell, found in the Derbyshire and Yorkshire coal fields.

MU'SITE. } A mineral, thus named
MU'SSITE. } from Mussa, in Piedmont, where it occurs. It is a white, or pale green, variety of augite.

MY'A. (from μυών, a muscle, Gr.) A genus of bivalves belonging to the family Myacidæ. Animal an ascidia. Shell transverse, oval, thick, gaping at both ends; ligament internal. Hinge with broad, thick, strong, patulous tooth, seldom more than one, perpendicular to the valve, and giving attachment to the ligaments.

MY′LIOBATES. A genus of fossil Rays. They are abundant in the London clay and in the crag.

MY′OLOGY. (from μυς, ˙μυος, a muscle, and λογος, discourse.) A description of the muscles.

MYO′PORA. A genus of bivalve conchifera, established by Lea, belonging to the family Arcacea.

MYRIAN′ITES. A genus of the Nereidina of Mac Leay; body linear, very narrow, and formed of very numerous segments, with indistinct feet and short cirri. One species, Myrianites Mac Leaii, so named by Sir R. Murchison, and described in his Silurian system, has been found in the older rocks.

MYOCO′NCHA. A genus of fossil bivalve shells, belonging to the Cardiacea; they are oval, equivalve, oblique; the umbones are terminal, the hinge has an external ligament, and an oblique elongated tooth in the left valve, the general form of this shell is that of Mytilus; the only known species is the M. Crassa of the Great and Inferior Oolite.

MYRIA′PODA. ⎫ (from μυρια, ten thous-
MY′RIAPODS. ⎭ and, and πους, ποδος, a foot, Gr.) A class of insects, commonly called Centipedes, possessing a number of feet, from six to some hundreds. The Myriapoda, in general, resemble little serpents, or Nereides, their feet being closely approximated to each other throughout the whole extent of the body. Myriapods exhibit the following general characters. Animal undergoing a metamorphosis by acquiring in its progress from the egg to the adult state several additional segments and legs. Body without wings, divided into numerous pedigerous segments, with no distinction of trunk and abdomen. Head with a pair of antennæ; two compound eyes; a pair of mandibles; under-lip connate with the maxillæ. Myriapoda constitutes the second class of the sub-kingdom Annulosa, and comprises two orders, Chilopoda and Chilagnatha.

MYTILA′CEA. In Cuvier's arrangement, the second family of the order Acephala Testacea. All belonging to this family are bivalves, having a foot which they use in crawling. Mytilacea comprises, in Lamarck's system, Modiola, Mytilus, and Pinna.

MY′TILUS. (*mytilus*, Lat.) A genus of the family Mytilacea. The muscle. A rough, longitudinal, bivalve; with equal, convex, and triangular valves; the anterior, and longest side of the shell, allowing passage of the byssus. The *Mytilus* is a littoral shell, moored to rocks, stones, crustaceans, &c. The foot of the Mytilus edulis, or common muscle, can be advanced to the distance of two inches from the shell, and applied to any fixed body within that range. By attaching the point to such body, and retracting the foot, this animal drags its shell towards it; and by repeating the operation successively on other points of the fixed object, continues slowly to advance. Some Mytili produce pearls.

# N

**NA´CRE.** (*nacre*, Fr. *nácchera*, It.) A sort of mother-of-pearl. The fossil ink-bags of belemnites found in the lias are surrounded by *nacre*.

**NA´CREOUS.** Glistening; silvery; irridescent. Having the appearance of mother-of-pearl. Many membraneous shells exhibit a nacreous appearance on their internal surface, as the Haliotis, or sea-ear; Anodon, or fresh-water muscle, &c.

**NA´CRITE.** (from *nacre*.) A mineral so called in consequence of its pearly lustre. The Talcite of Kirwan. Nacrite occurs in reniform masses, composed of extremely minute spangles, or glittering scales. Colour pearly grey, with a tinge of red or green. It fuses easily before the blow-pipe. When rubbed between the fingers it leaves a pearly gloss. Unctuous to the touch. Its constituents are, silex 56·0, alumine 18·25, potash 8·50, lime 3·10, iron 4·20, water 6·0.

**N´AGLE-FLUE.** The name given to a conglomerate of great thickness, occuring in Switzerland.

**NA´IAS.** A genus of plants of the order monandria, class dioecia, and in the natural method ranking with those of which the order is doubtful. The male calyx is cylindrical and bifid; the corolla quadrifid; there is no filament, nor is there any female calyx or corolla; there is one pistil; the capsule is ovate and unilocular. To this genus belong the several species of zosterites, found fossil in the cretaceous group.

**NA´KED.** (*nackt*, Germ.)
1. In botany, applied to flowers having no calices; to stems without leaves; also to leaves when perfectly smooth, and quite destitute of hairs.
2. In zoology, applied to mollusca, when the body is not defended by a calcareous shell.

**NA´PHTHA.** (νάφθα, Gr. *naptha*, Lat. *naphte*, Fr.) A variety of bitumen, thin, volatile, fluid, and inflammable; unctuous to the touch, and constantly emitting a strong odour. Colours yellowish-white and yellowish-grey; transparent. Specific gravity from 0·70 to 0·85. It is highly combustible, igniting even on the approach of a lighted taper. It burns with a white or bluish flame, produces a dense smoke, and yields no residuum. It is insoluble in alcohol. When long exposed to the air, it becomes yellow and then brown; its consistence is increased, and it passes into petroleum. Naphtha consists of carbon 82·2, hydrogen 14·8. Springs of it exist in many countries, particularly in the neighbourhood of volcanoes. The finest varieties of naptha are found on the shores of the Caspian. The soil is sandy and marly, and the surrounding minerals are calcareous. The naptha is constantly rising in the state of an odorous inflammable vapour. The inhabitants of the town of Baku, a port on the shores of the Caspian sea, are supplied with no other fuel than that obtained from the naphtha and petroleum with which the neighbouring country is highly impregnated. Mr. Coxe estimates the produce of the naphtha springs at Rangoon, Pegu, at 92,781 tons per annum. Naphtha is burnt in lamps instead of oil.

**NARC´ODES.** The name assigned to a genus of ichthyolites of the old red sandstone.

**NA´RWAL.** } (*narwhal, narwall*, Germ.
**NA´RWHAL.** } the sea unicorn.) The

*Monodon monoceros* of Linnæus. Placed by Cuvier in the family Cetacea ordinaria, order Cetacea. The tusk of this animal is sometimes ten feet long, and spirally furrowed. Portions of the skull of the narwahl have been found in the Lewes levels, in Sussex.

NA′SSA. A genus of fossil univalve shells separated from Buccinum by the tooth-like projection which terminates the columella; ten fossil species are recorded in the English tertiary deposites, and two from the green sand of Black Down, but the genus of the latter is perhaps rather doubtful.

NA′TATORY. (from *natator*, Lat. a swimmer.) Enabling to swim. Certain organs possessed by many animals are *natatory* organs. Several of the cephalophods and pteropods, and other molluscans, have *natatory* appendages.

NA′TICA. A genus of nearly globose, umbilicated, univalves, belonging to the family Neritacea. Aperture entire and semicircular; columella transverse, without teeth, and callous externally. These shells, though strongly resembling Neritæ, may be distinguished from those of that genus by their being always umbilicated, and the columella never dentated. The recent Natica is found in estuaries and tidal rivers, in mud and sandy mud, at depths varying to forty fathoms. It is also found fossil.

NA′TROLITE. The Natrolith of Werner; Natrolithe of Haüy; considered by some mineralogists to be a variety of prismatic zeolite. Occurs in small, reniform, rounded, or irregular masses, composed of very minute fibres; the fibres are divergent, or even radiate from a centre, and are sometimes extremely minute and close. Colours yellow, yellowish brown, and brown, with striped-colour delineations. Translucent at the edges. Sp. gr. 2·16, to 2·20. Before the blow-pipe it fuses readily into a white glass. In nitric acid it is reduced, without any effervescence, into a thick jelly. It derives its name from containing soda. Its constituents are silex 48·20, alumine 24·50, soda 16·10, oxide of iron 1·75, water 9·0. It occurs principally at Roegau, in Suabia, imbedded in amygdaloid.

NA′TRON. (*natron*, Fr.) The Soude carbonatée of Haüy; L'alkali mineral natif of Brochant; the Natron of Kirwan. A carbonate of soda occurring massive and crystallised, the principal supplies coming from lakes in Egypt and Hungary. In Egypt, the lakes which yield natron abundantly are called the Lakes of Natron. These are six in number, to the westward of the Nile, not far from Terrana, in a valley, surrounded by limestone.

NA′TURAL HISTORY. This extensive science has for its object the enquiry into the being of natural bodies, and their thorough investigation in reference to their various qualities, and the relative functions of their component parts. Understood in this extent, it presents us with a distinct unique entirety, which treats the natural body as complete, but gradually perfected; and at the same time seeks to discover the means whereby it attained its completion and perfection. Natural history, therefore, is no mere description of form,—no description of nature, as it has been, latterly, very incorrectly considered, but a true and pragmatical history, developed from its own fundamental principles.—*Burmeister*.

NAU′LAS. A genus of ichthyolites of the old red sandstone.

NAUTILA′CEA. A family of Polythalamous cephalopods, in the arrangement of Lamarck. This family comprises Discorbites, Nautilus,

Nummulites, Polystomella, Siderolites, and Verticialis.

NAUTI'LIDÆ. A family of chambered shells belonging to the order Tetrabranchiata; of this family the nautilus is a genus.

NAU'TILITE. A fossil nautilus.

NAU'TILUS. A **genus of** shells belonging to the **family** Nautilidæ. A spiral, polythalamous, discoidal univalve with smooth sides. The **turns contiguous,** the outer side covering **the** inner. The chambers separated by transverse septa, which are concave outwards, and perforated by a tube passing through the disk. Three or four recent species are known. "It is a curious fact," says Dean Buckland, "that although the shells of the nautilus have been familiar to naturalists, from the days of Aristotle, and abound in every collection, the only authentic account of the animals inhabiting them, is that by Rumphius, in his history of Amboyna." At the present day the nautilus **is** an inhabitant of tropical **seas, but** its fossil remains are found **in** formations of every age. The organ of locomotion in the nautilus appears to have been a foot, resembling that of the snail. This organ is expansive, and surmounts the head. The **oral** organs are much more complicated and numerous than those **of the cuttle-fish, and are** furnished **with no suckers. Its** tentacles **are retractile within four** processes, each pierced by **twelve** canals protruding an equal number of these organs, so that, in all, there are forty-eight. In fact, the whole oral apparatus, except the mandibles and the lip, is formed upon a plan different from that **of** the cuttle-fish, as likewise from that of **the** carnivorous trachelipod molluscans, and indicate very different **modes of** entrapping and catching **their prey**; being deprived of suckers, **they** seem destitute of any powerful means of prehension and detention. The eye, also, is reduced to the simplest condition that the organ of vision can assume, without departing altogether from the type of the higher classes, so that it appears not far removed from that of the proper molluscs. The nautilus has only a single heart, the branchial one being absent. The nautilus resides in the capacious cavity of its first, or external, chamber; and it is now well ascertained that this animal is not a piratical parasite, occupying the shell of another animal, which it has murdered, but that it lives, and sails, in a skiff of its own building. A siphuncle connects the body of the nautilus with the air chambers, passing through an aperture and short projecting tube in each transverse septum, till it terminates in the smallest chamber at the **inner** extremity of the shell. These internal chambers contain only air, and have no commnnication with the outer chamber but by one small aperture in each septum through which the siphuncle passes. No water can by any possibility pass into these chambers, between the exterior of the siphuncle and the siphonic apertures of the transverse plates, because the entire circumference of the **mantle in** which **the siphuncle** originates, is firmly attached to the shell by a horny girdle, impenetrable by any fluid. The number of chambers varies greatly, according to the age of the animal. Dr. Hook states that he has found in some shells as many as forty. The siphuncle, as appears from Prof. Owen's statement, terminates in a large sac surrounding the heart of the animal; if we suppose this sac to contain a pericardial fluid, the place of which is alternately changed from the pericardium to the siphuncle, we shall find in these organs an hydraulic apparatus for varying the specific gravity of the shell; so that it sinks when the pericardial fluid is forced into the siphuncle, and becomes buoyant, when the same fluid returns to the pericardium. The substance of the siphuncle is a thin and strong membrane, surrounded by a coat of muscular fibres, by which it could contract or expand itself, in the process of admitting or ejecting any fluid to or from its interior. When the arms and body are expanded, the fluid remains in the pericardium, and the siphuncle is empty, and collapsed, and surrounded by the portions of air that are permanently confined within each chamber; in this state the specific gravity of the body and shell together is such as to cause the animal to rise, and be sustained floating at the surface. When, on any alarm, the arms and body are contracted, and withdrawn into the shell, the retraction of these parts, causing pressure on the pericardium, forces its fluid contents into the siphuncle, and as the quantity of matter within the shell is thus increased, without any increase of magnitude, the specific gravity of the entire animal is increased, and it begins to sink. Rumphius states that, at the bottom of the sea, the nautilus creeps with his boat above him.

Fossil remains of the nautilus are found in strata from the mountain limestone upwards. In some of these the siphunculus is beautifully preserved. But while, as a genus, the nautilus occurs in formations of every age, from the transition series upwards, yet certain species appear limited to particular geological formations. The eocene, miocene, and pliocene has each its particular nautili.—*Buckland, Kirby. Owen. Parkinson. Sowerby.*

NAU'TILUS SY'PHO. The name given to a very beautiful, camerated, siphuncled fossil shell, found in the tertiary strata at **Dax**, near Bourdeaux. This fossil presents deviations from the usual characters of the nautilus, whereby it approximates to the ammonite.

NAU'TILUS ZIC ZAC. A fossil, camerated, siphuncled shell, found in the London clay. This and the nautilus sypho appear to form connecting links between the genera Nautilus and Ammonite.

NECRO'PHAGOUS. (from νεκρὸς, dead, and φαγεῖν, to eat, Gr.) Animals which devour dead substances. The unclean animals, with respect to their habits and food, were divided into two classes; namely, *zoophagous* animals, or those which attack and devour living animals; and *necrophagous* animals, or those which devour dead ones, or any other putrescent substance.

NE'CROMITE. (from νεκρὸς, dead, Gr.) A mineral occurring in small masses in limestone; found near Baltimore. When struck, it exhales a fetid odour, resembling that of putrid flesh; from this quality it obtained its name.

NE'CTARY. That part of a flower which secretes and contains the honey, (an almost universal fluid in flowers) and is either a part of the corolla, or an organ distinct from it, and variously formed, as in the monks'-hood, black hellebore, &c.; or it is a tubular elongation of the calyx, or of a petal; or, an assemblage of glands.

NEE'DLE ORE. The Nadelerz of Werner. Colour steel-gray. Amorphous, or in acicular hexaëdral prisms, which are occasionally invested with a yellowish or greenish crust. Fracture uneven and metallic. Specific gravity 6·8. Constituents, bismuth 43·6, lead 24·50, copper 12·12, sulphur 11·60, nickel 1·58, tellurium 1·32.

NEMATOIDE'A. The fourth order of

the class Scolecidæ; the threadworm.

NEE'DLE STONE. A variety of zeolite, of a yellowish-white colour, found in Iceland.

NEMATU'RA. A genus of shells belonging to the family Turbinacea. Recent and fossil.

NEMERTI'NA. The name assigned by Dr. Mac Leay to an order of annelida of the group apoda. He thus describes them; "animals aquatic, without eyes or antennæ. Body not externally setigerous. Articulation indistinct. The nemertinæ are white-blooded animals, like some of the leeches."

NEMERTI'TES. A genus of nemertina, thus named by Mr. Mac Leay.

NEOCO'MIAN. (from *Neocomiensis*, the Roman name for Neufchatel.) The name given by the French geologists to a formation synchronic with the Wealden. Dr. Fitton considers that the Neocomian strata are but the equivalents of the English greensand system.

N'EOLOGY. The introduction of a new word or words, or of new names into a language. Dr. Mc. Culloch has, notwithstanding his great dislike to *neology*, enriched our nomenclature more than any other British geologist, with new names. —*Dr. Boase.*

NEOZO'IC. A term proposed by Prof. Forbes to include the Mesozoic and Kainozoic epochs; or to group the whole of the great series of stratified rocks, and divide the whole lapse of past geologic time into two great epochs only, namely Palæozoic and Neozoic.

NE'PHELINE. (from νεφέλη, a cloud, Gr.) The Sommite of Jameson; Nephelin of Werner. A mineral found only in the cavities of lava at Mount Somma, from which circumstance it has been called Sommite. Occurs generally in small, regular, six-sided prisms, associated with mica, hornblende, and idiocrase. Specific gravity 3·27. Colour greyish-white, or greenish-grey. It is translucent, and sometimes transparent. Before the blow-pipe it fuses, with difficulty, into a transparent glass. Its constituents are silex 46, alumine 49, lime 2, oxide of iron 1.

NE'PHRITE. (from νεφρίτης, ab νεφρὸς, a kidney, Gr.) A mineral, formerly worn from an absurd notion that diseases of the kidney were relieved by so doing. It is a sub-species of jade, possessing the hardness of quartz, combined with a peculiar tenacity which renders it difficult either to break, cut, or polish. It is unctuous to the touch; fracture splintery and dull; translucent. Colours green, grey, and white. Sp. gr. from 2·9 to 3·1. Constituents, silex 53·80, lime 12·75, soda 10·80, potash 8·50, alumine 1·55, oxide of iron 5·0, oxide of manganese 2·0, water 2·30. Nephrite is brought from India, China, and Persia; it is found also, in primary rocks, in Germany and Egypt. It is worked into handles for sabres, knives, daggers, &c.

NEPTU'NIAN THE'ORY. That theory which attempted to prove that all the formations have been precipitated from water, or from a chaotic fluid.

NE'PTUNISTS. The supporters of the Neptunian theory; they were opposed by the Vulcanists. Werner taught that all the various formations had been, each in succession, precipitated over the whole earth from a common menstruum, or elastic fluid. His disciples supported his opinions to their full extent, maintaining that even obsidian was an aqueous precipitate.

NEREIDI'NA. Red-blooded, many-legged worms, resembling elongated centipedes. According to Mr. Mac Leay, animals belonging to the class Annelida, having a distinct head, provided with either eyes or antennæ, or both. Mr. Mac

Leay observes, "these are the most perfect in their structure of all Annelida, as they possess numerous organs, and have a distinct head. Some of them, after the manner of Serpulina, inhabit tubes, which tubes are membranaceous, and formed by a transudation from their body; but in general, the Nereidina are naked, and they are always agile animals freely moving about in search of prey."

NEREI'TES. A genus of Nereidina, which, says Mr. MacLeay, "comes very near to Savigny's genus Lycoris in its external appearance, only the segments of the body are here perhaps more slender and in proportion longer than usual." Sir R. Murchison has established two species, namely, N. Cambrensis, and N. Sedgwickii, both discovered in the older rocks of the Silurian system.

NERITA'CEA. A family of Trachelipods, including the genera Natica, Nerita, Neritina, Navicella, and Janthina.

NERITI'NA. A genus of fresh-water univalves, belonging to the family Neritacea. Shell thin, semiglobose, obliquely oval, smooth, flattish in front; spire short; aperture semicircular; outer lip thin; columella lip broad, flat, denticulated. Differs from Nerita in the minuteness of the denticulation of the columella.

NERI'TA. A genus of marine univalves, included in the family Neritacea. The Nerita is a littoral shell, creeping on rocks and sea-weeds. A semi-globose univalve, depressed beneath, and having no umbilicus; aperture entire and semicircular. The aperture is generally large in comparison with the shell, but it is furnished with an operculum which completely closes it.

NERINEA. A genus of fossil turricated univalve shells consisting of numerous whorls; the aperture has a fold upon the columella, one on the outer lip, and one on the inner lip at the edge of the body whorl. A longitudinal section of one of these shells has a singular appearance produced by the three folds; the oolitic formations contain numerous species.—*Lycett.*

NERITO'PSIS. A genus of univalve shells, separated from nerita by the character of the columella lip, which is smooth, flat, with a notch in the centre of its inner edge. The oolitic formations contain several species.

NEURO'PTERA. (from νεῦρον, a nerve, and πτερὸν, a wing, Gr.) Nerve-winged insects. Neuroptera, in Cuvier's arrangement, constitutes the eighth order of Insecta. The Neuroptera have four membranous wings, usually reticulated by numerous nervures, but having no sting, or ovipositor. The Neuroptera are mostly bold, rapacious, and sanguinary; perpetually chasing and devouring other insects. The libellula, or dragon-fly, is a familiar example.

NEURO'PTERIS. A genus of fossil terrestrial plants found in the coal measures.

NEW RED SANDSTONE. (Called also Red Marl, Red Rock, and Red Ground. The Grès Bigarrés of the French, and Bunter Sandstein of the Germans.) A member of the red sandstone group, lying between the variegated marls and muschelkalk, and above the magnesian limestone, or zechstein: sometimes called variegated sandstone. The marl and sandstone are often red, but vary in their hue from chocolate to salmon colour; they are not unfrequently variegated, exhibiting streaks of light blue or verdigris, buff, or cream colour; this forms so prominent a character, that Werner denominated the formation "bunter sandstein," variegated sandstone. It is principally silicious and argillaceous, and is sometimes found to contain mica, and masses of rock-salt and gypsum. It affords a good and handsome stone for building purposes in some parts, when but slightly coloured. It does not abound in organic remains; among its conchiferæ may be enumerated Plagiostoma, Avicula, Mytilus, Trigonia, and Mya. Natica, Turritella, and Buccinum may be included in its genera of molluscs. By some, the new red sandstone has been divided into three series; the upper, the middle, and the lower beds. Sir R. Murchison proposes a four-fold division of what he terms the new red system, namely: 1. Saliferous marls, &c. 2. Red sandstone and quartzose conglomerate. 3. Calcareous conglomerate = magnesian limestone. 4. Lower red sandstone. Over a large part of England the new red sandstone rests unconformably upon the carboniferous group, showing that the latter was disturbed, dislocated, and partially removed, before the former was accumulated upon it; there is, however, reason to believe that in other parts of the European area deposits still continued quietly to be thrown down upon undisturbed parts of the carboniferous series, so that no real line of separation can be well established between them. It is a very extensive deposit, stretching, with but little interruption, from the northern bank of the Tees, in Durham, to the southern coast of Devonshire. It is almost needless to observe, when we contemplate the red sandstone series as a whole, and consider that it is in great measure composed of matter which must have been deposited from water, where it was, for the time, mechanically suspended, that great variations should be expected at the same geological levels; here clay or marl being found, there sandstone or conglomerate, while, occasionally, calcareous matter should be dispersed among it, under favourable circumstances, in sufficient abundance to constitute numerous beds of limestone.

When this deposit appears as a sandstone, its characters differ greatly in different places; it is occasionally calcareous, and sometimes of a slaty texture. Above all, this extensive deposit is remarkable for containing masses of gypsum, and the great rock-salt formation of England occurs within it, or is subordinate to it. "Whether considered in its central or in its lower member," says Sir R. Murchison, "there is no system of rocks, which occasionally offers greater difficulties for determining its real laminæ of deposit than the new red sandstone. Besides the joints or fissures, the diagonal lines of false stratification are sometimes so prevalent, that is only by tracing at wider intervals the true laminæ of deposit as marked by herbage or moss, that we can correctly ascertain the real dip of the strata."

A very remarkable discovery was made in 1828 of the foot-marks of some unknown quadruped in strata of new red sandstone, three miles from Lochmaben in Dumfries-shire. They were found forty-five feet under the present surface; the strata are inclined thirty-seven degrees.

NEW RED SANDSTONE GROUP. This includes all those deposits found below the lias group, and above the carboniferous group. It contains the red or variegated marls (marnes iriseés, keuper), the muschelkalk, the new red or variegated sandstone (grès bigarré, bunter sandstein), the zechstein or magnesian limestone, and the red conglomerate (rothe todte liegende, grès rouge). The whole is considered as a mass of conglomerates, sandstones, and marls, generally of a red colour, but most frequently variegated on the upper parts.

NEWER PLIOCENE PERIOD. The term "Newer Pliocene" has been supplanted by that of Pleistocene, but as many readers may find in works on geology, written only a few years since, the term "Newer Pliocene," I have retained it in this edition. Lyell refers the European tertiary strata to four successive periods, each characterized by containing a very different proportion of fossil shells of *recent* species. These four periods he termed Newer Pliocene, Older Pliocene, Miocene and Eocene; the etymology of these terms will be fully explained under the several words. The Newer Pliocene period is the latest of the four periods, and immediately precedes the recent era. Nevertheless, the antiquity of some Newer Pliocene strata, as contrasted with our most remote historical eras, must be very great, embracing perhaps *myriads* of years. Out of 226 fossil species brought from the Sicilian beds, M. Deshayes found that no fewer than 216 were of species still living.

NI'CKEL. (*nickel*, Germ.) A metal of considerable hardness, nearly equal to that of iron, of a colour intermediate between silver and platina. When polished, it has a high lustre. Specific gravity 8·93. It is both ductile and malleable, and may be hammered into very thin plates. It is difficult to be purified. In common with iron, it is magnetic, capable of acquiring polarity, and may be formed into permanent magnetic needles; this property is destroyed by an alloy with arsenic. Nickel unites in alloys with gold, copper, tin, and arsenic, which metal it renders brittle. With silver and iron its alloys are ductile.

Nickel was discovered as a distinct metal by Cronstadt, in 1751. Its solution in nitric acid is nearly grass green. Nickel is found in all meteoric stones.

NI'GRINE. (from *niger*, Lat. black.) A variety of ferruginous oxide of titanium, occurring in grains, or rolled pieces. Colour black, or brownish black. It consists of titanium 84, oxide of iron 14, oxide of manganese 2.

NIO'BIUM. One of the sixty simple or elementary bodies, its symbol being N B.

NIPADI'TES. The name given by Mr. Bowerbank to a group of fossil fruits of the London clay. Brongniart had named them Pandanocarpum, but Mr. Bowerbank observes "the resemblance existing between the whole of the species of *Nipadites*, both as regards their external form and their internal structure, with those of Nipa, is so close as to leave scarcely a doubt of their being members of the same genus. I have therefore thought it advisable to reject M. Adolphe Brongniart's name of Pandanocarpum, and to apply that of Nipadites, as more expressive of their true relation to their recent analogue." These fossil fruits are found in great abundance in the Isle of Sheppey, and are known, by the women and children who collect them, by the name of figs. Many species have been distinguished.

NI'TRATE. A compound of nitric acid with a salifiable base.

NI'TRE. (νίτρον, Gr.) Nitrate of potash; saltpetre. The potasse nitratée of Haüy; natürlicher salpeter of Werner. Nitre, or nitrate of potash, is found native in all countries, where there are circumstances favourable to its production. It frequently effloresces on the soil; but never exists at a greater depth than that of a few yards beneath the surface. It occurs, naturally, either in masses, or in thin irregular crystals; it is white, semi-transparent, and brittle; salt and cold in taste. When thrown on hot coals, it burns with a sparkling bright light, and with a crackling noise. It crystallizes in six-sided prisms, terminated **by a** dihedral summit, and **retains no** water of crystallization. **The** crystals are permanent, **and** soluble in seven parts **of water at 60°**, and in less than **their** own weight at 212°. The principal supply of nitre **is** from India. One of the most remarkable localities of nitre in Europe, is in the Pula, or cavity of Molfetta, in the kingdom of Naples. This cavity, which is about one hundred feet deep, contains several grottoes or caverns, in the interior of which is found nitre in crusts, attached to compact limestone. When these crusts are removed, others appear in about a month. The various sources of native nitre **not** being sufficient to supply the great demand there exists for it, it is manufactured wholesale, in the following manner. Rubbish, consisting of lime, mortar, plaster, and earth, is mixed up in heaps, under sheds, with decaying vegetables and refuse matter, and left to rot; the masses being occasionally moistened with animal fluids, as urine, **blood, &c.** The nitrogen, disengaged **from the** corrupting mass, **unites with the** oxygen of the atmosphere, **and** forms **nitric** acid ; this, combining with **the** potash furnished by the vegetable substances, produces an impure nitre. The salt is collected, and afterwards washed and purified.

NI'TROGEN. (from νίτρον, nitre, and γεννάω, to produce, Gr.) Called **also azote.** Nitrogen was discovered in 1772 by Prof. Rutherford, of Edinburgh: it may be obtained **by** several processes, the object **of** most of which is to take away the oxygen gas from atmospheric air, of which nitrogen constitutes above four-fifths, or eighty per cent., the rest being principally oxygen. In its pure state, nitrogen is remarkable for its negative qualities; that is to say, for the difficulty with which it enters into combination with other matters. Thus, it is neither combustible, nor a supporter of combustion; it **is** neither acid, nor alkaline; possesses neither taste, colour, nor smell; nor does it directly combine with any known substance. Yet when made by peculiar management to unite with oxygen, hydrogen, or carbon, nitrogen forms some of the most energetic compounds we possess; thus, *mixed* with oxygen, it forms atmospheric air; *united* with oxygen, it forms aquafortis, the most corrosive of liquids; *united* with hydrogen, **it** forms **the** *volatile alkali*, or **ammonia**, likewise an energetic compound, but of an opposite nature; while *united* with carbon and hydrogen, it forms *prussic acid*, **the** most virulent poison in **existence.**

The absorption of *nitrogen* during respiration, was one of the results Dr. Priestley deduced from his experiments; and this fact, though often doubted, appears, on the whole, to be tolerably well ascertained by the inquiries of Davy, Pfaff, and Henderson. With regard to the respiration of cold blooded animals, it has been satifactorily established by the researches of Spallanzani, and more especially by those of Humboldt and Provençal, on fishes, that nitrogen is actually absorbed. A confirmation of this result has been obtained by Macaire and Marcet, who have found that the blood contains a larger proportion of *nitrogen* than the chyle, from which it is formed.

Nitrogen has been recently found, by Dr. Daubeny, to be contained very generally in the waters of mineral springs. The king's bath, at Bath, evolves 96·5 per cent. of nitrogen, 3·5 oxygen, and some carbonic acid. The hot-well at Bristol evolves 92 per cent. nitrogen, and 8 oxygen. The springs at Buxton, Bakewell, and Stony Middleton, Derbyshire, evolve nitrogen only. Those of the Spas in Germany, yield various proportions, as do the other thermal springs in other parts of the globe.

Its specific gravity is 0·9722. The combining power of nitrogen is variously estimated, some chemists making it 14·15, while others consider it to be only half of that number. Its symbol is N.

NOA'CHIAN. Pertaining to the great deluge related by Moses, from which Noah and his family were saved, and thus called after Noah. Cuvier says "that if there be any one fact thoroughly established by geological investigations, it is the certainty of the low antiquity of the human race, and present state of the surface of the earth, and the circumstance of its having been recently overwhelmed by the waters of a transient deluge."

NODE. (*nodus*, a knot, Lat. *nodus*, Fr. *nado*, It.) 1. A hard knot or swelling, a bump, or rising.

2. In astronomy, the nodes of a satellite's orbit are the points in which it intersects the plane of the orbit of a planet. The ascending node is the point through which the body passes in rising above the plane of the ecliptic, and the descending node is the point through which the body passes in sinking below the plane of the ecliptic.

NOCTI'VAGANT. } (from *nox*, the night,
NOCTI'VAGOUS. } and *vagor*, to wander, Lat.) A name given to such animals as wander, in search of prey, during the night.

NODOSA'RIA. A genus of orthocerata, found only fossil. They are polythalamous univalves.

NO'DULAR. In the form of a nodule or small lump; having irregularly globular elevations.

NO'DULAR IRON ORE. A variety of argillaceous oxide of iron; occurring in masses, varying from the size of a walnut to that of a man's head. Their form is spherical, oval, or nearly reniform, or sometimes like a parallelopiped with rounded edges and angles. They have a rough surface, and are essentially composed of concentric layers. These nodules often contain, at the centre, a kernel or nucleus, which is sometimes moveable, and always differing from the exterior in colour, density, and fracture. The texture of the exterior is compact and solid; but the density gradually diminishes to the centre, which has an earthy texture. Specific gravity, about 2·57. Its constituents are oxide of iron 77·0, silex 6·0, oxide of manganese 1·0, alumine 0·5, water 13·5. These nodules have also been called Œtites and Eaglestones, from an opinion that they were found in eagles' nests, where, it was supposed, they prevented the eggs from becoming rotten.

NO'DULE. (from *nodulus*, a little knot, Lat.) A rounded, irregular-shaped mineral mass. Ironstone forms regular layers of round nodules, sometimes as much as a foot or eighteen inches in diameter. These nodules, when broken open, are often found to be traversed by cracks in all directions, more or less filled up with crystalline spar, together with crystals of galena, blende, iron pyrites, and other minerals.—*Jukes.*

NO'GROBS. A fossil resembling a belemnite.

NO'MENCLATURE. (*nomenclatura*, Lat. *nomenclature*, Fr. *collection des mots qui sont propres aux différentes parties d'une science ou d'un art.*) The names of things in any art or science, or the whole vocabulary of technical terms which are appropriated to any particular branch of art or science.

The imposition of a name on any subject of contemplation is an epoch in its history of great importance. It not only enables us readily to refer to it in conversation or writing, without circumlocution, but, what is of more consequence, it gives it a recognized existence in our own minds, as a matter for separate and peculiar consideration. How important a good system of nomenclature is, may at once be seen, by considering the immense number of species presented by almost every branch of science of any extent, which absolutely require to be distinguished by names. Thus, the botanist is conversant with from 80,000 to 100,000 species of plants; the entomologist with, perhaps, as many, of insects. And the same as regards chemistry, astronomy, &c.

Nomenclature, then, is, in itself, an important part of science, as it prevents our being lost in a wilderness of particulars, and involved in inextricable confusion. Happily, in those great branches of science, where the objects of classification are more numerous, and the necessity for a clear and convenient nomenclature most pressing, no very great difficulty in its establishment is felt. The facility with which the chemist, the botanist, or the entomologist, refers by name to any individual object in his scinece, shows what may be effected in this way when characters are themselves distinct.

Nomenclature, in a systematic point of view, is as much, perhaps more, a consequence than a cause of extended knowledge. Any one may give an arbitrary name to a thing, merely to be able to talk of it; but to give a name which shall at once refer it to a place in a system, we must know its properties; and we must *have* a system, large enough, and regular enough, to receive it in a place which belongs to it and to no other.

There is no science in which the evils resulting from a rage for nomenclature have been felt to such an extent as in mineralogy. The nomenclature of most minerals is at present so encumbered with synonyma, that it has become extremely perplexing to the student. This may be illustrated by the example of Epidote. This mineral, which is called *epidote* by Haüy, is named *pistazit* by Werner, *thallite* by Lemetherie, *akanticone* by Dandrada, *delphinite* by Saussure, *glassy actinolite* by Kirwan, *arendalit* by Karsten, *glassiger strahlstein* by Emmerling, *la rayonnante vitreuse* by Brochant, *prismatoidischer augitspath* by Mohs, &c., &c.

In all subjects where comprehensive heads of classification do not prominently offer themselves, all nomenclature must be a balance of difficulties, and a good, short, unmeaning name, which has once obtained a footing in usage, is preferable to almost any other; Fabricius maintained "optima nomina, quœ omnino nil significant." When the composition is unknown, those names, which are altogether unmeaning in regard to any property of the thing, are, perhaps, the least objectionable; at all events, they cannot lead to error.

Linnæus was the first to introduce systematic names into natural history. By the introduction of these scientific, fixed, and universally valid names, Linnæus has undoubtedly acquired his greatest merit in science, and if every thing else which he has done should be forgotten, this, which is wholly his work, will secure his name from forgetfulness. — *Herschel. Cleaveland. Burmeister. Lyell.*

NON'RUMINANTIA, The order artiodactyla, or those animals possessing an even number of toes, is divided into **two families**, nonruminantia and **ruminantia**; the hippopotamus, pig, &c., are examples.

NO'RKA. The name given by Cronstadt to an aggregate of quartz, mica, and garnet. This aggregate is included by Kirwan in the granatines.

NO'RFOLK CRAG. An English tertiary formation belonging to the older pliocene. It is observed to rest on the chalk and on the London clay. It consists of irregular beds of ferruginous sand clay, mixed with marine shells. According to an account of Mr. S. Woodward, if a line be drawn from Cromer, on the northern coast of Norfolk, to Wayburn, about six miles west, and from thence extending in a southerly direction towards Norwich, about 18 miles, it will comprise all the regular beds of Norfolk crag.

NO'RITE. The name assigned by Esmarck to a rock not yet determined; some of its characters appear to belong to diorite, and some to gabbro.

NOTHO'CERAS. M. Barrande has lately established a new genus of Cephalopod under this designation. It is intermediate between Nautilus and Goniatite, and is of the Upper Silurian age.

NOTHO'SAURUS. A reptile of lacertian or lizard-like character, of the new red sandstone era.

NOVA'CULITE. (from *novacula*, a razor, Lat.) The Wetz schiefer of Werner; argile schisteuse novaculaire of Haüy. Honestone. See *Hone.*

NU'CLEOLITES. A genus of radiaria, twelve species of which have been determined as occurring in the chalk formation, and six species as occurring in the oolite.

NU'CULA. A genus of marine bivalve shells belonging to the family Arcacea. An inequilateral, equivalved, transverse, subtrigonal bivalve; covered with an epidermis. The hinge linear, bent at an angle formed by numerous, alternately inserted teeth; muscular impressions, two, simple; beaks approximating, and turned backwards. The recent species of this genus are found in estuaries, and in the ocean, at depths varying to sixty fathoms, in mud and sand. Several fossil species are described.

NU'DIBRANCHIATA. The second order of the class Gasteropoda. The nudibranchiata have no shell whatever; neither are they furnished with any pulmonary cavity, their branchiæ being exposed on some part of their back, from which circumstance they have obtained their name. The triton, doris, &c., are examples.

NU'MMULITE. (from *nummus*, money, Lat. and λίθος, a stone, Gr.) The nummulites compose a fossil extinct genus of multilocular cephalopods, presenting, externally, a lenticular figure, without any apparent opening, and, internally, a spiral cavity, divided by septa into numerous chambers; they do not possess a siphuncle, but their chambers communicate by means of small foramina with each other. They have obtained their name from their supposed resemblance to pieces of money. It is of stone composed of *Nummulites* that the pyramids of Egypt are constructed. Nummulites have been named *Helicites*, from their spiral structure; *Phacites*, from their resemblance to a lentil; and *Salicites*, from the supposed resemblance of their sections to the leaf of the willow. Pliny is supposed to refer to them under the name of Daphnias when he mentions that Zoroaster employed these substances for the cure of epilepsy. They have been also termed Lentes lapideæ, Lapides cumini, circulares, numismales, &c.

Nummulites vary in size from less than an eighth of an inch, or even microscopic minuteness, to an inch and a half in diameter. Their surface is, in some, nearly smooth, in others rough and scabrous, with numerous small projecting knobs, or undulating lines. Their colour varies from nearly white to brown and red, and sometimes nearly blue. The number of spiral turns seems to depend on the age and size of the animal; in those of a quarter of an inch in diameter, being three or four, while in those of the largest size the number of whorls is frequently upwards of twenty.

Nummulites occupy an important place in the history of fossil shells, on account of the prodigious extent to which they are accumulated in the later members of the secondary, and in many of the tertiary strata. They are often piled on each other nearly in as close contact as the grains in a heap of corn. Entire calcareous hills are, in some instances, composed of fossil nummulites.

NUMMULI'TIC. Containing nummulites; composed of nummulites. The nummulitic limestone of the Alps, a formation which extends through all the countries surrounding the Mediterranean, and thence through Lower Asia into India, and which is many thousands of feet in some places, has been shown by Sir R. Murchison to be clearly referrible to the eocene period.— *Jukes.*

NUTA'TION. (from *nutatio,* a nodding, Lat. *nutation,* Fr.) A tremulous or vibratory motion of the earth's axis, by which its inclination to the plane of the ecliptic is continually varying, being, in its annual revolution, twice inclined to the ecliptic, and as often returning to its former position. Both the celestial latitudes and longitudes are altered to a small degree by nutation. In consequence of this real motion in the earth's axis, the pole-star, forming part of the constellation of the Little Bear, which was formerly 12° from the celestial pole, is now within 1° 24′ of it, and will continue to approach it till it is within $\frac{1}{2}$°, after which it will retreat from the pole for ages; and 12,934 years hence, the star *a* Lyræ will come within 5° of the celestial pole, and become the polar star of the northern hemisphere.

NUT-GALL. An excrescence which grows on some species of oaks. These excrescences are produced by the Cynip quercus folii, of Linnæus, a small insect which deposits its egg in the tender shoots of the quercus infectoria, a species of oak abundant in Asia Minor. When the maggot is hatched, it produces a morbid excrescence of the surrounding parts, and it ultimately eats its way out of the nidus thus formed, and makes its escape. The best galls are imported from Smyrna and Aleppo.

# O

OB'CONICAL. Of the shape of a reversed cone.

OBCO'RDATE. In botany, an epithet for an inversely heart-shaped leaf, petal, or legume.

O'BLATE. (*oblatus,* Lat.) Flattened or depressed at the poles; generally applied to spherical bodies, flattened at the poles; of the shape of an orange.

OBLA′TE SPHE′ROID. A spheroid flattened at the poles is called an *oblate* spheroid: such is the form of the earth and planets. When, on the contrary, a spheroid is drawn out at the poles instead of being flattened, it is called a *prolate* spheroid.

OBLI′QUE. (*obliquus*, Lat. *oblique*, Fr. *obblico*, It.) Not direct; not perpendicular; not parallel.

In botany, applied to the position of leaves, and implies that one part of the leaf is horizontal and the other vertical.

In conchology, applied to the whorls of spiral univalves which commonly are in an oblique direction in reference to the axis of the shell. The term is also applied to bivalves when they slant off from the umbones.

O′BLONG. (*oblongus*, Lat. *oblong*, Fr.) Having greater length than breadth; longer than broad.

In botany, applied to leaves several times longer than broad. The term is chiefly used to discriminate a leaf whose form does not accurately come under the denominations oval, linear, or round.

O′BLONG O′VATE. Oblong egg-shaped; between oblong and egg-shaped.

OBO′VATE. In botany, applied to leaves having the form of an egg, with the broad end forming the base, and the pointed apex of the leaf.

OBSI′DIAN. The Obsidienne of Brongniart; Lave vi′treuse obsidienne of Haüy. Vitreous lava, a volcanic production, of a dark green colour approaching to black. An analysis of obsidian from Mount Hecla, by Vauquelin, gives the constituents as follows, silica 78, alumina 10, potash 6, lime 1, soda 1·6, oxides of iron and manganese 1.

Obsidian has been divided into two kinds, the vitreous and pearly; these may be distinguished by their fracture, which is either vitreous or pearly. Professor Jameson divides obsidian into two subspecies, namely, transparent obsidian and translucent obsidian.

Vitreous obsidian bears a strong resemblance to the glass of wine-bottles. Its fracture is conchoidal, showing frequently large cavities. Lustre vitreous. Specific gravity from 2·34 to 2·90. It generally occurs in large amorphous masses, when it appears almost black; it is sometimes found in rounded grains.

Pearlstone, the Obsidienne perleé of Brongniart; this variety has a granular structure, and is traversed by fissures in all directions. It is consequently very brittle. Its fracture is uneven or granular, and, as before mentioned, *pearly*. When moistened by the breath it frequently returns an argillaceous odour. It occurs amorphous only.

Before the blow-pipe both varieties intumesce, but the vitreous alone fuses into a globule. Obsidian bears indisputable characters of having once been in a state of fusion.

OBTU′SE. (*obtusus*, Lat. *obtus*, Fr.) An angle which is more than ninety degrees, or that of a right angle.

O′BVOLUTE. In botany, applied to leaves, when their margins alternately embrace the straight margin of the opposite leaf.

OCCI′PITAL. (from *occiput*, Lat.) Pertaining to the back part of the head.

O′CCIPUT. (*occiput*, Lat.) The back part of the head: the fore part is called *sinciput*.

O′CEAN. (ὠκεανὸς, Gr. *oceanus*, Lat. *océan*, Fr. *oceáno*, It.) That vast body of water which covers more than three-fifths of the earth's surface. The average depth of the ocean has been very variously estimated. Laplace considered, in order to account for the height of the tides according to the laws of gravitation, the depth to average ten miles; others rate it at five miles. The present cannot be considered as having always been the bed of the ocean; on the contrary, what are now the most elevated portions of the earth's crust were once submerged, and over them the ocean for ages rolled its majestic waves. This is not an invention of modern geologists, Ovid declares the same:—

Vidi factas ex æquore terras;
Et procul a pelago conchæ jacuere marinæ;
Et vetus inventa est in montibus anchora summis;
Quodque fuit campus, vallem decursus aquarum
Fecit; et eluvie mons est deductus in æquor:
Eque paludosa siccis humus aret arenis;
Quæque sitim tulerant, stagnata paludibus hument.
*Metamorph.* lib. xv.

OCEA'NIC DE'LTA. A delta formed at the mouth of rivers where they enter the ocean, as distinguished from either lacustrine or mediterranean deltas.

OCELLA'RIA. A genus of fossils, thus described by Mr. Parkinson: "a lapideous polypifer, expanded in a membranous form; variously convoluted and rather infundibuliform, with an arenaceous surface, porous on both sides; pores cylindrical, in quincunx order, with a solid axis in a raised centre."

O'CELLATED. (*ocellatus*, Lat.) In conchology, applied to shells, when marked with little eye-like spots.

OCHRA'CEOUS. Of a brown yellow colour, resembling ochre. Alternating with an *orchraceous* ironstone.

O'CHRE. (ὤχρα, Gr. ochra, Lat. ocre, Fr.) Red iron ore; it yields good malleable iron. Colours red, yellow, and brown. It occurs in dull earthy masses, nearly or quite friable, which soil the fingers. Its constituents are oxide of iron 83, silex 4, water 12.

OCRY'NIAN. The *Ocrynian* formation is not only distinguished by containing shorl, but also by a great excess of felspar, which forms the principal constituent in nearly all its rocks.

OCTAË'DRAL. } Having eight sides
OCTAHE'DRAL. } all equal.

OCTAË'DRITE. } Octaëdral oxide of titanium; the Titane
OCTOË'DRITE. } anatase of Brongniart; Octaëdrit of Werner; Octaëdrite of Jameson; the Oisanite of Lameth. A pure oxide of titanium, crystallized in acute, elongated octaëdrons, consisting of two pyramids, whose faces are isosceles triangles, and whose bases are squares. Colours blue, blackish-blue and brown. Lustre splendent and adamantine. Fracture foliated; easily broken. It scratches glass. Specific gravity 3·8. Before the blow-pipe it is infusible by itself, but with borax it fuses into a glass. It occurs in veins in Dauphiny, Norway, Spain, and Brazil.

OCTAË'DRON. } (ὀκταέδρος, from ὀκτώ,
OCTAHEDRON. } eight, and ἕδρα, a side, Gr. *octaèdre*, Fr.) The solid angles of an octaëdron are formed by four equal and equilateral plane triangles; consequently it is formed by two equal spare pyramids joined together at their bases, the sides whereof are equilateral triangles. The octahedron (unlike some forms which are not susceptible of variation, as the die or cube, a solid invariably bounded by six square surfaces or planes) is susceptible of variation; it is sometimes flat and low, and, at others, acute and high.

O'CTOPUS. (from ὀκτώ, eight, and ποὺς, a foot, Gr.) A genus of sepiæ. The *octopus* was the animal denominated polypus by Aristotle. It has eight arms, all of equal length, and contains in its interior two very small rudimental shells, formed by the inner surface of the mantle.

O'CTODENTATE. (from *octo*, eight, and *dentatus*, toothed, Lat.) Having eight teeth, and no more.

O'CTOFID. (from *octo*, eight, and *findo*, to cleave, Lat.) Eight-cleft. In botany, an epithet for a calyx divided into eight segments.

OCTOLO'CULAR. (from *octo*, eight, and

*loculus*, a cell or pocket, Lat.) Eight-celled.

OCTONO′CULAR. (from *octo*, eight, and *oculus*, an eye, Lat.) Having eight eyes.

OCTOPE′TALOUS. (from ὀκτώ, eight, and πεταλὸν, a petal, Gr.) Having eight petals or flower-leaves.

OCTOSPE′RMOUS. (from ὀκτώ, eight, and σπέρμα, seed, Gr.) Eight-seeded; having eight seeds.

ODONTACA′NTHUS. A genus of ichthyolites of the old red sandstone formation, previously known as Ctenoptychius, but for which Odontacanthus has been substituted. Two species have been described by Agassiz.

ŒSO′PHAGUS. (*œsophage*, Fr.) The gullet, or passage leading from the mouth to the stomach, through which the food passes. In the structure of the œsophagus, we may trace an adaptation to the particular kind of food taken in by the animal. When it is swallowed entire, or but little changed, the œsophagus is a very wide canal, capable of being greatly dilated. Serpents, which swallow animals of greater circumference than themselves, have an œsophagus admitting of great dilatation; the food in such cases remaining a long time in the canal, before it reaches the stomach. Grazing animals, who carry their heads close to the ground while feeding, have the œsophagus strengthened by thick muscular coats, whereby the food is propelled towards the stomach, the direction being contrary to that of gravity.

OGY′GIA. The name given by Brongniart to a genus of trilobites; he thus defines the generic characters: "Corps très déprimé, en ellipse allongée, non contractile en sphère. Bouclier bordé; un sillon peu profond, longitudinal, partant de son extrémité antérieure. Point d'autres tubercules que les oculiformes. Lobes longitudinaux peu saillans. Huit articulations à l'abdomen." In establishing this genus, M. Brongniart observes that although its appearance is very different from that of most other genera of the family of trilobites, yet it is not always easily separated from them.

OGY′GES. The name given by Guettard to a species of trilobite, from its being found among the most ancient rock formations, containing vestiges of organic life.

O′KENITE. A bi-silicate of lime, with two equivalents of water.

OLD RED SANDSTONE. (The Grès rouge intermediare of the French; the Jüngeres Grauwackengebirge, and Alter rother sandstein of the German geologists.) The lowest member of the carboniferous group, extensively developed in the counties of Shropshire and Herefordshire, in England; Brecknockshire, in Wales; and Dumfriesshire and Forfarshire, in Scotland. The old red sandstone strata lie between the carboniferous series and the silurian rocks, or grauwacke group. It consists of many varieties and alternations of silicious sandstones and conglomerates of various colors, red predominating. The old red sandstone of some countries graduates into grauwacke, and it is the opinion of most continental geologists that it should be considered as forming the upper portion of the grauwacke series. The old red sandstone is a course-grained, micaceous sandstone, evidently of mechanical origin, constituted apparently of abraded quartz, felspar and mica, and containing fragments of quartz, clay-slate, flinty-slate, &c.; sometimes passing into the state of a quartzose conglomerate, sometimes possessing a structure coarsely schistose, and sometimes, particularly towards its lower regions, becoming finely schistose, and passing into a fine-grained micaceous sandstone slate. In some situations the old red sandstone attains a thickness of 2000 feet and upwards.

"It was a prevalent belief among geologists," says Sir R. Murchison, "that few or no animal remains existed in the old red sandstone of England. I first undeceived myself on this point by observing shells in the lower group, or tilestones, in Caermarthenshire. I afterwards discovered similar fossils in the great outlier of Clun forest. The old red system may now fairly be said to be characterized throughout by ichthyolites peculiar to it. The rocks known to English geologists under the name of the old red sandstone, consist of various strata of conglomerate, sandstone, marl, limestone, and tilestone, the *youngest* beds of which dip conformably beneath the carboniferous deposits, whilst the oldest repose upon and pass into certain grey-coloured rocks. These last form the upper part of the Silurian system."

OLD RED SYSTEM. Sir R. Murchison has applied this term to what has heretofore been known as the old red sandstone. "Being convinced," he says, "that the old red sandstone is of greater magnitude than any of the overlying groups, I venture for the first time, in the annals of British geology, to apply to it the term *system*, in order to convey a just conception of its importance in the natural succession of rocks, and also to show, that as the carboniferous system, in which previous writers have merged it (but from which it is completely distinguishable, both by lithological characters and zoological contents,) is surmounted by one red group; so it is underlaid by another, this lower group being infinitely thicker than the upper." Sir R. Murchison proposes to divide the old red system into—1. Quartzose conglomerate and sandstone. 2. Cornstone and marl. 3. Tilestone. He considers the united thickness of the old red system, at a moderate calculation, to amount to nine or ten thousand feet.

"It will be found," says Hugh Miller, "that this hitherto neglected system yields in importance to none of the others, whether we take into account its amazing depth, the great extent to which it is developed both at home and abroad, the interesting links which it furnishes in the zoological scale, or the vast period of time which it represents. There are localities in which the depth of the old red sandstone fully equals the elevation of Mount Etna over the level of the sea, and in which it contains three distinct groups of organic remains, the one rising in beautiful progression over the other." See *Tilestone*.

OLDER PLIOCENE. Sir C. Lyell has sub-divided the tertiary epoch into four periods; namely, the newer pliocene or pleistocene, the older pliocene, the miocene, and the eocene. The term pliocene he derived from the two Greek words πλείων, more, and, καινὸς, recent. The older pliocene formations lie between the miocene and the newer pliocene or pleistocene. Of fossil shells examined by M. Deshayes, the older pliocene contained from thirty-five to fifty per cent. of *recent* fossils.

OLE′CRANON. (ὀλέκρανον, Gr. from ὠλενη, the ulna, and κάρηνον, the head.) A process of one of the bones of the fore arm, the ulna, forming part of the elbow joint.

OLE′FIANT GAS. (from *oleum*, oil, and *fio*, to become, Lat.) A gas devoid of colour and taste, deriving its name from the property it possesses of forming an oil-like liquid with chlorine.

O′LIGOCLASE. Soda Spodumene. A feldspathic mineral, whose constituents are silica 63·37, alumina 25·86, soda 11·77.

O′LIVA. (*oliva*, an olive, Lat.) So named from the oblong and elliptical shape of the shell. A marine subcylindrical univalve; aperture narrow, long, and emarginated opposite to the spire, which is short; the plicæ of the columella are numerous, and resemble striæ; whorls sulciform. The shells of this genus are very beautiful, and display a great variety of rich markings and splendid colours. Recent olivæ are found at depths varying to twelve fathoms, in mud, sandy mud, coarse sand, &c. They are also caught by fishing lines. Fossil olivæ are found in the calcaire grossier, and London clay. Several species have been described.

OLI′VENITE. An ore of copper of an olive-green colour. It consists of oxide of copper 63·0, **phosphoric acid** 28·6, water 8·4.

O′LIVINE. The prismatischer chrysolithe of Mohs; peridot of Haüy; olivin of Werner. A mineral, generally of an olive-green colour, from which circumstance it obtains its name: it is sometimes of an asparagus green, or yellowish green. Occurs in distinct granular concretions, or in rounded masses. Structure foliated. Fracture imperfectly conchoidal. Lustre shining, translucent, and, sometimes, transparent. Its constituents are, silex 50·0, magnesia 87·5, oxide iron 12·0, lime 0·5. It is found in basalt, and is a constituent of many lavas.

O′MBRIA. (from ὄμβριος, rain, Gr.) Fossil echini, to which the name of ombria has been given, from a supposition that they fell from heaven in the midst of heavy rain; they are of a rounded form, and have been compared to turbans.

OME′NTUM. (*omentum*, Lat.) The caul.

OMNI′VOROUS. (from *omnis*, all, and *voro*, to devour, Lat.) Animals which eat food of all kinds.

O′MOPLATE. (from ὦμος, the shoulder, and πλατὺς, broad, Gr.) The scapula, or shoulder-blade.

O′NCHUS. A genus of sharks, belonging to the sub-family of Hybodonts. The genus onchus, says M. Agassiz, is easily distinguished. It embraces certain cartilaginous fishes, the dorsal spines of which only have been discovered. The dorsal fins are large at their base, very much bent backwards, narrowing rapidly towards the superior end, furrowed along the whole of their surface by grooves parallel to the posterior margins, between which pretty strong ribs rise up, which so terminate as to give the anterior edge a toothed aspect. The onchus arcuatus is the species thus described, and is the largest species of the old red system, though inferior in size to some species of the carboniferous limestone.—*Murchison.* Teeth of the onchus have been found in the lias, at Lyme Regis.

ONTO′LOGY. The science of organic beings; divided into zoology, the science of animals; botany, the science of vegetables; or again specifically divided into Cainontology, the science of now living organic beings, and Palæontology, the science of extinct organic bodies.—*Jukes.*

O′NYX. (from ὄνυξ, Gr. a nail, *onyx*, Fr.) The quartz agathe onyx of Haüy. A variety of calcedony having different colours arranged in distinct parallel stripes or zones, and consisting of alternate layers of white and brown calcedony. The on**y**x is used by lapidaries in the formation of cameos.

O′OLITE. (from ὠὸν, an egg, and λίθος, a stone, Gr. *oolites*, Fr.) The Calcaire de Jura, Calcaire Jurassique of the French; the Oolithenbildung, Jurakalk of the Germans.

A group of strata, whose order of superposition is below the Purbeck and above the lias: called also the Jura limestone from the extensive chain of the Jura mountains being principally composed of lias and the oolitic series. The two lowest members of **this** group, or those immediately **above** the lias, are called **the great** oolite, and the inferior oolite. All the members **of the** group **are** marine deposites. The oolite has been thus named from its being composed of spherical granular concretions, supposed to resemble the roe, or eggs, of a fish: it is a mere term of convenience, like those of carboniferous, red sandstone, &c., for many limestones in other groups are oolitic, while, on the other hand, this mineralogical character is found only in an insignificant part of the rocks known as the oolite formation in England and France. The oolite is an accumulation of sands, sandstones, marls, clays, and limestones, ranging across our island from Dorsetshire on the south-west, to Yorkshire on the north-east. The central members of this group occupy the high districts of Oxfordshire and Gloucestershire. A very striking zoological feature of this group is the immense abundance of **ammonites and** belemnites which **must have existed** previous to, and during **its deposit; for,** notwithstanding the usual chances of destruction to which **we may** suppose they were exposed, myriads of their shells have been found entombed entire, and not unfrequently the animal must have been in them. One hundred and seventy-three species of ammonites, and sixty-five species of belemnites have been enumerated as discovered in the oolite. There can be little doubt that this group, greatly expanded in thickness, and mixed with sandstones, marls, and slates, possessing

a very different aspect from the equivalent rocks in a large portion of Western Europe, extends over **various** parts of Eastern Europe. The aggregate average thickness of the oolite may be estimated at 1200 feet. In some instances, the spherical granular concretions, which are imbedded in many of the strata, attain the size of a pea, and this **variety** has obtained the name of *pisiform* oolite. Some oolites have been used for building-stone, but they **are said not** to be durable. Somerset House is built of oolite. The vertebrated animals, whose remains are found in oolite, are fishes and reptiles of the same genera as those discovered in the lias. Mammalia have been found at Stonesfield in Oxfordshire, where **there are the remains of more than one species of Didelphis. Pterodactyles have been discovered at Solenhofen (where there would appear to be many species), at Lyme Regis, in Dorsetshire, and at Banz, in** Bavaria. Some strata of this group **are** composed, almost entirely, **of** madreporites, and these have been called "coral ragg." Other strata abound in the remains of fossil alcyonia and sponges, and with congeries of minute millepores and madrepores. In England, the limestone of the oolite has a yellowish-brown, or ochreous colour, by **which it may at once be** distinguished from the lias; and the fossils partaking of the colour of the limestone, renders it easy to separate them from the fossils of the lias. The oolite has been divided into three formations, the upper, the middle, and the lower. These formations, in England, occupy a zone having nearly thirty miles in average breadth, extending across the island from Yorkshire on the north-east, to Dorsetshire on the south-west. Between the lower and the middle division of oolites, there occurs a bed of dark blue clay, called Oxford, or clunch, clay, the thickness of which has been stated to be 200 feet. Between the middle and upper also, there is found a thick bed of clay, called Kimmeridge clay, of a thickness exceeding, in some parts, 100 feet. The uppermost members of the oolite group are the Portland beds, lying immediately under the Purbeck beds.

Oolite has been also called roe-stone, from a supposition of the older geologist that the globules contained in it were the petrified roes of fishes. In the lithographic limestone of Solenhofen, belonging to one of the upper members of the oolite, a great variety of organic remains is found; and in the museum of Count Munster, there are not fewer than seven species of flying lizards, six saurians, three tortoises, sixty species of fishes, forty-six species of crustacea, and twenty-six species of insects, taken from that deposit.

The oolitic tracts of England present a broad band of dry limestone surface, rising westward to elevations of from 800 to 1,400 feet, with escarpments commanding very extensive prospects over the undulating plains of lias and red marl. The whole tortuous line of oolitic escarpment from the Humber to the Avon, may be regarded as the wasting effects of water on the subjacent red marls and lias clays.—*Bakewell. De la Beche. Lyell. Cleaveland. Mantell. Phillips. Conybeare.*

OOLIT'IC. Composed of oolite; resembling oolite. The name of a large group of strata commencing with the Portland beds above, and terminating in the inferior oolite below.

OOLI'TIC PER'IOD. Called also the Jurassic period. The Oolitic period, is comprised in the secondary or Mesozoic epoch, and is placed between the cretaceous and triassic periods. The series of rocks deposited in the British isles during the Oolitic period is so complete both petrologically and palæontologically, that they serve for a type for those known all over the world.

OOLITI'FEROUS. Producing oolite, or roe-stone.

OPA'CITY. (*opacitas*, darkness, Lat. *opacité*, Fr. *opacità*, It.) Opaqueness; darkness. The quality of opacity is not a *contrary* or *antagonist* quality to that of transparency, but only its extreme lowest degree.

OPAL. The quartz résinite of Werner; untheilbarer quartz of Mohs; a sub-species of indivisible quartz. Of this there are many varieties, the principal of which are,— 1. The precious opal, a milk-white variety, with a beautiful play of various rich colours. The play of colours is supposed to be caused by numerous minute rents that traverse this mineral; thin layers being contained in them, which have the property of reflecting the prismatic colours. The precious opal is sometimes imperfectly imitated by artificial glasses; and substances which resemble the opal in its play of colours are said to opalesce. 2. Fire-opal; a transparent variety, brought from Mexico, with a carmine-red and apple-green iridescence of great beauty. Found also in Hungary and Cornwall. 3. Common opal; a variety differing but little from the precious opal in many of its characters, but not presenting that effulgence, or play of colours, by which the precious opal is distinguished. Its colour is white, shaded with grey, green, or yellow, sometimes milk-white. When viewed by transmitted light, the milk-white and greenish varieties often change their colours. These varieties constitute the Opal resinite Girasol of the French mineralogists. 4. Semi-opal; a feebly translucent variety, having a conchoidal fracture; colours white, grey, and brown. Prof. Ehrenberg states that nodules of *semi-opal*, which occur in the Poliers-chiefer, are composed of silex derived from infusorial remains that have been dissolved and formed into silicious concretions, having dispersed through them numbers of infusorial shields, partially dissolved, together with others that are unaltered. Ehrenberg also thinks that he has found indications of microscopic organic bodies of a spherical form in *semi-opal* from Champigny, and also in *semi-opal* from the dolerite of Steinheim, and in precious opal from porphyry of Kaschan. 5. Menilite; a variety occurring in small, irregular, roundish masses, often tuberose, or marked with little edges on the surface. The exterior is often bluish or striped, but the interior has a brown or dark grey appearance. Fracture conchoidal. It is translucent. 6. Mother of pearl opal or Cacholong. 7. Jasper opal. 8. Wood opal. For details of these varieties, see the several words. These varieties consist of silex in various proportion, from 86 to 95 per cent., combined with oxide of iron and water. Specific gravity 2·09. Not sufficiently hard to give fire with steel. The semi-opal contains about 3 per cent. of alumina.

OPALE'SCENT. Resembling opal; exhibiting a play of various colours; displaying iridescence.

O'PALIZED. Converted into a substance resembling opal.

O'PALIZED WOOD. This has the form and texture of wood; the vegetable matter having gradually given place to a silicious deposite possessing the characters of semi-opal. Its texture is fibrous; fracture conchoidal, with a moderate lustre. It does not strike fire with steel. Specific gravity between 2·0 and 2·6. Colours white and grey, often shaded with yellow or red, and passing into yellow or brown. Translucent at the edges.

OPAQUE. } Not transparent; not permitting the passage of rays of light.
OPAKE. }

OPE'RCULAR. Having a lid, or cover, or operculum.

O'PERCULE. } (*operculum*, Lat.)
OPE'RCULUM. }

1. A lid, by means of which many of the molluscous animals close the aperture of their shells. It is in some animals testaceous; in others, horny or cartilaginous. It is affixed to the animal. The operculum of multivalves is composed of two or four pieces. The operculum is calculated for the protection of the animal when it retires within its dwelling, of which it may be termed the door; it is adapted to the shape of the aperture, which it closes completely. The cartilaginous operculum of the common periwinkle is a familar example.

2. The flap which covers the gill, or organ of respiration in fishes.

OPHI'DIA. (from ὄφις, a serpent, Gr.) The third order in the class Reptilia, in Cuvier's arrangement, comprising three families, Anguina, Serpentia, and Nuda. In the structure of the skeleton of the serpents, the first of the true reptiles, we may observe a beautiful illustration of the simple means employed in organic structures to accomplish the most numerous and diversified ends, and of the resources of nature in adapting the forms of bones, in all their essential and common parts, to the various uses the animal is to make of them in the living state. We have here animals destitute of anterior and posterior extremities, destitute of arms and legs, of hands and feet, yet capable of a great variety of those active movements which we see in animals the most gifted with those parts. We see them as if running on all fours, pursuing their prey, rapidly winding through the turf, and through the low vegetables that cover extensive plains. If the prey, to escape from danger, betake itself to the trees, imagining there to be in safety, we find these serpents winding round the tree, and almost without any apparent motion of any portion of their trunk, gliding, as if they were sticking by suckers, up the trunks of the trees they climb, till within reach, and then with a velocity, like an elastic spring let loose, they dart forward and twine round their prey. If their prey should even rise from the ground into the air, we see these serpents, as if they were gifted with wings, spring with velocity from the ground, dart upon the bird and seize it, or if the animal be a quadruped, and plunge for safety into the water, the serpents still pursue it in that element, swimming like fishes. Yet, when we examine the condition of the skeleton, we find it simply to consist of a vertebral column and ribs; and with that simple condition of the solid internal frame-work, we see all those varied movements effected. The spine of serpents is formed of a great number of vertebræ; in the rattle-snake there are about two hundred, and in the coluber natrix above three hundred have been counted. These vertebræ are all united by ball and socket joints, the posterior rounded eminence of each vertebra being received into the anterior surface of the next. Serpents swallow their prey entire; and it is well ascertained that they will swallow animals having ten times the diameter of that of their own neck. The loose connexion of all the bones surrounding the mouth of serpents, enables them to distend their jaws and mouth to receive undivided prey, and thus, so far as food is concerned, to dispense with arms to grasp it, and assist in its sub-division. Neither are their teeth suited for mastication, being conical, slender, sharp, osseous, and recurved.

Venomous serpents, or those with isolated fangs, have their organs of manducation constructed on a very peculiar plan. Their superior maxillary bones are very small, attached to a long pedicle, and are very moveable; in them is fixed a sharp-pointed pervious tooth, through which flows a liquor which, poured into the wound made by the tooth, produces effects according to the species of the reptile secreting it. This tooth, when the animal does not wish to use it, is concealed in a fold of the gum, and behind it are several germs destined to replace it, in the event of its being broken.

All those venomous serpents, whose mode of production is well ascertained, bring forth living young ones, as their eggs are hatched without being laid. In more modern classification ophidia constitutes the seventh order of Reptilia, comprising Coluba, and the extinct genera Palæophis and Paleryx.

OPHI′DIOUS. Belonging to the order Ophidia.

OPHI′OLITE. (from ὄφις, a serpent, and λίθος, a stone, Gr.) Another name for mineral serpentine.

OPHIOMO′RPHA. The fourth order of the class Amphibia, comprising Cecilia, or the blind worm.

O′PHITE. (ὀφίτης, *lapis in modum serpentis maculosus, ab* ὄφις, a serpent, Gr.) Green porphyry, or serpentine. A green-stone, varying from blackish-green to pistachio green. It contains greenish-white crystals of felspar, which on the polished surface often shew themselves in parallelograms, and are sometimes cruciform. It occurs massive and disseminated. Lustre glistening and resinous. Fracture conchoidal, and often splintery.

OPHIU′RA. A genus of fossil radiaria, of the order Ophiuridæ, found in the lias of Yorkshire and in the inferior oolite sands near Bridport; in the oolite of Germany, in the muschel kalk, in the red sandstone group, in the upper lias, and in the upper chalk.

OPHUI′RIDÆ. The third order of the class Echinodermata.

OPI′STHOBRANCHIATA. The sixth order of the class Cephalophora, comprising Tornatella, Bulla, Doris, Aplysia, &c.

OPO′SSUM. A genus of quadrupeds belonging to the order Marsupialia. The opossums are peculiar to America, and are remarkable for their number of teeth, amounting in all to fifty. They approach the quadrumanes, by having the thumb of their hind foot opposed to the fingers, whence they have been called pedimanes: the thumb is not armed with a nail.

ORBI′CULA. (from *orbis*, an orb, Lat.) A genus of bivalve shells, fossil and recent, belonging to the family Brachiopoda. The orbicula is a very small inequivalved flat bivalve; the lower valve very thin, and adherent to other bodies.

ORBI′CULAR. (*orbiculaire*, Fr. *orbiculáre*, It.) Spherical; circular; roundish and flat. In botany, leaves are so called when their length and breadth are equal, and their form nearly circular.

O′RBIT. (*orbita*, Lat. *orbite*, Fr. *órbita*, It.) The line described by the revolution of a planet; the path of a planet, or of a comet. The mean distance of a planet from the sun is equal to half the major axis of its orbit. A planet moves in its elliptical orbit with a velocity varying every instant, in consequence of two forces, one tending to the centre of the sun, and the other in the direction of a tangent to its orbit, arising from the primitive impulse, given at the time when it was launched into space.

ORBULI′TES. A genus of foraminated polypifers, differing from lunulites in having pores on both sides. Mr. Parkinson thus describes the genus: "a free, circular, stony polypifer; flattish; pores on both sides, or at the margin; resembling nummulites. Set with minute pores, sometimes scarcely visible, regularly disposed, but not in spiral order." The species are nearly all fossil but some are recent."

ORCHI′DEÆ. (from ὄρχις, Gr. *orchis*, Lat.) An order of monocotyledonous plants. Perianth superior, sepals three, usually coloured, the odd one uppermost, from the twisting of the ovarium; petals three, usually coloured, of which two are the uppermost, while the third, called the labellum, is usually lobed, and differs in figure, colour, or size, from the other two, and is often spurred; stamens three, united in a central column, the two lateral generally abortive, the central one perfect; anther persistent or deciduous; pollen either powdery or cohering in granular or waxy masses; ovarium one-celled, with three parietal placentæ; style forming part of the column of the stamens; stigma a viscid space in front of the column; fruit usually a capsule, dehiscing by three valves, sometimes baccate; seeds numerous, testa loose and reticulated, no albumen, embryo a solid undivided fleshy mass; herbaceous plants, either stemless, or forming a kind of tuber above ground; or sometimes with a true stem; leaves simple and entire, sometimes articulated with the stem; inflorescence terminal or radical spikes, racemes or panicles, occasionally solitary.

Natives of all countries, except very cold or very dry. There are thirty-seven British, and, probably, altogether not fewer than fifteen hundred species.

The flower of the orchideæ is very peculiar; the calyx and corolla consist of three pieces each, and one of those forming the latter, differs very greatly in size and form from the other two; it is called the labellum, or little lip, and is often spurred. In many species, this resembles an insect, and hence they have received the name of bee, fly, spider, &c., &c.

ORCHI'DEOUS. Belonging to the order Orchideæ; parasitical plants.

O'RDER OF SUPERPOSI'TION. That arrangement of strata in which they are invariably found. The order of superposition is never inverted. Strata are frequently absent, but the order of superposition of such as are present is invariably the same.

ORE. (*erz*, Germ.) A metallic compound. Metals are found usually combined with other substances: the compounds they thus form are called *Ores*, when the metal exists in them in sufficient quantities to form a considerable portion of the mass.

ORGA'NIC. (ὀργανικὸς, Gr. *organicus*, Lat. *organique*, Fr. *orgánico*, It.) Consisting of various parts co-operating with each other; consisting of natural instruments of action or operation.

ORGA'NIC BO'DIES. Such as possess natural instruments of action; on the action of each, and their co-operation together, depend the growth and perfection of the body.

ORGA'NIC REMAI'NS. The relics of what were once living bodies: generally applied to the fossil remains of animals or plants.

"Chemical analysis," says Mr Parkinson, "has been called in to the aid of the naturalist, in order to account for the perfect state of preservation observable in remains organized with the most exquisite delicacy, and which there is every reason for supposing to have been decomposable in their recent state. From this investigation we learn the manner in which these memorials of the old world, so interesting and so frail, have been preserved. Some have been impregnated with calcareous matter, others with silicious, and others with iron and copper pyrites."

ORISK'ANY SANDSTONE. A member of the North American Devonian rocks, attaining a thickness of 200 feet, and by its fossils proved to be contemporaneous with the Lower Devonian group of the Rhine.

ORISMO'LOGY. (from ὁρισμὸς, definition, and λόγος, discourse, Gr.) Called also terminology. In entomology, orismology contains the various technical terms used in explaining the perceptible differences in the body of an insect, and at the same time acquaints us with its exterior visible parts in the several periods of its existence, until its full and perfect development. It is the very first requisite of a precise orismology to apply an exclusively proper term to each constantly distinct and peculiar part.

ORNITHI'CKNITES. (from ὄρνις, a bird, and θίγω, to touch, Gr.) The footmarks of birds found in different formations. Some recent discoveries of ornithicknites are very remarkable; the footsteps appear in regular succession, on the continuous track of an animal in the act of running or walking, with the right and left foot always in their relative places. An account of these has been published by Prof. Hitchcock, in the American Journal of Arts and Sciences: they were discovered in the new red sandstone of the valley of Connecticut. The most remarkable were those of a gigantic bird, twice the size of an ostrich, whose foot measured fifteen inches in length, exclusive of the largest claw, which measured two inches. The discovery of these ornithicknites is exceedingly interesting to the palæontologist, as proving the existence of birds at the early epoch of the new red sandstone formation.

ORNI'THOLITE. (from ὄρνις, a bird, and λίθος, a stone, Gr.) A fossil bird. Stones of various colours and forms, bearing the figures of birds. Specimens of this kind may be obtained at Matlock, in Derbyshire, and at other places where the water is surcharged with lime. The gypsum quarries of Paris contain the debris of birds in great abundance. The feet are the most remarkable part in all the ornitholites, the feet of birds being composed in a peculiar way, not resembling those of any other animals.

ORNITHO'LOGY. (from ὄρνις, a bird, and λόγος, discourse, Gr.) That department of natural history which treats of birds; describes their structure, external and internal; and teaches their economy and their uses.

ORNITHO'LOGIST. One versed in that branch of **natural history** which treats of the **habits, structure and** uses of birds.

ORNITHORHY'NCHUS. (from ὄρνις, a bird, and ῥύγχος, a beak, Gr.) The platypus of Shaw. The duckbill; an animal indigenous to New Holland, and found in no other country. In this anomalous animal, we have a quadruped clothed with fur, having a bill like a duck, with four webbed feet, suckling its young, and most probably ovoviviparous: the male is furnished with spurs. The mouth of the ornithorhynchus has a form of construction between that of quadrupeds and birds, being furnished, like the former, with grinding teeth at the posterior part of both the upper and lower jaws, but they are of a horny substance; the mouth is terminated in front by a horny bill, greatly resembling that of the duck, or the spoon-bill. It has also small cheek-pouches. Membranes unite the toes of the fore and hind-feet; in the fore-feet it extends beyond the nails, in the hind-feet it terminates at the root of the nails. It has also a flattened tail. It inhabits the rivers and marshes.

O'RPIMENT. (from *auripigmentum*, Lat. *orpiment*, Fr. *orpimento*, It.) The Arsenic sulfuré jaune of Haüy; Arsenic sulfuré orpiment of Brongniart. Yellow sulphuret of arsenic, an ore of arsenic combined with sulphur. Its colour is usually lemon-yellow, which is often shining and beautiful. It occurs in laminated or lamellar masses; in **concretions;** and sometimes in minute crystals. It is principally volatilized before the blow-pipe, with a white smoke, and with the odour of both sulphur and arsenic, leaving a small earthy residue. According to Thenard it is composed of arsenic 57, sulphur 43.

O'RTHIS. (from ὀρθός, straight, Gr.) A genus of fossil shells, constituting a division of spirifer, but distinguished from spirifer by the long narrow hinge and circular flat form of the striated shells. Twenty-six species of the genus orthis are enumerated by Sir R. Murchison as occurring in the Silurian rocks.

O'RTHITE. A mineral found in the mine of Finbo, in Sweden, and thus named from its being always found in straight layers.

ORTHO'CERAS. } (ὀρθός, straight,
ORTHOCE'RATITE. } and κέρας, a horn,

Gr.) An extinct genus of polythalamous, or many chambered, cephalapods, which inhabited straight shells. The orthoceratite resembles an ammonite unrolled, having its chambers separated by transverse septa, concave externally, and convex internally; the septa being pierced by a siphuncle. Orthoceratites are abundantly found in the transition strata, appearing to have been early called into existence, and at an early period to have been consigned to almost total destruction. At the close of the Triassic period the Orthoceras appears to have become extinct. It is regarded as characteristic of the Palæozoic strata. Part of the pavement of the palace at Hampton Court, and that of the hall of University College, Oxford, are composed of marble containing remains of orthoceratites. Some species, found in the carboniferous limestone of Closeburn, in Dumfries-shire, are nearly the size of a man's thigh.

ORTHOCŒ'LA. A group of animalcules, thus named from the straight course of the intestine.

O'RTHOCLASE. Called also potash-feldspar; a mineral consisting of silica 65·35, alumina 18·06, potash 16·59.

ORTHO'PTERA. (from ὀρθὸς, straight, and πτερὸν, a wing, Gr.) An order of insects, the sixth in Cuvier's arrangement. The coverings of the wings, instead of being of a horny texture, are soft and flexible. The wings themselves, being broader than their coverings, are, when not in use, folded longitudinally like a fan.

O'RTHOSE. The name of a feldspathic mineral.

ORYCTOLO'GICAL. (from ὀρυκτὸς, a fossil, and λόγος, discourse, Gr.) Pertaining to that part of physics which treats of fossils.

ORYCTO'LOGIST. One who studies, or is versed in, that part of physics which treats of fossils.

ORYCTO'LOGY. (from ὀρυκτὸν, a fossil, and λόγος, discourse, Gr.) Oryctology is that branch of mineralogy which has for its object the classification of fossils; or, in other words, it consists in the description of fossils, the determination of their nomenclature, and the systematic arrangement of their different species. Oryctology has also been defined as the science which enquires into the nature, origin, and formation of those bodies which possess the figures, markings, or structure of vegetables, or animals, whilst their substance evinces their having been preserved through many ages, by certain changes effected in subaqueous or subterranean situations.

OSCILLA'TION. (*oscillatio*, Lat. *oscillation*, Fr. *oscillazione*, It.) Vibration; the act of swinging to and fro; a movement to and fro, like the swinging of the pendulum of a clock, or waves in water. The tides are oscillations of the sea.

OSMERO'IDES MANTE'LLII. The name given by Dr. Mantell to an ichthyolite of the chalk formation discovered in the Lewes chalk-pits. It is closely related to the smelt.

O'SMIUM. (from ὀσμή, odour, Gr.) A metal discovered by Tennant in crude platinum, and deriving its name from the strong odour emitted by some of its compounds. Its symbols are O. S.; specific gravity 10·00.

O'SSEOUS BRE'CCIA. A mass of fragments of the bones of animals, cemented together by a calcareous gangue, and commonly found in fissures and caves.

O'SSICLE. (*ossiculum*, a little bone, Lat.) A small bone.

OSSI'FEROUS. (from *os*, a bone, and *fero*, to produce, to bear, or contain, Lat.) Yielding bones or fragments of bones; containing bones. Thus we have ossiferous gravel, ossiferous clay, ossiferous strata, ossiferous caves, &c., &c.

OSTEOCO'LLA. (from ὀστέον, a bone, and κόλλα, glue, Gr.) Vegetables of the most delicate texture, when immersed in waters containing carbonate of lime, become incrusted, still preserving their form even to their most **delicate** ramifications. These incrustations somewhat resemble the bone of an animal, and the property has been absurdly attributed to them of facilitating the union of fractured bones.

OSTEOLE'PIS. (from ὀστέον, a bone, and λεπίς, a scale, Gr.) The name assigned by Mr. Pentland and M. Valenciennes to a genus of fossil fishes, found in the old red sandstone of Scotland. Six species have been described. Sir R. Murchison states that this genus has not hitherto been discovered in the old red sandstone of England. The osteolepis had a shell as naked as its teeth, the bone being merely covered by a hard shining enamel; its toes also were of bare enamelled bone. The enamelled teeth were placed in jaws which presented outside a surface as naked and as finely enamelled as their own. The entire head was covered with enamelled osseous plates, furnished inside, like other bones, with their nourishing blood vessels. The fins had a sort of bird-wing construction. The mouth opened below the snout, but not so far below it as in the purely cartillaginous fishes. It was thickly furnished with slender and sharply-pointed teeth. The gills opened, as in the osseous fishes, in continuous lines, and were covered by large bony opercules. While the head and its appendages resembled, in some points, the bony fishes, the tail differed in no respects from the tails of cartilaginous ones. In size, **the** osteolepis varied from six to twelve inches.

OSTEO'LOGY. (from ὀστέον, a bone, and λόγος, discourse, Gr.) A description of the bones; that part of anatomy which treats of the bones. That branch of science named fossil osteology has assumed a most important character, for which we are mainly indebted to the labours of Cuvier. From a single fragment of a bone, certainly of a characteristic part, the order and genus of the animal may be determined with a precision amounting almost to mathematical certainty.

OSTEO'LOGICAL (from Osteology.)

OS'TRACITE. A fossil oyster.

OSTRA'CODA. The fifth order of the class Crustacea, comprising Cypris and Cythere.

OSTRÆ'A. } The oyster. A rough, adherent, inequivalved bivalve; the hinge without a tooth. One muscular impression in each valve. The oyster is found both fossil and recent. Of this genus, one hundred and thirty-seven species have been described in Turton's Linné. The most extraordinary shell of this genus for size, says Mr. Parkinson, is the large fossil oyster, the recent analogue of which, from Virginia, appears to be depicted by Lister. Some attain to the length of twenty inches. An under valve in Mr. Parkinson's possession weighed four pounds, being thirteen inches in length and three in thickness.

OU'TCROP. A term, used by miners, to express the exposure at the surface of a stratum or strata.

OU'TLIER. A portion of a stratum detached from the principal mass, and lying detached at some distance from it.

O'VA ANGUI'NA. A species of fossil cidaris or echinus.

O'VAL. (from *ovum*, an egg, Lat, *ovale*, Fr. *ovále*, It.) A rounded surface, its two right-angular diameters being of an unequal length, so that its longest transverse diameter does not pass through the centre of its longitudinal diameter, but lies nearer to one end: egg-shaped.

OVA′RIUM. ⎫ (*ovaire*, Fr. *ovaja*, It.)
O′VARY. ⎭ That part of the body which contains the ova, and in which impregnation is performed. In animals, it is only in the organs termed ovaries, that ova are formed.

In botany, that part of the flower which ripens into the fruit, and contains the seed.

O′VATE. (*ovatus*, made like an egg, from *ovum*, Lat.) Of the shape of an egg; egg-shaped.

OV′IDUCT. (from *ovum*, an egg, and *ductus*, a passage, Lat.) A canal, or duct, through which the ova pass, after impregnation, from the ovary to the uterus. In the human subject the oviducts are called the Fallopian tubes.

O′VIFORM. (from *ovum*, an egg, and *forma*, shape, Lat.) Of the form or shape of an egg.

OVI′PAROUS. (from *ovum*, an egg, and *pario*, to produce, Lat. *ovipare*, Fr. *oviparo*, It.) All animals which lay eggs, enclosed in a calcareous shell, are called oviparous. Oviparous production is thus characterized; the young animal is not attached to the parieties of the oviduct, but remains separated from it by its external envelope; its aliment being enclosed in a sac, which is attached to its intestinal canal.

OVO′-VIVIPAROUS. Producing young in an egg, which is hatched internally, and previously to exclusion by the parent. Some animals, such as the salamander and the viper, never lay their eggs, but these are hatched within the body of the parent; so that although originally contained in eggs, the offspring are brought forth in a living state. Such animals are termed ovo-viviparous.

OVIP′OSITOR. A name given to the terminal apex of the abdomen of insects.

O′VULE. ⎫ (dim. of *ovum*, an egg,
O′VULUM. ⎭ Lat.) In botany, the seed before it is perfected. The small bodies produced on the margins of the carpella in the pistil, are called ovula, or ovules; when perfected, they become the seed of the plant. The ovule is generally attached to the placenta of the ovarium by a very small stalk.

OVULI′TES. A genus of foraminated polypifers, said to be known only as fossil, of minute size, not exceeding two millimetres in length. Ovulites is a stony, free, oviform, or cylindrical, polypifer; hollow within; the extremities generally perforated, having minute pores regularly diffused over the surface.

O′XFORD CLAY. Called also clunch clay. A bed of dark blue clay, sometimes nearly six hundred feet in thickness, interposed between the lower and the middle oolites.

OXIDA′TION. That process by which metals, and other substances, are converted into oxides by their combination with oxygen.

O′XIDE. ⎫ A substance combined with
O′XYDE. ⎭ oxygen, without being in the state of an acid.

O′XIDIZED. Converted into an oxide, by combination with oxygen.

O′XYGEN. (from ὀξύς, acid, and γεννάω, to produce, Gr.) So called from its property of forming acids. One of the simple or elementary substances, and one of the five which exist as gas. So generally does oxygen enter into combination with metallic and non-metallic bodies, and in such large proportions, that it has been computed that one-half of the ponderable matter of the globe is composed of oxygen gas. Oxygen constitutes, by measure, 21 per cent, by weight 23 per cent. of the volume of the atmosphere; it forms a third part, by measure, of the gasses composing pure water; and is locked up to an immense amount in the various rocks, which are little else than a mass of oxidized substances. Plants give out oxygen, animals absorb it. It is to Dr. Priestley we owe the knowledge of the former of these two facts; and he it was who first discovered oxygen, in 1774. Oxygen has neither taste nor smell. It is a trifle heavier than atmospheric air, 100 inches weighing 33.88 grains. The combining proportion, or equivalen number of oxygen, that of hydrogen being taken as unity, is 8; its symbol, O.

OXYGENA'TION. "This word," says Dr. Ure, "is often used for oxidation, and frequently confounded with it; but it differs, in being of more general import, as every union with oxygen, whatever the product may be, is an oxygenation; but oxidation takes place only when an oxide is formed."

O'YSANITE. A name given by Lameth to pyramidal titanium, or anatase.

# P

PACHYDE'RMATA. (from παχὺς, thick, and δέρμα, skin, Gr.) Thick-skinned animals. The seventh order of the class Mammalia, in Cuvier's arrangement. This order Cuvier divided into two families, namely, Proboscidiana, or those pachydermatous animals, which have tusks and a proboscis, as the elephant and mastodon; and pachydermata ordinaria, in which are included the hippopotamus, anoplotherium, palæotherium, tapir, &c.

Several genera of the order Pachydermata have become extinct, their fossil remains alone proving that such ever existed. Amongst these are the mastodon, the anoplotherium, the palæotherium, the lophiodon, the anthracotherium, the cheropotamus, adapis, &c. Of these there are about forty species, all extinct, and to which there are none analogous in the living world, except two tapirs and a daman. Of the existing *genera* of pachydermata, many *species* which existed during the older and newer pliocene periods also seem to have become extinct, and, in fact, the living species bear no sort of proportion to the extinct. Those shades which approximate genera to each other, those intermediate forms, those steps from one genus to another, so common in the other families of the animal kingdom, are here wanting. It was reserved for the science of fossil osteology to recover them from the entrails of the earth, among the races which completed the grand system of animated nature, and whose destruction has produced such wide and striking intervals.

The pachydermata appear to be, as it were, only the remnants of a very extensive order, which formerly inhabited the earth, but have now almost entirely disappeared. They feed upon grass, but they do not ruminate. They are, for the most part, huge and unwieldly animals, with thick integuments; solidity and strength appear to be the objects chiefly regarded in their construction.

PACHY'MIA. A genus of fossil bivalve shells described by Mr. Sowerby as obliquely elongated, equivalve, thick, sub-bilobed, with beaks near the anterior extremity; the ligament partially immersed, attached to prominent fulcra. Its place in the conchological system has not been determined with certainty; the P. gigus is found in the green sand of Lyme Regis.—*Lycett.*

PACHYRI'SMA. A fossil genus of thick equivalve, inequilateral bivalves, belonging to the Cardiaceæ of Lamarck; the figure is transverse, oblong or cardiform, the umbones are large and incurved, the ligament external, placed upon prominent fulcra, the hinge is massive, having a large sub-conical prominent tooth in each valve, and a small accessory tooth in the right valve.—*Lycett.*

PACHYDE'RMATOUS. Thick-skinned; belonging to the order Pachydermata.

PA'DDLE. The swimming apparatus of the chelonian reptiles and of the marine saurians, has obtained the name of paddles.

PÆCILO'PODA. (from ποικίλος, and πούς, Gr. various footed.) The second order of the class crustacea; it comprises two families, Xysophura and Siphonostoma.

PALA'GONITE. A mineral; an amorphous highly hydrated scapolite.

PALÆO'LOGY. (from παλαιὸς, ancient, and λόγος, discourse, Gr.) The study of ancient things. This word is commonly written paleology.

PALÆONI'SCUS. The name assigned by De Blainville to a genus of fossil fishes or ichthyolites, approximating to the sturgeons. The name has also been given to a genus of insects, and in order to avoid confusion this latter should, as applied to insects, be altered.

PALÆONTO'LOGIST. (from *palæontology.*) One who studies, or is versed in, the history of fossil plants and animals.

PALÆONTO'LOGY. (from παλαιὸν, ancient, ων, being, and λόγος, discourse, Gr.) The history of fossil plants and animals; that branch of natural history which treats of fossil and extinct animals and plants.

PALÆOSAU'RUS. (from παλαιὸς, ancient, and σαῦρος, a lizard, Gr.) A genus of fossil saurians, now extinct, found in the magnesian limestone.

PALÆOTHE'RIUM. } (from παλαιὸς, an-
PALÆOTHE'RE } cient, and θηρίον, a wild beast, Gr.) An extinct genus of fossil quadrupeds, belonging to the order Pachydermata, having six incisors in each jaw, ranged in one and the same line; four canines, two in each jaw, conical, and so distant as to cross each other when the mouth was closed; fourteen molars in each jaw, the upper of a square form, with four roots and three crests on the outer side, leaving between them two channels; a furrow on the inner side; their coronal nearly anologous to that of the upper molars of the daman and rhinoceros. The lower molars show their enamelled outlines in the form of a double crescent. The general form of the head is like that of the tapirs, the nasal bones, short and slender. It possessed three toes to each foot, and had a short fleshy probocis. "The place of the genus Palæotherium," says Buckland, "is intermediate between the rhinoceros, the horse, and the tapir. Some of the discovered species were as large as a rhinoceros, others were from the size of a horse to that of a hare. These animals probably lived and died upon the margins of the then existing lakes and rivers."

PALÆOTHE'RIAN. Belonging to the genus Palæotherium, as palæotherian remains, &c.

PALÆO'SPALAX. (from παλαιὸς, ancient, and σπάλαξ, a mole, Gr.) "An extinct genus of insectivore, referable, says Prof. Owen, to the mole tribe, but as large as a hedgehog. The only specimen hitherto discovered is in the British Museum, and was met with in the lacustrine formation at Ostend, near Bacton, in Norfolk.

PALÆOTHRI'SSUM. The name assigned to a genus of fossil fishes discovered in the metalliferous schists of Mansfield, Thuringia, &c. This genus comprises several species, as the P. macrocephalum, P. magnum, P. parvum, and P. inæquilobum. Hitherto this genus has been found fossil only, and in strata anterior to the chalk.

PALÆOZOIC. Formerly termed primary. The palæozoic or primary epoch comprises six periods, namely, the Cambrian, the Lower Silurian, the Upper Silurian, the Devonian, the Carboniferous, and the Permian.

PA'LEA. (*palea*, chaff, Lat.) In botany, a term applied to the inner bracteæ of grasses: the paleæ are membranous or chaffy in texture.

PALLA'DIUM. (from the planet Pallas.) A metal of a greyish or bluish-white colour, discovered by Dr. Wollaston in 1803, in platinum. It is malleable, ductile, and flexible, but does not possess much elasticity. In hardness it surpasses all other metals, with the exception of tungsten, which it equals. Specific gravity 11·8. It is not oxidated by the action of the atmosphere. It is fusible only at a very high temperature. Symbols Pd.

PA'LLEAL. (from *pallium*, a mantle, Lat.) In conchology, the name given to the mark or impression observed in bivalves, formed by the muscular attachment of the mantle.

PA'LLIOBRANCHI'ATA. The same as Brachiopoda.

PALM. (from *palma*, Lat.) The palms constitute a natural order of monocotyledonous, or endogenous, plants. The flowers are hermaphrodite, or polygamous. Perianth six-parted, persistent. Stamens inserted into the base of the perianth, definite or indefinite. Ovary three-celled, or deeply three-lobed, with an erect ovule. Fruit baccate or drupaceous, with fibrous flesh. Albumen cartilaginous; embryo in a cavity at a distance from the hilum. Leaves terminal, large, pinnate, or flabelliform, plaited in vernation. Spadix enclosed in a valved spatha. Flowers small.

A palm tree affords an example of the mode of growth in endogenous plants. The stem of this tree is usually perfectly cylindrical, attains a great height, and bears on its summit a tuft of leaves. It is composed of an extremely dense external cylindric layer of wood; but the texture of the interior becomes gradually softer and more porous as it approaches the centre. It has neither medullary rays, nor central pith, nor true outward bark. The first stage of its growth consists in the appearance of a circle of leaves, which shoot upwards from the neck of the plant, and attain during the first year, a certain size. The following year another circle of leaves arises; but they grow from the interior of the former circle, which they force outwards as their vegetation advances, and as ligneous matter is deposited within them. As soon as the outer layer has become too hard to yield to the pressure from within, the growth of the inner layers is immediately directed upwards; so that they rise in succession by distinct stages, always proceeding from the interior; a mode of development which has been compared by De Candolle to the drawing out of the sliding tubes of a telescope. The whole stem, whatever height it may attain, never increases in diameter after its outer layer has been consolidated. A circle of leaves annually sprouts from the margin of the new layer of wood; these, when they fall off, leave traces, consisting of a circular impression round the stem. By the number of these circles the age of the tree may be ascertained. The existing family of palms is supposed to consist of nearly a thousand species, of which the greater number are limited to peculiar regions of the torrid zone.

PA'LMATE. ) *(palmatus,* Lat.) Web-
PA'LMATED. ) bed, like the feet of some water-birds; deeply divided into lobes like the fingers on the hand; resembling a hand; palmed or hand like. Applied to leaves which are divided, half-way, or more than half-way, down the middle, into several nearly equal segments, having a space between each.

PALMI'PEDES. (from *palmipes,* that hath its feet closed with a film or web, Lat.) The sixth order of birds in Cuvier's arrangement. The goose and duck are familiar examples.

PA'LPI. In entomology, the palpi, or feelers, are the auxiliary organs of a masticating mouth. Those upon the maxillæ are termed the palpi maxillares, or maxillary feelers; those placed laterally upon the labium, are designated the palpi labiales, or labial feelers.

PALUDI'NA. A genus of fresh-water univalves, belonging to the family Peristomata. Remains of paludinæ are very abundant in the Sussex marble, and the P. fluviorum is its common and characteristic fossil.

PAMPHRA'CTUS. A genus of ichthyolites of the old red sandstone, two species of which have been described by Agassiz. It appears to have become extinct at the close of the Devonian period.

PA'NCAKE. The name given by Klein to the Echinodiscus laganum, a species of fossil echinus, belonging to the division Catocysti.

PANDA'NEA. ) (from *pandus,* Lat. crook-
PANDA'NUS. ) ed.) The screw-pine, so named from the spiral arrangement of its leaves, is a monocotyledonous tree, growing only in the warmer zones, and principally near the sea. The pandanea, like the cocoa-nut palm, is generally the first vegetable colonist of the newly-raised coral islands. Its appearance is that of a gigantic pine-apple plant, with arborescent stems. The pandanus bears a large spherical, drupaceous fruit: the seed within each drupe being enclosed within a hard nut. From the pandanus growing near to the sea, its fruit frequently drops into the water, and is drifted by the waves and winds to distant shores: thus the elements of vegetation are transported to the emerging coral islands, where it vegetates. A fossil fruit of the pandanus was found by Mr. Page in the inferior oolite, and is in the Oxford museum. It is of the size of a large orange, and is covered by a stellated rind, or epicarpium, composed of hexagonal tubercles, forming the summits of cells, which occupy the entire surface of the fruit. Fruits of a genus, to which M. Adam Brongniart has given the name of Pandanocarpum, occur, together with cocoa-nut fruit, at an early period of the tertiary formations, in the London clay of the Isle of Sheppey.

PA'NGOLIN. A species of manis, or scaly lizard; called also the scaly ant-eater. Its armature is composed of separate, horny, moveable scales. It is destitute of teeth, has a very extensile tongue, and lives on ants and termites.

PA'NICLE. *(panicula,* Lat. a bunch or cluster.) A species of inflorescence, in which the flowers are scattered on peduncles, variously subdivided without any order, and more or less close. The oat affords a familiar example. When the middle branches of a panicle are longer than the others, it is termed a thyrsus.

PANNI'CULUS CARNO'SUS. (from *panniculus*, a cloth, and *carnosus*, fleshy, Lat.) A peculiar set of subcutaneous muscular bands which serve to erect the bristles, or armour of certain animals; as in the hedge-hog, porcupine, &c.

PANOPÆ'A. A genus of bivalve shells of the family Solenacea. The Panopæa is a transverse inequilateral bivalve, gaping at both extremities. The hinge similar in both valves, with an acute cardinal tooth in each, and, on the right valve, a little pit, which receives the tooth of the opposite valve. It is found both recent and fossil.

PA'PER COAL. A bituminous shale, to which the name has been given from its divisibility into extremely thin leaves.

PA'PER NAU'TILUS. Called also the Paper Sailor. See *Argonauta*.

PAPI'LIO. (*papilio*, Lat. a butterfly.) A genus of the family Diurna. belonging to the order Lepidoptera, The butterfly. The species are numerous. It has been well observed that the chrysalis is the tomb of the caterpillar, and the cradle of the butterfly.

PAPILIONA'CEOUS. Resembling a butterfly. In botany, the corolla is called papilionaceous when it consists of five petals of particular forms, of which the uppermost is generally the largest, and turned back; the two next resemble each other, and are called the alæ; the two lowermost are generally united by their lower edge, and form a keel-like figure, and are, from that circumstance, called the carina or keel; the two last, so united, contain, and protect, the internal organs.

PAPI'LLA. (*papilla*, Lat.) This word is generally used in the plural, papillæ. Malphigi first discovered this structure in the foot of the pig, and gave to it its name. The external surface of the skin presents a great number of minute projecting filaments; these are the papillæ. "It is probable," says Dr. Roget, "that each of these papillæ contains a separate branch of the nerves of touch, so that we may consider these papillæ as the principal and immediate organs of touch. The papillæ are much more easily perceived on some parts than others, but no where are they more perceptible than on the tongue, where, more especially in a morbid condition of the body, they are frequently much elevated."

PAPI'LLOSE. In botany, a term applied to stems covered with soft tubercles; also to leaves covered with fleshy dots or points.

PA'PPUS. (*pappus*, Lat. thistle-down, πάππος, Gr.) The feathery appendage that crowns many seeds which have no pericarpium; a particular form of calyx, of which we have a familiar example in the dandelion.

PARA'BOLA. (*parabola*, Lat.) One of the five conic sections: thus, if a cone be cut by a plane parallel to one of its sloping sides, the section will be a parabola.

PARADO'XIDES. The name given by Brongniart to a genus of trilobites. The Olenus of Dalman. "This genus," says Sir R. Murchison, "may always be recognised by the ends of all the lateral ribs terminating in deflected points, some of which extend in spikes beyond the tail." In Pictet's system, Paradoxides constitutes a family comprising the genera Remopleurides, Paradoxides, Hydrocephalus, Jao, Arionellus, Ellipsocephalus, Olenus, Conocephalites, Peltura, and Triarthrus.

PARALLE'LOGRAM. (from παράλληλος and γράμμα, Gr. *parallélograme*, Fr. *parallelogrammo*, It.) In geometry, a right-lined quadrilateral figure, whose opposite sides are parallel and equal.

PARALLEL'OPIPED. *(parallelipipede, Fr. terme de géométrie. Corps solide terminé par six parallélogrammes dont les opposés sont parallèles entre eux.)* A solid figure contained under six parallelograms, the opposites of which are parallel and equal; or it is a prism whose base is a parallelogram; it is always triple to a pyramid of the same base and height. The term *parallelopiped* includes, in mineralogy, all those solids whose bounding planes are parallel two and two; as for instance, all the varieties of the rhomboid; and all the prisms of which the terminal planes are rhombic; and all the square and rectangular prisms which do not possess the precise proportions of the cube.

PARALLELO'PIPEDAL. Having the figure of a parallelopiped.

PARALLELISM. Nothing is more remarkable in volcanic dikes than the *parallelism* of the opposite sides, which usually correspond with as much regularity as the two opposite faces of a wall of masonry.

PARAMO'UDRA. (The derivation of this word, says Buckland, I could trace to no authentic source, but shall adopt because I find it thus appropriated.) A singular fossil body, found in the chalk of the North of Ireland. Among the organic remains of the chalk in the north of Ireland, are large silicious bodies of a very peculiar character. These singular fossils are found in many of the chalk pits from Moira to Belfast and Larne, but are most numerous at Moira. They have, I believe, never yet been found in the chalk of England, except at Whitlingham, near to Norwich, and at some other places in the same neighbourhood, whence there is a good specimen in the collection of the Geological Society, about two feet long and one foot in diameter. The length of these bodies varies from one to two feet, their thickness from six to twelve inches. Their substance in all cases is flint. The termination of of these silicious bodies is distinct, and the separation of the flint from its matrix of chalk always clear and decided. Their outer covering has the appearance of a thin epidermis, smooth externally, and whiter than the mass of flint enclosed, which is usually of a dark grey colour.

In all cases these bodies seem to have had a central aperture passing into and generally through their long diameter. The breadth of this aperture varies in different specimens.

PARAMO'RPHISM. The name given to a particular kind of pseudo-morphism.

PARA'NTHINE. A rare mineral, thus named by Haüy, more commonly known as Scapolite.

PARASI'TA. *(parasita,* Lat. παράσιτος, Gr.) In Cuvier's arrangement, the third order of Insecta; they have six legs and are apterous. The Rev. Mr. Kirby observes, "the order of parasites, consisting of the most unclean and disgusting animals of the whole class, infest man, beast and bird, no less than four species being attached to man, may be divided into two sections, namely, those that live by suction, and those that masticate their food. To the first of these belong the human and the dog-louse, and to the other the various lice that inhabit the birds, of which almost every species has a peculiar one."

PARASI'TIC } 1. In botany, applied
PARASI'TICAL. } to plants which fix their roots into other plants, and from them, instead of from the earth, derive their nourishment: the mistletoe is a familiar example. 2. In zoology, a name given to certain insects which live upon the animals they infest.

PARENCHY'MA. (παρέγχυμα, Gr. *paren-chyme*, Fr.)
1. A spongy or porous substance forming the bulk of some of the viscera, as the parenchyma of the liver, &c.
2. In botany, a fine, **transparent,** membranous tissue, **lying immediately beneath the epidermis of plants; it is of a deep green colour, very tender, and succulent.** When **viewed with a microscope, it seems to be composed of fibres which cross each** other in every direction. In its simplest state, it appears like a mass of globules or vesicles, crowded together; these, from pressure, assume a six-sided, or hexagonal figure.

PARENCHY'MATOUS. **Consisting of parenchyma; spongy; porous.**

PA'RGASITE. **The name given to a variety of actinolite, from its being found in the Isle of Pargas, in Finland.** Colour forms the principal difference between **hornblende** and pargasite, the latter being somewhat translucent and of a lighter green, or more generally of a bottle-green hue.

PARIE'TAL. (from *paries*, a wall, Lat.)
1. The name given to certain bones of the skull, from their serving as walls to the brain.
2. In botany, a term used to express an adhesion of some part to the inner side of an organ; as when the seeds are attached to the placentæ, the latter are termed *parietal.*

PA'RIS BA'SIN. A large area, to which the name of Paris Basin has been given, about 180 miles in length, from north-east to southwest, and about ninety miles wide, from east to west. The country in which the capital of France is situated, is perhaps the most remarkable that has yet been observed, both from the succession of different soils of which it is formed, and from the extraordinary organic remains which it contains. Bones of land animals, of which the genera are entirely unknown, are found in certain parts; other bones remarkable for their vast size, and of which some of similar genera exist only in distant countries, are found scattered in the upper beds. Millions of marine shells, which alternate regularly with fresh-water shells, compose the principal mass. The strata composing the Paris basin rest upon chalk, lying, as it **were,** in a depression of the chalk. The depth of these strata varies from one to five hundred feet. MM. Brongniart and Cuvier divided the strata into the five following formations, commencing with the undermost. 1. First fresh-water formation; consisting of **plastic** clay, lignite, and first **sandstone.** 2. First marine formation; comprising the calcaire grossier. 3. Second fresh-water formation; containing silicious limestone, gypsum, with bones of animals, and fresh-water marls. **4. Second marine** formation; consisting of gypseous marine marls, upper marine sands and sandstones, and upper marine marls and limestones. 5. Third fresh-water formation; containing silicious millstone with shells, and upper fresh water marls. Subsequent observations have proved that this division, as well as many of the views entertained by MM. Brongniart and Cuvier, is not in accordance with facts with which they were unacquainted, and much modification of the above arrangement has been the consequence. The silicious limestone, with fresh water and terrestrial shells and plants, and the calcaire grossier, or first marine formation, often alternate, and are deemed by M. Constant. Prevost to be contemporaneous formations; and it is not improbable that while the waters in one lake or basin might be marine, those in another might be fresh, and thus two formations containing different organic remains might be deposited contemporaneously. However the rocks of this group may be eventually discovered to differ from this type, the labours of MM. Cuvier and Brongniart on the rocks of the Paris basin will not the less retain that place in the annals of geology, which by common consent has been assigned them. Nor will the zoological discoveries of Cuvier, constituting as they did such a brilliant epoch in the history of geological science, the less claim the gratitude of geologists in succeeding ages.—*De la Beche. Lyell. Bakewell.*

PA′RTICLES ELEME′NTARY. The final results of chemical analysis. Elementary particles are those of which integrant particles are composed; thus, while the latter remain invariable in the same body, the former must vary with the progress of chemistry. In bodies really simple, the integrant and elementary particles must be the same.—*Cleaveland.*

PA′RTICLES I′NTEGRANT. These are the smallest particles into which a body can be reduced without destroying its nature, or, in other words, without decomposing it. Only three forms of integrant particles have hitherto been discovered. They are the three most simple, geometrical solids; namely, a tetraëdron; a triangular prism; and a parallelopiped, including all solids of six sides, parallel two and two.

PAR′TICLES OF MATTER. The indefinitely small or ultimate atoms into which matter is believed to be divisible. Their form is unknown; but though too small to be visible, they must have magnitude.—*Mrs. Somerville.*

PA′RTITE. *(partitus,* Lat.) Divided. In botany, a partite leaf is one separated to the base.

PA′SSERES. The third order of the class Aves, or birds. The thrush, warbler, sparrow, swallow, crow, and lark, are comprised in this order.

PATE′LLA. *(patélla,* Lat. a little deep dish with a broad brim.) That bone of the leg commonly known as the knee-pan.

PATE′LLA. In conchology, the limpet shell. Animal a limax. A marine shell, univalve, subconic, shaped like a basin; without a spire. The patella is found both recent and fossil. Many species have been found fossil in the neighbourhood of Paris. Patellæ have also been obtained from the Shanklin sand, and from the Harwich cliffs.

PATE′LLIFORM. (from *patella,* a dish, and *forma,* form, Lat.) Of the form of a small dish.

PATE′LLITE. A fossil patella.

PA′TULOUS. *(patulus,* from *pateo,* Lat. to be open.) In botany, spreading, as a patulous calyx. In conchology, gaping; with a spreading aperture.

PAVO′NIA. (from *pavo,* Lat. a peacock.) "A genus of stony polypifer, fixed and frondescent; the lobes flat, subfoliaceous, erect or ascending, with stelliferous rugæ or grooves on each side. The stars lamellated, in rows, sessile, and rather imperfect." — *Parkinson.* There are several known species. A coral with a deep and isolated cell, such cell containing a large depressed polypus, very similar to the actinia, as regards both its structure and appearance.

PEA ORE. } The Pisiform iron-ore
PEA IRON ORE. } of Kirwan. Masses of *pea iron ore* are found on the sea beach near Scarborough. This species of ore, which is met with in Germany, France, Switzerland, &c., yields from thirty to forty per cent. of iron. Its colour is bluish-

grey, and the grains or globules of which the mass is composed, vary from the size of mustard seed to that of large peas.

PEA GRIT. A member of the inferior oolite; a pisolitic limestone, made up of flattened oval concretions rather larger than peas; it occurs near Cheltenham, and attains a thickness of forty feet.

PEACH. A provincial term for either chlorite or mica, if the latter have a tinge of green.

PEAR E'NCRINITE. The Apiocrinites rotundus, or Bradford encrinite. A species of crinoïdea abounding in the oolitic limestone in the neighbourhood of Bradford, near Bath. When living, their roots were confluent, and formed a thick pavement over the bottom of the sea, from which their stems and branches rose into a thick submarine forest, composed of these splendid zoophytes. This bed of beautiful remains has been buried by a thick stratum of clay. The body of the pear encrinite was of a pyriform shape, from which circumstance it has been thus named. The pear encrinite is confined to the middle oolite.

PEARL. *(perle*, Germ. *perle*, Fr. *pérla*, It.) A spherical concretion consisting of concentric coats of the same substance as that which forms the mother-of-pearl of the shell. It is produced by the extravasation of a lapidifying fluid, secreted in the organs of the animal, the pearl oyster, and filtered by its glands. The animal that produces pearls in the greatest abundance, of the purest nature, and of the highest value, has been formed by Lamarck into a genus named Meleagrina; Linné classed it with the muscles. It inhabits the Persian gulf, the coast of Ceylon, &c. It attains perfection no where but in the equatorial seas. The pearl fishery off the island of Ceylon is the most productive of any; the oyster-beds extending over a space thirty miles long by twenty-four broad. The oysters at the greatest depths yield the largest pearls, which are situated in the fleshy part, near the hinge. For one pearl that is found perfectly round and detached between the membranes of the mantle, hundreds of irregular ones occur attached to the mother-of-pearl; these are sometimes in such numbers as to prevent the animal from closing its valves, and thereby cause its destruction.

The pearl is supposed by some writers to be the effect of disease; it is a formation forced upon the oyster by some extraneous substance within the shell, which it covers with mother-of-pearl. Sir Everard Home considered that the abortive eggs of the animal were the nuclei upon which the pearls were formed.

To collect the pearl oysters, divers are employed; these men, provided with baskets, descend to the bed at the bottom of the sea, and during their stay there, which does not exceed two minutes, generally a minute and a-half, collect into their baskets every thing they can grasp, when they are rapidly, at a signal given, hauled up to the surface. When the bed is richly stored, a diver will collect 150 oysters at one dip, and a single diver will, in one day, bring up from 1000 to 4000 oysters.—*Kirby*.

PEARL SI'NTER. Called also fiorite. A variety of silicious sinter of a white or grey colour, found in volcanic tuff.

PEA'RLSTONE. An igneous or volcanic rock with a mother-of-pearl lustre. The Perlstein of Werner; Obsidienne perlée of Brongniart; Lave vitreuse perlée of Haüy. Pearlstone is a variety of obsidian, occurring in globular and concentric lamellar, irridescent, translucent concretions. It scratches glass. Specific gravity from 2·20 to 2·55. When breathed upon, it frequently gives out an argillaceous odour. Its constituents are silica 77·0, alumina 13·0, lime and natron 2·6, potash 1·4, oxides of manganese and iron 2·0, water 4·0.

PEA STONE. A variety of limestone, called also pisolite.

PEAT. (derived by some from the German word *pfutze*, a pool, or standing water.) An intermediate substance between simple vegetable matter and lignite, the conversion of peat into lignite being gradual, and brought about by the action of water. This substance, arising sometimes from the subversion of forests covered by the sphagnum palustre, and other mosses, and sometimes from the growth of various maritime and semi-maritime plants on the marshes bordering the coasts, is found among the most modern alluvia, generally covering them; often containing works of human art imbedded, and in many instances still in the act of progressive increase. Peat is composed of the remains of many different plants, but generally a great portion is derived from the Sphagnum palustre, and the process by which these vegetables are thus converted is clearly seen in the sphagnum palustre. As the lower extremity of the plant dies, the upper sends forth fresh roots, thus furnishing a perpetual supply of decomposing vegetable matter. Dr. Maculloch states, "where the living plant is still in contact with the peat, the roots of the rushes, and ligneous vegetables, are found vacillating between life and death, in a spongy half decomposed mass. Lower down, the pulverized carbonaceous matter is soon mixed with similar fibres, still resisting decomposition. These gradually disappear, and at length a finely-powdered substance alone is found, the process being completed by the total destruction of all the organised bodies. The generation of peat, when not completely under water, is confined to moist situations, where the temperature is low, and where vegetables may decompose without putrifying." While in the upper portions of a peat bed we may perceive the fibres of the vegetables, whence it has originated, in an almost unchanged state; in the middle portions the texture presents itself gradually obliterated, and the mass is found passing into a compact peat; in the lowest part this change is carried still farther, and substances analogous to jet are met with: in some instances beds of peat alternate with beds of mud or sand deposited in lakes, or of silt and sand formed in the estuaries of rivers; in these cases they represent an imperfect and unmatured coal formation.

Sir H. Davy states that one hundred parts of dry peat contain from sixty to ninety-nine parts of matter destructible by fire. One-tenth of the whole of the surface of Ireland is stated to be peat. At the bottom of peat-mosses there is occasionally found a cake or pan of oxide of iron; whence this is derived does not appear to be clearly understood. The preservative property of peat is very remarkable; bodies of persons who have perished in peat-bogs have been kept free from putrefaction for many years.

PECOP'TERIS. A genus of fossil terrestrial plants found in the coal measures.

PE'CTEN. (*pecten*, Lat. a comb.) A genus of marine bivalves, belonging to the family Ostracea. The pecten is a fossil as well as a recent shell, many species being found in our seas. It is a regular, eared, longitudinally ribbed, inequivalved bivalve, with contiguous beaks, having a triangular auricle on each side of the umbones. Hinge toothless; pit trigonal. One muscular impression. Fossil pectens are found in the neighbourhood of Paris, and in many parts of England; in the Harwich cliff; in the green-sand of Wiltshire; near Thame in Oxfordshire; in Gloucestershire; and in Sussex.

PEBBLE. } A roundish stone of
PEBBLE STONE. } any kind from the size of a small nut to that of a man's head. A general term for water-worn minerals. A mineral distinguished from a flint by its variety of colours.

PE'CTINATE. } (from *pecten*, a comb,
PE'CTINATED. } Lat.)
1. In conchology, resembling a comb; cut into regular, straight segments like a comb.
2. In botany, applied to a pinnatifid leaf, whose segments are extremely narrow, resembling the teeth of a comb.

PECTINIBRANCHIA'TA. The sixth order of mollusca, in Cuvier's arrangement.

PE'CTOLITE. A mineral, a silicate of lime and soda.

PECTU'NCULUS. A genus of orbicular sub-equilateral marine bivalves, with an arched hinge; numerous teeth, alternately inserted in a single row; of the family Arcadidæ. Pectunculi are easily recognized by their rounded or lenticular form. Muscular impressions two, and strongly marked. Recent pectunculi are found at depths varying from five to seventeen fathoms, in sandy mud and sands. Fossil pectunculi are met with in the London clay and calcaire grossier, and in the Bognor sandstone.

PE'DATE. (*pedatus*, Lat. from *pes*, a foot.) In botany, applied to leaves in which a bifid petiole connects several leaflets on the inside only; also to a peculiar kind of ternate leaf, its lateral leaflet being compounded in the fore part: the black hellebore is an example.

PE'DICEL. In botany, a partial flower-stalk, or a subdivision of the general one, each subdivision being termed a pedicel.

PEDIPA'LPI. The second family of Arachnidans. They have very large palpi terminated by a forceps or claw. The principal animals among the pedipalps are the scorpions, possessing powerful organs for seizing their prey, and having a tail terminating in a deadly sting. The other pedipalps are not armed with a sting.

PE'DUM. (*pedum*, Lat. a shepherd's crook.) A genus of marine bivalvular shells, found attached by a byssus to rocks. It is an eared inequivalved bivalve, gaping at the lower valve, and having its beaks separated: hinge toothless; ligament exterior; inferior notch grooved.

PE'DUNCLE. (*pedunculus*, Lat. from *pedo*, a splay foot.)
1. In botany, the stalk that bears the flower and fruit.
2. In conchology, a sort of stem by which the shells of the second division of lepas are attached to wood, &c.

PEDU'NCULATE. } Attached to objects
PEDU'NCULATED. } such as wood, rocks, &c., by a peduncle; having a peduncle.

PE'GMATITE. A name given by the French mineralogists to a variety of granite composed of granular quartz and felspar. When in granite, the mica disappears, and only quartz and feldspar are left, we have Pegmatite, or Graphic Granite.

PELA'GIAN. } (*pelagus*, Lat. the sea.)
PELA'GIC. } Belonging to the sea. Lyell says, "belonging to the *deep* sea."

PELA'GIAN FORMATIONS. Oceanic accumulations; deposits by currents, or from other causes, at the bottom of the sea.

PEL'DON. We can sometimes trace sandstone becoming finer and finer, till we arrive at a rock for which we have no very accurate or distinctive name,—a flinty or silicious rock, with a perfectly smooth compact texture, a conchoidal fracture, without anything we could call a grain. A rock of this kind, when met with in the coal-measure of S. Staffordshire, is there called a *peldon*, and I venture to propose that designation for the acceptance of my brother geologists.—*Jukes*.

PE'LIOM. (from πελίωμα, Gr. blueness, or of a lead colour.) A blue coloured mineral resembling iolite, of which it is a variety. It is found in Bavaria.

PE'LLICLE. (from *pellicula*, Lat.) A film; a thin crust or covering.

In botany, a membranous or mucilaginous covering, closely adhering to the outside of some seeds, so as to conceal their proper surface and colour.

PELTA'TE. (from *pelta*, Lat. a target.) In botany, a term applied to leaves which have their footstalk inserted in the middle of the leaf, and not joined to the edge; the nasturtion is a familiar example.

PE'LVIS. (*pelvis*, Lat. from πέλυς, Gr. a basin.) The lower part of the trunk of vertebrated animals.

PE'NCIL. A name given to the belemnite.

PE'NNATE. } (*pennatus*, Lat. from *penna*, a wing.) Winged; feathered.
PE'NNATED. }

PENNA'TULA. Called, commonly, the Seapen. A polypus with a calcareous axis or stem, having a double set of branches extending in the same plane from both sides, like the vane of a quill. Pennatulæ are not fixed by any attachment to the ground, but float about in the waters of the ocean, carried hither and thither as the current may direct them.

PE'NNIFORM. (from *penna*, a feather, and *form*.) Having the form of a feather or quill. In anatomy, muscles in which the muscular fibres pass obliquely outwards on either side from a tendinous centre, are termed penniform. The rectus femoris affords an illustration of a penniform muscle.

PENTACA'PSULAR. (from πέντε, Gr. five, and *capsular*, Lat. a cell or capsule.) Having five cavities, capsules, or cells.

PENTA'CEROS RETICULA'TUS. A fossil species of asteria, or stella marina.

PENTA'CEROS DENTIGINO'SUS. A fossil species of asteria or stella marina, found in the chalk and in the London clay.

PENTACRI'NEA. A tribe of stone-lilies, comprising the living genus Pentacrinus (including Chladocrinus and Extracrinus) and the extinct genus Isocrinus.

PENTA'CRINITE. The fossil Pentacrinus.

PENTACRI'NUS. (from πέντε, five, and encrinite.) The five angled lily-shaped animal. A genus of the family Crinoïdea. Miller thus describes the generic character of pentacrinus. "An animal with a column formed of numerous pentangular joints, articulating by surfaces with pentapetalous semistriated markings. Superior columnar joint supporting a pelvis of five joints, on which five first costals rest, succeeded by five second costals and five scapulæ, from which ten arms proceed, having each two hands, composed of several tentaculated fingers. Column long, having numerous auxiliary side arms. Base unascertained." The arms, when expanded, resemble a star of five (or six) rays; and when they converge, a pentapelous or hexapetalous flower. The whole animal, when alive, is supposed to be invested with a gelatinous muscular integument. From the circumstance of pentacrinites abounding in the lias and lower strata of the oolite, and disappearing entirely in the uppermost strata, geologists were disposed to limit their existence to certain periods, and to conclude that the pentacrinite furnished the remains of an extinct genus. Subsequent discoveries, however, prove that the pentacrinus does still exist, and thus, probably, it may be with many genera which, in ignorance, geologists describe to be extinct, merely because they have not met with living or recent specimens. The calcareous joints which compose the fingers of the Pentacrinus Europæus, are capable of expansion and contraction in all directions; now spreading outwards, like the petals of an expanded flower, and again rolled inwards towards the mouth in the form of a closed bud. These organs serve to seize, and convey to the mouth, the food of the animal. The number of bones in each animal is computed at thirty thousand. Dr. Buckland says that the number of bones in the Briarean Pentacrinite exceeded a hundred and fifty thousand. The number of bones in the fingers and tentacula amount at least to a hundred thousand, and fifty thousand more, (which is considerably under the real number), may be added for the ossicula of the side arms. Each bone requiring at least two fasciculi of fibres, one for expansion, the other for contraction, the Briarean Pentacrinus must have had three hundred thousand fasciculi of fibres equivalent to muscles.

"The pentacrinites differ considerably," says Mantell, "in their form and structure from the encrinites, the name of encrinite being given to the species in which the bones of the column are circular or elliptical, that of pentacrinite to those which have an angular or pentagonal stem. The lias of Lyme Regis abounds in the remains of these animals; and large slabs often have the whole surface covered with the plumose tentacula of pentacrinites, converted into pyrites."

"There are instances," says Miller, "of the column of this genus having sometimes a tetragonal, or hexagonal form, these I consider as monstrous varieties."

PENTADA'CTYL. (from πέντε, five, and δάκτυλος, a finger, Gr.) Having five fingers. Applied also to leaves.

PE'NTAGON. (from πέντε, five, and γωνία, an angle, Gr.) A figure having five sides and five angles.

PENTA'GONAL. Having five angles and five sides; quinquangular.

PENTAGONA'STER SE'MILUNATUS. A fossil species of the stella marina.

PENTAGONA'STER REGULA'RIS. A fossil asterite, or species of the stella marina.

PE'NTAGYN. (from πέντε, five, and γυνή, a woman, Gr.) A plant which has five pistils.

PENTAHE'DRAL. Having five equal sides.

PENTAHE'DRON. (from πέντε, five, and ἕδρα, a base, Gr.) A figure of five equal sides.

PENTA'MERUS LIM'ESTONE. The name assigned to a rock of the Helderberg group. The Upper Pentamerus beds, or May Hill sandstone, or Upper Llandovery beds, attain a thickness of 1000 feet, being a formation of the Upper Silurian period.

PENTA'MERUS. A genus of fossil Brachiopodous shells, described as equilateral, inequivalve, one valve being divided by a central septum into two parts, the other by two septa into three parts; the umbones are incurved and imperforate; five species of Pentamerus are found in the Silurian system of rocks in England and Wales.—*Lycett.*

PENTA'NDER. (from πέντε, five, and

ἀνήρ, a man, Gr.) A plant having five stamens.

**PENTA'NDRIAN.** Having five stamens.

**PENTAPHYLLOI'DAL.** (from πέντε, five, φύλλον, a leaf, and εἶδος, resemblance, Gr.) Appearing to have five leaves; resembling five leaves. The Placentæ are all ornamented with a pentaphylloidal flower.

**PENTELA'SMIS.** (from πέντε, five, and ἔλασμα, a plate, or layer, Gr.) A species of Anatifa, or of the Lepas of Linnæus.

**PE'NTREMITES.** An extinct genus of fossil encrinites, established by Say, of the order Blastoidea.

**PEPERI'NO.** The name given by Italian geologists to a particular form of volcanic tuff, composed of basaltic scoriæ.

**PERA'MELES.** A genus of marsupiala. The following description is from Col. Mitchell's Australia. "The most remarkable incident of this day's journey was the discovery of an animal, of which I had seen only the head amongst the fossil specimens of Wellington valley. This animal was of the size of a wild young rabbit, and of nearly the same colour, having a broad head, terminating in a long and very slender snout, like the narrow neck of a wide bottle; and no tail. The feet, and especially the fore legs, were singularly formed, the latter resembling those of a hog, and the marsupial opening was downwards, and not upwards, as in the kangaroo and others of that class of animals."

**PERE'NNIAL.** (perennis, Lat. perénne, It.) In botany, applied to plants that live many years, bearing flowers and fruit frequently.

**PERFO'LIATE.** In botany, applied to leaves when the stem appears to pass through their substance. The common hare's-ear is an example.

**PE'RIANTH.** (from περί, about, and ἄνθος, a flower, Gr.) The calyx is so called when it is united with the corolla, so as to form only one floral envelope.

**PERICA'RDIUM.** (from περί, round, and καρδία, the heart, Gr. péricarde, Fr. pericárdio, It.) The membrane which envelopes the heart.

**PE'RICARP.** (from περί, round, and καρπὸς, fruit, Gr. péricarpe, Fr.) The ovarium, when ripened into fruit, is called the pericarp; this consists of three parts, which in some fruits, as the peach and plum, are easily separable. The outer skin is called epicarp; the fleshy part, the sarocarp; the stone, or shell, the endocarp. There are ten different kinds of pericarps, namely, drupe, pome, berry, follicle, silique, silicle, legume, capsule, nut, and strobile.

**PE'RICHLIN.** A feldspathic mineral, an albite, in which part of the soda has been replaced by potash.

**PERICLI'NIUM.** The name given by foreign botanists to a kind of involucre.

**PE'RIDOT.** The name given by Haüy to prismatic chrysolite.

**PERIGE'E.** (from περί, round, and γῆ, the earth, Gr. périgée, Fr.) A term used to denote that point where the sun is nearest to the earth. The perigee of the lunar orbit is the point where the moon is nearest to the earth.

**PERI'GYNOUS.** (from περί, about, and γυνή, a woman, Gr.) Inserted around the pistil. When the stamens grow out of the corolla, calyx, or perianth, or are not in any way joined to the seed-vessel, they are said to be perigynous.

**PERIHE'LION.** The point of an orbit nearest to the sun.

**PERISSODA'CTYLA.** Odd-toed; having one or three toes. This order of mammalia is divided into two families, namely, Solipedia, comprising the horse, and Pachydermata.

**PERIO'STEUM.** (from περί, around, and ὀστεὸν, a bone, Gr. perioste, Fr.)

The membrane which covers the bones; that, however, which covers the bones of the skull is called the pericranium.

PERIPHORA'NTHIUM. When bracts are collected into a whorl, as in umbelliferous plants, they are said to form an involucre, which, if very small, receives the diminutive name of involucel. This kind of organ is very remarkable in compound-flowered plants, appearing as if it constituted a calyx common to many flowers; and hence it used to be called a common calyx. It, however, does not differ from the involucre in any thing more than its bracts being more numerous, more closely packed, and parallel with each other, instead of diverging. Foreign botanists have given the names *Periphoranthium* and *Periclinium* to this kind of involucre.

PER'MEABLE. *(permeo,* Lat. from *per* through, and *meo* to pass or glide; *perméable,* Fr.) That may be passed through without rupture or displacement of its parts. Applied to substances that admit the passage of fluids. All *permeable* strata receive rain-water at their surface, whence it descends until it is arrested by an impermeable subjacent bed of clay, causing it to accumulate, and to form extensive reservoirs, the overflowing of which on the sides of valleys constitute the ordinary supply of springs and rivers.

PER'MIAN. The newest of the primary or palæozoic formations, consisting of the following deposits in ascending order. 1. Red sandstone and marl. 2. Conglomerates and breccia. 3. Lower red sandstone and marl. 4. Magnesian limestone. The term Permian is derived from the district of Perm in Russia, where the formation in question is very extensively developed.—*Lycett.*

PE'RNA. A genus established by Bruguiéres. A flat, irregular, subequivalve, compressed, foliaceous, marine, bivalve. Several parallel cavities across the hinge opposed to each other in the two valves, and lodging as many elastic ligaments: anterior margin with a passage for a byssus. Recent and fossil.

PE'ROXIDE. When oxygen is combined with any substance, in the highest proportion it is capable of, whatever that may be, it is called a *peroxide*.

PERSI'STENT. (from *persisto,* Lat. to abide.) In botany, opposed to deciduous. Not withering and falling; remaining. 2. Unbroken; continuing. In almost every formation the individual strata are rarely *persistent* for a great distance.

PE'RSONATE. A term applied to a monopetalous flower of an irregular form, the border of the corolla having an oral appearance with the labia closed.

PE'TAL. ($\pi\epsilon\tau\alpha\lambda o\nu$, Gr. *pétale,* Fr.) The name given to each leaf of the corolla, or flower of a plant.

PE'TALITE. (from $\pi\epsilon\tau\alpha\omega$ and $\lambda\iota\theta o s$, Gr.) A mineral, of a reddish, or greyish-white colour, which has only been found in Sweden. It consists of silica 77·0, alumina 17·0, lithia 6·0; or according to others, of silica 76·21, alumina 17·22, lithia 5·76. It occurs in a mine of black iron ore, associated with spodumene, felspar, tourmaline, mica, and quartz. Its fracture is foliated; scratches glass; brittle, and translucent at the edges. Specific gravity 2·62.

PETALO'IDEÆ. In botany, the second order of the class Endogens, comprising the banana, orchis, palms, lilies, &c.

PE'TIOLE. (from *petiolus,* Lat. the stalk of fruits.) In botany, a foot stalk; the stalk, or stem, which connects the leaf with the branch.

PETR'AIA. A genus of fossils found in the Silurian formation, of which many species have been described.

PETRI'COLA. (from πέτρος, a stone, Gr. and *colo*, to inhabit, Lat.) A genus of transverse, inequilateral, bivalve, marine shells, belonging to the family Lithophagi. Two muscular impressions; two hinge-teeth on one valve, and a bifid one on the other; anterior side rounded, posterior side more or less slightly gaping; ligament external. Petricolæ are found at depths varying to ten fathoms; they inhabit cavities, of their own working, in rocks and shells. They may be distinguished from saxicava by the regularity of their form, and by the teeth on the hinge, which in saxicava become obsolete when the animal is full grown.

PETRIFA'CTION. *(pétrifaction,* Fr.)
1. A substance converted into stone. Substances, either animal or vegetable, converted into stone by the infiltration, or incrustation, of silicious matter.
2. That process by which organic remains are mineralized, and their external form, and more or less of their internal structure, preserved.

PETRE'SCENT. (from *petrescens*, Lat.) Becoming stone; growing into stony matter.

PE'TRIFIED NAIL. A local name given to a bone of the Asterolepis, a gigantic ganoid of the old red sandstone. This nail-like bone formed a characteristic portion of the Asterolepis.—*Hugh Miller.*

PETRO'LEUM. *(pétrôle,* Fr.) A mineral oil, rather thicker than tar, and of a reddish-brown colour; it has obtained its name from the circumstance of its oozing out of rocks like oil. In the East it is burnt as oil. It is unctuous to the touch, and exhales a strong and unpleasant odour. It is lighter than water, its specific gravity being 0·87. In the island of Zante, petroleum is at the present time obtained from the same spot, and in the same manner, as in the days of Herodotus. The most powerful springs producing petroleum are on the Irawaddi, in the Burman empire. In one locality there are said to be 520 wells, yielding annually 400,000 hogsheads of petroleum. It occurs in most countries where coal is found.

In describing the coal-field of Coal Brook Dale, Sir R. Murchison says, "Petroleum is of constant occurrence in the upper as well as lower measures; the chief source of this mineral at Coal Port, which formerly afforded one hogshead a-day, being in a thick-bedded sandstone of the upper measures. This supply has, however, much decreased with the opening of new coal works. Other wells have been discovered in the lower coals at Prior's Lee. In some pits, the petroleum exudes in such quantities that the works are necessarily boarded up to prevent its infiltration upon the workmen."

PETRO'LOGY. The study of rock masses; or the examination of those characters, structures, and accidents of rocks which can only be studied on the large scale, and only be observed in the field.—*Jukes.*

PETRO-PHILOIDES. The name assigned by Mr. Bowerbank to a genus of fossil fruits found in the London clay: he thus describes the generic characters. "Fruit, a strobilus. Squamæ usually confluent, rarely separate. Seed bilobate, without a comose or foliaceous appendage." Mr. Bowerbank describes seven species, and in speaking of the genus he says, alluding to their approximation to Petrophila, "I have thought it better to call them Petro-philoides, rather than designate them by any other name which would not serve as an indication of their nearest affinity with genera existing at the present period."

PETROSI'LEX. A fusible variety of hornstone; according to some authors, the same as clinkstone; to others, compact felspar.

PETU'NSE. ⎫
PETU'NTZE. ⎬ The **felspathe** pétuntzé of Brongniart. A variety of felspar, used in the manufacture of porcelain. It is called Petuntze by the Chinese.
PETU'NZE. ⎭

PE'TWORTH MA'RBLE. Called also Sussex marble, occurs in layers varying **from a** few inches to a foot and upwards in thickness, separated from each other by seams of clay. The Petworth or Sussex marble is a limestone of various shades of colour, occurring in the Weald clay; it is composed of the remains of fresh-water univalves, shells of the paludina, and crusts of the cypris faba, united into a compact marble by a gangue of calcareous cement. The more compact varieties bear a beautiful polish, and are elegantly marked, when cut into slabs, by the section **of the** contained shells.

PHA'COPS. A genus of fossils found in the Silurian rocks, of which **there** are many species.

PHANEROGA'MIC. ⎫
PHANEROGA'MOUS. ⎬ (from φανερὸς, manifest, and γάμος, marriage, Gr.) Plants in which the stamens and ovarium are distinctly **visible**; plants having the reproductive **organs** visible. In all the *phanerogamous* plants, the whole of the double apparatus required for reproduction is contained in the flower. The term is used in contra-distinction **to** *cryptogamous*.

PHARMA'COLITE. The name given by Brochant to arseniate of lime.

PHASCOLOTHE'RE. ⎫ An extinct genus
PHACOLOTHE'RIUM. ⎬ of entomophagous marsupials, discovered in the Stonesfield slate. I am indebted to my friend, Mr. Lycett, of Minchinhampton, for the following observations:—In remarking upon the little Marsupial Mammal *Phascolotherium Bucklandi*, portions of which have occurred in the great oolite of Stonesfield, Oxon, Hugh Miller speculates upon the possibility of there having existed along with them extinct species of Thylacinas and Dasyurus, (two recent genera of Australian carnivorous Marsupials) being led to such a train of thought by the analogy which the fauna and flora of recent Australia presents to those of the ancient oolitic formations, but no fossil Thylacinus or Dasyurus has been found in Britain. Bones of extinct species of Dasyurus, Macropus, Phascolomys, Phalangista, Diprotodon, (N. G. Owen) and Hypsiprymnus were collected in the bone caves of Wellington Valley, Australia, by the late Sir Thomas Mitchell.

PHASIANE'LLA. (from *phasianus*, Lat. a pheasant.) A genus of shells belonging to the family Trochoida; or, according to Lamarck's arrangement, Turbinacea. It is a solid ovate or conical univalve; opening longitudinal, ovate, and entire; lip thin; columella smooth, with an attenuated base. Recent phasianellæ are inhabitants of the Indian ocean; they are found on the coasts **and in** estuaries, at small depths only.

PHI'LLIPSITE. A mineral, found accompanying Herschelite. It is a species of Harmotome or crossstone, containing lime and potash instead of baryta.

PHO'LAS. (φωλὰς, from φωλέω, Gr. to lie concealed.) A genus of marine bivalves, belonging to the family Inclusa; or, according **to** Lamarck's arrangement, the family Pholadaria. A transverse gaping shell, composed of two principal valves, with several small accessory pieces placed on the ligament, or at the hinge. A long curved tooth protrudes in each valve from beneath the umbones. Pholades are found at depths varying to nine fathoms; they pierce wood, rocks, indurated clay, &c. They are much sought for in consequence of their delicious flavour.

PHO'LADITE. A fossil or petrified pholas.

PHO'NOLITE. (from φονη, sound, and λίθος, a stone, Gr.) Another name for clinkstone. A felspathic rock, sonorous when struck with a hammer, from which circumstance it derives its name.

PHO'SPHATE. A salt formed by the union of phosphoric acid with a salifiable base.

PHO'SPHATE OF LIME. This is found in the bones of animals, and constitutes their base; as well as in the mineral kingdom. It consists of lime 59·0, phosphoric acid 41·0. It is destitute of taste, insoluble in water, and not affected by exposure to the atmosphere. Exposed to a very high temperature it becomes soft, and is converted into a white semitransparent enamel, or rather porcelain. Human bones, according to Berzelius, contain 51·04 of phosphate of lime, and the enamel of teeth, according to Mr. Pepys, is composed of 78 per cent. of it. Sulphuric, nitric, muriatic, fluoric, and several vegetable acids are capable of more or less decomposing phosphate of lime.

PHOSPHORE'SCENCE OF THE SEA. This luminous appearance of sea-water arises from the presence of immense numbers of microscopic medusæ which people every region of the ocean, and, being specifically lighter than the sea-water, float in incalculable multitudes on its surface.

PHOSPHORE'SCENT. Emitting light in the dark without sensible heat.

PHO'SPHORITE. Amorphous phospate of lime. The chaux phosphatée terreuse of Haüy; phosphorit of Werner; phosphorite of Jameson. A variety of apatite, with commonly an earthy aspect: it occurs in masses whose surface often displays mammillary projections. Fracture dull and earthy. Colour white or grey, often marked with spots or zones of a brownish tinge. Specific gravity from 2·8 to 3·2. Before the blow-pipe it is infusible, but its powder thrown upon live coals emits a yellowish-green phosphorescent light. In Spain it forms whole mountains; it is also met with in Germany. According to Pelletier, it contains lime 59, phosphoric acid 34, and the remainder consists of the carbonic, fluoric, and muriatic acids, with a trifling portion of silex and oxide of iron.

PHO'SPHORUS. (φωσφόρυς, Gr. phosphorus, Lat. phosphore, Fr.) One of the simple or elementary substances, and belonging to that subdivision termed non-metallic. Phosphorus is principally known as entering into the chemical composition of animals. As a component part of minerals, phosphorus is rare; but there must be some amount of it entombed in fossiliferous rocks. Phosphorus is never found pure in nature, and is only to be obtained from organic matter by elaborate chemical processes. It is yellow, and semi-transparent; resembling wax in softness, but more cohesive and ductile. Specific gravity 1·77. Its affinity for oxygen is so great that it burns spontaneously in the atmosphere. It should always be kept in bottles filled with water, and well corked.

PHOTO'METER. (from φῶς, gen. φωτὸς, and μετρον, measure, Gr.) An instrument for measuring the intensity of light.

PHYLLA'DE. The name given by D'Aubuisson, and the French geologists, to clay-slate. Under the name of Phyllade, Dr. Boase includes a genus of rocks comprising four species. Phyllade is largely found in Cornwall and Devon. The rocks comprised in this genus consist in great measure of very thick beds, which are closely lamellar, and even compact and jointed, like the felspar rocks; indeed, they seem to be principally composed of felspar, the most compact and silicious becoming weathered on exposure to the atmosphere. Phyllade differs from corneanite in having a granular instead of a compact basis, and it always contains minute spangles, like scales of mica.

PHY'SALITE. A variety of prismatic topaz, of a greenish-white colour. It occurs in coarse granular concretions, having a low degree of lustre. Edges feebly translucent. It consists of alumina 57·75, silica 34·30, fluoric acid 7·81.

PHY'LLITE. (from φύλλον, a leaf, and λίθος, a stone, Gr.) A petrified leaf.

PHYLLO'PHAGOUS. An animal that feeds on leaves; the silk-worm is an example.

PHYLLOLE'PIS. The name assigned to a genus of ichthyolites of the old red sandstone.

PHY'SICAL. *(physique,* Fr. *fisico,* It.) Relating to nature or to natural philosophy; not moral; pertaining to material things.

PHY'SICS. (from φύσις, nature, Gr.) Taken in its most enlarged sense, comprehends the whole study of nature; but in the usual acceptation of the word, that branch of science which treats of the properties of natural bodies, and includes natural history and philosophy.

PHYTI'VOROUS. (from φυτὸν, Gr. a plant, and *voro,* Lat. to devour.) Feeding on plants.

PHY'TOLITE. (from φυτὸν, a plant, and λίθος, a stone, Gr.) A petrified or fossilized plant.

PHYTO'LOGY. (from φυτὸν, a plant, and λόγος, discourse, Gr.) That department of science which treats of the nature, habits, qualities, &c., of plants.

PHYTO'PHAGOUS. (from φυτὸν, a plant, and φαγεῖν, to eat, Gr.) Feeding on plants; devouring plants; feeding on vegetable substances.

PHYTOSAU'RUS. A fossil saurian discovered in the saliferous formation, and thus named by Jaeger.

PHYTOZO'A. (from φυτὸν, a plant, and ζῶον, an animal, Gr.) Plant-like animals. Another, and more modern, as well as appropriate name for zoophytes.

PI'CROSMINE. A silicate of magnesia, having nearly the same composition as serpentine.

PIGME'NTUM NI'GRUM. A dark brown, or nearly black substance, which covers the surfaces of the choroid membrane of the eye, and gives to it its colour.

PI'LEUS. The name given to a genus of fossil echini; pileus is another name for conulus.

PI'LLAR. In conchology, the columella, or perpendicular centre, which extends from the base to the apex, in most of the spiral shells.

PI'LLAR-LIP. In conchology, a continuation of the glossy process with which the aperture of shells is lined, expanded on the columella.

PI'LOSE. *(pilosus,* Lat.) In entomology, covered with dispersed, long, and bent hairs.

PI'MELITE. A variety of steatite, coloured by chrome or nickel.

PI'NEAL GLAND. (from *pineus,* Lat. a pine.) The name given to a gland of the brain from its supposed resemblance to a pine. This gland was at one time supposed to be the seat of the soul.

PI'NNA. *(pinna,* Lat. the fin of a fish.) A genus of marine bivalves, belonging to the family Mytilacea. A cuneiform, longitudinal bivalve, with an acute base, the upper part gaping; hinge without a tooth, lateral, and very long; valves coalescent. Recent pinnæ are found in the ocean at depths varying to seventeen fathoms; they are moored by a long silky byssus, which has been manufactured into stockings and gloves. Pinnæ have been found fossil in the limestone of Gloucestershire, Wiltshire, and Somersetshire.

PI'NNIGRADA. A division of the order Carnivora, of which the Seal and Morse are examples.

PI'NITE. A silicate having one equivalent of silicate of potash, or protoxide of iron, to one of bisilicate of alumina, and also containing water. Sp. gr. 2·8. Slightly translucent; colourless before the blowpipe, and fuses at the edges to blistered glass.

PI'NNITE. A fossil pinna.

PI'NNATE. } *(pinnatus,* Lat. winged.)
PI'NNATED. } In botany, applied to compound leaves, composed of many leaflets, placed on each side of the petiole; these are placed in pairs opposite to each other, and in some cases, an odd leaflet at the termination, or, in others, a tendril: of the former the rose furnishes an example; of the latter, the vetch.

PINNA'TIFID. In botany, applied to leaves cut transversely into several deep, oblong, parallel segments, the incisions reaching nearly to the midrib, and dividing the leaf into irregular forms, termed lobes. The groundsel affords a familiar illustration.

PIPE-CLAY. The Pfeifenthon of Werner. This is the purest kind of potter's clay, and is called pipeclay, from its being manufactured into tobacco pipes. It is of a grey or greyish white colour; is infusible; and on exposure to a strong heat becomes white. It is abundant in Devonshire and Staffordshire; a stratum, in a horizontal position above the chalk, extends from Headfast Point to beyond Corfe Castle in Dorsetshire. It forms the basis of the Queen's ware pottery.

PISCI'VOROUS. (from *piscis,* a fish, and *voro,* to devour, Lat.) Feeding on fishes; devouring fishes; subsisting on fishes.

PI'SIFORM. (from *pisum,* a pea, and *forma,* form, Lat.) Of the form of a pea; having a structure resembling peas.

PI'SIFORM IRON ORE. } The pea ore
PI'SIFORM IRON STONE. } of Jameson; pisiform iron stone of Kirwan. A variety of argillaceous oxide of iron, occurring in small masses or grains, nearly or quite spherical, and often equal in size to a pea, or even larger. These globules are composed of thin, concentric layers, which decrease in density as they approach the centre. The exterior layers are compact, and present an even, glistening fracture with a resinous lustre, whereas the centre of the grain is almost always friable, and has a dull earthy fracture. They are easily broken, and may be cut by a knife. Specific gravity 3.40. These grains, sometimes solitary, are generally united by a ferruginous cement, either calcareous or argillaceous, which adheres to their surface. This variety is composed of oxide of iron 71·5, water 14·5, silex 7·5, alumine 3·5, oxide of manganese 0·5. It is abundant in France, Switzerland, and Germany: occurring in secondary rocks.

PI'SOLITE. (from πίσον, a pea, and λίθος, a stone.) So called from its resembling an agglutination of pease. The pisolithe of Brongniart; the peastone of Jameson. Pisolite is nearly or quite opaque, and consists of small rounded masses, composed of concentric layers, each concretion having a grain of sand for its nucleus, or centre. These concretions, agglutinated by a calcareous cement, form masses of considerable magnitude, and sometimes continuous beds.

PISOLI′TIC. Composed of pisolite; containing pisolite; resembling pisolite.

PI′STACITE. ⎫ Another name for epidote.
PI′STAZITE. ⎭

PI′STIL. (from *pistillum*, Lat. *pistil*, Fr.) In botany, the female organ of the plant; situated in the centre of the flower, and forming the rudiments of the fruit. A perfect pistil is composed of three parts, the ovarium, the style, and the stigma. Each modified leaf which forms the pistil, is called a carpellum : the carpella are so folded that the margins of the leaf are next to the axis, or centre; and from these a species of bud is produced, which is the seed. The form of the pistil must depend on that of the carpella, on their number, and on their arrangement.

PITCH-STONE. A vitreous lava, of a blackish-green, or a nearly black colour; a semi-vitreous substance having the lustre and appearance of pitch, and containing a portion of bitumen. Specific gravity from 2·29 to 2·64. Before the blowpipe it whitens, tumesces, and fuses into a porous, whitish enamel. It consists of silex 73, alumine 14·5, soda 1·75, lime 1, oxides of iron and manganese 1·1, water 8·50. Pitch-stone occurs in veins and in beds, sometimes forming whole mountains.

PIT COAL. The coal usually consumed in our houses, and thus called from its being dug out of pits.

PLACE′NTA. (*placenta*, Lat. a cake, *placenta*, Fr.)
1. In anatomy, the medium of communication between the mother and the fœtus.
2. In botany, that part of the ovarium to which the seeds are attached.
3. In fossilogy, the name given by Klein to a section of catocysti, from the shells being flat, like a cake. They are all ornamented with a pentaphylloid flower. The mouth is in the middle of the base, and the anal orifice near the margin. Placentæ are divided by Klein into three genera, mellita, laganum, and rotula.

PL′ACENTALIA. (from *placenta*, a cake, Lat.) The name given to one of the two great sections into which the class mammalia is divided. The orders comprised in this section have been thus named from a predominating feature in the reproductive system of the animals comprehended in them. Fœtus ad uterum maternum annexus per medium placentæ veræ, diversé conformatæ, diversis mammalibus; vel, inloco illius, membranâ villosa, et vasculari, χωρίον dictâ.—*Martin*.

PLACOI′DEA. (from πλάξ, a broad plate, and εἶδος, form, Gr.) One of the orders into which M. Agassiz divides the class of fishes. The placoidians are distinguished by their skin being irregularly covered with plates of enamel. In this order are comprised all the cartilaginous fishes of Cuvier, the sturgeon only excepted.

PLACOTHO′RAX. A genus of ichthyolites of the old red sandstone.

PLAGIA′ULAX. (an abbreviation for Plagiaulacodon, from πλάγιος, oblique, and αὖλαξ, a groove, having reference to the diagonal grooving of the premolars.) The name assigned to a genus of fossil herbivorous marsupials, found in the Purbeck series.

PLAGIO′STOMA. (from πλάγιος, oblique, and στόμα, a mouth, Gr.) A genus of bivalve shells, first named by Llwhyd and established by Mr. Sowerby in his mineral conchology for a generic form presumed to be known only in a fossil state. Certain differences of aspect induced Mr. Sowerby to divide them into two sections, one of which has been shewn by M. Deshayes to pertain to the Lima, the other, from which have been constructed the three genera Podopsis, Pachytes, and Dranchora, has likewise been proved by M. Deshayes to be live spondyli; the genus Plagiostoma is therefore useless and should expunged. The well known Plagiostoma Spinosune of the chalk is a Spondylus.—*Lycett.*

PLANO'RBIS. (from *planus*, flat, and *orbis*, an orb, Lat.) A genus of discoidal, fresh-water univalves, resembling the ammonite, but not chambered. Planorbis belongs to the family Pulmonea in Cuvier's arrangement, and to Lymnacea in Lamarck's and Blainville's. Planorbes may be distinguished from Helices by the slight increase of the whorls of their shell, by the convolutions being nearly in one plane, and by the aperture being wider than it is high. All the shells of this genus are reversed: they abound in pools and ditches.

PLAN'TIGRADE. A term applied to the feet of animals when so constructed as to allow the sole, from the heel to the toes, to be fairly applied to the surface of the ground.

PL'ANULITE. The name assigned by Lamarck to a genus of univalve fossils formerly considered as a species of ammonites. The planulite differs from the ammonite in not being articulated.

PLA'SMA. (from πλασμα, Gr. image, this stone having been formerly used for engraving.) A grass-green variety of rhombohedral quartz. Fracture conchoidal; lustre feeble and resinous. It occurs in beds associated with common chalcedony. It is brought from Italy and the Levant: it was worn by the Romans, and formed into ornamental articles of dress. By some mineralogists, plasma is considered to be a variety of calcedony. Specific gravity 2·55. Before the blow-pipe it is infusible, but parts with its colour, and becomes whitish. According to Klaproth, it consists of silica 96·75, alumina 0·25, iron 0·50.

PLA'STER OF PA'RIS. A sub-species of gypsum. See *Gypsum.*

PLA'STIC CLAY. (from πλαστικὸς, Gr. fit for the art of fashioning, *plastique*, Fr.) The argille plastique of the French geologists. A name given to one of the beds of the eocene period, from its easily receiving and preserving the forms given to it. The plastic clay comprises the lowest formation of deposits of the Eocene period. The plastic clay of the Paris basin is described as sometimes consisting of two beds, separated by a bed of sand. The lower bed is properly the plastic clay. It is unctuous, tenacious, contains some silicious but no calcareous matter, and is absolutely refractory in the fire, when it has not too great a portion of iron. It varies much in colour, being very white, grey, yellow, grey mixed with red, and almost pure red. This clay is employed, according to its quality, in making a coarse and fine pottery and porcelain. The French sands are of great variety of colours. A species of imperfect coal occurs in the lower strata of the Paris basin. The sands of the formation in this country are of almost every variety of colour, and this is more particularly seen at Alum Bay, in the Isle of Wight, where they are employed in forming ornamental articles resembling landscapes, by a tasteful arrangement of them in glass cases. The plastic clay and the London clay are deemed by some geologists as one formation, and, although separated by others, the line of separation appears to be quite arbitrary. In some parts of England, more particularly the western, the *plastic clay* has suffered considerable destruction. Two isolated patches of it may be observed near Weymouth. one on Came Down, on the Ridgeway, and the other at Black Down. Viewed on an extended scale, the *plastic clay* is composed of an indefinite number of sand, clay, and pebble beds, irregularly alternating, the distribution of the organic remains, like the alternation of the strata, being exceedingly variable: sometimes they occupy the clay; at other times the sand or pebbles; and very frequently are altogether wanting in both.

The Druid sandstones, grey weathers, sarsenstones, and puddingstones, scattered in loose blocks over many of the chalk downs around the London Basin, are believed by Mr. Prestwich to be consolidated portions of the sands and gravels of the plastic clay series.

PLA'STRON. (*plastron*, Fr.) A name given to the sternum of reptiles.

PLATE. A provincial term for shale, or slate clay. This *plate* or shale is found in the coal and associated beds, intercalated with them.

PLA'TINA. } (*platina*, Spanish, from
PLA'TINUM. } *plata*, silver.) A metal; one of the simple or elementary bodies. Platinum was not known in Europe till Mr. Wood brought some of it from America in 1741. When pure, it is of a white colour, like silver, but not so bright. It has neither taste nor smell. It is exceedingly malleable and ductile; it may be hammered into plates of extreme thinness, and Dr. Wollaston succeeded in drawing out a wire of this metal to the fineness of 1-10,000dth of an inch. Platinum is one of the most infusible of metals, not yielding before the utmost heat of the furnace; it is soluble in chlorine and nitro-muriatic acid. It was first obtained from Choco and Santa Fé, in South America; it has since been discovered in the Brazils, Spain, and in the Ural mountains, in Siberia. In the ore of platinum four new metals have been discovered, namely, iridium, palladium, osmium, and rhodium.

PLATYCRINY'TES. (from πλατὺς, broad, and κρίνον, a lily, Gr.) A genus of the family crinoïdea, of the division, according to Miller's arrangement, crinoïdea inarticulata. That author thus describes the generic characters of Platycrinites:— "A crinoidal animal, with an elliptic, or (in one species) pentagonal column, formed of numerous joints, having a few side arms at irregular distances. Pelvis, saucer-shaped, formed of three unequal pieces, from which five large plate-like scapulæ proceed. Base provided with numerous fibres for attachment. The want of costæ, supplied by the large plate-like scapulæ, gives the superior part of these animals a pentagonal appearance, and furnishes so conspicuous a character, that they are readily distinguished from all other genera."

PLATYGNA'THUS. A genus of ichthyolites of the old red sandstone. A jaw, in the possession of Dr. Traill, that of an Orkney species of Platygnathus, does not exceed in bulk the jaw of a full grown cod fish.— *Hugh Miller.*

PLEC'TRODUS. The name given by M. Agassiz to a genus of fossil fishes discovered in the Upper Ludlow rock. The name has been chosen from the circumstance of the teeth being bristled with sharp points, like the spurs of a cock. The teeth, says M. Agassiz, cannot be referred to any species already known, and constitute a genus, the fishes of which were without doubt the pirates of the seas of that period. If there is but one species, it might be named Plectrodus mirabilis; if there are two, that with the greatest number of points may be Plectrodus pleiopristris. — *Murchison's Silurian System.*

PLEISTO'CENE. A sub-division of the Pleiocene division, implying that it contains the maximum proportion of the recent organisms, being from 90 to 95 per cent. of existing forms. The Pleistocene deposits may be defined as those in which more then three-fourths of the fossils are of existing species.

PLEROSAU'RUS. A fossil saurian of the lias and oolite.

PLESIOSAU'RUS. (from πλησίον, near to, and σαῦρα, a lizard, Gr.) We are indebted to the researches of the Rev. W. Conybeare for our earliest acquaintance with the plesiosaurus. A genus of extinct amphibious animals, nearly allied to the Ichthyosaurus. Cuvier says this inhabitant of the ancient world, is perhaps the most heteroclite, and appears to merit the name of monster above all others. "Cet habitant de l'ancien monde est peut-être la plus hétéroclite et celui de tous qui paroît le plus mériter le nom de monstre." It united the teeth of a crocodile to the head of a lizard; its neck was of enormous length, exceeding that of its body, and resembling the body of a serpent; it possessed a trunk and tail of the proportions of an ordinary quadruped; to all these were added the ribs of a cameleon, and the paddles of a whale. The teeth were conical, very slender, curved inwards, finely striated on the enamelled surface, and hollow throughout the interior. Five or six species of the plesiosauri are known; they appear to have lived in shallow seas and estuaries, and, in the opinion of some, they swam upon or near the surface, having the neck arched, like the swan, and darting it down at the prey within reach. Prodigious numbers of remains are found in the lias. Vertebræ and teeth are found in the Hastings beds. Some of the plesiosauri were upwards of twenty feet long. The plesiosaurus is supposed to have been oviparous. Of all the species yet discovered the most extraordinary is the plesiosaurus dolichodeirus, or long-necked plesiosaurus: the neck of this animal is equal to half the entire length of the body and tail united, and is composed of 35 vertebræ; the back of 27, and the tail of 28; altogether 90 vertebræ. The head does not exceed one-fifth of the neck in length.—*Geological Transactions.*

PLE'TA. An orthocerite limestone, of a dull red and dingy-grey; an earthy and slightly consolidated limestone; with thin seams of clay. It is placed amongst the lower silurian rocks of Russia.

PLEUROCY'STI. The third class of echini.

PLEURO'TOMA. A genus of spiral univalves, which have a narrow aperture and long channel: the outer lip is distinguished by a transverse slit, about the middle or upper part. Many species have been found fossil in the tertiary deposits, more especially in the Paris basin, and in the London clay.

PLEUROTOMA'RIA. A fossil genus of turbinated, spiral, univalve shells belonging to the family Turbinacea. They are found only fossil, and occur in the inferior oolite.

PLI'CATED. (from *plico*, Lat. to fold.) Plaited; folded.

PLI'OCENE. } (from πλείων, more, and
PLEI'OCENE. } καινὸς, recent, Gr.)
The name given by Sir C. Lyell to a division of the supracretaceous group, or tertiary strata. The tertiary series Sir C. Lyell divided into four principal groups, namely, the eocene, the miocene, the older pliocene, and the newer pliocene, each characterized by containing a

very different proportion of fossil recent species. The newer pliocene, the latest of the four, contains from ninety to ninety-five per cent. of *recent* fossils; the older pliocene contains from thirty-five to fifty per cent. of *recent* fossils; the miocene contains eighteen per cent. of *recent* fossils; the eocene contains only three and a-half per cent. of *recent* fossils. The newer pliocene period is that which immediately preceded the recent era; the older pliocene period is that which intervened between the miocene and the newer pliocene. The newer pliocene formations occur in Sicily and Tuscany; the older pliocene at Nice, Perpignan, Norfolk, Suffolk, and near Sienna. Both the newer pliocene and the older pliocene exhibit marine as well as fresh-water deposites.

PLUMBA'GO. *(plumbago*, Lat.) Graphite. Commonly called black-lead.

PLU'MULE. (from *plumula*, Lat. a little feather.) In botany, that part of the seed which grows into the stem and axis of the future plant. In the bean, horse-chesnut, &c., the plumule is distinctly visible, but in plants generally, it is scarcely perceptible without the aid of a magnifying glass; and in many it does not appear till the seed begins to germinate. The first indication of development is the appearance of the plumule, which is a collection of feathery fibres, bursting from the enveloping capsule of the germ, and which proceeds immediately to extend itself vertically upwards.

PLUTO'NIC. (from *Pluto*, one of the heathen deities.) A name given to certain rocks elaborated in the deep recesses of the earth, from the opinion they were formed by igneous action at great depths; whereas the volcanic, although they may have risen up from below, have cooled from a melted state upon or near the surface.

PLUTO'NIC ROCKS. Unstratified crystalline rocks, such as granites, greenstones, and others, of igneous origin, formed at great depths from the surface. Plutonic rocks are distinguished from those which are called volcanic, although they are both igneous; Plutonic rocks having been elaborated in the deep recesses of the earth, while the volcanic are solidified at or near the surface. Plutonic rocks differ from volcanic, not only by their more crystalline texture, but also by the absence of tuffs and breccias, which are the products of eruptions at the earth's surface. They differ also by the absence of pores or cellular cavities, which the entangled gases give rise to in ordinary lava.—*Lyell.*

PNEUMA'TICS. (from πνευματική, Gr.) That branch of science which relates to the equilibrium or movements of aërial fluids under all circumstances of pressure, density, and elasticity. The weight of the air, and its pressure on all the bodies on the earth's surface, were quite unknown to the ancients, and only first perceived by Galileo, on the occasion of a sucking-pump refusing to draw water above a certain height. The manner in which the observed law of equilibrium of an elastic fluid, like air, may be considered to originate in the mutual repulsion of its particles, has been investigated by Newton, and the actual state of the law itself, as announced by Mariotte, "that the density of the air, or the quantity of it contained in the same space, is, cæteris paribus, proportioned to the pressure it supports," has been verified by direct experiment. This law contains the principal of solution of every dynamical question that can occur relative to the equilibrium of elastic fluids, and is therefore to be regarded *as one of the highest axioms* in the science of pneumatics.

POCILLO'PORA. A genus of stony polypifers, thus named from the pocilliform shape of the cells, from which shape, as well as the margins having little or no projection, this genus is separated from madrepora. Pocillopara is plant-formed, ramose or lobated; the surface set with deep cellules, with porous interstices. There are several species, found principally in the South sea and Indian ocean.

PODOPHTHA'LMIA. The first order of the class crustacea, comprising the lobster, crab, crayfish, &c.

PODOPHTHA'LMA. The name assigned by Leach to a group of crustaceans comprising the orders Decapoda and Stomatopoda. The animals of this group have their eyes supported on moveable peduncles.

POIKILITIC. (from ποικίλος, Gr. various, variegated.) To the new red sandstone group, M. Brongniart has applied the name of Terrain Pœcilien. Conybeare has proposed to extend the term *Pœcilitic* to the entire group of strata between the coal formation and the lias, comprising the new red conglomerate, the magnesian limestone, the variegated sandstone, the shell limestone, and the variegated marl. Some common appellative, says Buckland, for all these formations has been long a desideratum in geology; but the word *pœcilitic* is, in sound, so like pisolite, that it may be better to adhere more literally to the Greek root, and apply the common name of *Poikilitic group* to the strata in question.

POLARIZED LIGHT. Light which by reflection or refraction, at a certain angle, or by refraction in certain crystals, has acquired the property of exhibiting opposite effects in planes at right angles to each other.

POLISHING SLATE. (Poherschiefer of Werner.) A variety of tripoli. Colour white, yellowish-white, or yellow. Structure slaty. Cross-fracture dull and earthy. It is very light, so as sometimes to swim on water, strongly adheres to the tongue, and is easily reduced to a fine dry powder. Before the blowpipe it hardens, but does not fuse. It principally occurs at Bilin, in Bohemia, and is regarded as a volcanic production.

PO'LLEN. *(pollen,* Lat. fine flour.) In botany, the fecundating powder or dust contained in the anther. In dry and warm weather the anther bursts, and the pollen is thrown out.

POLYCH'ŒTA. The first order of the class annulata, comprising the Nereis, serpula, and lob-worm.

POLYGA'STRICA. (from πολὺς, many, and γαστὴρ, a stomach, Gr.) So named from their possessing numerous internal digestive cavities. The lowest class of animals, belonging to Diploneura, or Helminthoida.

When we place, says Professor Grant, a drop of any decayed infusion of animal or vegetable matter under a powerful microscope, and throw a light through that drop, and through the microscope to the eye, we discover in the drop of water various forms of living beings; some of a rounded, some of a lengthened form, and some exhibiting ramifications shooting in all directions, but all apparently of a soft, transparent, gelatinous, and almost homogeneous texture. In these minute animals there are numerous cavities or stomachs, in some of them being two hundred in number. There is every reason to believe that polygastrica exist in every drop of water. They form the food of other classes, more especially the zoophytes. Almost all the known genera of polygastric animalcules possess eyes: they are also found to possess an acute sense of taste; they distinguish, pursue, and seize their prey, and, although so excessively minute that five millions have been calculated as being contained in one drop of water, they avoid infringing on one another while swimming. All their movements appear to be as well directed, regular, methodical, and spontaneous, as those of the higher classes of swimming animals. These movements are effected by means of very minute, hair-like, tapering, transparent, vibratile filaments disposed frequently around the mouth, where they are generally largest and longest. There is no proper skeleton in the whole order polygastrica, nor any secretion of shell on the surface, yet there are parts destined to give support: no nervous system has hitherto been detected in the polygastric animals. Some of the polygastric animals exude on their surface a secretion which agglutinates foreign particles floating in the waters which surround them, and thus form for themselves a partial covering. In the majority of polygastric animals there is an alimentary canal, with an oral and an anal orifice, which traverses the body; in the simplest forms of animalcules, however, there is but one general orifice to the alimentary cavities, which is placed at the anterior extremity of the body, and is surrounded with long vibratile cilia, which serve both as organs of motion and tentacula. No teeth for mastication, nor any glandular organs to assist in digestion, have been discovered in them. Notwithstanding their extreme minuteness, they appear to be the most numerous, the most prolific, the most active, and the most voracious of all living beings. Some naturalists, like Lamarck, believed that these animals were without a mouth or any internal organs, and were nourished by superficial absorption, like marine plants. Lewenhoeck and Ellis, however,

observed that they possessed an internal cavity, and devoured each other. Spallanzani perceived them swallowing each other so avariciously that their bodies became distended with their prey.—*Lectures on Comp. Anatomy, passim.*

POLYHA'LLITE. A mineral found at Ischel, in Austria. It occurs in masses of a fibrous texture. Lustre pearly. Specific gravity 2·76.

POLYME'RIC ISOMO'RPHISM. The replacement of an equivalent by a multiple of another, is termed polymeric isomorphism.—*Jukes.*

POLYMI'GNITE. (from πολὺς, many, and μίγνυμι, to mix, Gr.) A mineral, thus named in consequence of the variety of its constituent parts. It consists of titanic acid, zirconia, lime, yttria, the oxides of iron, cerium, and manganese, with minute portions of magnesia, potash, silica, and oxide of tin. It is of a black colour; crystallized in small prisms; scratches glass; specific gravity 4·8. Fracture conchoidal. Lustre almost metallic.

PO'LYPE. The name given to each tube, surrounded with its tentacula, of the Polypus.

POLYPA'RIA. } The fourth class of Radiata or Zoophytes; thus named from a supposed resemblance to an Octopus, called Polypus by the ancients, this resemblance arising from the arrangement of the tentacula around the mouth.

The polypi constitute the second class (the Infusoria forming the first) of Lamarck, and form a large family of animals, in which we trace a gradual process to a more complicated organization. They, according to Lamarck, have a gelatinous, contractile body, a distinct mouth, surrounded by tentacula, or branching arms, and a simple alimentary canal or stomach, showing no vents; they increase by separation, or internal spontaneous production, and show no generative organs. The greater part of the species adhere one to another, and may be considered as animals depending on mutual support. Some of them approach closely to the Infusoria, as the Polypi Ciliati, whilst others are capable of attaching themselves by means of a pedicle, and in many instances are able to detach and affix themselves to new spots.

These animals are commonly known as *Corals*. From an idea which long prevailed that these animals are allied to marine plants, they also obtained the name of Zoophytes. The body is cylindrical or conical, sometimes possessing no viscus but its cavity; at others possessing a stomach, which is visible, and other organs. The greater number of Polyparia are inhabitants of the ocean, and from the ocean's depths they raise those immense reefs that at some future period may form a communication between the inhabitants of the temperate zones. Although Polypi abound in every part of the ocean, still it is in the warmer regions that they grow in greatest luxuriance. The tentacula of Polypi are exquisitely sensitive, and are frequently seen, either singly or altogether, bending their extremities towards the mouth when any minute floating body comes in contact with them. A question arises, says Dr. Roget, with regard to the constitution of these Zoophytes, similar to that which has been proposed with regard to trees, namely, what limit should be assigned to their individuality? Is the whole mass, which appears to grow from one root, and which consists of multitudes of branches, proceeding from a common stem, to be considered as one individual animal, or is it an assemblage, or aggregation of smaller individuals; each individual being characterised by having a single mouth, with its accompanying tentacula, and yet the whole being animated by a common principle of life and growth? The greater number of naturalists have adopted this latter view, regarding each portion as provided with a distinct circle of tentacula, as a separate animal, associated with its neighbours in the construction of a common habitation, and contributing its quota to the general nourishment of this animal republic.

POLYPHRA′CTUS. A genus of ichthyolites of the old red sandstone.

POLYPI′FERA. } That class of animals
POLYPI′PHERA. } commonly known by the name of Zoophytes. They are carnivorous, feeding upon living animalcules. These animals precipitate immense quantities of carbonate of lime, especially in tropical seas.

"These," says Prof. Grant, "are soft aquatic animals of a plant-like form, generally fixed, and supported by an extra-vascular axis of a calcareous or horny texture. Instead of the pores of the poriphora, the common fleshy mass of the body here developes small tubular digestive sacs called polypi, the margins of which are furnished with sensitive tentacula, and the sides of the tentacula are almost always furnished with sensitive, or prehensile, or vibratile cilia."

POLYPI′FEROUS. Animals which have polypi; zoophytes.

POLYPE′TALOUS. (from πολὺς, many, and πέταλον, a leaf of the corolla, Gr.) In botany, a term applied to a corolla which has the petals separate.

POLYSE′PALOUS. (from πολὺς, many, and *sepal*, the name given to the parts of which the calyx is composed.) In botany, a term given to a calyx which has its sepals separate from each other.

POLYPOTHE'CIA. A genus of spongeous zoophytes found in flints, and thus named by Miss Bennett, of Norton House, near Warminster.

POLY'PTERUS. A genus of fishes found in the Nile and in the rivers of Senegal. The Polypterus and Lepidosteus are the only known genera of living representatives of the sauroid fishes.

POLYSP'ERMOUS. (from πολύς, many, and σπέρμα, seed, Gr.) In botany, a term applied to the ovarium and fruit, when they contain many seeds.

POLYTHA'LAMOUS. (from πολύς, many, and θάλαμος, a chamber, Gr.) Having many cells or chambers, as *polythalamous* shells; multilocular; camerated.

POLYZO'A. Called also Bryoza: the second order of the class Molluscoidea, comprising flustra, retepora, eschara, and other zoophytes.

PO'RCELAIN EARTH. } For a description of this, See *China Clay*.
PO'RCELAIN CLAY. }

PO'RCELAIN JAS'PER. A variety of jasper; called also Porcellanite. see *Porcellanite*.

PORCELANA'CEOUS. } Resembling porcelain. Shells
PORCELA'NEOUS. } have been divided into two classes. The first are of a compact texture, have an enamelled surface, and are generally beautifully variegated; the shells of this class have been termed *porcelanaceous*, or *porcelaneous* shells; they contain but a small proportion of soft animal matter.

PORC'ELAINOUS. The name given to a species of protogine. Porcelainous Protogine, or China Stone, is composed of felspar quartz and talc, and is of a greenish-yellow colour. It affords, by decomposition, the china clay of commerce.

PORCE'LLANITE. The Porzellan Jaspis of Werner; Thermautide porcellanite of Haüy; Porcellanite of Kirwan. This is a variety of Jasper, according to some mineralogists; Prof. Jameson places it as a sub-species. Mr. Allan, however, says "it is merely clay indurated by heat;" and according to Werner, it is slate-clay converted into a kind of porcelain by the action of the heat of the volcano. "A mineral of various colours, from grey to nearly black, occurring in amphorous masses or fragments, which are often rifted. Porcellanite sometimes resembles a brick which has undergone a slight vitrification. Its fracture is imperfectly conchoidal or uneven, more or less glistening, and often has the aspect of certain porcelains. It is opaque, very brittle, and less hard than quartz. An analysis of of it yielded silex 60·75, alumine 27·25, potash 3·66, magnesia 3, oxide of iron 2·50. It is most likely an alteration of some variety of argillaceous slate by pseudo-volcanic fires: it does not constitute a distinct species. It is found in large masses near the pitch-lake in Trinidad, and occurs usually in the vicinity of coal mines."—*Cleaveland*.

POR'CELAIN SPAR. A mineral combination with a chloride, consisting of four equivalents of the double silicate of lime and alumina with one of chloride of sodium.

PORI'FERA. } (from *porus*, a pore,
PORI'PHERA. } and *fero*, to bear.) A classs of animals belonging to Cyclo-Neura, or Radiata. Periphera constitutes the second lowest class of animals, coming between Polypiphera and Polygastrica. They form the various species of sponge which are met with in such multitudes on every rocky coast of the ocean, from the shores of Greenland to those of Australia. Their surface is porous; those pores lead to canals which ramify through all parts of their texture; and those canals anastomosing into larger and

larger trunks, lead, again, to orifices on the surface, from which there issues constant streams of water. The poriferous animals present various and remarkable forms in the skeleton; and the simple gelatinous body of the animal is supported by a skeleton composed of different kinds of earth: in one group the earth is silica; in another it is the carbonate of lime; in another it is a horny substance. The skeleton, thus composed, has been called the axis of the animal. The material of which the fleshy portion is composed is of so tender and gelatinous a nature, that the slightest pressure is sufficient to tear it asunder, and allow the fluid parts to escape, and the whole soon melts away into a thin oily liquid.

The surface of a living sponge presents two kinds of orifices; the larger of a rounded shape with raised margins, which form projecting papillæ; the smaller, minute and numerous, constituting the pores of the sponge. The porifera present a digestive system, which, by its form and simplicity, approaches the nearest to that of plants. The cellular tissue of their body is permeated in all directions by anastomosing and ramifying canals, which begin by minute superficial pores, closely distributed over every part, and terminate in larger orifices variously placed, according to the form of the entire animal. The pores are provided with a gelatinous network and projecting spicula, to protect them from the larger animalcules and floating particles. The internal canals, like the venous system, leading from capillaries to trunks, are bounded by a more condensed portion of the general cellular substance of the body, and are incessantly traversed by streams of water, passing inwards through the minute pores, and discharged through the larger orifices or vents, but no polypi or cilia have been discovered in those parts, although from analogy we might consider them necessary. From the incessant streams that are conveyed through the bodies of these animals, it appears that all parts of these interior perforations, as well as the general external surface of this cellular structure, serve for the conveyance of nutritious matter into the interior substance of the body. On watching the streams of water which issue from the fœcal orifices, there may be seen minute flocculent particles that are incessantly detached and thrown out, which appear as if they were the residue of digestion, or pellicles excreted from the body, and thrown off from the surface of internal canals.

No nervous filaments have been detected in the soft gelatinous bodies of poriphera. Their ciliated gemmules, however, are endowed with remarkable living properties, and powers of spontaneous motion. They have an evident object in their motions; they can accelerate, retard, or cease, at pleasure, the vibrations of their cilia; they can change the direction of their course in the water, perceive each other's vicinity, revolve round each other, distinguish the most suitable place for the fixing of each species, or bound forward suddenly from a state of rest. They appear in this state of freedom to be sensible to light, and to shun it.

Although sponges, or poriferous animals, are permanently attached to rocks, and other solid bodies in the ocean, and are consequently destined to an existence as completely stationary as that of plants, yet such is not the condition of the earlier, and more transitory stages of their development. On the gemmule the power of locomotion is conferred, until it has found for itself a proper habitation; this chosen, it there fixes itself and there continues for the remaining period of its existence.

PORI'PHEROUS. Possessing pores; animals belonging to the class poriphera. The alimentary apparatus of *poripherous* animals approaches the nearest to that of plants.

PO'RPHYRY. (from πορφύρα, purple, Gr.) Porphyry has been so called in reference to the purple, or reddish, colour so commonly perceptible in it. Generally, any form of rock in which one or more minerals are scattered through an earthy or compact base. Porphyry has generally a compact texture. Sometimes it is composed of tabular, columnar, or globular distinct concretions; and not unfrequently it is traversed by numerous seams and rents. There are many varieties of porphyry, named according to the base of each, as Petrosiliceous Porphyry, Felspar Porphyry, Clinkstone Porphyry, Argillaceous Porphyry, &c. Porphyry occurs in enormous masses; at the head of Glen Ptarmagan, a cliff of porphyry fifteen hundred feet in height, in shape resembling an oblique truncated pyramid, passes through granite. In some instances porphyry is, beyond all question, a volcanic formation. Near Christiana, in Norway, an immense mass of porphyry, from 1600 to 2000 feet in thickness, covers beds of gneiss, limestone, and greywacke. Dykes of porphyry cutting through the subjacent rocks indisputably prove the volcanic character of this immense mass.

PORPHYRI'TIC. ⎱ Resembling porphyry; containing porphyry; composed of a compact homogeneous rock, in which distinct crystals or grains
PORPHYRA'CEOUS. ⎰

are imbedded: the compact stone is called the base, and sometimes the paste. The base, or paste, is generally felspar.

PORI'TES. A genus of stony polypifers, fixed; ramified; or lobated and obtuse; the outer surface everywhere stellated.

PO'RTLAND BEDS. A marine formation, occurring in the Isle of Portland and in Wiltshire. These beds consist of coarse shelly limestone, fine grained white limestone, and compact limestone, all possessing an oolitic structure; and beds of chert. The Portland beds lie immediately under the Purbeck beds, and above the Kimmeridge clay. They constitute the uppermost members of the oolite group, and abound in ammonites, trigoniæ, &c.

PO'RTLAND L'IMESTONE. ⎱ One of the members of the Portland beds; a marine oolitic formation, obtained principally from Portland, whence the name, and used in building. The Portland limestone abounds in trigoniæ, ammonites, pernæ, pleurotomariæ, and other marine shells.
PO'RTLAND OOLITE. ⎰

PO'RTLAND SANDS. A name given to one of the divisions of the Portland oolite. These beds consist of white and green sand and sandstone, with concretionary masses of grit; the lower strata argillaceous.—*Mantell.*

POST. A north of England term for any bed of firm rock; generally applied to sandstone.

POST PLE'IOCENE. See *Quaternary.*

PO'TASH. An alkali obtained by the incineration of vegetables, or the woody parts of plants that do not grow near the sea. The water in which the ashes are washed is evaporated in iron pots, from which circumstance it was called *potash.* There are few, if any, of the inferior stratified rocks without potash; and, viewing them in the mass, potash may be considered as constituting five or six per cent. of the whole. Potash may be regarded as constituting between six and seven per cent. of granites, greenstones, and rocks of that class.

POTA′SSIUM. A metal discovered by Sir H. Davy in 1808. At a temperature of 32° potassium is hard and brittle, with a crystalline texture; at 50° it becomes malleable, with a lustre like that of polished silver; and at 150° it fuses. Potassium is lighter than water, its specific gravity being 0·85; to preserve it unchanged, it should be kept in a phial with pure naptha.

POTERIOCRINI′TES. (from ποτήριον, a cup, and κρίνον, a lily, Gr.) Vase-like, lily-shaped animal. A genus of Crinoïdea, belonging to the division crinoïdea semiarticulata, according to Miller's arrangement. That author thus describes the generic characters of Poteriocrinites. "A crinoidal animal, with a round column composed of numerous thin joints, having in their centre a round alimentary canal, and articulating by surfaces striated in radii. Round auxiliary side arms, proceeding at irregular distances from the column. Pelvis formed of five pentagonal plate-like joints, supporting five hexagonal intercostal plate-like joints, and five plate-like scapulæ, having on one of the intercostals an interscapulary plate interposed. An arm proceeding from each of the scapulæ. Base, probably fascicular, and permanently adhering."

PO′TSTONE. The Lapis ollaris of Pliny. A variety of steatite, nearly equal in hardness to common steatite; it is, however, more tenacious, and though it may be turned with the lathe, it breaks with difficulty. It is smooth and unctuous to the touch. It is usually of a greenish-grey colour, with various shades, and often spotted. Its fracture curved, and, sometimes, almost foliated. Specific gravity from 2·8 to 3·2. It emits an argillaceous odour when breathed on. From its being formed into culinary vessels it has obtained its name. From an analysis by Weigleb, the potstone of Corno, in Lombardy, where it occurs in great abundance, consists of magnesia 38, silica 38, alumina 7, iron 15, carbonate of lime 1, fluoric acid 1.

PO′TTERS' CLAY. A variety of clay, of a reddish or grey colour, which becomes red when heated. That used in our potteries for making coarse red ware comes chiefly from Devonshire. It is exceedingly infusible, and contains a large proportion of alumine. Potter's clay, mixed with ground flints, is employed in the manufacture of the finer kinds of pottery in Staffordshire.

POZZUOLA′NA. Scoriæ or volcanic ashes, brought from Pozzuoli, a town in the bay of Naples, and named therefrom. Pozzuolana is used, mixed with lime, for making Roman, or water-setting, cement.

PRASE. (from πράσον, Gr. a leek; so called from its colour.) The Prasem of Werner; Quartz hyalin vert obscur of Haüy; Quartz prase of Brongniart. A leek-green translucent variety of rhombohedral quartz; lustre vitreous; fracture splintery. Specific gravity 2·5. Prase appears to be common quartz, coloured by actynolite or epidote.

PRA′SINOUS. (prasinus, Lat.) Of a light green colour, inclining to yellow.

PRE′HNITE. A mineral thus named after Colonel Prehn, who brought it from the Cape of Good Hope. Prehnite is of a green, grey, or white colour. It occurs crystallised; in granular, scopiform, and stellular fibrous distinct concretions; massive and reniform. Its texture is foliated. Fracture uneven. Internal lustre pearly. It scratches glass, though feebly, and gives sparks with steel. Specific gravity from 2·60 to 2·94. Hardness 6 to 7. Veins of prehnite occur in Cornwall; they are very irregular, both in size and direction. Asbestus and stilbite have been found in the veins with the prehnite.

PRI'MARY. 1. In astronomy, the planet about which a satellite revolves. The planetary bodies which revolve round the sun, and not round any other body, are called primary planets, or primaries. Some of the planets are accompanied by satellites or moons, which revolve round their *primary*, in a manner similar to the revolution of our moon.

2. A term applied to rocks or strata, because it was supposed, from the absence of fossil remains, that they were formed before animals and vegetables; as well as that they were the first rocks formed. Sir C. Lyell proposes to substitute the word *hypogene* for *primary*.

PRI'MARY FORM. In mineralogy, that form to which minerals may be reduced by cleavage, and which is no longer changed by continued cleavage. If a mineral can be cleaved in directions which produce only one particular form that form is denominated its primary. The whole number of primary forms are comprised in the regular tetrahedron, the cube, the rhombic dodecahedron, the octahedron, the six-sided prism, and the parallelopiped.

PRI'MARY LIMES'TONE. This, says Jukes, instead of primary, ought to be simply called altered limestone. It is always highly crystalline, generally has lost all appearance of bedding, and is a granular crystalline carbonate of lime.

PRI'MARY STRATA. The primary strata are defined above by the old red sandstone; and when that is absent, by the carboniferous limestone; below, they usually rest, but sometimes unconformably, upon granite. They consist, in a great measure, of mechanical aggregates, comparable with sandstones and clays, but yet generally distinguishable by superior hardness, and somewhat of a crystalline structure in mass, or texture in detail, from the secondary rocks. In the secondary rocks there is more variety of arenaceous and calcareous members. In the tertiary strata loose sands, marls, and clays abound, while these scarcely occur at all among the primary rocks.—*Prof. Phillips.*

PRI'MITIVE. A term applied to certain rocks, from the circumstance of no fossil remains of animals or vegetables, nor any fragments of other rocks, being found in them. The term has given way to what is considered a more appropriate one, namely, primary, and primary has given way to Palæozoic.

PRISM. ($\pi\rho\iota\alpha\mu\alpha$, Gr. *prisme*, Fr. *prisma*, It.) A solid figure, the ends whereof are parallel, equal, and similar plane figures, and the sides which connect the ends are parallelograms. Prisms take particular names from the figure of their bases or ends, namely, triangular, square, rectangular, pentagonal, hexagonal, &c.

"A prism is rarely found having only three, very commonly four, six, eight, or even more sides; the sides, or lateral planes, surround its axis, which is an imaginary line passing down the middle of the prism, from the centre of the upper terminal plane to the centre of the lower; the terminal planes being also called the bases. But prisms are found both very long and very short; when long, and the crystals slender and curved, they are termed *capillary*; when straight, *acicular*; when the prism is short, the crystal is said to be *tabular*."—*Phillips.*

PRI′STIS. The fossil saw-fish.

PROBOSCI′DIA. A family of quadrupeds belonging to the order Pachydermata, or, according to more recent classification, the sixth order of the sub-class Placentalia, class Mammalia. The proboscidians have five toes to each foot; they possess no canine teeth, but two tusks, which project from the mouth, and frequently attain to an immense size. The nostrils are continued out into a proboscis, which is exceedingly flexible, possesses great sensitiveness, and terminates in a finger-like appendage. The proboscis may be considered as the hand of the elephant.

PROBO′SCIS. (προβοσκίς, Gr. from βόσκω, to feed, and πρὸ, before.) A lengthened tube, snout, or trunk belonging to certain animals. The proboscis of the elephant is of great length, serving the purposes of a hand, conveying to the mouth anything it desires to swallow. It is an instrument of most delicate touch, of scent, and breathing, and of prehension as adroit as that of a hand. In insects, when the instrument for suction extends for some length from the mouth, it is called a proboscis; such is the apparatus of the butterfly, the moth, the gnat, the house-fly, &c.

PRODU′CTA. } An extinct genus of
PRODU′CTUS. } equilateral inequivalve, striated bivalves. The name Producta has been, by some, objected to, and that of Leptæna has been assigned by Dalman. The genus Producta has not hitherto been found in any deposit more recent than the magnesian limestone or zechstein.

PROTE′OLITE. (from πρωτευς and λίθος, Gr.) The name assigned by Dr. Boase to a genus of schistose rocks. He says, "in many parts of Cornwall, the schistose rocks adjoining the granite, have a basis more granular than that of cornubianite; are much softer, and more prone to disintegrate: following, therefore, the plan adopted in the instance of curite and felsparite, these may be referred to a distinct genus which, on account of the various appearances it assumes, may be called *proteolite*. It may be divided into two families; mica being the accessory mineral in the one; shorl in the other. Proteolite, like cornubianite, consists of a basis of compact felspar with mica or shorl, but its ingredients are differently aggregated, and its felspathic basis is softer, granular, and more prone to disintegration."

PRO′TEUS. (προτευς, Gr. *proteus*, Lat.) The name given to a genus of the order Batrachia. One species only has been hitherto discovered, namely, the Proteus Anguinus. A subterranean saurian, which never makes its appearance on the earth's surface, but is always concealed at a considerable depth below it, being found in subterraneous lakes and caves two or three hundred feet below the surface of the ground. The following particulars are extracted from Sir H. Davy's Consolations in Travel:—"Independently of the natural beauties found in Illyria, and the various sources of amusement which a traveller, fond of natural history, may find in this region, it has a peculiar object of interest in the extraordinary animals which are found in the bottom of its subterraneous cavities, namely the Proteus anguinus, a far greater wonder of nature than any of those which the Baron Valvasor detailed to the Royal Society a century and a half ago, as belonging to Carniola.

At first view, you might suppose this animal to be a lizard, but it has the motions of a fish. Its head, and the lower parts of its body, and its tail, bear a strong resemblance to those of the eel; but it has no fins; and its curious bronchial organs are not like the gills of fishes; they form a singular vascular structure, almost like a crest, round the throat, which may be removed without **occasioning** the death of the animal, **who is** likewise furnished **with** lungs. With this **double** apparatus for supplying **air to the** blood, it can either **live below** or above the water. Its **fore** feet resemble **hands, but they** have only three claws or fingers, and are too feeble to be of use in grasping or supporting the weight of the animal; the hinder feet have only two claws **or** toes, **and** in the larger specimens these are found so imperfect as to be almost obliterated. It has small points instead of eyes, as if to preserve the analogy **of nature. It** is of a fleshy whiteness **or transparency** in its natural state, but when exposed to light, its skin gradually becomes darker, and at last gains an olive tint. Its nasal organs appear large and it is abundantly furnished with teeth, from which it may be **concluded** that it is an animal of prey; yet in its confined state it has never been known to eat, and it has been kept alive for many years by occasionally changing the water in which it is placed. The proteus was first discovered in Illyria by the late Baron **Zöis**; but it has been reported **that some** individuals of the same species have **been** recognized in the calcareous strata **in** Sicily."

The proteus has been found **of** various sizes, from that of the thickness of a quill to that of the thumb, but its form of organs has always been the same. It is a perfect animal of a peculiar species, and it adds one instance more to the number already known, of the wonderful manner in which life is produced and perpetuated in every part of the globe, even in places which seem the least suited to organized existences. And the same Infinite Power and Wisdom which has fitted the camel and the ostrich for the deserts of Africa, the whale for the Polar seas, and the morse and the white **bear for** the Arctic ice, has given the proteus to the deep and dark subterraneous lakes of Illyria,—an animal to whom the presence of light is not essential, and who can live indifferently in air and in **water, on the** surface of the rock, **or in the** mud. The organization of the spine of the proteus is analogous to that of one of the sauri, the remains of which are found in the older secondary strata. The problem of the reproduction of the proteus, like that of the eel, is not yet solved, but ovaria have been discovered in animals of both species.

2. This name is also given to a species of infusoria. Of these, the most singular, says Dr. Roget, is the Proteus, which cannot, indeed, be said to **have any determinate shape; for it seldom remains the** same for two minutes together. It looks like a mass of soft jelly, highly irritable and contractile in every part; at one time wholly shrunk into a ball, at another stretched out into a lengthened riband; and again, at another moment, perhaps, we find it doubled upon itself like a leech. If we watch its motions for any time, we see some parts shooting out, as if suddenly inflated, and branching forth into star-like radiations, or assuming various grotesque shapes, while other parts will, in like manner, be as quickly contracted. Thus the whole figure may, in an instant, be completely changed, by metamorphoses as rapid as they are irregular and capricious.

PROTHE′EITE. A mineral species, occurring in the valley of the Zillerthal, in the Tyrol.

Protococ′cus niva′lis. The name given to an extremely minute plant which grows upon the surface of the snow in certain regions, imparting to the snow a red appearance, from which circumstance it was at first supposed that the snow itself was red. Mr. Darwin, in his account of his passage of the Cordillera of the Andes, thus describes it: "On several of the patches of perpetual snow, I found the Protoccocus Nivalis, or red snow, so well known from the accounts of Arctic navigators. My attention was called to the circumstance from observing the footsteps of the mules stained a pale red, as if their hoofs had been slightly bloody. I, at first, thought it was owing to dust blown from the surrounding mountains of red porphyry, for from the magnifying power of the crystals of snow, the groups of these atom-like plants appeared like coarse particles. The snow was coloured only where it had thawed rapidly, or had been accidentally crushed. A small portion of it rubbed on paper communicated a faint rose tinge, mingled with a little brick-red. I placed some of the snow between the leaves of my pocket-book, and a month afterwards examined with care the pale discoloured patches on the paper. The specimens, when scraped off, were of a spherical form, with a diameter of the thousandth of an inch. The central part consists of a blood-red substance, surrounded by a colourless bark. When living on the snow they are collected in groups, many lying close together. The dried specimens placed in any fluid, as water, spirits of wine, or diluted sulphuric acid, were acted on in two different ways: sometimes an expansion was caused, at others a contraction. The central part after immersion invariably appeared as a drop of red oily fluid, containing a few most minute granules; and these probably are the germs of new individuals."

Pro′togine. A term applied by the French to the talcose granite of the Alps. The name given to a granite composed of felspar, quartz, and talc or chlorite; the talc supplying the place of mica. Protogine is distinguished from granite not only in the substitution of talc for mica, but also in its great tendency to disintegration. Dr. Boase divides protogine into five species, namely, porcelainous, shorlaceous, fluoric, porphyritic, and quartzose. These are all found in Cornwall.

Protozo′a. A sub-kingdom of animals comprising the two classes, Stomatoda and Astomata.

Protozo′ic. (from πρῶτος, first, and ζῶον, animal, Gr.) A name proposed by Sir R. Murchison to be given to those rocks which are supposed to contain the remains of the first formed animals or vegetables; the first or lowest formations in which animals or vegetables occur. "That there is a limit in the descending scale of formations, beneath which no traces of life have been discovered, is now pretty generally recognized; and looking merely to this fact, geologists may agree to use the term *Protozoic*, however they may differ in their interpretation of the phenomenon." —*Sir R. Murchison.*

Pru′inose. (from *pruina*, Lat. a frost, or rime.) In entomology, applied to the clothing of insects when covered with a minute dust, scarcely discoverable by the lens.

Psammo′bia. A small bivalve found in the cyclas limestone of Burwash.

Psammo′steus. A genus of ichthyolites of the old red sandstone, of which Agassiz, in his Poissons Fossiles, describes four species.

Pseu′do. (ψεῦδος, Gr. false.) A term generally used as a prefix to, and in composition with, other words; it implies a sense of spuriousness, as, for example, a pseudo-tuber is a false tuber; pseudo-galena is false galena; a pseudo-volcano is a coal mine in a state of combustion.

PSEUDOMO'RPH. (from ψεῦδος, false, and μορφή, form, Gr.) A Pseudomorph is one mineral occurring in the crystalline form of another.

PSEUDOMO'RPHISM. That process by which the constituents of minerals are changed, and the place of the removed constituent is supplied by the presence of another of the same form.

PTERA'SPIS. A generic name proposed by Dr. R. Kner, and accepted by Mr. Salter, for certain species of Cephalaspis, namely C. Lewisii, and C. Lloydii.

PTERIC'HTHYS. (from πτέρυξ, a wing, and ἰχθύς, a fish, Gr.) The winged fish, an ichthyolite of the lower old red sandstone. Hugh Miller thus graphically describes the pterichthys, "on a ground of light coloured limestone, lay the effigy of a creature fashioned apparently out of jet, with a body covered with plates, two powerful-looking arms articulated at the shoulders, a head as entirely lost in the trunk as that of the ray or the sun-fish, and a long angular tail. This ichthyolite in my large specimens does not much exceed seven inches in length." Agassiz has distinguished nine species.

PTERO'CERES. A genus of winged zoophagous univalves; they are turrited, oval, ventricose, and tuberculated, the spire is short, the aperture oval, terminating in a lengthened canal at both extremities; the outer lip is thickened, expanded into horn-shaped, hollow, thickened spines with an anterior sinus apart from the caudal canal. Several fossil species are recorded from the Jurassic and Cretaceous rocks of France and Germany.

PTERODA'CTYLE. } (from πτερόν, a
PTERODA'CTYLUS. } wing, and δάκτυλος, a finger, Gr.) An extinct genus of winged reptiles, belonging to the family Iguanida. It is found fossil in the Jura limestone formation, in the lias at Lyme Regis, and in the oolitic slate of Stonesfield. "The structure of these animals," says Buckland, "is so exceedingly anomalous, that the first discovered pterodactyle was considered by one naturalist to be a bird; by another, a species of bat; and by a third, a flying reptile. The form of its head and length of neck, resembled that of birds, its wings approached to the proportion and form of those of bats, and its body and tail approximated to those of ordinary mammalia. The pterodactylus forms an extinct genus of the order Saurians, in the class of reptiles; adapted by a peculiarity of structure to fly in the air." The pterodactyles are considered by Cuvier to rank among the most extraordinary of all the extinct animals that have come under his consideration. Species have been discovered, of sizes varying from that of a snipe to that of a cormorant. They had a short tail, an extremely long neck, and a very large head; their eyes were of enormous size, apparently enabling them to fly by night; the beaks were long, like those of a crocodile, and furnished with sixty sharp-pointed teeth. Their most remarkable character consisted in the excessive elongation of the second toe of the fore-foot, which was more than double the length of the trunk, and, in all probability, served the purpose of supporting some membrane which enabled the animal to fly. The fingers terminated in long hooks, like the curved claws of the bat. The form and size of the foot, leg, and thigh, show that this extraordinary ani-

mal was capable either of standing firmly on the ground, or of perching upon the branches of trees. It is deemed probable that the pterodactyle had the power of swimming.

PTE'ROPOD. } (from πτερὸν, a wing,
PTERO'PODA. } and πούς, a foot, Gr.) The second order of the class Cephalophora. A class of molluscs possessing organs adapted for either swimming or sailing. The genera belonging to this class have the sac formed by the mantle closed on every side; a structure rendering it necessary that the gills should be placed externally, as regards the sac; and they are found spreading out like a pair of wings on each side of the neck. This position of the gills, causing them to resemble the wings of an insect, suggested to Cuvier the name which he assigned to the class.

PTERO'PODOUS. Belonging to the class Pteropoda; wing-footed.

PTERYGO'TUS. The name given by M. Agassiz to a genus of fossil fishes discovered in the Upper Ludlow Rock. These singular winged bodies have been called Seraphims by the Scotch quarrymen. M. Agassiz says, "the more I know of this creature, the more I am tempted to believe that it was a fish; but how absolutely decide upon it, when we have neither discovered head nor tail, but only large wings."—*Murchison. Silurian System.*

PTILO-DI'CTYA. (from πτίλον, pluma, and δίκτυον, rete, Gr.) The name assigned by Mr. Lonsdale to a genus of corals found in the Silurian rocks, from the feather-like arrangement of the middle and lateral cells, and their net-like union. "This fossil," says Mr. Lonsdale, "is considered by Goldfuss to be a *flustra*, but it is placed by Milne Edwards among the doubtful species of that genus. It differs essentially from *Flustra* in the thickening of the external crust. From *Eschara* it differs in not having a central partition, and in the surface of the cells not being convex, but depressed as in Flustra." For a description of the generic characters the reader is referred to Sir R. Murchison's Silurian System. One species, P. lanceolata, has been established by Mr. Lonsdale.

PTYCHACA'NTHUS. The name of a genus of ichthyolites of the old red sandstone.

PUBE'SCENCE. (from *pubesco*, Lat. to grow mossy, or hairy.) The downy substance on plants, resembling fine silken short hairs.

PUBE'SCENT. Clothed with fine short hairs or down.

PU'DDING-STONE. A conglomerate composed of rounded stones imbedded in a paste. Pudding-stone is distinguished from breccia by the form of the contained pebbles: in the latter they are sharp angular fragments; in the former they are rounded nodules.

PULVINI'TES. (from *pulvinus*, Lat. a cushion.) A fossil bivalve found in the baculite limestone of Normandy.

PU'MICE. } (*pumex*, Lat. *pómice*,
PU'MICE-STONE. } It.) The Bimstein of Werner: Ponce of Brongniart: La pierre ponce of Brochant: Lave vitreuse pumicée of Haüy. A light, spongy, fibrous lava, produced by the action of gases on trachytic and other lavas. The island of Lipari contains a mountain entirely formed of white pumice: when viewed from a distance, the appearance is that of a mountain completely covered with snow; it is called Il Campo Bianco. Klaproth gives the analysis of a specimen from Lipari as follows: silex 76·5, alumine 17·5, soda and potash 5·0, iron 1·75. Immense quantities of pumice are sometimes ejected from volcanoes. Werner and Kar-

sten have divided pumice into three sub-species, namely, Glassy pumice, Common pumice, and Porphyritic pumice. Glassy pumice is distinguished from the other sub-species by its darker colours, its vitreous and conchoidal cross-fracture, and its greater hardness and translucency. Porphyritic pumice is distinguished by its containing crystals of felspar, quartz, and mica, thus forming a sort of porphyry. Abich divides pumice into two groups; the cellular being dark green, poorer in silica and richer in alumina, derived from clinkstone, trachyte, or andesite; the filamentous white, containing more silica, and derived from trachytic porphyry.—*Jukes.*

PU'NCTATED. (from *pungo*, Lat. to prick.) Marked with small dots or punctures; full of small holes.

PU'NCTULATED. Some of the ganoid scales of Burdie House present surfaces similarly punctulated.—*Hugh Miller.*

PU'PA. (*pupa*, Lat.)
1. In conchology, a genus of cylindrical univalve land shells, belonging to the family Colimacea.
2. In entomology, the chrysalis; one of the states of existence of such insects as undergo metamorphoses.

PU'RBECK BEDS. (The Calcaire Lumachelle Purbeckien of Brongniart.) A fresh-water deposit, consisting of various kinds of limestones and marls. The Purbeck beds constitute the lowest members of the Wealden group, lying below the Hastings sands, and immediately above the Portland beds. The Purbeck limestone abounds in organic remains, and the marble is a congeries of small fresh-water snail-shells (paludina), intermixed with the minute crustaceous coverings of a species of cypris. The Purbeck beds consist of many thin strata of argillaceous limestone, alternating with schistose marls, and forming an aggregate of upwards of 300 feet in thickness.

PU'RBECK STRATA. The lowest deposits of the Wealden group.

PU'RPLE CO'PPER ORE. The Cuivre pyriteux hepatique of Haüy; the buntkupfererz of Werner. A species of sulphuret of copper. This ore occurs in masses, or plates, or disseminated; it is characterized by its lively and variegated colours, from which circumstance it is frequently called variegated pyritous copper.

PU'RPURA. (*purpura*, Lat. the shell-fish from which purple cometh.) A genus of marine univalves belonging to the family Purpurifera. The shells of this genus are found at depths varying to twenty-five fathoms, and the greater number of the species are littoral. The Purpura is an ovate univalve, its surface being rather rough with spines or tubercles; aperture notched, and slightly channelled in the lower part; the columella naked, flat, depressed, and terminating in a point at the base. The animal resembles that of a true buccinum. The species are very numerous; the animal secretes a purple liquor, which was formerly used in dyeing. Fossil and recent.

PURPURI'FERA. A family of Trachelipoda in Lamarck's system, comprising the genera Buccinum, Cassidarea, Cassis, Concholesas, Dolium, Eburna, Harpa, Monoceros, Purpura, Ricinula, and Terebra.

PURPURO'IDEA. A genus of turbinated and tuberoolated univalves found in the great oolite; it is ventricose, with one or more circles of spines or tubercles: the aperture is large ovate, the base widely notched, the columella is smooth, rounded and curves upwards at its base: three species are found at Minchinhampton.—*Lycett.*

PUTA'MEN. *(putamen,* Lat. a shell of a nut.) In botany, another name for the endocarp, stone, or shell of certain fruits.

PYCNODO'NTIDÆ. According to some naturalists, a family of extinct fishes of the order Ganoidea Rhombifera. This family comprises many genera, namely, Pycnodus, Gyrodus, Phyllodus, &c., &c.

PY'CNODONTS. (from πυκνός, thick, and ὀδούς, a tooth, Gr.) Thick-toothed fishes. An extinct family of fishes which prevailed extensively during the middle ages of geological history. Their leading character consists in a peculiar armature of all parts of the mouth with a pavement of thick, round, and flat teeth, the remains of which, under the name of bufonites, occur most abundantly throughout the oolite formation.

PYCNO'DUS TRIGONUS. A genus of thick-toothed fishes, belonging to the family of Pycnodonts.

PY'CNITE. A mineral of a yellowish-white colour, found principally at Altenberg, in Saxony.

PYRA'LLOLITE. (from πῦρ, fire, ἄλλος, and λίθος, a stone, Gr., in allusion to its change of colour when exposed to the action of fire.) An earthy mineral discovered by Nordenskiold in the lime quarries, at Pargas, in Finland. It is a trisilicate of magnesia mixed with hydrate of the same. Its analysis, according to Nordenskiold, gives silica 56·62, magnesia 23·28, alumina 3·41, lime 5·56, proxide of iron 0·99, protoxide of manganese 0·99, water 3·58. Specific gravity, 2·55; hardness, 3·5. It occurs massive, and crystallized in flat rhombic prisms, much resembling those of tremolite, and is divisible parallel to the planes of a rhombic prism: the prisms are generally above an inch in length, have occasionally a greenish tinge, but by exposure become of a pale yellow, and are then soft and friable.—*Phillips.*

PY'RAMID. (πυραμίς, Gr., from πῦρ, fire; *pyramide,* Fr. *pyramide,* It.) The name given to a certain figure, from its resembling the shape of flame. A solid figure, whose base is a polygon, and whose sides are plain triangles, their several points meeting in one. In mineralogy, a pyramid is formed by the meeting of three or more planes at a point, which is termed the apex, each plane being bounded by *edges;* considered separately, a pyramid is supposed to have a base, which is the case in regard to the tetrahedron; but in respect of such other forms, it is only imaginary, as in the instance of the octahedron, which is often termed a double four-sided pyramid; and also the dodecahedron with triangular faces, which is frequently denominated a double-sided pyramid.

PYRAMIDE'LLA. A genus of marine univalves, belonging to the family Plicacea. It is found on coral reefs, sands, and sandy mud, at depths varying to twelve fathoms. The pyramidella is a turriculated univalve; opening entire and semioval; columella projecting, with three transverse folds, and perforated at its end.

PY'RENAITE. } A greyish-black dode-
PY'RENEITE. } cahedral, opaque variety of dodecahedral garnet. It consists of silica 43, alumina 16, lime 20, oxide of iron 16, water 4. It is found in the French Pyrenees, from which circumstance it has obtained its name, in limestone.

PY'RGOM. (Called also Fassaite.) An earthy mineral of a dingy green colour; a variety of augite.

PYRI'TES. (πυρίτης, Gr. *pyrites,* Lat. *pyrite,* Fr.) Sulphuret of iron. The fersulfure of Haüy, who has described sixteen modifications of the primitive form of its crystals. The colour of pyrites is usually bronze yellow, passing to brass yellow, and sometimes to brown. It possesses the hardness of quartz, striking fire with steel, and emitting an odour of sulphur. Specific gravity from 4·10 to 4·80. Before the blow-pipe it exhales a strong odour of sulphur, and yields a brownish globule. It is composed of sulphur and iron; iron 47·85, sulphur 52·15. It occurs in almost every rock, stratified or unstratified. The shining yellow streaks so common in our coals afford familiar examples of pyrites.

PYRI'TIFEROUS. Yielding or containing pyrites (sulphuret of iron). "These beds are underlaid by sandy *pyritiferous* shale."—*Murchison*.

PY'ROCHLORE. The pyrochlor of Werner. Octahedral titanium ore.

PYRO'GENOUS. (from πῦρ, fire, and γεννάω, to produce, Gr.) Produced by fire, as rocks of igneous origin.

PYROLU'SITE. Another name for the common ore of manganese, as prismatic manganese ore.

PYRO'METER. (from πῦρ, fire, and μέτρον, measure, Gr.) An instrument for measuring intense degrees of heat. The most celebrated pyrometer is that invented by Mr. Wedgewood, the extremity of his scale reaching to 240° Wedgewood, or 32277° Fahrenheit.

PYRO'PE. (from πυρωπὸς, Gr. *pyropas*, Lat.) The Pyrop of Werner; Grenat, rouge de feu, granuliforme, of Haüy; Grenat pyrope of Brongniart; Bohemian garnet of some authors. A deep blood-red variety of dodecahedral garnet. It consists of silica 40·0, alumina 27·6, oxide of iron 16·0, magnesia 10·0, lime 3·5, chromic acid 2·6, oxide of manganese 0·3. It occurs in small masses or grains. It is generally transparent, with a splendent, vitreous, conchoidal fracture. Specific gravity from 3·7 to 3·9. It refracts double, and is sufficiently hard to scratch glass. Its fine blood-red colour is supposed to be owing to the presence of chromic acid. It is found imbedded in serpentine, wacke, and trap-tuff in Ceylon, and different parts of Germany. It is also found in some alluvial soils, accompanied by sapphires and hyacinths. As a gem it is highly prized.

PYRO'PHORUS. (from πῦρ, fire, and and φορὸς, bearing, Gr. *pyrophore*, Fr.) A substance which has the property of igniting merely on exposure to the air. It is an artificial production, and may be variously prepared. M. Gay Lussac formed a pyrophorus of one part of lamp-black and two parts of sulphate of potash.

PYROPHY'LLITE. An earthy mineral, described by Hermann of Moscow, occurring in the Ural mountains. It is of a light green colour, has a pearly lustre, and may be divided into thin and transparent laminæ. It is found in fibrous radiating masses, and small elongated prisms, sometimes with terminations, whose form is not however ascertained.

PYROPHY'SALITE. A variety of topaz. The Pyrophysalith of Hisenger. This mineral which some authors consider to be intermediate between topaz and schorlite occurs principally at Finbo, near Fahlun, in Sweden. It is of a greenish-white and mountain green colour, translucid or opaque, and sometimes prismatic. On burning coals it phosphoresces, in consequence of the fluate of lime with which it is mixed; from this property it has obtained its name, which is derived from the Greek. It occurs massive and crystallised, and may be regarded as a coarse opaque variety of topaz. Specific gravity 3·451. It scratches glass, but is itself scratched by quartz. It contains alumina, silica, fluoric acid, and a trace of lime and iron.

PYROSO'MA. (from πῦρ, fire, and σῶμα, a body, Gr.) From the phosphorescent character which they possess they have obtained their name. A floating polyp, differing from the coral in being locomotive. Peron observed that when irritated its phosphorescence was augmented.

PYROXE'NE. (from πῦρ, fire, and ξένος, a stranger, Gr.) The name given by Haüy and Brongniart to augite.

PY'RULA. (from *pyrum*, a pear, Lat.) A genus of marine univalves, belonging to the family Canalifera. Recent pyrulæ are found at depths varying to nine fathoms, in mud and sand sand. The pyrula is a somewhat pyriform univalve, swelling in the upper part, with no variciform sutures, caudated, canaliculated, spire short, aperture wide, outer lip thin, and not slit, columella smooth. Six species have been found fossil by Lamarck. Pyrulæ are also found in the London clay, and in the Bognor sandstone.

# Q

QUA'DRATE. (*quadratus*, Lat.) Square; having four equal and parallel sides.

QUADRICA'PSULAR. In botany, having four capsules to a flower.

QUADRIDE'NTATE. Four-toothed.

QUADRILO'BATE. (from *quatuor*, four, and *lobus*, a lobe, Lat.) A term applied in botany to a part having four lobes; as a quadrilobate leaf.

QUADRILO'CULAR. (from *quatuor*, four, and *loculus*, a cell, Lat.) Four-celled; having four cells.

QUADRIPHY'LLOUS. (from *quatuor*, four, Lat. and φύλλον, a leaf, Gr.) In botany, having four leaves; four-leaved.

QUADRIPLI'CATED. (from *quatuor*, four, and *plica*, a fold, Lat.) A term in conchology; having four plaits or folds.

QUADRIVA'LVULAR. (from *quatuor*, four, and *valva*, a valve, Lat.) A term in botany; having four valves.

QUADRU'MANA. (from *quatuor*, four, and *manus*, a hand, Lat.) The second order of Mammalia, including the monkeys, lemurs, and onisites. All the animals of this order have the toes of the hind feet free and opposable to the others, and all the toes are as long and flexible as the fingers. The great character which distinguishes the members of this order is the possessing a moveable thumb on their lower extremities opposed to the fingers.

QUADRU'MANOUS. Having four hands; four-handed.

QUA-QUA-VE'RSAL. (*quaquà*, Lat.) on every side, and *versus*, inclined, Lat.) Inclined towards every side; facing all ways: a term applied to the dip of a bed which is inclined, facing all sides.

QUARTZ. (*quarz*, Germ. *quartz*, Fr. *mot emprunté de l'Allemand.* The meaning of the word quartz does not appear to be known.) Silex in its purest form. Quartz is a hydrate of silicon, or silex with some water of crystallization; it is a compound of a metallic basis, silicium, and oxygen. The usual colours of common quartz are white and grey. Quartz is found in every variety of form, although in its composition it varies but slightly. When crystallised, it usually occurs in six-sided prisms, terminated by pyramidal points; but it also occurs in many derivative forms, its primitive being a rhomboid. The primitive crystal of quartz, says Phillips, is considered to be an an obtuse rhomboid, of which the angles are given by Haüy in his Tableau Comparatif, as being 94° 24′ and 85° 36′. A minute examination of several specimens afforded perfect coincidences of 94° 15′ on the one angle, and 85° 45′ on the other, the former differs from that given by Haüy, in being 9′ less; the latter in being 9′ more. It scratches glass, of which it is an essential ingredient; it gives sparks with steel in great abundance. Hardness = 7. Specific gravity = 2·5, to 2·7. Its lustre ranges from splendent to glimmering, and is vitreous. Fracture conchoidal. Brittle, and easily frangible. It exhibits double refraction, which must be observed by viewing an object through one face of the pyramid and the opposite side of the prism. It is infusible before the blow-pipe, even when the flame is excited by oxygen. Before the compound blow-pipe a fragment of rock crystal fuses instantly into a white glass. Quartz is very generally distributed, and, as far as our knowledge extends at present, appears to be the most abundant mineral in nature. It occurs in every rock, from granite to the newest secondary formation; it is found in every district of the globe. When mixed with alkalies quartz melts easily, and forms glass. It is not acted upon by any acid but the fluoric. Quartz often encloses, or is mixed up with, foreign substances, from which circumstance it assumes great varieties of colour. Many of the precious stones consist of simple quartz combined with some colouring matter. Among these we may place the amethyst, cats'-eye, opal, Bristol diamond, Scotch topaz, &c. Professor Jameson divides quartz into six sub-species, namely amethyst, rock crystal, rose or milk quartz, common quartz, prase, and cats'-eye. Quartz exists in veins intersecting mountains, and it sometimes forms large beds, and even entire mountains, which are composed of this mineral in grains united without a cement, called granular quartz. Combined with alumine and iron, quartz loses its translucency and passes into jasper. When rubbed, quartz yields a peculiar odour, and a phosphorescent light. Quartz is a constituent of granite.

QUA′RTZOSE. } Containing quartz; composed of quartz; resembling quartz; having the properties of quartz.
QUA′RTZY.

QUARTZOSE CONGLOM′ERATE. } The name given to the upper division of the New Red System of Murchison. When the coverings of turf or bog are removed from the old red sandstone, "a conglomerate," says Sir R. Murchison, "composed of pebbles of white quartz in a red matrix, forms the upper member of the old red sandstone, or substratum of the carboniferous limestone."
OLD RED CONGLOMARATE.

QUARTZ ROCK. } A compact, fine-grained but distinctly granular rock, very hard, frequently brittle, and often so divided by joints as to split in all directions into small angular but more or less cuboidal fragments.—*Jukes*.
QUART′ZITE.

QUATE′RNARY. (*quaternarius*, Lat. *quaternaire*, Fr.) Consisting of four; the number four. In geology, a term applied to the upper tertiary strata, or those which are supposed to be of later formation than any of the strata in the Paris or London basins. The faluns, or marls of Tourraine and the Loire, are *quaternary* formations.

2 c

A group of rocks more recent than the Pleistocene, and by some authors called "Post-pleiocene," in which all the fossil shells are still found living on the globe, though not always in the immediate neighbourhood of the places where they are found fossil. The quaternary, or Post-pleiocene formations consist of irregular deposits of clay, sand, and gravel, with which, in many cases, are associated huge blocks of rock, that have been transported sometimes from vast distances.

QUI'CKLIME. This may be obtained by exposing chalk, limestone, or any calcareous substance, for a length of time, to a full red heat, whereby the carbonic acid and the water, which were previously in combination, are expelled.

QUI'NATE. In botany, applied to compound leaves when composed of five leaflets.

QUINQUA'NGULAR. (from *quinque*, five, and *angulus*, an angle, Lat.) Having five corners or angles.

QUINQUECA'PSULAR. (from *quinque*, five, and *capsula*, a capsule, Lat.) In botany, applied to a flower having five capsules.

QUINQUEDE'NTATE. (from *quinque*, five, and *dens*, a tooth, Lat.) Five-toothed.

QUINQUEFO'LIATED. (from *quinque*, five, and *folium*, a leaf, Lat.) Having five leaves.

QUINQUELO'BATE. } (from *quinque*, five,
QUINQUELO'BED. } and *lobus*, a lobe, Lat.) In botany, applied to parts which are divided to the middle into five distinct parts or lobes.

QUINQUELO'CULAR. (from *quinque*, five, and *loculus*, a cell, Lat.) Five-celled; having five cells.

QUINQUEVA'LVULAR. (from *quinque*, five, and *valvæ*, doors or valves, Lat.) Having five valves.

# R

RACE'ME. (*racemus*, Lat. a bunch of berries.) A term used in botany to express a kind of inflorescence, when all the buds of a newly-formed axis unfold into flowers, each having stalk; a raceme consists of numerous flowers, rather distant, each on its own proper stalk, and all connected together by one common peduncle.

RADIA'TA. The name given to the fourth great division of the animal kingdom, in consequence of the radiated form of the body which is so apparent in some of the classes which compose it. This division comprises five classes, namely, Echinoderma, Acalepha, Polyphera, Poriphera, and Polygastrica. The radiata are amongst the most frequent organic remains in the transition strata, and they present numerous forms of great beauty. Radiata have also been called zoophyta. They form the lowest division of the animal kingdom, having skeletons as various as the forms of the animals. "The muscular system, so essential to all the voluntary and involuntary movements of the higher classes, entirely disappears in the class of radiated animals, and with it every trace of nervous system, which we are apt to suppose essentially connected with all the motions and functions of animals. In the majority of radiated animals the blood-vessels also cease to exist, and with them all circulating motion of the blood."

"In the lowest tribes of animals the internal organization relates almost solely to digestion, and the food consists almost entirely of the simplest forms of animal matter. The alimentary canal has often but one orifice, it is seldom provided with masticating organs, and scarcely a trace of any glandular organ is yet observed to assist in the process of assimilation."—*Prof. Grant.*

Ra′diate. In botany, a corolla consisting of a disk, in which the corollets are tubular and regular, and of a ray, in which the florets are irregular.

Ra′diated. (*radiatus*, Lat.)
1. Adorned with rays.
2. Belonging to the division Radiata.
3. In mineralogy, when the fibres are broad and flat, and diverging as from a centre.

Ra′diated iron pyri′tes. A variety of sulphuret of iron of a pale bronze yellow, more or less inclining to steel-grey, or to brass-yellow. When its form is spherical, the fibres diverge from the centre; when nearly cylindrical, from the axis. Its constituents are iron 46·03, sulphur 53·97. It occurs more particularly in the chalk deposit.

Ra′diated qua′rtz. A variety of common quartz occurring in masses having a crystalline structure and composed of imperfect prisms, closely applied to each other, and sometimes terminating in pyramids at the surface.

Ra′dicated. (*radicatus*, Lat.) In conchology, when the shell is fixed by the base, or by a byssus, to some other body.

Ra′dical. (from *radix*, Lat. a root.) Primitive; original. In botany, radical leaves are such as spring from the root; the dandelion is a familiar example.

Ra′dicle. (*radicula*, a little root, from *radix*, Lat. *radicule*, Fr.) In botany, that part of the embryo which grows downwards and becomes the root. The primary object of vegetable structures appears to be the establishment of the functions of nutrition; and we find that while the plumule, bursting from its enveloping capsule, proceeds to extend itself vertically upwards, at the same time, slender filaments, or *radicles*, shoot out below to form the roots.

Ra′dii vecto′res. Imaginary lines joining the centre of the sun and the centre of a planet or comet, or the centres of a planet and its satellite.

Radioli′tes. A genus of irregular inequivalved fossil shells obtained from that part of the Pyrenees which is named Les Corbieres. They are striated externally. The inferior valve is in the form of a reversed cone; the superior valve convex. They have neither hinge nor cartilage.

Ra′dius. (*radius*, Lat.)
1. The semi-diameter of a circle. The mean radius of the earth is intermediate between the distances of the centre of the earth from the pole and from the equator.
2. One of the bones of the forearm, or that part of the upper extremity which extends from the elbow to the wrist: the fore-arm contains two bones, the *radius* and the *ulna*.

Ragg.            ) Called also Rowley ragg,
Ra′gstone. ) or Dudley basalt. A fusible silicious stone, of a dark grey colour, occurring near Dudley.

Ra′mose. ) (from *ramus*, a branch,
Ra′mous. ) Lat.) Branched; full of branches. Applied also to flowers growing on the branches; to peduncles proceeding from a branch; and also to leaves growing on branches when they differ from those on the stems. Minerals

having a branched appearance are described as ramose.

RANE'LLA. A genus of marine univalves, belonging to the family Canalifera. Recent ranellæ are found principally in the Indian seas, at depths varying to eleven fathoms. Some fossil species have been discovered in the London clay.

RAPTO'RES. The first order of the class Aves, or birds: this order comprises the eagle, hawk, vulture, and owl.

RAPTO'RIOUS. (from *rapio*, Lat. to snatch.)
1. The name given to animals which dart upon and seize their prey.
2. The name given to certain parts of insects. The legs are called *pedes*; when adapted to the seizing of prey, they are called *pedes raptorii*, not arms.

RA'VIN. } (*ravin*, Fr.) A deep, hollow, narrow excavation formed by the force of running water. The French use the two words ravin and ravine in different significations. The word *ravin* is used to express a place which has been hollowed out by a stream of running water, as "*passer un ravin profond.*" Ravine is employed to denote a torrent of water, "espèce de torrent formé d'eaux qui tombent subitement et impétueusement des montagnes, ou d'autres lieux élevés, après quelque grande pluie." Sometimes, however, *ravine* is used to signify the place worn by the torrent.

RAY. (from *raja*, Lat. *raie*, Fr.) A sea fish. The rays form a genus of the order Chondropterygii; they may be known by their flattened body, which is in the form of a disk, from the union of the body with the very broad and fleshy pectorals, which are joined to each other before or on to the snout, and which extend behind the two sides of the abdomen as far as the base of the ventrals. The rays have no ribs. The phalanges of the carpus are very numerous, and each is subdivided into several pieces by regular articulations; they are arranged close to one another in one plane, and form an effectual base of support to the integument which covers them. Both the anterior and posterior extremities are supported by arches of bones, forming a sort of belt. In this genus are included the skate, torpedo, and stingray. Kirby, quoting from Lacepede, says, an individual of a species of this tribe, called by the sailors the sea-devil, taken at Barbadoes, was so large, as to require seven pairs of oxen to draw it on shore; he very judiciously, however, adds two notes of admiration to this marvellous story. Fossil rays are abundant throughout the tertiary formation, they occur also in the jurassic limestone.

RAY. 1. A name given to the fins of certain fishes. 2. In botany, the florets composing the margin of a compound flower.

RE'ALGAR. (*realgar*, Fr.) Red sulphuret of arsenic. Arsenic combined with sulphur forms *realgar* and *orpiment*, which are found as natural ores; the realgar is of a beautiful and variable red. Realgar is a bi-sulphuret of arsenic, consisting of arsenic 69, sulphur 31. Specific gravity 3·27. It occurs in regular crystals, in compact masses, in concretions, or in crusts, which are sometimes earthy. Before the blow-pipe it melts easily, burns with a blue flame and garlic smell, and soon evaporates. Nitric acid deprives it of its colour. It occurs in veins in primary, transition, secondary, and volcanic rocks; in Sicily, Germany, America, and other parts.

RE'CENT FORMA'TION. Any formation, whether igneous or aqueous, which can be proved to be of a date posterior to the introduction of man is called *recent*.

RE'CENT PE'RIOD. That period of time commencing with the introduction of man upon this earth.

RECE'PTACLE. (*receptaculum*, Lat.) In botany, the basis or point, upon which all the parts of the fructification rest. The receptacle is distinguished into *receptaculum proprium*, or receptacle appertaining to one fructification only; and *receptaculum commune*, or common receptacle, connecting several distinct fructifications.

RECLI'NED. In botany, applied to leaves when the point is lower than the base; also to stems when curved towards the ground.

RED CO'RAL. A branched zoophyte, somewhat resembling in miniature a tree deprived of its leaves and twigs. It seldom exceeds one foot in height, and is attached to the rocks by a broad expansion or base. It consists of a bright red, stony axis, invested with a fleshy, or gelatinous substance of a pale blue colour, which is studded over with stellular polypi.

RED CRAG. The Crag of Suffolk consists of two groups, the Red Crag and the Coralline Crag. The Red Crag, which is the later formation, is composed of red quartzose sand, and accumulations of rolled shells.

RED MARL. Another name for the new red sandstone.

RED OR VARIEGATED MARLS. The uppermost division of the red sandstone group. The Marnes Irisées of the French, the Keuper, Bunte Mergel of the German geologists.

RED SNOW. See *Protococcus Nivalis*.

RE'DDLE. The roethel of Werner; the crayon rouge of Brochant; the argile ocreuse rouge graphique of Haüy. Red chalk; a species of argillaceous iron-stone ore. The best specimens are brought from Germany. It occurs in opaque masses, having a compact texture. In hardness, it differs but little from chalk. It is dry, and rough to the touch, adhering to the tongue, and yielding an argillaceous odour. Specific gravity from 3·10 to 3·90.

RED SANDSTONE GROUP. This group is placed below the oolitic group, and above the carboniferous group; and in Sir H. De la Beche's table constitutes the sixth group of the superior stratified or fossiliferous rocks. The red sandstone group comprises— 1. The variegated or red marl; 2. The muschelkalk; 3. The red sandstone; 4. The zechstein or magnesian limestone; 5. The red conglomerate.

REEF. (*riff*, Germ.) A range of rocks lying generally near the surface of the water.

REFLEC'TING GONIO'METER. An instrument invented by Wollaston, and improved by Sang, for measuring the angles of crystals. By the reflecting goniometer the smallest crystals may be accurately measured, for, provided the surface be perfect and brilliant, the 100th part of an inch in length and breadth will suffice.

REFLE'CTION. The act of throwing back; the act of bending back; that which is reflected. By the laws of optics, the angle of reflection is equal to that of incidence, whatever the reflecting surface may be, and however obliquely the light may fall upon it.

REFRA'CTION. (*réfraction*, Fr. *refrazióne*, It.) The incurvation, or change of determination in the body moved, which happens to it whilst it enters or penetrates any medium; in dioptricks, it is the variation of a ray of light from that right line, which it would have passed on in, had not the density of the medium turned it aside. When light passes through a drop of water or a piece of glass, it obviously suffers some change in its direction. These bodies have therefore exercised some action, or produced some change upon the light, during its progress through them. The power which thus bends or changes the direction of a ray of light is called *refraction*, and the amount of this refraction varies with the nature of the body. All the celestial bodies appear to be more elevated than they really are; because the rays of light moving through the atmosphere in straight lines, are continually inflected towards the earth. Light passing obliquely out of a rare into a denser medium, as from vacuum into air, or from air into water, is bent or refracted from its course towards a perpendicular to that point of the denser surface where the light enters it. The denser the medium the more the ray is bent.

REFRA′CTION DOUBLE. When a ray of light passes obliquely from one medium to another of a different density, it is refracted, or bent from its original direction. Still, the image of any object, seen through a refracting medium, usually appears single. There are, however, some transparent minerals, which have the remarkable property of causing objects to appear double; that is, they present two images of any object seen through them. In this case it is evident that the ray must be divided into two portions after entering the refracting medium, and that each portion presents an image of the object. This is a distinctive character of very considerable value in some minerals, not depending on any accidental circumstances, but on the nature of the mineral. Double refraction is exhibited most strikingly in Iceland spar.

REFRA′CTORY. A term applied to minerals when they possesss the property of resisting the application of strong heat; it is also used for such as are so tough as to withstand repeated blows.

REFRIGERA′TION. (*réfrigération*, Fr.) The act of cooling down; the state of being cooled

It is the opinion of some geologists that the whole of this planet was formerly in an incandescent state, and that the process of *refrigeration* has been constantly proceeding; that the crust of the earth has cooled down to its present temperature, but that the centre of the earth is still a molten mass. Professor Whewel, speaking of Fourier's arguments on the subject, says, "it results from Fourier's analysis that at a depth of twelve or eighteen miles the earth may be actually incandescent, and yet that the effect of this fervent mass upon the temperature at the surface may be a scarcely perceptible fraction of a degree. The slowness with which any heating or cooling effect would take place through a solid crust is much greater than might be supposed. If the earth below twelve leagues depth were replaced by a globe of a temperature five hundred times greater than that of boiling water, 200,000 years would be required to increase the temperature of the surface one degree."

RE′NIFORM. (from *ren*, a kidney, and *forma*, form, Lat.) Kidney-shaped.

RE′PTILE. (*reptilis*, Lat. *reptile*, Fr. *rettile*, It.) A vertebrated, cold-blooded, animal. The body covered with a shell, or with scales, or entirely naked. Possessing neither hair, mammæ, feathers, nor radiated fins; breathing through the mouth and nose by means of lungs; oviparous, but never hatching its eggs, and amphibious.

REPTI′LIA. Reptiles form the third class of vertebrated animals, which is subdivided into four orders, namely, Chelonia, or tortoises, Sauria, or lizards, Ophidia, or serpents, and Batrachia, or frogs. Reptilia have red and cold blood, with true lungs; they are oviparous; the heart is trilocular, that is possesses three cavities, namely, two auricles and one ventricle. The brain of reptiles is characterized by the position of the thalami behind the hemispheres; **neither is** there any arbor vitæ in the cerebellum. Some naturalists **divide the** class reptilia into **seven orders,** namely, Chelonia, Crocodilia, Lacertilia, Ophidia, *Dinosauria, Enaliosauria,* and *Pterodactylia,* the three last being extinct, and found fossil only.

REPU′LSION. (*répulsion.* Fr.) That property possessed by bodies which causes their particles to recede from one another, or to avoid coming in contact. In air and in liquids, the most perfect freedom of motion **of** the parts among each other subsists, and from this, and other considerations, it has been concluded that the several parts do not touch, but are kept asunder at determinate distances from each other, by the constant action of the two forces of attraction and *repulsion,* which are supposed to balance and counteract each other.

RE′SIN. (*resina,* Lat. *résine,* Fr.) Called also rosin. A yellowish-white coloured substance, which exudes **from** many **trees,** more particularly the different species of fir. It is somewhat transparent, is hard and **brittle, of a** disagreeable taste, and **may be collected in** considerable quantities. Resin may be distinguished **from other** substances by the following properties. It is more or less concrete, and possesses a certain degree of transparency. Its taste is sometimes hot and disagreeable, but not unfrequently it is tasteless, or nearly so. Colour generally between pale yellow and brown. Specific gravity from 1·0 to 1.3. It is electric and a non-conductor of electricity. It is insoluble in water, and by this may be distinguished, as well as separated from gum. It is soluble in alcohol, ether, and in the volatile oils. There is scarcely a plant which does not contain some kind of resin.

RESPIRA′TION. (*respiratio,* Lat.) **The** function of breathing; the act of inhaling air into and exhaling it from the lungs. Respiration consists of two parts, inspiration, which, in a healthy condition, takes place about twenty-six times in a minute, thirteen cubic inches **of air** being the average quantity taken in at each inspiration; and expiration, which alternates with inspiration. It is by respiration that the blood becomes freed of its carbon and, at the same time, obtains fresh supplies of oxygen.

RESU′PINATE. (*resupinatus,* Lat.) In botany, a term applied to leaves, when the under surface is **turned** upwards.

RE′TE MUCO′SUM. (from *rete,* a net, and *mucosum,* mucous, Lat.) A tissue lying immediately under the epidermis, or scarfskin, and above the cutis vera, or true skin. The colour of negroes depends upon a black pigment, situated in this substance.

RETEPO′RA. The name given by Lamarck to a genus of fossil corals.

RE′TINA. (from *rete,* a net, Lat.) The net-like expansion of the optic nerve placed at the back of the eye, and which has been called one of the membranes of that organ. The *retina* is an exceedingly thin and delicate layer of nervous matter, supported by a fine membrane. No nerve but the optic nerve, and no part of that nerve but the *retina,* is capable of giving rise to the sensation of light.

RE′TINASPHALTUM. A sub-species of bitumen. An opaque, ochre-yellow, and brittle substance found in Bovey coal and fossil wood. Fracture vitreous, and imperfectly conchoidal. Specific gravity 1·13. It consists of resin 55, asphaltum 41, earthy matter 3.

RETROFLEC'TED. (from *retro*, back, and *flecto*, to bend, Lat.) Bent in different directions, usually in a distorted manner.

RETROMI'NGENTS. In zoology, a class of animals, whose characteristic is, that they void their urine backwards.

RETU'SE. (*retusus*, Lat. blunted.)
1. In conchology, a shell ending in an obtuse sinus is termed a *retuse* shell; bluntly notched.
2. In botany, leaves are called *retuse*, when ending in a broad shallow notch, or sinus.
3. In entomology, when the terminal margin has an obtuse impression.

REVE'RSED. (*reversus*, Lat.) Turned upside down; turned side for side.
1. In conchology, a reversed shell is one, the volutions of which are the reverse way of the common corkscrew.
2. In botany, when the upper lip of the corolla is larger and more expanded than the lower.

RE'VOLUTE. (*revolutus*, Lat.) In botany, applied to leaves, when the margins are rolled backwards towards the under surface.

RE'YGATE STONE. } (So named
RE'YGATE FIRE-STONE. } from having been formerly quarried almost exclusively at, or in the neighbourhood of, Reygate, in Surrey. Webster says that Reygate stone is identical with the green sandstone. The Reygate stone is situated below the chalk marl and above the ferruginous sand. It is now procured principally at Merstham, three miles from Reygate.

RHACHEOSAU'RUS. A fossil saurian of the lias and oolite.

RHINO'CEROS. (ῥινόκερως, Gr. from ῥίν, a nose, and κέρας, a horn.) A genus of thick-skinned mammalia included by Cuvier in the family Pachydermata Ordinaria. The rhinoceros is found, at the present day, in India, Java, Africa, and Sumatra. It is a large animal, having three toes, but the feet not cloven. The bones of the nose support a solid horn in two species of this genus; but the other species possess two horns.

Fossil remains of the rhinoceros are found in Siberia and Germany. The entire carcass of a fossil rhinoceros was discovered, in frozen sand, on the banks of the Wilaji, in Siberia. Bones and teeth of the rhinoceros are found in this country in superficial gravel and loam.

RHIPI'PTERA. A new order of insects, established by Kirby under the name of *Strepsiptera* (twisted wings), named Rhipiptera by Latreille, includes only two genera, namely Stylops and Xenos. These insects are remarkable for their anomalous form, and the irregularity of their habits. The tegmina are fixed at the base of the anterior legs; they are both long and narrow, and appear to be incapable of protecting the wings. The wings are large, membranous, divided by longitudinal and radiating nervures, and fold longitudinally, after the manner of a fan.

RHIZO'MA. (ῥίζωμα, a root, Gr.) A species of creeping stem which grows under-ground.

RHIZO'PODES. A class of animals of lower degree than the Radiata, possessing a power of locomotion by means of minute tentacular filaments; comprising the animals which construct the miliola, and some microscopic foraminiferous shells.

RHO'DIUM. A metal of a white colour, having a metallic lustre, brittle and hard: specific gravity 11. Discovered by Wollaston in 1803, in the native ore of platinum.

RHODOCRINI'TES. (from ῥόδον, a rose, and κρίνον, a lily, Gr. So named from the rose-like figure of its alimentary canal.) A genus of fossil crinoidea occurring in the mountain limestone.

RHOMB. (ῥόμβος, Gr. *rhombus*, Lat.) In geometry, an oblique-angled parallellogram, or quadrilateral figure, whose sides are equal and parallel, but the angles unequal, two of the opposite sides being obtuse, and two acute.

RHO'MBOID. (from ῥόμβος, a rhomb, and εἶδος, form, Gr.) A figure which has its opposite sides equal to one another, but all its sides are not equal, nor are its angles right angles.

RHOMBOHE'DRON. A solid contained by six plane surfaces, the opposite planes being equal and similar rhombs parallel to one another; but all the planes are not necessarily equal or similar, nor are its angles right angles.

RHOMBOI'DAL. Having the form of a rhomboid.

RHOMB SPAR. Called also bitter spar, a variety of magnesian carbonate of lime. This variety occurs crystallised, most of its crystals being rhombs, sometimes truncated, and sometimes with rounded edges. Fracture foliated, the foliæ having a shining or splendent lustre, more pearly than that of calcareous spar. Specific gravity from 2·48 to 3. It is generally of a greyish colour to pale yellow, or yellowish brown. It appears to pass by imperceptible shades into dolomite.

RIBB'ON JAS'PER. A variety of jasper, presenting green, red, and yellow colours, of various shades.

RIBBON AGATE. Called also Striped Agate. A kind of agate which, on being cut at right angles with the layers of which it is composed, presents alternate colours, sometimes straight, at others curved, but always parallel with each other. For a description see *Agate*.

RI'MA. In conchology, the interstice between the valves, when the hymen is removed.

RI'MULA. A genus of small patelliform shells separated from Emarginula on account of the position of its fissure which approaches the margin but does not reach it. The great and inferior oolite of Gloucester have species.—*Lycett*.

RI'NGENT. (from *ringo*, to grin, Lat.) In botany, applied to a monopetalous corolla, the border of which is usually divided into two lips, which gape like the mouth of an animal. A corolla with two lips is called bilabiate; when these present an appearance like the mouth of an animal, the *corolla* is called *ringent*.

RINGI'CULA. A genus of fossil univalves; a species of which occurs in the Suffolk Crag, and also at Bordeaux, and which was formerly attributed to Auricula.

RI'SSOA. A genus of small turrited univalve shells, placed by Sowerby near to the Scalariæ. The aperture is expanded anteriorly, and the outer lip thickened. Four species are known fossil in the great and inferior oolite of Gloucestershire, and in the great oolite at Ancliff. *Lycett*.

RO'ACH. A provincial term given to the upper bed of the Portland oolite, "it is," says Mantell, "a congeries of casts of trigoniæ, pernæ, terebræ, ammonites, lucinæ, &c."

ROCK. (*roc* and *roche*, Fr. *rocca*, It.) Rocks, and the substances they enclose, lie beneath the superficial accumulations, and constitute the crust of the earth. The term "rocks" is apt to mislead beginners; for under this title geologists rank clay, sand, coal, and chalk, as well as limestone, granite, slate, and basalt, and other hard and solid masses, to which the use of the term is generally restricted.

The rocks of which the mineral crust of the globe is composed are

2 D

divided into those of aqueous and igneous origin, from the two agents known to us as capable of their production. Rocks are also divided into primary, transition, secondary, tertiary, &c.

ROCK BUTTER. Native alum. It occurs in the cavities or fissures of argillaceous slate in soft masses; it is of a yellowish-white colour; a little unctuous to the touch; massive, tuberose, or stalactical. It is mingled with clay and oxide of iron.

ROCK CRYSTAL. Called also Mountain Crystal. The Berg Crystal of Werner. This, which is only the most perfect variety of quartz, has, when crystallized, received the name of *rock crystal:* the same name has been extended to coloured crystals, when transparent. Rock crystal is usually limpid, colourless, and as transparent as glass; it is, however, occasionally met with coloured, passing from white into yellow, brown, red and brownish black. The finest specimens of *rock crystal* are found in Dauphiné, in the Alps, in Madagascar, &c.; but it abounds in every country, in all parts of the world, in chasms or clefts of the oldest rocks. The primitive form is a rhomboid of 94° 15′ and 85° 45′; the secondary forms are an equiangular six-sided prism, rather acutely acuminated on both extremities by six planes, which are set on the lateral planes; a double six-sided pyramid; an acute simple six-sided pyramid, and acute double three-sided pyramid. Splendent; fracture perfect conchoidal; gives double refraction when viewed through a pyramidal and lateral plane at the same time. It is perfectly infusible before the blow-pipe, and is sufficiently hard to scratch glass. Its specific gravity is from 2·5 to 2·8. By friction it exhales a peculiar odour, and some varieties also phosphoresce in the dark. An analysis by Bergman gave silex 93, alumine 6, lime 1. Bucholz states it to consist of silica 99¾, with a trace of ferruginous alumina.

ROCK WOOD. A variety of asbestus of a brown colour; in its general appearance greatly resembling fossil wood.

ROCK CORK. Called also Mountain cork; a white or grey-coloured variety of asbestus. Its specific gravity varies from 0·68 to 0.99; this, and its fibrous structure, have obtained for it the name of cork. Its constituents are silex 56·3, magnesia 26·2, lime 12·4, alumine 2, iron 3·1. It occurs in France, Germany and some other countries.

ROCK SALT. Common salt. This is found in vast solid masses or beds, in different formations, but most extensively in the new red sandstone formation. Although rock salt, salt mines, and salt springs are most frequently found to occur in the new red sandstone formation, (from which circumstance many geologists have applied to it the name of saliferous formation) yet are they not confined to them. The salt mines of Wieliezka and Sicily are in tertiary formation; those of Cardona are in the chalk series; some are in the oolitic deposits; and in the county of Durham salt springs occur in the carboniferous series.

The saliferous strata of Northwich, form two beds of great thickness, one being 120, the other 110 feet in thickness. The origin of these beds does not appear, to the present time, to be satisfactorily understood or explained.

RODENT. An animal belonging to the order Rodentia, or gnawers.

RODENTIA. (from *rodo*, Lat. to gnaw.) The fifth order of Mammalia: called also Gnawers. The order contains many genera, some of which are familiar to us, namely, the squirrel, the rat, the mouse, the hare, the rabbit, &c., &c. From the characters of their teeth, which are adapted neither for seizing nor tearing their food, but merely to nibble and gnaw it, they have received their name of *Rodents*, or gnawers.

ROE STONE. A name given to the oolite, a variety of limestone, from its being composed of small rounded particles, resembling the roe or eggs of a fish.

ROMA'NZOVITE. A mineral of a brown, brownish yellow, and blackish-brown colour, named after Count Romanzoff. According to the analysis of M. Julin, it consists of silica 41·22, lime 24·78, alumine 24·08, oxide of iron 7·02, magnesia and oxide of manganese 0·92.

The name given by Nordenskiold to a variety of cinnamon-stone, occurring at Kimito in Finland. See *Cinnamon-stone*.

ROSA'CEÆ. An order of plants, including the genera potentilla, fragaria, rosa, rubus, spiræa, brayera, dryas, &c., &c. The whole order is innocent.

ROSA'CEOUS. Belonging to the order Rosaceæ. Applied to polypetalous corollas, consisting of four or more petals, spreading like a rose.

ROSE QUARTZ. (The Quartz hyalin rose of Haüy.) A subspecies of quartz of a rose-red colour, as its name implies. This mineral occurs massive only; it differs from quartz only in its colour, which colour is attributed to the presence of manganese. See *Milk quartz*.

ROSTELLA'RIA. A genus of marine univalves, recent and fossil. A slightly turretted or fusiform univalve, terminating at its base in a lengthened canal, similar to a sharp beak. The lip whole or dentated, and dilated with age; with a groove contiguous to the canal. Recent Rostellariæ are found in the Indian seas.

RO'STRATED. (*rostratus*, Lat. beaked.) 1. In botany, a term applied to plants when the fruit has a beak-like proccess.
2. In conchology, applied to shells having a beak-like extension of the shell, in which the canal is situated.

RO'STRUM. (Lat.) A beak or bill; the beak or bill of a bird.

ROTA'LIA. A genus of microscopic shells found in flint and in chalk.

ROTALI'TES. A genus of shells existing only in a fossil state, and found at Grignon. The shells of this genus are convex, conical, spiral, multilocular univalves, slightly radiated beneath; aperture marginal, trigonal, and inclined downwards.

RÖTHE TODTE LIEGENDE. The name given by the Germans to the lower beds of the new red sandstone group.

ROTI'FERA. The first order of the class Infusoria. "Minute, transparent, soft, aquatic animals with distinct muscular and nervous systems, provided with eyes, lateral maxillæ, an intestine with distinct buccal and anal openings."—*Prof. Grant*. The researches of Ehrenberg show that a group, formerly believed to belong to the class of the most minute animalcules, possess an organization extremely complex. Ehrenberg has called them "rotatoria," but the term rotifera is more generally used. They are distinguished by their circles of cilia, sometimes single, sometimes double, which, through the microscope, appear like revolving wheels.

The object of the rapid gyration of this wheel or wheels is to create a vortex in the water, whose centre is the mouth of the animal; a little charybdis, bearing with it all the animalcules or molecules that come within its sphere of action.

RO′TTEN STONE. Another name for Tripoli. See *Tripoli.*

RU′BELITE. ⎫ The Tourmaline rubel-
RU′BELLITE. ⎭ lite of Brongniart; Tourmaline apyre of Haüy. Red tourmaline. A mineral of a red colour, of various shades; in the form of its crystals it resembles schorl, as well as in its power of acquiring opposite electricities by heat. It is translucent, sometimes transparent. Specific gravity 3·07. Its crystals are cylindrical or acicular, and aggregated in groups. Before the blowpipe it becomes white, but does not fuse. Its constituents are silex 42·0, alumine 40·0, soda 10·0, oxides of manganese and iron 7·0. In the British Museum there is a specimen of Rubellite, from the kingdom of Ava, valued at £1000. Some mineralogists deem rubellite to be merely a variety of tourmaline, it was however arranged as a distinct subspecies by Karsten and Steffens. Rubellite has also obtained the names of Daourite and Siberite.

RU′BBLE. A term applied to loose angular gravel, or a slightly compacted brecciated sandstone.

RU′BY. (from *rubeo*, Lat. to be red, *rubis*, Fr. *rubino*, It. *rubin*, Germ.) The Spinell of Werner; Spinelle rubis of Brongniart. A transparent red variety of rhombohedral corundum. The ruby is a variety of the same mineral genus as the sapphire, but differs from it in containing rather more silex, and in being less hard. Rubies are found in alluvial soil in Ceylon, Pegu, and other countries in the East. The ruby ranks next to the diamond in value. When a specimen is fine, and free from flaws, a ruby of large size will sell for from ten to fifteen thousand pounds.

RUB′ISTES. A family of the order Conchifera Monomyaria of Lamarck, containing the genera Crania, Discina, Birostrites, Radiolites, Sphœrulites and Calceola.—*Lycett.*

RUGO′SA. The name of an extinct order of four-starred corals, of the class Actinozoa.

RUGO′SE. ⎫ (*rugosus*, Lat.)
RU′GOUS. ⎭

1. In conchology, applied to shells which are rugged and full of wrinkles.
2. In botany, applied to leaves when the veins are more contracted than the disc, so that the disc rises into little inequalities; the primrose and sage afford examples.
3. In entomology, when longitudinal elevations are placed irregularly on the surface, resembling coarse wrinkles.

RUMINA′NTIA. ⎫ The eighth order of
RU′MINANTS. ⎭ Mammalia, or those animals that chew the cud. These animals possess the singular property of returning the food to the mouth after it has been swallowed, that it may undergo the process of a second mastication. This property depends upon the structure of the four stomachs which these animals possess, the three first being so arranged that the food may pass from the œsophagus into either of them. Cuvier divides this great order into those that have horns, and those that have none. " Fossil ruminantia," says Pidgeon, " are found in the depositions of many different eras. The ruminants were clearly coeval with the other mammifera of the ancient world; and they existed in a numerical proportion sufficiently great to produce an abundance of their bones in various depositions."

RU′NCINATE. (from *runcina*, a saw, Lat.) In botany, a term applied to leaves whose edges are cut into teeth turning backwards, like a scythe.

RUTHE′NEUM. One of the noble metals, its symbols being Ru. Specific gravity 8·60.

RU′TILE. Red oxide of titanium. The titane oxydé, of Haüy; titane ruthil of Brongniart; rutil of Werner; peritomes titan-erz of Mohs. A brown, red, yellow, and sometimes nearly velvet-black ore. Occurs regularly crystallised, massive, disseminated, in angular grains, and in flakes. External lustre considerable, and sometimes metallic. Opaque or translucent. Scratches glass. Specific gravity = 4·2 to 4·4. It is infusible before the blow-pipe unless a flux be employed. It is found in Scotland, in the granite of Cairngorm.

RYA′KOLITE. A name given to glassy felspar.

# S

SACCHA′RINE LIMESTONE. When limestone is fine-grained and regular, it resembles the crystalline structure of loaf sugar; it is hence called saccharine limestone, or marble.

SA′CCHAROID. (from σάκχαρ, sugar, and εἶδος, form, Gr.) Resembling white, loaf, or crystallized sugar. A term applied to rocks which have a texture resembling that of loaf-sugar.

SA′CRUM. (Lat.) The bone which forms the basis of the vertebral column.

SADDLE-SHAPED STRATA. When strata are bent on each side of a mountain, without being broken at the top, they are called saddle-shaped.

SA′GENITE. Another name for rutile, or red oxide of titanium.

SAGI′TTATE. (from *sagittatus*, Lat.) Arrow-shaped.

SAH′LBANDE.  } By this word is under-
SAALB′ÄNDE. } stood a slip or band interposed between the veins and the rock which forms the body of the mountain. The saalbände is generally of a different nature from the substance to which it is contiguous. It is not found in all veins, and those from which it is absent are said to be adherent. The term saalbände, for which we have no corresponding scientific expression, is frequently denominated in some of the mining districts of this country, *pasting* or *sticking*.—*Geological Transactions.*

SA′HLITE. (The sahlit of Werner: the malacolith of Brongniart.) A variety of augite, discovered in a silver mine at Sahla, in Sweden, from which it takes its name. It is of a green or greenish-grey colour of various shades. It occurs in straight, lamellar, and granular concretions sometimes crystallized with a shining, vitreous, or pearly lustre; translucent at the edges. It is soft to the touch, scarcely scratches glass, and is easily scratched by a knife. Specific gravity 3·2. Before the blow-pipe it melts, with some ebullition, into a porous glass. It consists of silica 53, magnesia 19, lime 20, alumina 3, iron and manganese 4.

SALAMA′NDER. (σαλαμάνδρα, Gr. *salamandra*, Lat.) A genus of reptiles belonging to the order Batrachia. The salamander possessses the general form of the lizard, and is placed by Linnæus among the lizards; but its characters are those of the Batrachians. Its body is elongated; it has four feet, and a long tail. When arrived at an adult state, its respiration is performed in the same manner as in frogs and tortoises. Aristotle and Pliny state that if the salamander passes through fire, the fire is immediately extinguished, and that it emits a milky saliva which is depilatory. Bose says that it emits from its skin a milky fluid when annoyed, and when put into the fire, it sometimes happens that this fluid sufficiently extinguishes it to permit the animal to escape. This secretion of a milky fluid appears to be exceedingly acrid; produces, if applied to the tongue, a very painful sensation; is an excellent depilatory, and destroys small animals. Spallanzani has discovered that the salamander has the power of reproducing lost or mutilated organs, so that if its legs or tail be cut off, or its eyes plucked out, these organs will, in the course of a few months, be reproduced. It has been found fossil.

ST. CASSIAN BEDS. A large mass of beds occurring in the Austrian Alps: called also Hallstatt Beds. In these beds are found fossils of an intermediate character between those occurring in the palæozoic and mesozoic rocks.

SA'LIENT A'NGLE. A projecting angle. In a zig-zag line the upper are the salient, the lower the re-entering angles.

SALI'FEROUS. (from *sal*, salt, and *fero*, to bear, or produce, Lat.) Containing salt; yielding salt. Thus, in geology, we have saliferous deposites; saliferous rocks; saliferous strata, &c. &c.

SALI'FEROUS SYSTEM. The New Red Sandstone system of some authors; the Poikilitic system of Conybeare. The saliferous system comprises the new red sandstone and the magnesian limestone formations. In Germany and France there is added to the series of strata which we possess in this country, a member which is called muschelkalk; this, though extensively developed in Germany, has never been discovered in England. Organic remains of this system, says Prof. Phillips, though few in number, are exceedingly interesting to the naturalist and geologist, from the strong testimony they offer of the successive changes of the living creation, according to the new circumstances of the land and sea. The fossil plants, shells, fishes, and reptiles of the saliferous system, appear to partake both of the character of those in the older carboniferous, and the newer oolitic, deposites. Calamites, resembling those of the coal formation, are mingled with cycadeæ, like those of the oolites. Fishes of the genus palæoniscus are here found for the last time; while the remains of oviparous quadrupeds, the phytosaurus and protorosaurus, are first discovered. Regarding it according to its mineral characters, it forms one great series of deposites, which were thrown down at a period when a decided change in the conditions of the globe was taking place. The manner in which the group rests upon the carboniferous group in England is such as to show that the latter was disturbed, dislocated, and partially removed before the former was accumulated upon it; nevertheless, in other parts of the European area, there is reason for supposing that the new red sandstone was quietly deposited upon the carboniferous series, no real line of separation being established between them. The saliferous system, commencing with the keuper, or variegated marls, lies immediately under the lias, and, terminating in the red conglomerate, rests upon the carboniferous series. Its depth in some parts has been estimated at eight or nine hundred feet.

SALIFI'ABLE. (from *sal*, salt, and *fio*, to become, Lat.) That may become a salt by combination with some other body; capable of combining with an acid to form a salt.

SA'LIVARY GLANDS. Organs which secrete the saliva. All animals that masticate their food are provided with salivary glands, which pour the saliva into the mouth as near as possible to the grinding surfaces of the teeth. Fishes and the cetacea, performing no mastication, have no salivary glands.

SALT. (*sal*, Lat. *salz*, Germ.) In an impure state, one of the most abundant productions of nature. It occurs in two forms, either as a solid mineral, or in solution, in the waters of the ocean, and of lakes and springs in inland districts. The waters of the ocean contain about one-thirtieth of their bulk in solution. The **uses of salt are numerous**, putting **aside its great importance**, or absolute necessity, as a matter of food. It is employed in glass-making, enamelling, glazing, and bleaching. It is a valuable manure; and is used in the making of bread. Although the most frequent position of rock-salt is in strata of the new red sandstone formation, yet it is not exclusively confined to them. The salt mines of Wieliezka and Sicily are in tertiary formations; those of Cardona **in cretaceous;** some are found in **the oolite; while others** occur in **the coal formation.**

SAND. (*sand*, Germ.) **Flint or quartz** broken fine by the **action of water,** but not reduced to powder. **Very** small particles of silicious matter not cohering together, nor softened by water.

SA'NDSTONE. An aggregate of silicious grains. Any stone composed of grains of sand agglutinated together. The grains of sandstone are sometimes so fine as scarcely to **be** distinguished by the unaided eye; at others their magnitude is equal to that **of** a walnut or an egg,

as in the coarse sandstones known as conglomerates, pudding-stones, breccias, &c. The cement which agglutinates the silicious particles of sandstones may be calcareous, argillaceous, or silicious: when silicious the sandstone sometimes resembles quartz. Sandstones are close, porous, and vesicular, with every intermediate gradation, from perfectly loose sand to the hardest sandstone. They vary in colour, from white to red or brown, but their most common colour is grey or greyish white: sometimes their colour is uniform, at others it is variegated.

SA'PPARE. A mineral first described by Saussure. The Cyanit of Werner, and Disthene of Haüy.

SA'PPHIRE. (from $\sigma\acute{a}\pi\phi\epsilon\iota\rho\sigma\varsigma$, Gr. *sapphirus*, Lat.) A precious stone, exceeding all others in hardness except the diamond. It occurs crystallized, in six-sided prisms variously terminated; the crystals yield readily to cleavage in one direction, presenting a most brilliant surface; cleavage fourfold; **fracture** conchoidal. It consists of nearly pure alumina, with a little oxide of iron, with some silex or lime; but the sapphire contains upwards of ninety-eight per cent. of alumina. Its specific gravity is from 3·70 to 4·30. It possesses double refraction, and varies from opaque to transparent. Its colours are blue, red, green, white, grey, yellow, brown, and black. There are several varieties of the sapphire, **as** the white, blue, or oriental sapphire, the *oriental* amethyst, the *oriental* topaz, and the *oriental* emerald. Some varieties of sapphire exhibit particular kinds of opalescence, and these have obtained the name of girasol sapphire, chatoyant or opalescent sapphire, and asteria or asteriated sapphire, the last, when cut *en cabochon*, presents a silvery star of six rays, in a direction perpendicular to the axis. The finest sapphires are found in alluvial soil in Ceylon and Pegu. Lately the sapphire has been employed in the formation of small lenses for microscopes; it is also employed, in addition to its use as an ornament, for jewelling the pallets of escapements, and the holes of wheel pivots in **astronomical** clocks and watches. The red sapphire is the most highly esteemed, its value being sometimes equal to that of a diamond of the same size: a single stone has been estimated at the value of one thousand guineas.

SARCI'NULA. A genus of lamellated polypifers, thus described by Parkinson. "A stony polypifer, formed in a free, simple, thick mass, by tubes united together. The tubes numerous, cylindrical, parallel, and vertical, accumulated in bundles by intermediate and transverve septa. Radiated lamellæ within the tubes." Sarcinula differs from Tubipora in its tubes being lamellated, and from Stylina in having no central style.

SA'RCOCARP. The fleshy part of certain fruits, placed between the epicarp and the endocarp. That part of fleshy fruits which is usually eaten.

SA'RCOLITE. (from $\sigma \grave{\alpha} \rho \xi$, flesh, and $\lambda \acute{\iota} \theta o s$, stone, Gr.) A variety of analcime, found at Mount Somma, and obtaining its name from the flesh colour of ts crystals, which are cubo-octahedral.

SARD. The best specimens are brought from Sardinia, whence its name. A variety of chalcedony, of a deep rich reddish-brown colour.

SA'RDONYX. ($\sigma \alpha \rho \delta \acute{o} \nu \upsilon \xi$, Gr. *sardonyx*, Lat.) The Quartz agathe sardoine of Haüy; Silex sardoine of Brongniart. A variety of calcedony differing from carnelian only in its colour, which is reddish yellow, or nearly orange, with occasionally a tinge of brown.

SA'SSOLIN. } So called from having
SA'SSOLINE. } been found near the warm spring of Sasso, in Tuscany. Native boracic acid.

SA'TIN SPAR. A fibrous variety of calcareous spar. It is susceptible of a fine polish, and exhibits **the** lustre of satin, from which circumstance it has obtained its name. Its colours are grey and pale rose-red. Very fine specimens are met with in Cumberland.

SAU'RIA. (from $\sigma \alpha \hat{v} \rho o s$, a lizard, Gr.) The second order in the class Reptilia. This order, according to Cuvier's arrangement, includes six families, namely, Crocodilia, Lacertinida, Iguanida, Geckotida, Chamæleonida, and Scincoide.

SAU'RIAN. A reptile belonging to the order Sauria. The species of fossil saurians are exceedingly numerous, attaining in many instances a magnitude unknown among the living orders of that class, and which seems to have been peculiar to those middle ages of geological chronology that were intermediate between the transition and tertiary formations. It is in the oolitic period, between the eras of the red sandstones and the greensands, that the gigantic saurians existed in greatest abundance about the shores, in the rivers, and on the land, in these now cold regions of the globe. Some of the saurians were exclusively marine; others amphibious; others were terrestial, ranging in marshes and jungles, or basking on the margins of estuaries, lakes, and rivers. Even the air was tenanted by flying lizards, under the dragon form of pterodactyles.

SAU'ROBATRACHIA. The third order of the class Amphibia, comprising the proteus, siren, &c., and the extinct genus Andrias.

SAUROCE'PHALUS. A fossil saurian, found in the oolite, and by Agassiz ranked among fishes. Its form was adapted for swimming.

SAU′ROID. (from σαῦρα, a lizard, and εἶδος, form, Gr.) The name given to a group of fishes found in great abundance in the carboniferous and secondary formations. The *sauroid* fishes occupy a higher place in the scale of organization, than the ordinary forms of bony fishes. In the tertiary strata the sauroids almost disappear, and are replaced by less complex forms. The sauroid, or lizard-like fishes, combine in their structure, both in the bones and some of the soft parts, characters which are common to the class of reptiles.

SAU′SSURITE. A combination of crystallized serpentine with jade or felspar. The jade de Saussure of Brongniart. In its external characters it differs but little from nephrite, but in its composition it by no means resembles it. It was first noticed by Saussure near the lake of Geneva, scattered about in rounded pieces and loose blocks. Its colours are green, greenish grey, or white with a slight tinge of green or blue. Its specific gravity is about 3·35. Before the blow-pipe it fuses. It consists of silex 49·0, alumine 24·0, lime 10·5, soda 5·5, magnesia 3·75, oxide of iron, 6·5. By many mineralogists, Saussurite is included under nephrite.

SAXICA′VA. A genus of bivalves, belonging to the family Lithophagi, or stone-eaters.

SAXICA′VOUS. Animals which make holes in the rocks, either by boring them, by means of some auger-like process they possess; or by dissolving the rock, by some acid which they secrete.

SAXIGENOUS. Producing stone. This term is more particularly applied to those polypi which produce the reefs and islands of coral, so abundant in the Pacific Ocean and Indian seas.

SCA′BROUS. (from *scabrosus*, Lat.)
1. In entomology, applied to the surface of an insect when covered with small and slight elevations.
2. Applied to shells, when rough, rugged, harsh, or like a file.
3. In botany, applied to stems that are rough, from any little inequalities or tubercles.

SCA′GLIA. (Ital.) A kind of chalk, of a red colour. In an interesting paper, by Prof. Sedgwick and Sir R. Murchison, published in the Philosophical Magazine for June 1829, on the relations of the secondary and tertiary strata on the southern flanks of the Tyrolese Alps, the tertiary strata are described as forming a vast series of beds resting on *scaglia* or chalk. The *scaglia* occurs in beds nearly vertical; the upper ones contain nodules and layers of flints; their colour is red, and their structure fissile. The *scaglia* contains in some parts ammonites and belemnites.

SCALA′RIA. A genus of marine turreted univalves, with acute longitudinal raised ribs. The aperture nearly circular; the margins uninterrupted, bordered, and reflected.

SCA′NDENT. (from *scandens*, climbing, Lat.) A term applied to plants which climb upon some support, attaching themselves by fibres or tendrils.

SCANSO′RES. The second order of the class aves, or birds; the woodpecker, cuckoo, and parrot are examples.

SCAPE. (from *scapus*, Lat.) In botany, an herbaceous stalk, springing from the root, and bearing the flower and fruit, but not the leaves. The hyacinth is an example.

SCA′PHITE. (from *scapha*, a boat.) So named from its supposed resemblance to a boat. The scaphite resembles an ammonite partly unrolled. Scaphites are found in the chalk, and in the greensand; they are believed to be altogether extinct.

SCAPH'OID. (from σκαφη, a boat, and ειδος, form, Gr.) The name given to one of the carpal bones, from its fancied resemblance to a boat, or hollow oblong vessel.

SCA'POLITE. Pyramidal felspar. The scapolith of Werner; scapolithe of Brochant; paranthine of Haüy. A rare mineral, occurring massive, and in long prismatic crystals. It consists of silica 43·83, alumina 35·43, lime 18·96. It is of a grey, yellowish, greenish-white, or silver white colour. Specific gravity from 3·68 to 3·71. Before the blowpipe it intumesces, and fuses into a shining, white enamel. It is found in beds of magnetic ironstone and iron pyrites, at Arendal, in Norway. Jameson divides this species into three subspecies, namely, radiated scapolite, foliated scapolite, and compact scapolite. Its name is derived from the prismatic form of its crystals.

SCA'PULA. The shoulder-blade.

SCA'PULAR. Pertaining to the shoulder-blade or scapula.

SCARF-SKIN. The cuticle or external covering of the body; called also the epidermis.

SCA'RBROITE. An earthy mineral occurring near Scarborough, whence its name, of a pure white colour, consisting of alumina, silica, peroxide of iron, and water.

SCHAA'LSTEIN. Shell stone. The schaal-stone of Jameson; pierre calcaire testacée of Brochant; spath en table of Haüy. Called also tabular spar or table spar. A substance of a grey or pearly-white colour, usually occurring in masses, composed of thin laminæ, collected into large prismatic concretions. It is very rare, and has been found chiefly at Dognatska, in the Bannat.

SCELIDOTHE'RIUM. (from σκελις, hind leg, and θηριον, beast, Gr.) A genus or subgenus distinct from Megatherium and Megalonyx, but of the same natural family; discovered by Mr. Darwin in Patagonia, and so named by Prof. Owen. The characters of this genus are taken from the modifications of the teeth and of the bones of the extremities, especially of the astragalus.

SCHEE'LIUM. Another name for Tungsten.

SCHI'LLER SPAR. (from schillerm, Germ.) The schiller spath of Mohs. A genus of spars comprising four varieties, namely, common schiller spar; bronzite, or hemiprismatic schiller spar; hypersthene, or prismatoidal schiller spar; and anthophyllite, or prismatic schiller spar. The characters of the genus are, the cleavage monotomous, perfect. Metallic pearly lustre. Hardness 3·5 to 6·0. Specific gravity 2·6 to 3·4.

SCHIST. (σχιστος, Gr.) A term synonymous with slate. A rock, of a fissile character, which may easily be split.

SCHISTO'SE. } Slaty; fissile.
SCHI'STOUS. }

SCHISTO'SE MI'CA. The name given by Kirwan to mica slate.

SCHNEIDE'RIAN MEMBRANE. Thus named from Schneider, who first described it. The lining membrane of the nostrils.

SCHORL. (from Schorlan, a village in Saxony.) Some mineralogists comprise in one family, which they term the Schorl family, topaz schorlite, pyrophysalite, euclase, emerald, iolite, schorl, epidote, zoisite, and axinite. Others consider tourmaline and schorl as one mineral. Under the word schorl we shall describe, not the family, but the mineral known as the Schorl of Werner, Le Schorl of Brochant, the Tourmaline of Haüy. Jameson divides schorl into Precious Schorl and Common Schorl. It is a mineral, in general, easily recognised. It most frequently occurs in long prismatic crystals, more or less regular, whose lateral faces are almost always longitudinally striated. These prisms usually present six, nine, or twelve sides, and are terminated at both extremities by three principal faces. The principal form, not easily obtained by mechanical division, is an **obtuse rhomb of** which the plane **angle at the** summit is 115° 34', or according to some an obtuse **rhomboid** of 133° 50' and 46° 10'. Phillips terms it a combination of silica, alumina, oxide of iron, and lime, with small proportions of magnesia, potash, soda, and boracic acid, and the analysis of Gmelin give all these constituents. An attention to its electric powers, its vitreous, conchoidal fracture, and fusibility **into an** enamel, will generally prevent it from being confounded **with other** minerals, **more** particularly from hornblende, which it **most** resembles.

SCHORLA'CEOUS. Resembling schorl; containing schorl. The granite of Dartmoor **is** very frequently porphyritic from the presence of large crystals of felspar, and here and there *shorlaceous*. The schorl not unfrequently occurs in radiating nests of variable size and abundance.

SCHORL ROCK. A binary compound of schorl and quartz, in which the first considerably predominates. The schorl rock of Cornwall, says Jameson, is probably very intimately **connected with topaz rock**; its geognostic relations and characters are not well ascertained.

SCHO'RLITE. The Pycnite of Haüy and Brongniart; the Schorlous beryll of Jameson. A mineral of a straw colour, occurring at Altenburg, in Saxony, in a rock of quartz and mica.

SCI'ENCE. (from *scientia*, knowledge, Lat. *science*, Fr. *siénza*, It.) The knowledge of many, orderly and methodically arranged and digested, so as to become attainable by one. The knowledge of reasons and conclusions constitutes *abstract*, that of causes and their effects, and of the laws of nature, *natural science*.

SCITAMI'NEÆ. One of Linnæus's natural orders of plants; they are all natives of warm climates: the ginger, plantain, &c. are examples. *Scitamineous* plants have been found fossil in the strata of the Isle of Sheppey.

SCLER'ODUS. (from σκληρὸς and ὀδοὺς, Gr. rough-tooth. A genus of fossil fishes, discovered in the upper Ludlow rock, and thus named by M. Agassiz. It may be distinguished from Psammodus by the raised pustules on the surface of the teeth.

SCLERO'TICA. ⎫ (from σκληρὸς, hard,
SCLERO'TIC. ⎭ Gr.) The outermost coat, tunic, or membrane of the eye. It is exceedingly dense and firm, and does not pass over more than about four-fifths of the ball of **the** eye, its place in front being supplied by a transparent membrane, the *cornea*, **to permit the** passage **of** light.

SCO'LEZITE. A mineral having one equivalent of hydrated silicate of lime, to one of silicate of alumina, and two of water.

SCO'PIFORM. (from *scopa*, a besom, and *forma*, form, Lat.) Of the form of a besom or broom. In mineralogy, if **a number** of minute crystals or fibres be closely aggregated into a little bundle, with the appearance of diverging slightly from a common centre, they are said to be scopiform.

SCORBI'CULATE. In conchology, pitted; having the surface covered with hollows.

SCO'RIA. (σκωρία, Gr. *scoria*, Lat.) The dross or scum of metals; the cinders of volcanic eruptions (then used plurally, scoriæ); the recrementitious matter of metals in a state of fusion.

Scoria′ceous. Resembling scoria; containing scoria.

Sco′rious. Drossy; cindery; **excrementitious.**

Scorpion. (σκορπίος, Gr.) A genus of arachnidans, belonging to the family Pedipalpi, order Pulmonariæ. The scorpion is remarkable, not only for the powerful organs by the aid of which it is able to seize its prey, but also for its jointed tail, terminating in a deadly sting. The palpi are very large, with a forceps at the extremity resembling a hand; the tail is composed of six joints, the last joint terminating in an arcuated and exceedingly acute point or sting, which allow the exit of a poisonous fluid, contained in an internal reservoir. The scorpion is provided, on each side of the thorax, with four pulmonary cavities, into each of which air is admitted by a separate external opening. The eyes are compound, accompanied by stemmata. A fossil scorpion has been discovered by Count Sternberg in the ancient coal formation at Chomle, near Radnitz.

Screw stone. The name of a fossil resembling, at first sight, a screw; if, however, the marks be carefully examined they will be found to be circular, and not spiral.

Scrobi′culated. (*scrobiculus*, from *scrobs*, a furrow, Lat.) Furrowed; having small ridges and furrows.

Scru′bstone. A provincial term for a species of calciferous sandstone.

Scy′phia. A genus of zoophytes. Fossil scyphiæ occur in the oolitic and cretaceous groups. In the oolitic, forty one species have been determined by Goldfuss and Munster; in the cretaceous group twelve species have been determined by Goldfuss. Five species are mentioned as having been found in the grauwacke.

Seam. A term used to designate a thin stratum; as a seam of coal.

Scu′tum. (*scutum*, Lat. a shield or buckler.) A species of Echinite. The name given by Klein to the third section of the class Catocysti; the *scutum* of Klein is the Echinanthus of Leske.

Sea pen. A purple-coloured zoophyte.

Se′condary forma′tion. ⎫
Se′condary strata. ⎬ By secondary
Se′condary rocks. ⎭ rocks are meant those stratified rocks older than the tertiary, which contain certain distinct organic remains, and are now commonly called Mesozoic. The principal groups of the secondary formations are as follows:—1. The cretaceous group, beginning with the Maestricht beds, and terminating in the lower greensand. 2. The Wealden group, commencing with the Weald clay, and closing with the Purbeck beds. 3. The oolite, or Jura limestone group, beginning with the Portland beds, and ending with the inferior oolite. 4. The lias group. 5. The new red sandstone group, commencing with the Keuper and ending in the red conglomerate. 6. The carboniferous group, comprising the coal measures, the mountain limestone, and the old red sandstone. 7. The graywacke group.

The secondary strata cover a large portion of the habitable globe, and are the immediate subsoil of the most fertile districts of Europe. The secondary strata are composed of extensive beds of sand and sandstone, mixed occasionally with pebbles, and alternating with deposits of clay, marl, and limestone. The materials of most of these strata appear to have been derived from the detritus of primary and transition rocks; and the larger fragments, which are preserved in the form of pebbles, often indicate the sources from which these rounded fragments were supplied. Six substances are interstratified in this system: arenaceous, argillaceous, and calcareous rocks form the principal masses, and are associated with beds of chert, iron-stone, and coal.

It is in the strata belonging to the secondary formations that the bones of enormous reptiles are first discovered. The peculiar feature in the population of the whole series of secondary strata, was the prevalence of numerous and gigantic forms of Saurian reptiles.

"From the examination," says Mantell, "of the organic remains of the secondary formations, we arrive at the following results: that the seas, lakes, and rivers, during the geological epoch termed secondary, were peopled by fishes, mollusca, crustacea, radiaria, polyparia, and other zoophites; all of extinct species, and presenting as a whole, a greater discrepancy with existing forms, than those of the tertiary."

The commencement of the secondary epoch is a marked one, depending on a great change having taken place in the character of animal and vegetable life, in the interval between the formation of the last of the primary, or palæozoic rocks, and the first of the secondary. The discovery, however, of the St. Cassian beds, in the Austrian Alps, has tended to lessen that great interval, and to diminish the extent of the change; those beds having been found to contain fossils of an intermediate character.

With the cretaceous system ends the long series of deposits which are, by general consent, ranked as strata of the secondary periods of geology. Prof. Phillips says, "turning to the organic remains of the several secondary systems, it is apparent that, within the period of time which elapsed between the deposition of the primary and tertiary strata, two very distinct assemblages of terrestrial plants had flourished, and become extinct. The ancient and abundant flora of the carboniferous era, with its lepidodendræ, sigillariæ, and calamites, had been replaced by new races of zamiæ and cycadæ, which, in their turn, vanished from the northern zones of the globe before the cretaceous system. The marine zoophyta were changed. One total change had come over the crustacea, —not a single trilobite being known in the strata more recent than coal: the brachopodus conchifera, the gasteropodous and cephalopous mollusca were equally altered. Two large assemblages of fishes had vanished before the deposition of the chalk; and both on the land, and in the sea, gigantic reptile forms had come into being, reproduced themselves to a marvellous extent, and then all perished with the close of the secondary period."

"How, then, can they, by whom the magnificent truths of elapsed time and successive creations have been put in clear and strong evidence,—how can they be expected to yield to false notions of philosophy, and narrow views of religion, the secure conviction that, in the formation of the crust of the earth, Almighty Wisdom was glorified, the permitted laws of nature were in beneficent operation, and thousands of beautiful and active things enjoyed their appointed life, long before man was formed of the dust of the ancient earth, and endowed with a divine power of comprehending the wonders of its construction? It is something worse than philosophical prejudice, to close the eyes of reason on the evidence which the earth offers to the eyes of sense; it is a dangerous theological error to put in unequal conflict a few ill-understood words of the Pentateuch, and the thousands of facts which the finger of God has plainly written in the book of nature; folly, past all excuse, to suppose that the moral evidence of an eternity of the future shall be weakened by admitting the physical evidence for an immensity of the past."

SE'CTILE. (from *sectilis*, that may be easily cut, Lat.) A term in mineralogy, applied to minerals, when, being cut with a knife, the separated particles do not fly away, but remain on the mass. Sectile minerals are those which are midway between malleable and brittle. "A slice or portion cut from a sectile mineral is fragile, and the new surface on the mass is smooth and shining."—*Phillips.*

SECU'RIFORM. (from *securis*, a hatchet, Lat. and *form*.) Hatchet-shaped: a term applied to shells and to leaves.

SEDIME'NTARY ROCKS. Rocks which have been deposited by water.

SEED VE'SSEL. In botany, the pericarp.

SE'LENITE. (σεληνίτης, Gr. *sélénite*, Fr.) Sulphate of lime, or crystallized gypsum. A transparent and highly crystallized variety of gypsum. The crystals of selenite are frequently united, or collected into groups of various forms. Selenite consists of lime 33·0, sulphuric acid 44·8, water 21·0. It is found abundantly in the gypsum and salt formations. The primitive form of its crystal is a dodecahedron, which may be conceived as two four-sided pyramids, applied base to base, and which, instead of terminating in pointed summits, are truncated near the bases; so that the sides of the pyramids are trapeziums, each terminating in a rhomb. It causes double refraction. Before the blowpipe it melts into a white enamel.

SELE'NIUM. (from σελήνη, the moon, Gr.) One of the simple or elementary bodies, and a non-conductor of electricity. According to Prout, it appears to constitute the connecting link between sulphur and the metals. When in mass, selenium has the aspect of lead and a metallic lustre, but pulverized it displays a deep red colour. Specific gravity 4·3. Its equivalent is 39·6: symbol Se.

SE'MI-O'PAL. A variety of opal. The Halbopal of Werner; La demi-opale of Brochant; Quartz résinite commune of Haüy. The colours of semi-opal are white, grey, green, red, brown, and blue. Fracture imperfectly conchoidal. Specific gravity 2 to 2·5. It is infusible before the blow-pipe. It consists of silica 85, alumina 3, oxide of iron 1·74, carbon 5, ammoniacal water 8, with a fracture of bituminous oil: or, according to another analysis, of silica 82·7, water 10, oxide of iron 3, alumina 3·5.

SENSO'RIUM. (*sensorium*, Lat. *sensorium*, Fr.) That part of the brain where the senses transmit the impressions or perceptions to the mind.

SE'PAL. This word was invented by botanists to distinguish the several parts of the calyx from those of the corolla.

SE'PIA.
1. The name given by Linnæus to the cuttle-fish. See *Cuttle-fish*.
2. The ink of the cuttle-fish. This has been found in a beautiful state of preservation in fossil ink-bags of sepiæ in the lias at Lyme Regis. The common sepia used in drawing is from the ink-bag of an oriental species.

SEPIOSTA'IRE. The name given by Blainville to the internal bone of the sepia or cuttle-fish. The absence of a siphuncle renders the *sepiostaire* an organ of more simple structure, and of lower office, than the more compound shell of the belemnite.

SEPTA'RIA. (from *septa*, partitions, Lat.) Spheroidal concretions, varying from a few inches to a foot in diameter, and divided into cells or chambers of irregular form; sometimes they are nodules of clay, having the chambers filled with spar; they are usually found in argillaceous strata. Masses of argillaceous limestone, traversed interiorly by cracks passing in different directions, and containing calcareous spar. Septaria were at one time considered to be confined to the London clay deposit, and to be characteristic of it, but subsequent discoveries have proved the incorrectness of that opinion.

SE′PTUM. (*septum*, Lat.) A partition. The plates dividing the chambers of multilocular shells are termed septa; a partition separating certain portions of the brain is called the septum; and the cartilaginous partition of the nostrils is called the septum of the nose.

SE′RAPHIM. The name given by the workmen to impressions of a sort of fossil lobster or gigantic crustacean in the old red sandstone.

SE′ROLIS. A genus of crustaceans, affording the nearest approach among living animals to the external form of the trilobite. The greatest difference between the serolis and trilobite consists in the former possessing a fully developed series of crustaceous legs and antennæ, whilst the trilobite does not display any traces of either of these organs.

SE′RPENTINE. A mineral substance deriving its name from its spots and variegated colours, supposed to resemble the skin of the serpent; its colours and their peculiar arrangement are in great measure characteristic. It sometimes forms whole rocks. It differs from hornblende in containing a larger portion of magnesia and a smaller quantity of iron. There is, however, an intimate connection between serpentine and hornblende, as the latter is observed, in some situations, to be changed into serpentine by contact with limestone. Specific gravity from 2·5 to 2·7. Before the blow-pipe it hardens but does not fuse. Its constituents are magnesia 34·5, silex 28·0, alumine 23·0, lime 3·5, water 10·5, oxide of iron 4·5 = 101. There are two varieties, the precious and the common serpentine. When serpentine is found intermixed with patches of crystalline white marble, it constitutes a stone denominated verde antique. Some crystallised varieties have obtained the name of diallage, or Schiller spar.

SE′RPULA. A genus of the order Tubicola, class Annulata. The animal a terebella; shell univalve, tubular, generally adhering to other substance; often separated internally by septa at uncertain distances. Serpulæ are generally littoral, attached to rocks, stones, shells, crustaceans, corals, and other marine bodies; sometimes several species are found on one stone or shell. Seapulæ may commonly be seen upon the shells of lobsters, crabs, oysters, &c., to which they adhere by the lower surface, looking like small worms creeping upon them. Wherever the sea is or has been, they abound either in a recent or fossil state.

SERPULI′NA. An order of the annelida of MacLeay, who thus describes them "sedentary animals without eyes or antennæ. They live in tubes which are either a natural transudation of their body, and are membranaceous or calcareous, or the tubes are semifactitious, composed of agglutinations of sand, &c."

SERPULI′TES LONGI′SSIMUS. The name assigned by Sir R. Murchison to a fossil shell of the Upper Ludlow Rock. The following description is taken from his Silurian System. "Very long, hardly diminishing in diameter, compressed, smooth, slightly tortuous, composed of numerous thin layers of shell containing much animal matter. In structure, this fossil resembles the Serpula compressa, but it does not diminish so rapidly. Width half an inch. It is found near Ludlow very abundantly, and generally throughout the Upper Ludlow Rock of Salop, Hereford, Radnor, &c."

SE′RRATE. ⎱ (*serratus*, Lat.) Jagged;
SE′RRATED. ⎰ notched.

1. In botany, applied to leaves, the margins of which resemble a saw, the teeth pointing towards the extremity of the leaf.

2. In entomology, applied to the bodies of insects, the margins having jagged incisions, like the teeth of a saw.

SE′RRULATE. ⎱ (from *serrula*, a little
SE′RRULATED. ⎰ saw, Lat.) When the edges of leaves or margins of shells are very finely jagged or notched, they are said to be serrulated, and not serrated.

SERTULA′RIA. A genus of aborescent corals belonging to the family Tubularii.

SE′RUM. (*serum*, Lat. *serum*, Fr.) The thin, watery, transparent part of the blood.

SE′SAMOID. (from σεσάμη, an Indian grain, and εἶδος, resemblance, Gr. *sesamoide*, Fr.) The name of some exceedingly small bones found at the root of the thumb or great-toe. It is the opinion of some naturalists that sesamoid bones are rather to be regarded as depositions of bony matter, in parts disposed to its reception, as ligament, cartilage, &c., than as intrinsic portions of the skeleton; they are larger and more numerous in the hands and feet of such as take laborious exercise.

SE′SQUI. (A contraction of *semisque*, Lat. signifying and a half.) A prefix to many words signifying the quantity and a half more. A sesqui-oxide is therefore a combination with one equivalent and a half of oxygen.

SE′SSILE. (from *sessilis*, seated, Lat.)

1. In botany, applied to flowers when placed directly on the branch or stem; also to leaves when they grow directly from the stem, branch or root, without **any** footstalk: any part of a plant which commonly is borne on a stalk, **is** said to be sessile when it has no stalk.

2. Applied to animals. No truly *sessile* animal is provided with sight.

SE′VERITE. An earthy mineral consisting of alumina, silica, and water. It is found in small masses, white, without lustre, and slightly translucent. It was discovered near St. Sever, in France, whence its name.

SETA′CEOUS. (from *seta*, a bristle, Lat.) Bristle-shaped; bristly.

SHALE. (*schale*, Germ. Schiefer Thon of Werner: argile schisteuse of Brochant.) Shale occurs masive only; its usual colour is grey, sometimes blueish, blackish or yellowish. Often easily cut by the knife. Specific gravity about 2·64. Adheres to the tongue, yields an argillaceous odour when breathed upon, absorbs water considerably, and often falls gradually to pieces in that liquid, but never forms a paste. Slate clay; indurated slaty clay.

Fracture slaty, sometimes nearly earthy, and is dull unless it contains casually imbedded mica, which renders it glimmering. It is fusible before the blow-pipe. It is found resting upon, as well as interposed between, beds of coal, which it invariably accompanies. Some shales are highly bituminous and burn with a bright flame. Shales frequently contain impressions of fishes, reeds, and ferns.

SHA′NKLIN-SAND. (The Glauconie Sableuse of the French geologists.) Called also the lower green sand.

A marine deposit of silicious sands and sandstone of various shades of green, red, brown, yellow, feruginous, grey, and white, with subordinate beds of cherts and silicious limestones, constitute the formation called the **Shanklin sand**, or the lower green sand. It is the lowest member of the cretaceous group, intervening between the gault above and the weald clay below. The beds consists of an aggregation of sand, with comminuted shells and fragments of corals, impregnated with iron, and containing the remains of myriads of shells, polyparia, &c. Dr. Fitton divided the Shanklin sand into three distinct parts: the first or uppermost consisting of sand, with irregular concretions of limestone and chert, sometimes disposed in courses oblique to the general direction of the strata.

The second consists chiefly of sand, but in some places is so mixed with clay, or with oxide of iron, as to retain water: it is remarkable for the great variation in its colour and consistency.

The third and lowest group abounds much more in stone; the concretional beds being closer together and more nearly continuous. The Shanklin, or lower green, sand, is separated from the upper by the galt or Folkstone marl.

SHARK. (from καρχαρίας, Gr.) The squalus of Linnæus. A genus of fishes belonging to the family Selachii, order Chondropterygii Branchiis Fixis. The shark is a phosphoric fish. That tribe of sharks, called by the French Reguins, which is thought to be synonymous with the carcharias of the Greeks, and one of which was probably the monster that swallowed Jonah, are stated to exceed thirty feet in length. The genus of sharks may be considered as one of the most universally diffused, and most voracious, of modern fishes. Several rows of teeth are lodged in each jaw, but one only of these rows projects, and is in use at the same time; the rest lying flat, but ready to rise in order to replace those that have been broken off, or worn down. The shark is oviparous or ovo-viviparous, according to circumstances. The vertebral column is prolonged into the upper lobe of the tail, and the tail is of great service in enabling the shark to turn its body so as to bring the mouth, which is placed downwards beneath the head, into contact with its prey. Sharks appear to have existed throughout every period of geological history. M. Agassiz has separated the sharks into three sub-families, each containing forms peculiar to certain geological epochs, and which change simultaneously with the other great changes in fossil remains. The first of these sub-families, or the Cestracionts, commence with the transition strata, appearing in every subsequent formation till the commencement of the tertiary. Of the Cestracionts, one representative only now remains, namely, the Cestracion Philippi, or Port Jackson shark. The Cestracionts possessed large polygonal, obtuse, enamelled teeth, covering the interior of the mouth with a kind of tesselated pavement. The second sub-family, or Hybodonts, commenced with the muschel-kalk, is found throughout the whole of the oolitic deposits, and disappears at the commencement of the cretaceous group. The teeth of the Hybodonts were intermediate between the blunt polygonal teeth of the Cestracionts and the sharp-edged cutting teeth of the Squaloids. The third sub-family, termed Squaloids, appeared at the commencement of the chalk deposits, and continues downwards to the present period. In the

Squaloids, the teeth are smooth on the outer side, and plicated on the inner; sometimes the edge is serrated.

SHELL. (*schale*, Germ.) The hard covering of anything; the covering of a testaceous or crustaceous animal. The crustaceous coverings of animals, as of echini, crabs, lobsters, cray-fish, &c., are composed of the same ingredients as bones; but the proportion of carbonate of lime far exceeds that of phosphate of lime in shells.

The process employed by nature for the formation and enlargement of the shells of the mollusca was very imperfectly understood prior to the investigations of Reaumur. His experimental enquiries have established these two general facts; first, that the growth of a shell is simply the result of successive additions made to its surface, and, secondly, that the materials constituting each successive layer are supplied by the organized fleshy substance called the mantle, and not by any vessels belonging to the shell itself. The connexion between the animal and the shell may be regarded as mechanical rather than vital; for whatever portion of vitality it might have possessed when first deposited, all trace of that property soon disappears.

Shells are found fossil in the most ancient strata of the transition period that contain any traces of organic life, and many of these agree so closely with the existing species, that we infer their functions to have been the same, and that they were inhabited by animals of form and habits similar to those which fabricate the living shells most nearly resembling them. The most prolific source of organic remains has been the accumulation of the shelly coverings of animals which occupied the bottom of the sea during a long period of consecutive generations. A large proportion of the entire substance of many strata is composed of myriads of these shells, reduced to a comminuted state by the long-continued movements of water. Minute examination discloses occasionally prodigious accumulations of microscopic shells, no less surprising by their abundance, than their extreme minuteness; the mode in which they are sometimes crowded together, may be estimated from the fact, that Soldani collected from less than an ounce and a half of stone, found in the hills of Tuscany, ten thousand four hundred and fifty-four microscopic chambered shells. Of several species of these shells, four or five hundred weigh but a single grain; of one species, a thousand individuals would scarcely weigh one grain; and great numbers of them will pass through a paper in which holes have been pricked with a needle of the smallest size.

SHELL MARL. A deposit of calcareous earth and clay containing shells.

SHI'NGLE. (*schindel*, Germ.) The loose, water-worn, pebbles on the sea-shore.

SHORLA'CEOUS. A name given by Dr. Boase to certain primary rocks which contain shorl; containing shorl.

SHORL ROCK. A genus of primary rocks. Dr. Boase says, "Shorl-rock is composed of quartz and shorl, with the occasional addition of a third mineral. Its species generally possess the structure of granite: but sometimes the constituents are so intimately blended that they cannot be distinguished, forming homogeneous masses which, however, are so well characterized, that they cannot easily be mistaken for any other rock. Shorl-rock occurs in distinct beds, but more commonly in immense veins; in both forms it is found in granite and protogine, and also among the schistose rocks. The following species are enumerated: granular; large grained; fine grained; granitic; quartzose; compact; zoned; agate-like; porphyritic. They all occur in Cornwall.

Si'BERITE. Another name for Rubellite, or red tourmaline.

Si'DERO-CA'LCITE. The name given by Kirwan to brown spar; the braun spath of Werner.

Sy'ENITE. } (from Syene, a city of
Sy'ENITE. } Egypt, where this rock occurs in abundance, and whence the Romans obtained it for architectural and other purposes.) Werner gave the name of sienite to aggregates composed of felspar, hornblende, and quartz; or of felspar, hornblende, quartz, and mica. Sienite is the roche feldspathique of Haüy. It often bears the general aspect of a granite. Felspar and hornblende may be deemed its two constant and essential ingredients, but it frequently contains quartz and mica, and occasionally talc and epidote. It is the presence of hornblende, as a constituent part, which distinguishes this rock from certain granites, that accidentally contain hornblende. The structure of sienite is sometimes slaty, commonly granular. Greenstone and sienite are essentially composed of the same ingredients, namely, felspar and hornblende; from granitic greenstone there is a transition to sienite, and from sienite to true granite. The colour of sienite is usually grey, but this is affected by the ingredients entering into its composition. Prof. Delesse says, "the rose-coloured syenite of Egypt is formed of quartz, orthose, oligoclase, mica, and frequently also of hornblende."

SIENI'TIC. Containing sienite; resembling sienite; possessing some of the characters of sienite. Sienitic granite contains hornblende.

Sienitic porphyry is fine-grained sienite containing large crystals of felspar.

SIGARE'TUS. A genus of marine univalve shells belonging to the family Macrostomata. It is a depressed, oval, nearly auriform shell, with a short spiral columella: the aperture entire, wide, spread out towards the summit of the right lip, and longer than wide. It is a Tuscan fossil, and exceedingly rare. The living sigaretus is found in sand at depths varying from five to fifteen fathoms.

SIGILLA'RIA (from *sigillum*, Lat.) The name given by Brongniart, to certain large, and, in modern vegetation, unknown forms of plants discovered in the coal formation: the name has been assigned from the peculiar impressions on the stems. The stems are of various sizes from a few inch s to upwards of three feet in circumference, and of great length. A stem found in Craigleith Quarry, near Edinburgh, was forty-seven feet in length, the bark being converted into coal. Stems nearly as long, and four feet and a half in diameter, have sometimes been found in the coal series in the north of England. They are scattered throughout the sandstones and shales that accompany the coal, and may occasionally be seen in the coal itself. These stems are inclined in all directions, and some of them are nearly vertical: they are supposed to have been hollow, like the reed, and with but little substance. Brongniart has enumerated nearly fifty species.

Si'LEX. (*silex*, Lat. flint.) An oxide of silicon, constituting the greater part of all the rocks of which the crust of the earth is composed. Pure silex is perfectly white, it has neither taste nor smell, and its spec. gr. is 2·26. Silex consists of oxygen in the proportion of 50 per cent. united with the base silicium; it is found in the greatest purity in quartz or rock crystal.

SI′LICA. A peroxide of Silicon. One hundred parts of silica contain 48·4 of silicium, and 51·6 of oxygen. It is white; specific gravity 2·6; it is fusible.

SI′LICATE. A union **of silica with** some other substance as a **base,** and by uniting in a double or treble proportion, becomes a bisilicate or trisilicate. The principal *silicates* are those of alumina, potash, soda, magnesia and lime.

SILICICA′LCE. The quartz agathe calcifère of Haüy: silex silicicalce of Brongniart. A substance occurring in amorphous masses, in thin beds, under strata of compact limestone in Provence. It is of a grey or brown colour, sometimes nearly black. It effervesces with nitric acid, and before the blow-pipe fuses into a white scoria. It is a mixture of flint and carbonate of lime.

SILICIOUS SINTER. The Kiesel-sinter of Werner and Klaproth. Quartz agathe concretionnée thermogene of Haüy. Quartz hyaline concretionée of Brongniart. An earthy mineral, arranged by some mineralogists as a subspecies of quartz. Its usual colours are white, greyish-white, reddish-white, yellowish-grey, and yellow. According to the analysis of Klaproth it consists of silica 98, alumina 1·5, iron 0·5. Specific gravity 1·81. It is infusible before the blow-pipe. It is found in the neighbourhood of thermal springs containing silex in solution, and is deposited from them; more especially by the hot springs of Iceland.

SILIC′IFEROUS. (from *silex,* flint, and *fero,* to bear, yield, or contain, Lat.) Yielding or containing silex; as the pink *siliciferous* oxide of manganese.

SILI′CIFY. To convert into flint; to petrify.

SILICIFI′CATION. Called also petrifaction. The conversion of any substance into stone by the infiltration of silicious matter.

SILI′CIUM. } Silicon was discovered by
SI′LICON. } Berzelius in 1824. It is of a dark nut-brown colour, without any metallic lustre; incombustible in air or in oxygen gas; but oxidizable by certain methods, by which it is converted into silica or silicic acid. Its equivalent is 7·5; its symbol Si.

"Silicon is now shown not to be a metal, but to be nearly allied to carbon in some of its properties. It will combine with the metals like carbon, especially with aluminum, forming cast aluminum, as carbon and iron form cast iron."— *Jukes.* The hitherto undecomposed base of silica or silex. Of the metallic bases of the alkalies and earths, silicium is the most abundant on the surface of our planet, silica entering so largely into the composition of both chemical and mechanical rocks.

SILI′CULOUS. Having small pods or husks.

SI′LIQUA. (Lat.) A pod; a long seed vessel of two valves, separated by a linear receptacle, on whose edges the seeds are ranged alternately.

SILIQUA′RIA. A genus of marine univalves, found both fossil and recent. It is a tubular shell, spiral at its beginning, continued in an irregular form; divided laterally, through its whole length, by a narrow slit, and formed into chambers by entire septa. Recent siliquariæ have been found in sponges; they may be distinguished from serpulæ by the longitudinal slit. Cuvier places the genus in the order Tubulibranchiata.

SI′LIQUOSE. } Bearing pods. A term
SI′LIQUOUS. } applied to plants having that sort of pericarp denominated a pod or legume.

SILL. A provincial term signifying

a bed or stratum; as the whinstone sill. The great bed of basalt in Northumberland is called the Whinstone Sill.

SI′LLIMANITE. A dark grey or brown mineral, composed of silica 42·6, alumina 54·1, oxide of iron 1·9, water 0·5, discovered at Saybrook, in Connecticut, and named after Prof. Silliman.

SILT. The deposit of running water; mud.

SILT. To silt up, to fill with mud, sand, or other matter deposited by running water. The verb is commonly used with the word up immediately following, as the silting up of rivers; estuaries known to have been silted up, &c.

SI′LVAN. The name given by Werner to the metal tellurium.

SI′LVER. (*silber*, Germ.) One of the simple or elementary bodies, and included in the subdivision termed metals. When pure, it is nearly white. It is superior to gold in lustre, and inferior to it in malleability; it is however so malleable that it may be beaten out into leaves not exceeding 100,000th of an inch in thickness. It is very ductile, surpassing gold in tenacity, but inferior to iron, copper, and platinum. It may be drawn out into wire of greater fineness than human hair. It is harder than gold, but **softer** than copper, and may be easily cut **by** a knife. Its specific gravity **is 10·47.** Silver fuses at a temperature **of about 1,000 degrees Fahr. It is not** oxidated by exposure to the atmosphere, but becomes tarnished by sulphureous vapours. It is tasteless, and free from smell. It **is** soluble in nitric acid. Silver for domestic purposes, as well as that made into coin, is rendered harder by an alloy of copper. The stand**ard** silver of this country consists of eleven ounces two pennyweights of pure silver and eighteen penny-weights of copper. Silver has been known from the earliest ages. It is found native and in ores of several kinds. The ores of silver occur in metallic veins, traversing primary rocks. There are many ores which yield silver that are not, strictly speaking, ores of silver. Although the mines of Europe yield considerable quantities of silver, yet it is to Mexico and Peru that we are indebted for the main supplies. The mines of Potosi have paid a royal duty on silver valued at £234,700,000 sterling.

SILU′RIAN. The name given by Sir R. Murchison to an upper subdivision of the sedimentary strata found below the old red sandstone. He assigned this name to these strata from their being best developed in that portion of England and Wales formerly included in the ancient British kingdom of the Silures. The Silurian rocks are divided into upper and lower: the upper Silurian rocks comprise the Ludlow formation and the Wenlock formation; the Ludlow formation consisting of the upper Ludlow rock, the Aymestry limestone, and the lower Ludlow rock; the Wenlock formation consisting of the Wenlock limestone and the Wenlock shale: the lower Silurian rocks are subdivided into the Caradoc formation and the Llandeilo formation; the Caradoc formation consisting of flags, sandstones, grits, and limestones; the Llandeilo formation, of calcareous, dark-coloured flags, with sandstone and schist. The whole of the Silurian rocks attain a thickness in some parts of seven thousand five hundred feet. They are all marine deposites. The Silurian rocks, though ancient, are not the most ancient of the fossiliferous strata. They are but the upper portion of a succession of early deposits, which it may hereafter be found necessary to describe under one comprehensive name. For this purpose Sir R. Murchison proposes the term Protozoic, thereby to imply the first or lowest formation in which animals or vegetables occur. "Acting upon the principle which guided William Smith in subdividing the oolitic system of our island, I have named the rocks of the Silurian system from places in England and Wales, where their succession and age are best proved by order of superposition and imbedded organic remains, and have termed them, in descending order, the Ludlow, Wenlock, Caradoc, and Llandeilo formations. Each subdivision is characterized by a corresponding suite of organic remains, while the whole was formerly considered to be one assemblage, without definite sequence, and was included under the names greywacke or transition limestone" The Silurian rocks consist of a complete succession of fossiliferous strata, interpolated between the old red sandstone and the oldest slaty rocks. In the lower or Cambro-Silurian rocks we know of no plants, with the exception of a few sea-weeds: the animal remains are those of zoophyta, polyzoa, brachiopoda, conchifera, gasteropoda, cephalopoda, echinodermata, annelida, and crustacea (chiefly trilobites). No unquestionable remains of any higher order of animals have as yet been discovered. In the upper Silurian rocks, many generic forms make their appearance, not met with in the lower series, and many appear to have existed during that period only. Two genera of fish have also been discovered, Onchus and Plectrodus; of the latter, three species have been distinguished; the Plectrodus disappears with this period.—*Jukes.*

Sɪ'MPLE. (*simplex*, Lat. *simple*, Fr. *semplice*, It.)

1. In botany, applied to roots, when undivided; to a leaf, when consisting of only one leaf, and not divided into leaflets, &c., &c.

2. In mineralogy, a term applied to elementary, or undecomposed substances: these are about 60 in number, and are described under the article "*elementary substances.*" The mineralogist and the geologist consider those minerals as *simple* and homogeneous, which present no difference of qualities to our senses throughout the mass, although the chemist may discover that such minerals are composed of two or more elementary substances.

Sɪɴɪ'sᴛʀᴀʟ. A term applied to shells, where, in consequence of the heart being placed on the right side, the turns of the spiral are made to the left. These shells are termed *sinistral*, or *reversed.*

Sɪ'ɴᴛᴇʀ. (*sinter*, Germ.) Calcareous sinter is a variety of carbonate of lime, and may be either stalactical, tuberose, reniform, globular, cylindrical, tubular, branched, or in large undulated masses. It is composed, whatever may be its form, of a series of successive layers, concentric, plane, or undulated, and nearly or quite parallel.

Silicious sinter is an opaline silica, deposited on the margins of some hot springs. Some mineralogists have established three sub-species, namely, common silicious sinter, opaline silicious sinter, and pearly silicious sinter.

Sɪ'ɴᴜᴀᴛᴇ. ) (*sinuatus*, Lat.) In botany applied to leaves
Sɪ'ɴᴜᴀᴛᴇᴅ. ) when the margins are cut into wide rounded openings, as in the leaf of the oak.

Sɪ'ɴᴜs. (*sinus*, Lat. a bag, *sinus*, Fr.)

1. In anatomy, a cavity or cell; a narrow passage.

2. In conchology, a groove or cavity.

Sɪ'ᴘʜᴏɴ. ) (*siphon* and *siphunculus*,
Sɪ'ᴘʜᴜɴᴄʟᴇ. ) Lat.) An hydraulic

apparatus belonging to chambered or polythalamous shells, passing through the several chambers, and terminating in a large sac, which surrounds the heart of the animal. The use of the siphunculus appears to be the enabling the animal to rise to the surface, or descend to the bottom of the water, by increasing its specific gravity.

SIPHO'NIA. A genus of zoophytes, species of which have been discovered in the chalk and in the greensand. One species also has been found in the oolite. "A fossil animal, with a polymorphous body, supported by a stem proceeding from a fusiform or ramose root-like pedicle; the original substance spongeous, and pierced by a bundle of tubes derived from the pedicle, passing through the stem, then ramifying and terminating on the surface of the body."—*Parkinson.*

SI'PHUNCLED. Possessing a siphuncle; formed with a siphuncle.

SI'TU. (*situs*, Lat.) "In situ" is a term used in mineralogy, when a mineral is in its natural position or place; in its native site.

SKI'DDAW SLATE. (from Mount Skiddaw.) The Skiddaw slates form the lower division of the Cambrian group. They are of a thickness of 6,000 feet.

SKO'LEZITE. An earthy mineral, nearly allied to Thomsonite. It occurs at Pargas, in Finland.

SKO'RODITE. (from σκόροδον, garlick, Gr.) A mineral of a leek-green or brown colour; an arseniate of iron. Before the blow-pipe it fuses, giving out a smell of garlick, from which circumstance it has obtained its name.

SKO'RZA. The epidote skorza of Brongniart; epidote arenacé of Haüy. A variety of epidote. See *Epidote.*

SLAG. (*schlacke*, Germ. *slagg*, Dan.) The drop or recrement of metal.

SLATE. A kind of clay, of a structure termed **schistose**, which admits of being split into thin layers of considerable extent. Slate is commonly of a bluish or greenish colour, with a silky lustre. It consists of silex 50·0, alumina 25·0, oxide of iron 11·3, manganese 1·6, potash 4·8, carbon 0·3, water 7·5. It is opake; may be scratched by the knife; and fuses into a blackish slag.

SLATE CLAY. See *Shale.*

SLA'TY. Resembling slate; containing slate; composed of parallel thin plates which admit of being separated by splitting.

SLATE SYSTEM. This group is subdivided into, 1st, the Plynlymmon rocks, consisting of grauwacke and grauwacke slate, with beds of conglomerates, the thickness of the whole being estimated at several thousand yards. 2nd, The Bala limestone, a dark limestone associated with slate, containing shells and corals. 3rd, The Snowdon rocks, consisting of fine-grained slates, of various shades of colour, and of fine and coarse grauwacke and conglomerate. In the strata of the slate system are found the most ancient organic remains.

SLI'CKENSIDES.

1. A provincial name for a variety of galena.

2. "It sometimes happens," says Mr. Phillips, "that the vein-stuff of each wall of a vein is nearly compact, both so completely occupying the vein, that they meet together in close contact in the middle. The two faces in contact appear as though they had been polished, and are ribbed, or rather fluted, horizontally; and the face of each is sometimes covered by a remarkably thin coating of lead ore; these planes, when separated, are the *slickensides* of the mineralogist. The edges of the strata on the sides of faults frequently present that polished appearance known to miners under the name of 'slickensides.'"

SLOTH. The Bradypus of Linnæus,

the only existing genus of Tardigrada.

SMALT. (*smalto*, It. *schmalte*, Germ.) Powder-blue, a vitreous substance obtained by melting zaffre, silex, and potash together,

SMARA′GDITE. The name given by Saussure to diallage, from its emerald-green colour.

SMA′RAGD. } (from σμάραγδος, Gr.)
SMA′RAGDUS. } The emerald. See *Emerald*.

SOAP ROCK. } Names given to a kind
SOAP STONE. } of steatite, in consequence of its soapy feel. Soap rock is so tender that it may be cut as easily as new cheese. Its colour is a pearly white or grey with red and blue veins, and when pure is semi-transparent. On coming out of the quarry, it may be kneaded like a lump of dough, but after having been exposed to the air for some time, it becomes friable; it possesses the soapy feel in the highest degree. It is used in the manufacture of porcelain. Klaproth gives the following as the analysis of soap rock: silica 48·00, magnesia 20·50, alumina 0·14, iron 0·01, water 15·50.

SO′DA. (*soda*, *sode*, *soude*, Germ.) Mineral fixed alkali, found native in some situations, but generally obtained from the combustion of marine plants, more particularly of the salsola soda. "Soda," says De la Beche, "is found in schorl, and certain hypersthene rocks, in some eurites, in trachytes, pitchstones, basalts, and some diallage rocks. It is found in greatest abundance diffused through the waters of the ocean."

SO′DA FE′LSPAR. Another name for albite. See *Albite*.

SO′DALITE. (from *soda*, and λίθος, Gr. stone.) A sub species of lapis lazuli. The name sodalite has been given to this mineral from the large proportion of soda which it contains, being 25 per cent. Its constituents are silex 36·0, alumine 32·0, soda 25·0, muriatic acid 6·7, oxide of iron 0·2. It is found in Greenland and Vesuvius. It occurs massive, and in dodecahedrons with rhombic faces. Colour green, of different shades. Structure foliated; fracture conchoidal. Specific gravity 2·37. It is infusible.

SO′DIUM. One of the simple or elementary bodies. Sodium is the metallic basis of soda, and, like potassium, was discovered by Sir H. Davy in 1807. It has the appearance of silver, or of lead, and is both ductile and malleable. Its specific gravity is 0·97, consequently it is lighter than water. When united with oxygen in the proportion of 23·3 by weight to 8 oxygen, it constitutes soda.

SOIL. (*sol*, Fr. *suolo*, It.) The name given to that superficial accumulation of various substances which lies upon the surface of the globe, and covers the rocks below; it is also called earth, mould, loam, &c. Its depth is irregular, from a few inches to several feet.

SOLA′NOCRINITES. A genus of Radiaria, three species of which have been determined by Goldfuss and Munster; these have been found in the oolitic group of Germany.

SOLA′RIUM. A genus of fossil and recent depressed, conical, nearly discoidal, umbilicated, marine, univalve shells, belonging to the family Turbinacea. Recent solaria are littoral shells, found on rocks and weeds, and belong to tropical seas.

SO′LEN. A genus of marine bivalves, found on sandy beaches, wherein it burrows vertically, and lies concealed at a depth of about six inches, when the tide leaves the beach dry. The shell is bivalve, oblong, equivalve, inequilateral, open at both ends; hinge with a subulate reflected tooth, often double, and not inserted in the opposite valve.

SOLENE′LLA. A genus of bivalve conchifera, established by Sowerby, belonging to the family Arcacea.

SO′LENITE. A fossil solen. Fragments of solenites are found in the Essex cliffs.

SOLENI′TES. A genus of plants of the Algæ tribe, found fossil in the Jurassic or Oolitic rocks.

SOLFATA′RA. The name of an extinct volcano near Puzzuoli, which constantly emits aqueous vapour, and sulphureous and muriatic exhalations. The word solfatara is now applied to any volcanic vent emitting sulphureous, muriatic, and acid vapours or gases. Solfataras are usually considered as semi-extinct volcanoes, emitting only gaseous exhalations and aqueous vapours; but there can be no certainty that they may not again become active. According to Dr. Daubeny, sulphuretted hydrogen and a small quantity of muriatic acid are contained in the steam which rushes out of the fumeroles at the solfatara near Naples. Solfataras, variously modified, are by no means rare in volcanic countries.

SOLIDU′NGULOUS. (from *solidus*, solid, and *ungula*, a hoof, Lat.) Having the hoof whole and undivided, as in the horse.

SO′LITARY. (*solitarius*, Lat.)
1. In botany, applied to peduncles when there is only one on the same plant, or when they stand singly in the same place; to seeds, when there is only one in a pericarp.
2. In conchology, applied to a single tooth.

SO′MMITE. The name given by Jameson to the mineral called by Haüy Nepheline.

SO′RDAWALITE. An earthy mineral of a greyish or bluish-black colour, massive, and without any traces of cleavage. It was first discovered by Nordenskoild, near the town of Sordawala in Finland, whence its name.

SPALACOTHE′RIUM. A genus of fossil mammifers, of the upper oolite beds, and so named by Prof. Owen.

SPAR. (*spath*, Germ. *spath*, Fr. *terme de minéralogie, emprunté de l'Allemand. Quelques uns disent*, **spar**.) In mineralogy, a name given to those earths which easily break into rhomboidal, cubical, or laminated fragments with polished surfaces. Spar constitutes the sixth order of the second class in the natural history system of mineralogy. Spar is not metallic; its streak is white, grey, brown, or blue. Hardness from 3·5 to 7·0. Specific gravity from 2·0 to 3·7. As the term spar is applied to stones of different kinds, without any regard to the ingredients of which they are composed, an additional term must necessarily be employed to express the constituent parts as well as the figure; for instance, calcareous spar, gypseous spar, adamantine spar, cubic spar, brown spar, &c. &c. The term spar is sometimes applied to quartz.

SPA′RRY I′RON. The fer oxidé carbonaté of Haüy; spath eisenstein of Werner; fer spathique of Brongniart; sparry iron-stone of Jameson; sparry iron-ore of Kirwan. Carbonate of iron. It is of a yellow, grey, brown, or black colour; occurring crystallised in rhombohedrons, or in laminated and lamellar masses. It is found in metalliferous veins, as well as in common veins, in primary, transition, and secondary rocks. It consists principally of protoxide of iron and carbonic acid; some specimens yielding manganese and lime; others, magnesia, oxide of manganese, and lime, but in very small proportions. Sparry iron is a valuable ore, from the facility with which it may be converted into excellent steel. It is, from the last circumstance, sometimes called *steel ore*.

SPATA'NGUS. A genus of echini, of the section Cor marinum, belonging to the class Pleurocysti. It is characterised by the bilabiated mouth being in the third region of the axis of the base, and the anus in the side of the truncated extremity. There are a great many species of the genus. The shell of the spatangus is oval, possessing a great number of spines, by the action of which it buries itself in the sand. Fossil spatangi are very abundant in the chalk formation.

SPATHE. (*spatha*, Lat.) In botany, a kind of bractea; a large coloured bractea which envelopes the principal axis of sessile flowers; it forms a sort of hood or sheath, opening longitudinally, at some distance from the flower; the arum, calla, &c. are examples.

SPA'THIC. In mineralogy, lamellar; foliated.

SPA'TULATE.
1. In botany, applied to leaves shaped like a spatula or battledore, having the upper part of a roundish figure, the base tapering and linear.
2. In conchology, applied to shells which are rounded and broad at the top, and becoming narrower below.
3. In entomology, applied to the figure of insects, when commencing with a narrow base, gradually widening by the lateral margins sloping out, and terminated at the extremity by a sudden straight line.

SPE'CIES. (*species*, Lat.)
1. That which is predicated of many things as the whole of their essence.
2. In mineralogy, a species may be defined, a collection of minerals, which are composed of the same ingredients, and combined in the same proportions.
3. In entomology, a group of natural bodies which agree together in all their essential, unchangeable characters. The idea of species comprises in it a congruency, that is to say, not a mere conformity, but also a resemblance of its individuals. Species is the lowest of all the systematic groups, and consequently, the most fixed and conformable.
4. In botany, according to Jussieu and others, a species is a combination of individuals alike in all their parts. De Candolle makes it "a collection of all the individuals which resemble each other more than they resemble any thing else; which can by mutual fecundation produce other individuals; and which reproduce themselves, by generation, in such a manner that we may from analogy suppose them all sprung originally from one single individual."

"What is a species? has long been a *questio vexata* among biologists. If, however, we adopt the idea of a species being the descendants of a single pair, successive generations of their species being possible among those descendants, and them only, we shall find the facts of the distribution of species harmonize well with this idea."—*Jukes*.

SPECIFIC GRA'VITY. As this is a term extremely common and in constant use in mineralogy, it may be desirable to define it. "The specific gravity of a body is its weight, compared with that of another body of the same magnitude. Thus, if a cubic foot of water weigh 1000 ounces, and a cubic foot of iron 7000 ounces, their comparative weights or specific gravity are as 1000 : 7000, or as 100 : 700 or as 10 : 70, or as 1 : 7."—*Phillips*.

SPHA'GNOUS. Mossy.

SPE'ETON CLAY. A member of the Lower Cretaceous or Neocomian series. Its exact position is regarded as still uncertain. Prof.

E. Forbes placed it at or below the base of the Lower Greensand.

SPECULAR. (*speculum*, Lat. a mirror.) Any body which presents a smooth and brilliant surface, and which reflects light.

SPHAGO'DUS. (slaughtering or murdering tooth, from σφαγή and ὀδούς Gr.) The name given by M. Agassiz to a genus of ichthyolites found in the Upper Ludlow rocks.

SPHA'GNUM PALU'STRE. A species of moss, generally constituting a large portion of the entire mass of peat bogs. The sphagnum palustre has the property of throwing up new shoots in its upper part, while its lower part decays, and from this circumstance it mainly contributes to the formation of beds of peat.

SPHERE. (*sphæra*, Lat. σφαῖρα, Gr.) A solid, generated by the revolution of a semicircle about its diameter, which remains fixed; a globe; an orbicular body; a body of which the centre is at the same distance from every point of the circumference; such a solid body that all lines drawn from its centre to its surface are equal. The lines are called radii.

SPHE'RICAL. (*sphérique*, Fr. *sférico*, It.) Round; orbicular; globular; Having all its diameters equal.

SPHERI'CITY. Roundness; globosity.

SPHE'ROID. (from σφαῖρα, a sphere, and εἶδος, likeness, Gr.) A solid body approaching to the form of a sphere. A spheroid may be either oblate or prolate; an oblate spheroid resembles an orange, having its poles flattened, such is the form of the earth and planets; a prolate spheroid has its poles drawn out, and its form somewhat resembles an egg.

SPHERO'IDAL. Having the form of a spheroid, whether oblate or prolate.

SPHENO'PTERIS. A very beautiful and delicate subgenus of fossil ferns or filicites, described by Brongniart.

SPHE'RULE. (*sphærula*, Lat.) A little globe; a globule.

SPHI'NCTER. (from σφίγγω, Gr. to contract.) The name given to certain muscles whose office it is to contract the part in all directions, drawing it together as the mouth of a purse is contracted by a circular string.

SPI'DER. The different species of spiders compose the genus Aranea, order Pulmonariæ, class Arachnides. The male spider possesses four pairs of legs, the female five, the additional pair enabling her to carry her eggs. The legs are composed of seven joints, the two first forming the hip, the third the thigh, the fourth and fifth the tibia, the sixth and seventh the tarsus. The feet are spread out in diverging rays, so as to include a wide circle, and afford an extensive base of support; they terminate in two, or, sometimes, in three hooks. In front of the head are placed members resembling feet, having affixed to them, or terminating in, a moveable hook, or pincers, flexed inferiorly, underneath which is a minute opening that permits exit to a venomous fluid contained in an adjoining gland. By the injection of this poisonous fluid the common spider of this country is able to kill a fly in a few minutes, and the large spider of South America can, by the same means, destroy the smaller vertebrated animals, and produce, even in man, severe constitutional disturbance. The greater number of species possess a curious apparatus for spinning threads, and for constructing nets, for the entanglement of flies and small insects. This net, or web, is as various as the species, each species constructing its own peculiar form of net; in addition to the principal web, which is spun out for the capture of small insects, the spider frequently constructs a smaller one, both as a residence and a place of ambush. Between these two constructions there is placed a thread of communication, and no sooner is the struggling insect involved in the meshes of the larger net than the vibrations of this communicating thread afford information to the concealed spider, who instantly rushes towards his victim and endeavours to destroy it, by piercing it with his dart and infusing into the wound his poisonous fluid. The web is produced by a double series of spines, opposed to each other, and planted on a prominent ridge of the upper side of the metatarsal joint, or that usually regarded as the first joint of the foot of the posterior legs next the abdomen. These spines are employed as a carding apparatus, the low series combing, or extracting the ravelled web from the spinneret, and the upper series, by the insertion of its spines between those of the other, disengaging the web from them.

Fossil remains of spiders exist in strata of very high antiquity.

SPIKE. (*spica*, Lat.) In botany, a species of inflorescence, in which the flowers stand sessile along a common peduncle, and are either placed alternately and crowded together, or in separate groups; the plantain, lavender, corn, &c., afford examples.

SPI′KELET. In botany, the term applied to a subdivision of a spike, forming, as it were, a small spike.

SPINE. (*spina*, Lat. *epine*, Fr. *spina*, It.)
1. In anatomy, the vertebral column or back-bone of vertebrated animals.
2. In botany, a sharp point, or thorn; the spines of plants differ from prickles, inasmuch as they proceed from the wood of the plant, whereas a prickle comes from the bark only, and they are distinguished by their woody vascular centre.
3. In zoology, a thin pointed spike. Some of the spines of fishes are simply imbedded in the flesh of the animal, and attached to muscles; others are articulated with bones which lie beneath them.
4. The word spine is occasionally used to signify a ridge.

The fossil spines of various **fishes** are found in strata from the greywacke series to the chalk inclusive; they have obtained the name of ichthyodorulites.

SPI′NELLE. } (*spinelle*, Fr.) A species
SPI′NEL. } of corundum both of an octahedral and dodecahedral form. Some mineralogists place spinel in the Ruby family. Its colours are red, black, blue, brown, yellow, and white. It occurs in regular crystals, and, occasionally, in rounded grains: when crystallized, it is found either in regular octahedrons, occasionally having their edges replaced, or in macles, presenting very different forms. It scratches quartz, its hardness being $= 8$. Its structure is usually foliated, with laminæ parallel to the faces of the octahedron. Specific gravity from $3·5$ to $3·8$. It is infusible before the blow-pipe, and intense heat does not even deprive it of its colour. It consists of alumina $82·47$, magnesia $8·78$, chromic acid $6.18$, loss $2·57$. Sometimes its colouring matter is oxide of iron instead of chrome; the red specimens containing chromic acid; the blue, protoxide of iron. The spinelle ruby is a subspecies, of a scarlet colour; the rose-red specimens are termed Balas rubies; the yellow, or orange-red, spinel is called Rubicelle; and the violet coloured, Almandine ruby. Spinel ranks as a precious stone, and is worked by the lapidary for ornamental purposes. As a species, spinel was first established by Romé de Lisle and Werner, and by them separated from

sapphire, with which it had previously been confounded. It is principally met with in Ceylon, Siam, Pegu, and other eastern countries; it also occurs in drusy cavities, together with Ceylanite, Vesuvian, &c.

SPINE'LLANE. A variety of dodecahedral zeolite. A mineral of a plum-blue or blackish-brown colour, found on the banks of the river Laach, near Andernach. It occurs in hexahedral prisms, terminated by three-sided summits, whose faces stand on alternate, but different, lateral edges at each extremity.

SPI'NY. } Having spines, thorns, or
SPI'NOUS. } points.

1. In botany, applied to plants possessing thorns or spines; also to leaves, the margins of which are beset with thorns.

2. In anatomy, applied to certain processes of bones.

SPIRE. (*spira*, Lat.) That part of a body which shoots up to a point.

In conchology, the spire of univalve shells consists of all the whorls except the lower one, which is termed the body. The spire is a prominent feature in univalve shells, and upon its being elevated, depressed, &c., depends much of the generic and specific definition. There are several kinds of spire; the depressed spire, when the spire is very **flat**; the involuted spire, when the whorls are concealed in the inside of the first whorl, as in the nautilus; the reversed spire, when the whorls turn in the contrary direction to a right-handed screw, &c.

SPI'RIFER. } (from *spira*, a spire, and
SPIRI'FERA. } *fero*, to bear, Lat.) A genus of bivalve shells, distinguished from terebratula by its very extraordinary internal spiral processes or cones, of which there were two, and from which it obtains its name. The genus *spirifer* appears to be met with, in the descending series, first in the lias. It is in the carboniferous limestone that Spiriferæ most abound, not fewer than twenty species having been determined as occuring in that deposit. Deshayes recommends the entire suppression of the genus *Spirifer*, stating that all the species may be referred either to Producta or Terebratula. Spiriferæ, which are met with in great abundance in the grauwacke and carboniferous series, appear to have become extinct during, or immediately after, the deposition of the lias, above which not any have been discovered.

SPIROLI'NA. A genus of microscopic foraminiferous multilocular univalves, described by Lamarck, who discovered several species of them in the fossils of Grignon.

SPIRO'RBIS. A genus of shells belonging to the family of the Serpulacea. A familiar example of spirorbis is afforded in the common, small, white, coiled shell so frequently seen upon the shell of the lobster. The spirorbis is found on sea-weed, shells, &c.

SPI'RULA. (from *spira*, Lat.) Both a recent and a fossil shell. A genus of multilocular shells, partly spiral and partly straight, the whorls being arranged in a discoidal form, and separate from each other; the last turn being elongated, and continued in a straight line. The siphunculus, instead of being membranous, is formed by one continuous shelly tube. It appears that the spirula, notwithstanding it possessed a siphuncle, was, altogether or in part, an internal shell. The living spirula is an inhabitant of tropical seas; it floats on the surface of the ocean.

SPI'RULITE. A fossil spirula. Spirulites are sometimes termed lituites, from their supposed resemblance to a bishop's pastoral staff.

SPLE'NDENT. (*splendens*, shining, Lat. *splendénte*, It.) 1. In mineralogy, a term applied to metals as regards their degree of lustre. A mineral is *splendent* when perceptible in full day-light at a great distance, as in highly polished metals.

2. Applied to any colour which possesses a metallic splendour.

SPLENT COAL. } An impure variety of
SPLINT COAL. } cannel coal, occurring in Scotland.

SPLI'NTERY. In mineralogy, a term applied to a particular fracture of minerals. The fracture is called *splintery* when the surface, produced by breaking a mineral, is nearly even, but exhibits little splinters or scales, somewhat thicker at one extremity than the other, and still adhering to the surface by their thicker extremities.

SPLI'NTERY HORNSTONE. The Splittriger Hornstein of Werner. Splintery hornstone is a subspecies of hornstone, occurring both massive and crystallized. It is infusible, *per se*, before the blow-pipe. Its colours are various, principally grey, red, and green. For a further description, see *Hornstone*.

SPO'DUMENE. (from σποδόω, Gr. *in cinerem redigo*.) The Triphane of Haüy; Spodumene of D'Andrada. A rare mineral of a greenish-white or gray colour; occurring massive, and in large granular concretions. Spodumene has been found in the iron-mine of Uton, in Sweden, and in primary rocks in Ireland. According to the analysis of Vauquelin, it consists of silica 64·4, alumina 24·4, potash 5, lime 3, oxide of iron 2·2.

SPO'NDYLUS. A genus of rough, slightly-eared, inequivalved, marine, bivalves, with unequal beaks; hinge with two recurved teeth, separated by a small hollow. Spondyli are found only in the ocean, attached to rocks, corals, &c.: they are remarkable for their spines, and the richness of colouring of the shells. They are eaten like oysters.

Lamarck has described one species as occurring fossil in the neighbourhood of Paris, and very fine fossils are found in Tuscany.

SPONGE. (σπογγία, Gr. *spongia*, Lat. *eponge*, Fr. *spúgna*, It.) This word is pronounced, and frequently written, spunge. A porous marine substance found adhering to rocks, formerly supposed to be a vegetable production, but now classed among the zoophytes: it is soft, light, porous, and easily compressible, readily imbibing fluids, and thereby distending. "The existence of fossil sponge in the transition or in the mountain limestone," says Parkinson, "has not been ascertained, or in the different beds of the lias formation; but the tenuity, in general, of its substance, and the nature of the matrices in which it has been sought, may perhaps occasion its concealment."

SPO'NGIFORM QUARTZ. The name given to a white or grey, porous, variety of quartz, so light as to swim on water, and called also *Floatstone*.

SPONG'IADÆ. An order of the class Astomata, sub-kingdom Protozoa, and divided by Pictet into three families; Spongidæ, Clionidæ, and Petrospongidæ.

SPO'NGIOLE. In botany, an organ situated at the extremity of the root, and thus named from its peculiar texture. It is by the spongioles that plants are enabled to absorb. They are constructed of common cellular spongy tissue, and they imbibe the fluids which are in contact with them, partly by capillary action, and partly, also, by a hygroscopic power.

SPO'RULE. (from σπορά, a seed, Gr.) The organ of reproduction in cryptogamic plants. Ferns are increased by minute bodies, called *sporules*; these are produced either on the backs, or in the axillæ of the fronds, and on other parts. The organs of reproduction in mosses consist of *sporules*, contained within an urn or theca, placed at the top of a thin stalk. In lichens, the organs of reproduction are *sporules*.

SPRINGS OF WA'TER. All permeable strata receive rain-water at their surface, whence it descends until it is arrested by an impermeable subjacent bed of clay, causing it to accumulate throughout the lower region of each porous stratum, and to form extensive reservoirs, the overflowings of which on the sides of valleys constitute the ordinary supply of springs and rivers. The water, however, which descended from the atmosphere in the form of rain, having passed through the various strata, does not re-issue in the same condition. Rain-water contains carbonic acid; in passing through the strata it absorbs oxide of iron, lime, &c., and on issuing in the form of springs it loses its excess of carbonic acid, and again deposits carbonate of iron, &c. "Springs are seldom or ever quite pure, owing to the solvent property of water, which percolating through the earth, always becomes more or less charged with foreign matter. Carbonate, sulphate, and muriate of lime, muriate of soda and iron, are frequently present in spring waters. Some are more highly charged with these and other substances, such as carbonate of magnesia and even silica, than others, and have hence obtained the name of mineral springs. Many thermal springs contain silica, though this substance is of exceedingly difficult solution. Springs usually possess one particular average temperature, generally identical with that of the ground through which the particular spring passes."

SQUA'LOID. (from *squalus*, a shark, Lat. and εἶδος, resemblance, Gr.) The squaloid, or third division of fossils of the family of sharks, appears for the first time in the chalk formations, and extends through all the tertiary deposits to the present period.

SQUA'LUS. The name given by Linnæus to the true shark.

SQUA'MOSE. } (*squamosus*, Lat.) Scaly; covered with scales.
SQUA'MOUS. }

SQA'RROUS. (from *squarra*, roughness of skin, Lat.)
1. In conchology, consisting of scales spreading every way, or standing upright, and not parallel with the plane.
2. In botany, applied to parts with scales widely divaricating.

STAGONOLE'PIS. A genus of ichthyolites of the old red sandstone.

STA'LACTITE. (σταλακτὶς, from σταλάζω, Gr. to drop or distil; *stalactite*, Fr.) A concretion of carbonate of lime pendent from the roof of a cavern, and produced by the percolation and dripping of water, holding in solution, or super-saturated with, carbonate of lime. The mode of formation of a stalactite resembles that of an icicle; the water, as it slowly drips from the roof, continually deposits upon the pendent stalactite a small quantity of its carbonate of lime, and thus the stalactite increases in length and bulk.

STA'LAGMITE. (from σταλάγμὸς, Gr. *stalagmite*, Fr.) A concretion of carbonate of lime produced by the dripping of water holding in solution carbonate of lime. The difference between a stalactite and a stalagmite is this: the former is attached to, suspended from, and formed at, the roof of a cave or grotto; the latter is formed upon the floor: the stalactite generally resembles a large icicle; the stalagmite is an unshapen mass upon the floor. It sometimes happens that from the stalactite lengthening downwards, and the stalagmite increasing upwards, the two become united, and thus form a column extending from the roof to the floor.

STA'MEN. (*stamen*, Lat.) A constituent part of a flower, situated within the corolla, consisting of two parts, the filament and the anther. The stamens are the male organs of the plant. In the plural the word is written either stamens or stamina.

STA'NDARD. In botany, the upper large petal of a papillionaceous flower; also called the banner.

STANNI'FEROUS. (from *stannum*, tin, and *fero*, to bear, Lat.) Containing tin; yielding tin; producing tin.

STAU'ROLITE. The name given by Kirwan to the harmotome of Haüy, the cross-stone of Jameson; kreutzstein of Werner. The composition of staurolite, as described by Cleaveland, differs entirely from that of staurotide. Staurolite contains, according to some analyses, upwards of twenty per cent. of baryta, and no oxide of manganese; whereas staurotide contains no baryta, but four per cent. of oxide of manganese. Prof. Cleaveland assuredly does not intend the same mineral in describing Staurotide and Staurolite, nevertheless, I find most mineralogists giving these two words as synonymous. The reader, however, must bear in mind that Harmotome, or Cross-stome is not to be considered as *Staurotide*, although by some authors Staurotide is regarded as synonymous with *Staurolite*.

STAU'ROTIDE. The name given by Haüy to prismatoidal garnet or grenatite. The prismatoidischer garnet of Mohs; the granatit of Werner; the grenatite of Jameson and Brochant. This mineral, of a reddish-brown colour, occurs crystallized, in four and six-sided prims, sometimes intersecting each other at right angles. Its primitive form, under which it sometimes appears, is a four-sided prism, whose bases are rhombs with angles of 129° 30′ and 50° 30′: the reflection goniometer gives the measurement of the angles 129° 20′ and 50° 40′. Its integrant particles are triangular prisms. Its colours are reddish-brown to blackish-brown. Specific gravity from 3·3 to 3·9. Hardness from 7 to 7·5. It feebly scratches quartz, but does not yield sparks with steel. Fracture uneven or imperfectly conchoidal. It consists of silica, alumina, lime, and the oxides of iron and manganese. It is found in primary rocks only. It may be distinguished from garnet by its form and infusibility.

STEA'RIN. ⎫ (from $\sigma\tau\epsilon\alpha\rho$, Gr.) The
STEARI'NE. ⎭ solid part of oil and fatty matter. Fat is composed of two constituent principles, which Chevreul distinguished by the terms *stearine* and *elaine*.

STEA'TITE. Veins of *Steatite* are very plentiful in the serpentines of the Lizard. Their colour varies from white to yellow, green, and purple. The veins of *steatite* have distinct walls, and are as regular as most true veins. Fragments of serpentine and calcareous spar are sometimes found in them. Dr. Berger considered steatite to bear the same relation to serpentine that kaolin bears to granite. Dr. Thomson considers steatite to be a portion of serpentine altered by the action of water, or some other body. Sir H. Davy regarded the veins of steatite as mechanical deposits. Steatite has four equivalents of silica to three of magnesia.

STEEL. (*stahl*, Germ. From the Chalybes, who had considerable iron and steel works, has steel derived its Greek name of $\chi\alpha\lambda\upsilon\psi$, though some have thought to derive the name of the people from their works. The word chalybs was adopted by the Romans from the Greeks, and is has passed into our own language in the adjective chalybeate.) Iron combined with carbon. The proportion of carbon has, perhaps, never been accurately ascertained; but steel containing one-sixtieth of its mass of carbon is said to have the maximum degree of hardness. The following are some of the properties of steel:— It is so hard as to be unmalleable when cold; it is brittle, resists the file, cuts glass, affords sparks with flints, and retains the magnetic virtue for an indefinite length of time. By being ignited, and afterwards slowly cooled, it loses its hardness. It fuses at 130° Wedgewood. When red-hot it is malleable. It is more sonorous than iron, and may be hammered out into much thinner plates. The conversion of iron into steel is effected by combining it with carbon. This combination is performed in three ways, by three different processes, and the products are distinguished by the names of *natural steel, steel of cementation,* and *cast-steel*. Of these, the most valuable is cast-steel, its texture being the most compact, and it admitting of the finest polish. The manufacture of articles of steel, says Mr. Babbage, affords a most striking example of the value conferred by human labour on the raw produce of nature. The value of a pound of crude iron is twopence. This pound of iron, after having been converted into steel, is manufactured into balance-springs for **watches**. One of these springs weighs $\frac{15}{100}$ of a grain and sells for twopence. After deducting for waste, a pound of iron will make fifty thousand springs, and the twopennyworth of iron becomes worth £416 13s. The value of the charcoal is too minute to be taken into the calculation.

STEI'NHEILITE. A variety of iolite, a mineral of a blue colour.

STE'LLA MARI'NA. The name employed by Linck, on the authority of Pliny, to signify the asteria or star-fish.

STE'LLATED. (from *stella*, Lat. **a star**.) Having the fibres diverging **all** round a common centre.

STELLE'RIDÆ. The name assigned by Lamarck to the Linnæan genus Asteria; of this genus Lamarck has formed a family comprehending four genera.

STELLE'RIDAN. An animal resembling an asteria. Fossil stelleridans have not been discovered in strata more ancient than the muschel-kalk.

STE'LLITE. A fossil asteria or star-fish.

STELLI'FEROUS. (from *stella*, a star, and *fero*, to bear, Lat.) Having stars, as some of the corallines.

STE'LLULAR. Having markings resembling stars. The surface of the tubipora, or organ-pipe coral, **is** covered with a green fleshy substance, studded with *stellular* polypi.

STENEOSAU'RUS. A genus of fossil Saurians, with long and narrow beaks; thus named by M. Geoffroy St. Hilaire. In the Steneosaurus the arrangement of the nostrils was nearly the same as in the gavial, opening upwards, and of nearly a semicircular form on either side.

STE'RNAL. (from *sternum*, the breast-bone.) Pertaining to the sternum or breast-bone.

STE'RNUM. (στέρνον, Gr. *sternum*, Lat.) The breast-bone. In the human subject the sternum is divided into three parts; in some vertebrated animals it is formed of nine elementary species, each proceeding from a separate centre of ossification. "Few subjects in comparative osteology," says Dr. Roget, "are more curious and instructive than to trace the development of these several elementary parts in the different classes of animals, from the rudimental states of this bone as it occurs in fishes, to its greatly expanded conditions in the tortoise and the bird, which severally exhibit the most opposite proportions of these animals."

STI'GMA. (στίγμα, Gr. *stigma*, Lat.) In botany, the apex of the pistil; the stigma is variously formed, being either a fine point, a round head, or lobed; generally downy, often hollow and gaping, and more or less moist. Sometimes, though there is only one style, there are two or more stigmas; when there is no style, the stigma is sessile on the ovarium.

STIGMA'RIA. A family of extinct fossil plants of the coal formations. The stigmaria was an aquatic plant, inhabiting swamps or lakes, and, as regards its external structure, resembled the euphorbiaceæ. Fragments of stigmariæ occur abundantly in the coal shales. It appears to have been dicotyledonous.

"Authors are by no means agreed," says Sir H. De la Beche, "as to what families certain genera of fossil plants should be referred. Thus M. Ad. Brongniart refers the genus *Stigmaria*, very common in the coal measures, to the family of Lycopodiaceæ; while Lindley and Hutton consider that, if any existing analogy must be found, it is with greater probability of accuracy referrible to the *Euphorbiaceæ* or *Cacteæ*, most probably to the former; a difference of considerable importance, as upon it depends whether the genus in question belongs to the class *cellurares*, or to the class *vasculares*. This difference of opinion arises, no doubt, from the obscurity of the subject, fossil botany being beset with very great difficulties; difficulties far beyond those which attend the study of fossil zoology, though the latter are by no means either small or rare."

STI'GMATA. Spiracles from which the tracheæ of insects commence.

STI'LBITE. (from στίλβω, to shine, Gr.) From the degree of lustre which it possesses. The radiated zeolite of Jameson; strahl-zeolite of Werner; prismatoidischer kuphon-spath of Mohs. Stilbite is of a white colour generally, sometimes pure, at others shaded with grey, yellow, or red. It occurs both crystallized and massive. It is splendent externally; internally shining and pearly. Translucent, and sometimes transparent. Specific gravity 2·2. Hardness 3·5 to 4·0. It does not scratch glass. It consists of silica 52·5, alumine 17·5, lime 11·5, water 18·5. It occurs in secondary trap rocks, in Scotland, Norway, and in the Faroe islands.

STINK STONE. (Called also Swine-stone.) The name given to a variety of limestone, from the fetid odour which it gives out on friction; a smell resembling rotten eggs. It is the chaux carbonatée fetide of Haüy; the stinkstein of Werner. It occurs in masses, either compact, or having a granular or foliated structure, frequently forming large beds, or even whole mountains.

STIPE. (*stipes*, Lat.) In botany, the stem or base of a frond; a species of stem passing into a leaf, or not distinct from the leaf; the name given to the stem of palm trees; it differs essentially in form, structure, and mode of growth, from the trunk, increasing in length only, and not in thickness. The stem of a fungus is also called a stipe, as is the thread, or slender stalk, which supports the down, and connects it with the seed.

STI'PULA. } (from *stipula*, Lat.) In
STI'PULE. } botany, a membranous leafy appendage, placed at that part of the stem whence the leaf or footstalk arises. Stipules vary in number, being solitary or in pairs; in situation, being either external with regard to the leaf or footstalk, or internal, the internal sometimes embracing the stem in an undivided

tube; in form, linear at the base, or crescentic; in attachment, connected directly with the stem, or with the petal; in direction, erect, or variously reflected. Stipulæ serve to protect the nascent leaves.

STOMATO'PODA. An order of the class crustacea; the animals belonging to this order have the antenniferous region of the head distinct from the thorax.

STONE GALL. The name given by the workmen to an oval or round mass of clay, occurring in variegated sandstone.

STONE-BORER. The name for a molluscous bivalve which mechanically perforates, or bores into, rocks.

STONE COAL. Another name for culm. See *Culm*.

STRAMI'NEOUS. Straw-coloured; of a yellow colour resembling that of straw.

STRATIFICA'TION. The arrangement of substances in strata or layers, like the leaves of a book, one upon another.

STRA'TIFIED. Arranged in layers or strata, one upon another, like the leaves of a book.

STRA'TUM. *(stratum,* Lat. from *sterno,* to lay out, to spread.) A layer of any deposited substance. The term *stratum* is of general signification, and independent of the absolute thickness of the mass; it need never be used as a special term of definition, but reserved for general reasoning. Bakewell observes, "though the word *stratum*, in its original language, and by general acceptation in speaking of rocks, denotes a bed, it is convenient to restrict the term bed to a stratum of considerable thickness; for such *beds* are often subdivided into several distinct minor *strata*, and we cannot well describe a *stratified stratum*." The true thickness of a stratum is measured by a line perpendicular to the upper and under surface, let its inclination be whatever it may.

STREAK. (*strich*, Germ.) In mineralogy, that appearance of a mineral which arises from its being scratched by a hard sharp instrument. The *streak* is said to be *similar* when the colour of the powder produced by scratching the mineral is the same with the colour of the mineral itself; when the colour varies, the streak is said to be *dissimilar*. Streak forms one of the physical or external characters of minerals, and by no means an unimportant one. The particular hue of the powder of a mineral is most easily obtained by rubbing or *streaking* the specimen on a slab of porcelain biscuit.

STREPSI'PTERA. (Kirby.) An order of insects, parasitic animals, that have two ample wings, forming the quadrant of a circle, and of a substance between coriaceous and membranous; and two elytriform subspiral organs, appendages of the base of the anterior legs. Latreille has given to this order the name Rhipiptera.

STREPTOSPO'NDYLUS. (Reversed spine.) The name assigned to an extinct crocodilian reptile, and so named because the vertebræ are arranged in the spiral column in a position the reverse of that which obtains in all other reptiles of the same osteological type; the convexity of the vertebræ being placed anteriorly, whereas in the crocodile, &c., it is in the opposite direction.—*Mantell*.

STRI'ATED. (*strié*, Fr.) Arranged in fine lines running parallel to each other, let the direction of the lines be what it may. Marked with fine thread-like lines, running parallel to each other.

STRIKE. (*streich*, Germ.) The line of bearing of strata. The strike or direction of strata is always at right angles to their dip; if a bed *dip* due north or south, its *strike* will be due east and west.

STR′IPED AG′ATE. (See *Ribbon Agate.*)

STR′IPED JA′SPER. The Band jaspis of Werner. Quartz-jaspe onyx of Haüy. Called also Ribbon jasper. This is a variety differing from common jasper principally in the arrangement of its colours, which are various, and arranged in stripes, veins, rays, and in curved, concentric zones. It occurs massive, in whole beds. Specific gravity from 2·44 to 2·55. Fracture flat conchoidal. Opake. Being susceptible of a high polish, this variety of jasper is used for ornamental purposes. It is found principally in the Uralian mountains of Siberia, in Saxony, in Devonshire, and in the neighbourhood of Edinburgh.

STRO′BIL. ⎱ (from *strobilus*, Lat.) In
STRO′BILE. ⎰ botany, a pericarp formed from an ament by the hardening of the scales; a catkin hardened and enlarged into a seed-vessel; a seed-vessel composed of ligneous scales.

STROMA′TEUS. An ichthyolite, exhibiting much analogy with Zeus or Chætodon, and met with in the metalliferous schists of Mansfield, Thuringia, &c.

STRO′MBITE. A fossil strombus; a petrified strombus. Strombites are very rare: in the mountains of Arragon, and in the Veronese, some specimens beautifully preserved have been discovered.

STRO′MBUS. The name given by Linnæus to a genus of univalve, spiral, marine, shells: aperture much dilated: lip expanding and produced into a groove. Lamarck has divided this genus into two subgenera, Strombus and Pteroceras. In some of the shells of this genus the spines are of great length, and are arranged round the circumference of the base, being at first tubular, and afterwards solid, according to the period of growth. In Turton's Linné, fifty-five species of strombi are described.

STRO′NTIA. ⎱ A mineral sometimes
STRO′NTIAN. ⎰ transparent and colourless, but generally with a tinge of yellow or green. Hardness = 5. Specific gravity from 3·4 to 3·9. Texture, generally fibrous; sometimes it occurs crystallized in slender prismatic columns of various lengths. Its taste is acrid and alkaline; it converts vegetable blues to green. The principal use of strontian is to communicate a beautiful red colour.

STRO′NTIANITE. Carbonate of strontian. The Strontian carbonatée of Haüy; Strontianit of Werner; Strontiane of Jameson; Stronthianite of Kirwan. A mineral first brought to Edinburgh from the lead-mines of Strontian, in Argyleshire, in the year 1787. It greatly resembles carbonate of barytes, but it is not poisonous.

STRO′NTIUM. The metallic base of strontia. Strontium greatly resembles barium in its appearance, although it is, indeed, a very different substance. Strontium is harmless, but barium and all its salts are poisonous. The salts of strontium communicate to flame a fine red tinge; those of barium, a yellow.

STRU′CTURE. (*structura*, Lat.)

1. A term used, in mineralogy, to denote one of the characters of minerals. The structure of a mineral depends on the shape, size, and arrangement of the minute parts of which it is composed. It is sometimes used synonymously with fracture, but it is not correct so to do, there existing a considerable difference between the two terms.

2. The manner of organization of animals and vegetables. The manner in which the parts of an organized body are arranged among themselves.

STU′FA. (*stúfa*, It.) A jet of steam issuing from a fissure of the earth.

In volcanic regions stufas are by no means uncommon. The name is taken from the Italians, who thus apply it.

STYLE. (*stylus*, Lat.) In botany, that part of the pistil which elevates the stigma above the germen. The style is not absolutely essential, and is sometimes wanting. The style is a continuation of the midrib, and constitutes a portion of the pistil.

STYLI′NA. The name of a genus of lamellated polypifers, thus described by Parkinson, "A stony polypifer, formed of simple thick masses echinated in the upper part. Numerous cylindrical fascicular tubes containing radiating lamellæ, with a solid axis; the solid styliform axes projecting beyond the tubes."

STYTHE. The name given by the colliers to fire-damp.

SUB-A′PENNINE. A term applied geologically to a series of strata, of the older pliocene period. The beds which have been termed subapennine are composed of sand, clay, marl, and calcareous tufa; they are all tertiary deposites, and abound in marine shells, of genera and species which prove some of them to be contemporaneous with the crag deposit, and others of a more ancient epoch: they rest unconformably upon the inclined beds of the Apennine range. Brocchi, an Italian geologist, gave to this group the name it bears. The subapennine beds have resulted from the waste of the secondary rocks which now form the Apennines, and which had become dry land before the older Pliocene beds were deposited.

SUBLIMA′TION. (*sublimation*, Fr. *sublimazióne*, It.) That operation by which solids are, by the aid of heat, brought into a state of vapour, and again condensed into a solid form. One of the hypotheses proposed to explain the filling of chasms in solid rocks with metallic ores, is by a process of sublimation from subjacent masses of intensely heated mineral matter.

SUBSE′SSILE. In botany, applied to leaves having very short footstalks.

SU′BSTANTIVE CO′LOURS. Dr. Bancroft divides colours into *substantive* and *adjective*: those which he termed substantive colours, communicate their colour without the intervention of some other substance: those which he called adjective, require the aid of a mordant or basis.

SUBTERPOSI′TION. The order of arrangement in which strata or rocks are placed below each other, as superposition is the order in which they are arranged above one another. Subterposition in the plutonic, like superposition in the sedimentary rocks, being, for the most part, characteristic of a newer age.

SU′BULATE. (from *subula*, an awl, Lat.) Awl-shaped.
1. In botany, applied to leaves, when thickest at the base, and gradually tapering towards the point.
2. In conchology, applied to shells tapering gradually to a point.
3. In entomology, a long thin cone softly bent throughout its whole course.

SU′CCINITE. (from *succinum*, amber, Lat.) A mineral of an amber-yellow colour, thus named by Bonvoisin. It occurs in small rounded masses about the size of a pea. Some mineralogists refer succinite to the idocrase; others, to the garnet; by some it is deemed an amorphous variety of topazolite.

SU′CCULENT. (*succulentus*, Lat. *succulent*, Fr. *sugósa*, It.) Juicy; full of juice. A term applied to plants with a soft and juicy stem, as distinguished from those called ligneous.

SUCTO′RIA. A class of animals belonging to the sub-kingdom Diploneura or Helminthoida. Cuvier placed Suctoria in his placoid order, but by the more recent classification of Müller and Owen they have been removed from that high order and carried down to the lower point in the scale, which their inferior standing so obviously renders the natural one.

SU′DES. (*sudes*, Lat. a spear.) A class of spines comprising several genera, as Sudes villarum, Sudes fortalitiorum, &c.

SU′FFOLK CRAG. A marine deposit of the older pliocene period. It consists of beds of sand and gravel, abounding in shells and corals.

SU′LCATED. (from *sulcus*, a furrow, Lat.) Furrowed; grooved.
1. In botany, applied to stems marked with broad deep lines; also to leaves having broad, deep, parallel lines.
2. In conchology, applied to shells that are deeply furrowed, or marked with ridges or broad furrows.

SU′LCUS. Plural sulci. A broad furrow or groove.

SU′LPHATE. (from *sulphur*.) A combination of sulphuric acid with any salifiable base.

SU′LPHITE. A combination of sulphurous acid with any salifiable base.

SU′LPHUR. (*sulphur*, Lat. *soufre*, Fr. *zólfo*, It.) One of the simple or elementary substances, and a non-conductor of electricity. It is of different shades of yellow, and occurs either in masses or crystallized. Specific gravity from 1·9 to 2·1. By friction it acquires negative electricity. It is very brittle and friable. Its equivalent number is 16·1; its symbol, S.

SU′LPHURET. A combination of sulphur with a metallic base; as sulphuret of iron.

SU′LPHURET OF IRON. Iron pyrites. This is very commonly found in irregular and subglobular nodules and masses, in the chalk formation.

SU′MMIT. (*summitas*, Lat. *sommet*, Fr.) In conchology, the most elevated point of the shell in which the hinge is placed.

SUPERFI′CIES. (*superficies*, Lat. *superficie*, Fr. *c'est longeur et largeur sans profondeur*.) The surface only of a body; the exterior part.

SUP′ERIOR ORDER. By some geologists it has been proposed to place all the formations above the chalk in an order to be named the "Superior Order." The superior order may be deemed synonymous with the tertiary, or newest floëtz rocks. In the arrangement thus proposed the division of all the formations will be as follows.
1. Superior order; the same as the tertiary. 2. Supermedial order, comprising the chalk and all other formations downwards to the magnesian limestone inclusive. 3. Medial order, comprising the coal measures, the carboniferous limestone, and the old red sandstone. 4. Submedial order, synonymous with the transition, comprising the roofing slate, &c. 5. Inferior order, consisting of mica, slate, gneiss, granite, &c., &c.

SUPERME′DIAL ORDER. This class includes, generally, all the secondary formations more recent than the great coal deposit, and between it and the tertiary, supra-cretaceous, or newest floëtz class. This series of strata comprises distinct groupes well marked from each other, and therefore entitled to the name of separate formations. The supermedial order admits of four principal divisions. 1. The chalk formation. 2. The series of ferruginous sands, including, and divided by, the Wealden. 3. The series of oolites, including and terminating with the lias. 4. From the lias downwards to the coal, comprising the new red sandstone and the magnesian limestone.

SUPERPOSI'TION. The order in which bodies are placed upon or above other bodies, as more recent strata upon those that are older; secondary rocks upon primary; tertiary upon secondary, &c., &c. The order of superposition of rocks is never reversed, unless it be by volcanic agency, when rocks may be forced from below and thrown, as it were, upon those which, in the usual order of superposition, would be above them. Beds, or strata, may be altogether wanting, but whereever similar beds occur together, the order of superposition is never inverted. Thus the Wealden deposits are never found above the chalk; the chalk is never found above the London clay; the London clay is never found above the crag: nor do we meet with the chalk under oolite, the lias under the red sandstone, or the coal under the greywacke. The order of superposition alone does not form a certain criterion of the ages of rocks, but when coupled with their zoological contents, the two connected may be considered the only safe criterion.

SUPRA-CRETA'CEOUS. (from *supra*, above, and *cretaceous*, chalky, Lat.) Above the chalk; formations more recent than those of the chalk. The supra-cretaceous deposits are very commonly termed tertiary. This is, however, a name exceedingly objectionable, as it would imply that there were three great classes of rocks possessing marked characteristic distinctions, and that the deposits above the chalk constituted the third of such classes. The supra-cretaceous rocks constitute a large portion of the dry land of Europe, among the lowest of which, in Western Europe, are those of the London and Paris basins. Bakewell considers the term supra-cretaceous, or, as he writes it, super-cretaceous, to be peculiarly inappropriate, and adds, "if a new name were necessary, *post*-cretaceous should have been chosen." The supra-cretaceous rocks are exceedingly various, and contain an immense accumulation of organic remains, terrestrial, freshwater, and marine. This group has lately been shown to approach, more closely than was supposed, to the existing order of things on the one side, and to the cretaceous group on the other.

SU'RTURBRAND. A name given to Bovey coal, or brown coal.

SU'SSEX MA'RBLE. A member of the Wealden group; occurring in layers varying from a few inches to upwards of a foot in thickness, the layers being separated by seams of clay, or loose friable limestone. It is a fresh-water deposit, and contains in great abundance shells of paludinæ, a genus of fresh-water univalves. It is of various shades of grey and bluish-grey, mottled with green and yellow; it bears a high polish, and is used extensively for architectural and ornamental purposes. There is historical proof, says Dr. Mantell, of its having been known to the Romans, and in the early Norman centuries it was much sought after, and applied, when cut into small shafts of pillars, which were placed in the *triforia*, or upper arcades, of cathedral churches, as at Canterbury, Chichester, &c. The archiepiscopal chair of Canterbury cathedral is formed of Sussex marble.

SU'TURE. (*sutura*, a seam, Lat.) A seam; the junction of the bones of the head by an irregularly jagged zig-zag line.

SWAL'LOW. The name given to a chasm or hollow, commonly occurring in the mountain or carboniferous limestone, in which a stream of water is engulphed. The stream or river reappears after a certain subterranean course.

SWI'LLEY. A provincial term for a

coal-field of very limited extent.

SWINE-STONE. The name given by Kirwan to fetid carbonate of lime, or stink-stone.

SY'LVANITE. Native tellurium.

SY'MON FAULT. A local term applied to the tapering off of coal seams. The coal not unfrequently tapers away and disappears amid the shales and sandstones, constituting what are locally termed "Symon" faults.

SY'MPHYSIS. (σύμφυσις, Gr. from συμφύω to grow together.) A term used in anatomy to denote a particular form of union of two bones. Bones united by symphysis have no manifest motion.

SY'NCHRONAL. ⎱ (from σύν, the same,
SY'NCHRONOUS. ⎰ and χρόνος, time or age, Gr.) Occuring at the same period of time; simultaneous; of the same age.

SY'NCLINAL LINES. Lines which form ridges and troughs running nearly parallel to each other.

SYNGENE'SIA. (from σύν, with, and γένεσις, generation, Gr.) In botany, the nineteenth class of plants in Linnæus's artificial system. The orders of this large class, five in number, are founded on the circumstance of the florets of the capitule being hermaphrodite and unisexual, variously combined in the disk and ray. The anthers are united into a tube, and the flowers are compound. The following are the five orders composing the class Syngenesia; they are determined by the arrangement of their flowers, and by the sex of their florets.

1. Polygamia æqualis, where each floret is perfect, being furnished with stamens and pistils, and capable of bringing its seed to maturity: the leontodon taraxacum, or dandelion, is a familiar example. 2. Polygama superflua. The florets of the disk perfect, those of the margin having pistils only: the anthemis nobilis, or chamomile, is an example. 3. P tranea. Florets of t or united, those o: neuter, or destitute well as of stamens. necessaria. The flo: are male, of the ra garden marygold a: tration. 5. Polyga Several florets, cit: compound, but with included within one the globe-thistle is a: these five orders, so: a sixth, namely, Mo: has the flowers sep crowded in heads.

SYNO'VIA. (from σύν, ὠόν, an egg, Gr. s; glairy lubricating 1 within the capsula joints, serving the p venting friction, and white of an egg, wh

SY'NTHESIS. (σύνθεσις τίθημι, to join toge Fr.) The act of jo: to analysis. Wate consist of oxygen a: analysis, that is, b water and ascertai: tuents; it may how to consist of oxyge: by synthesis, that is, relative proportions hydrogen.

SYRINGODE'NDRON. T by Count Sternberg of sigillaria, from th shaped flutings tha the top to the bo trunks. These trur joints, and many the size of forest tre

SYSTE'MIC CIRCULA'TIO: lation of the bloo body generally, as from that other ci: is confined to the res and the heart, or circulation.

SY'STOLE. (συστολή, (

*systole*, Fr.) A term used to signify the heart's contraction. The two movements of the heart are its systole and diastole; by the systole of the heart, or its contraction, the blood is pressed out of the heart and forced into the arteries; by its diastole, the blood is received from the veins into the heart.

SY'ZIGY. (from συζυγία, Gr. conjunction; *syzygie*, Fr.) A term applied to the conjunction or opposition of a planet with the sun. The syzygies of the sun and moon occur at the time of full and new moon; the tides are much increased, and are called spring tides, in the syzygies.

# T

TABASHEE'R. A silicious secretion, or concretion, found in the joints of the bamboo: by some it has been supposed to be the juice of the plant inspissated and hardened. It is remarkable on account of its peculiar optical properties; its refractive power is between air and water, namely 1·111. The finest varieties reflect a delicate azure colour, and transmit a straw-yellow tint, which is complemental to the azure. When it is wetted slightly with a wet needle or pin, the *wet spot instantly becomes milk-white and opaque*. The application of a larger quantity of water restores its transparency. The word is from the Persian. The tabasheer found in the *green* bamboo is perfectly translucent, soft, and moist; but on exposure to the atmosphere its moisture evaporates, and it becomes opaque, hard, and of a white or grey colour.

TA'BULAR. (from *tabularis*, Lat.) Formed in laminæ or plates; having a flat or square surface; in large plates.

TA'BULAR STRU'CTURE. This form of structure consists of parallel plates, separated by regular seams; it is the consequence of crystallization, and though closely allied to the columnar structure, is not uncommonly confounded with stratification.

TA'BULAR SPAR. (The schaalstein of Werner; spath en tables of Haüy; prismatischer augit-spath of Mohs.) Called also Wollastonite. A greyish-white mineral, occurring massive, and in granular concretions. Lustre pearly; fracture splintery; translucent. Specific gravity 2·7 to 3·2. Its constituents are silica 51·40, lime 45·0, oxide of iron a trace, water 4·0. It occurs in primary rocks in Norway, and other places.

TÆNI'ADA. The second order of Scolecida, according to Busk; it comprises the tape-worm.

TALC. } (*talk*, Germ, *talc*, Fr.) A
TALCK. } somewhat fibrous and very commonly foliated mineral, resembling mica in its lamellar structure, its laminæ not being elastic as are those of mica, but only flexible. Its colours are white, pale yellow, or greenish. Both its surface and powder are unctuous to the touch. It is so soft as to be easily scratched with the fingernail. Its lustre is often pearly, or inclined to metallic. It is translucent, and, when divided into thin laminæ, transparent. Specific gravity from 2.58 to 2·90. Before the blow-pipe it whitens, its laminæ

separate, and their extremities fuse into a white enamel. Its constituents are principally silex and magnesia, with small quantities of potash, alumine, oxide of iron, and water. Talc and chlorite are nearly allied, and pass, by insensible gradations, into each other. Talc is sometimes used as a **substitute for glass, and windows are formed of its laminæ, which are sometimes obtained of the size of twelve inches** square. There are two varieties of talc, namely, fibrous talc and indurated talc.

TA′LCITE. The nacrite of Brongniart; talc granuleux of Haüy; erdiger talc of Werner. A rare mineral, occurring in coats or reniform masses, composed of very minute shining spangles or scales. When rubbed between the fingers, it leaves a pearly gloss.

TALCOCA′LCITE. A name proposed, by Dr. Boase, to be given to a genus of rocks. Talcocalcite is a compound of granular felspar and talc: its species all effervesce with dilute muriatic acid. The genus is divided into four species, namely, compact, schistose, lamellar, and silicious: these all possess, **more or less, a** silky texture.

TALC-SCHIST. A genus of minerals in the classification of Dr. Boase, and thus described by him: "A basis of granular felspar, intermixed with, and laminated by, talc." It is a glossy slate, having the peculiar saponaceous feel of magnesian minerals. Dr. Boase enumerates four species: T. foliated; T. lamellar; T. crenulated; T. schistose. Talc-schist is intimately connected with euphotide and serpentine, by the change of diallage into talc.

TA′LUS. *(talus,* Lat. *talus,* Fr. *inclinaison que l'on donne à la surface latérale et extérieure d'un mur, de telle sorte que de haut en bas il aille toujours en s' épaississent. Il se dit aussi d'une terrasse sans murs, lors- que ses faces latérales s'elargissent de haut en bas.)*
1. A sloping heap. When, from disintegration, the fragments of a face of rock accumulate at its base and form a sloping heap, the heap is called a talus.
2. In anatomy, a name sometimes given to one of the bones of the tarsus, the astragalus.

TA′NTALITE. Called also Columbite. The ore of tantalum or columbium. When recently broken, tantalite is of a dark bluish-gray, or nearly iron-black colour. It occurs in octohedral crystals, and in small masses. Its specific gravity is from 5·9 to 7·9 It consists of oxide of columbium 80, oxide of iron 12, oxide of manganese 8.

TA′NTALUM. A metal extracted from tantalite; it appears to be the same substance as columbium, and identical with it. It is one of the simple or elementary bodies.

TA′PIR. A genus of mammalia, belonging to the family Pachydermata Ordinaria. The nose of the tapir may be compared to a small fleshy proboscis, the snout being lengthened and moveable. There are several species of this genus still surviving, but they are all natives of tropical climates. The general appearance of the tapir is that of a pig, but some of the species are as large as the ass. The fore feet have four toes each, all of equal size; the hind feet have only three each. The skin is of a dark colour, nearly black, with but few hairs.

Fossil tapirs are found in different parts of Europe. They appear to have far exceeded in size any of the living species. They belonged to the same era as the fossil elephants and mastodons; lived with them, and were destroyed by the same catastrophe; their bones being found in the same strata.

TAP-ROOT. In botany, the main root

of a plant, which passes directly downwards.

TARDIGRA'DA. } (from *tardigradus*,
TA'RDIGRADE. } slow-paced, Lat.) A family of quadrupeds of the order Edentata. These animals have obtained their name from the extreme slowness of their motions. The only existing genus is the Bradypus or sloth. Their nails are enormously long, compressed, and crooked. The molars are cylindrical; the canini sharp, and longer than the molars. They live in trees, and so great is their indisposition to locomotion, that they continue on the same tree till they have devoured every leaf to be found thereon. Some authors state that, to avoid the trouble of a regular descent, they tumble themselves down from the branch they happen to be on. When they have eaten their full, they can roll themselves into a ball, and take a long and reckless sleep.

TAR MI'NERAL. A variety of bitumen, bearing a great resemblance to petroleum, but more viscid, and of a darker colour.

TARN. A bog; a fen; a marsh; a pool; a lake basin. This word appears to be of Icelandic origin.

TA'RSAL. Pertaining to the tarsus or instep, as the tarsal bones, &c.

TA'RSUS. (ταρσος, Gr. *tarse*, Fr.)
1. The instep, or that part of the foot situated between the bones of the leg and the metatarsus. In the human subject the tarsus is composed of seven bones, namely, the astragalus, the os calcis, the os naviculare, the os cuboides, and three cuneiform bones. In the lower mammalia, the number of bones forming the tarsus, and the general figure of this part of the foot are very variable. In the simiæ, as in man, the number of the tarsal bones is seven; in the solidungulous anmimals it is six; in the ruminants generally five, although in the giraffe it is four, and in the camel six.

2. In entomology, the tarsus, or foot, of insects is the last division of the limb: it is divided into several joints, which have been supposed to represent the toes of quadrupeds.

TAX'ODON. The name given by Prof. Owen to a fossil animal discovered in South America, in consequence of the curvature of its teeth. He says, "judging from the portion of the skeleton preserved, the Taxodon, as far as dental characters have weight, must be referred to the rodent order. But from that order it deviates in the relative position of its supernumerary incisors, in the number and direction of the curvature of its molars, and in some other respects. It deviates also, in several parts of its structure, from the Rodentia and the existing Pachydermata, and it manifests an affinity to the Dinotherium and the Cetaceous order. From the development of the nasal cavity and the presence of frontal sinuses, it is extremely improbable that the habits of the Taxodon were so exclusively aquatic as would result from the absence of hinder extremities, and Prof. Owen, therefore, concludes that it was a quadruped and not a cetacean; and that it manifested an additional step in the gradation of mammiferous forms leading from the Rodentia, through the Pachydermata, to the Cetacea; a gradation of which the water-hog of South America (Hydrochærus capybara) already indicates the commencement amongst existing Rodentia, of which order it is interesting to observe this species is the largest, while at the same time it is peculiar to the continent in which the remains of the gigantic Taxodon were discovered."

TAXO'NOMY. (from τάξις, order, and νόμος, law, Gr.) The classification or arrangement of animals or plants, according to certain principles, in divisions and groups.

TE'GMEN. (*tegmen*, any sort of covering, Lat.) A covering of the body, as the cuticle, &c.: used in botany to denote one of the coats of the seed; in entomology, applied to the coverings of the wings of the order Orthoptera, or straight-winged insects.

TELEOSA'URUS. A genus of fossil saurians, thus named and arranged by M. Geoffroy St. Hilaire. The *teleosauri* have long and narrow beaks, the nostrils forming almost a vertical section of the anterior extremity of the beak. "One of the finest specimens of fossil teleosauri yet discovered was found in the year 1824, in the alum shale of the lias formation at Saltwick, near Whitby; its entire length is about 18 feet, the head one foot in breadth, the snout was long and slender, as in the gavial, the teeth one hundred and forty in number. Some of the ungual phalanges show that the extremities were terminated by long and sharp claws, adapted for motion upon land, from which we may infer that the animal was not exclusively marine."

TELEO'STI. The fourth order of the class Pisces, comprising the perch, cod, salmon, and other ordinary osseous fishes.

TELLY'NA. (*tellina*, Lat.) A genus of orbicular, or ovate transverse, equivalved, marine bivalves, with a fold on the anterior part, and short beaks, found in sands, at depths varying to fifteen fathoms. Hinge with usually three teeth, the lateral ones smooth on one side. The shells of this genus are chiefly known by the inflection, or irregular fold, on the fore part; in the one valve the fold is convex, in the other, concave. There are three families of tellinæ: ovate and thickish; ovate and compressed; suborbicular. Some conchologists have divided Tellina into three genera; Tellina, Cyclas, and Pandora. Many species of this genus are found fossil.

TELLY'NIDES. A genus of sub-equivalve, inequilateral, transverse, marine bivalves, found in sandy mud at depths varying from five to fifteen fathoms.

TE'LLINITE. A fossil tellina.

TELLU'RIUM. The name given by Klaproth to a metal first discovered by him or by Müller in 1782, from an ore of gold, with which metal it is found combined in the Transylvanian mines. Its colour is nearly a tin white, with a shade of blue. Structure foliated. Specific gravity 6·1. Before the blow-pipe it fuses easily, and is very volatile, giving out a pungent odour, compared by some to that of a radish. Tellurium is not used in any form. Symbols Te.

NA'TIVE TELLU'RIUM. Rhombohedral Tellurium. The tellure natif auroferrifère of Haüy; gediegen silvan of Werner; tellure natif ferrifère of Brongniart. This ore is never perfectly pure. It always contains a greater or less quantity of gold, and sometimes silver, lead, copper, and sulphur. It is found in Transylvania only, in veins, traversing greywacke. It is of a white colour; of a shining and metallic lustre; brittle and frangible. Its constituents are tellurium 92·6, iron 7·2, gold 0·2.

TE'MPERATURE. (*temperatura*, Lat. *température*, Fr. *temperatura*, It.) Constitution of nature; the constitution or state of the atmosphere, whether it be hot or cold, humid or dry; the condition of a body, as manifested by its influence on the thermometer.

A question of great importance, in the study of geology, arises as regards the existing and the former temperature of this planet. Whether the nebular hypothesis be correct, or otherwise, it certainly does appear that a very high temperature did once exist on this planet; and that such temperature has been gradually diminishing. Whether also the nucleus of the globe be in a state of incandescence or fusion is a question which probably never will be solved, till the heavens shall be rolled away as a curtain, and the elements shall melt with fervent heat. Still we well know that the temperature of the earth does increase in a fixed and certain ratio as we descend into its depths. The following are general results from the various facts observable on the earth's surface, and such depths thereof as man has hitherto been able to penetrate. 1. Numerous experiments in mines shew an increase of temperature from the surface downwards, that is, from those depths where the action of the solar rays ceases to produce a variable heat. 2. Thermal springs occur in all parts of the world, and among all varieties of rock. 3. The temperature of the water in Artesian wells is found to increase with the depth. 4. Terrestial temperature, at small depths, does not coincide with the mean temperature of the atmosphere above it. 5. Igneous matter has been ejected at all periods from the interior of the earth. 6. Active volcanoes occur widely spread over the surface of the world, and so closely resemble each other, that they may be considered as produced by a common cause, and that cause deep-seated. 7. Geological phenomena attest a great decrease of temperature on the surface of the globe. 8. A decreased temperature of the earth would, by radiation, produce the various mountain ranges and fractured strata found on the surface of our planet. When all these circumstances are taken into consideration, and we add the probability that heat counteracts the effects of gravity in the sun and certain planets, and that the free passage of the particles of terrestrial matter among each other was necessary to produce the figure of the earth, the evidence in favour not only of a central heat at present, but also of a heat of far greater intensity at remote geological epochs, becomes exceedingly strong; so strong, indeed, that there is some difficulty in resisting the impression that we have, by various means, made as fair an approximation to the truth, as the nature of the subject will admit. If the theory of central heat be founded on probability, the very general occurrence of tilted and fractured rocks is of easy explanation. From a series of experiments, it appears that the temperature of the earth's crust increases at the rate of 1° Fahrenheit for every forty-five feet of perpendicular descent.

TE'NDRIL. (*tendron*, Fr.) In botany, a spiral appendage to certain plants, its use being to clasp and wind round other bodies, by which means weak and climbing stems support themselves, and rise to a great height.

TE'NNANTITE. A variety of sulphuret of copper, of a lead-grey or blackish colour, thus named by Prof. Phillips. It occurs in copper veins in some of the mines of Cornwall. Its constituents are, according to the analysis of Prof. Phillips, copper 45·32, arsenic 11·84, iron 9·26, sulphur 28·74, silica 5·00. Hardness 4·0. Specific gravity from 4·3 to 4·4. It occurs massive, and crystallized in rhomboidal dodecahedrons, cubes, and octohedrons.

TE'NTACLES. } Feelers; exploring or-
TENTA'CULA. } gans. In its most restricted sense this term is understood to signify organs, appendages

of the mouth, which have no articulations; but, in a larger sense, the term has been applied also to all jointed organs in its vicinity, and used for a similar purpose, which indeed are the precursors of feelers and antennæ. It is to these organs that polypes are indebted for what constitutes their principal ornament, that resemblance of a plant or shrub in full blossom, adorned with crimson or orange-coloured flowers. In the fixed polypes, the tentacles are the only motive organs. The tentacles of the fresh-water polypes, forming the locomotive genus hydra, are not, as those of the fixed marine ones, shaped like the petals of a blossom, but are long hair-like flexible arms, somewhat resembling the branches of a chandelier. Amongst the Radiaries, tentacles exist in some genera, and not in others. In the Stelleridans and Echinidans, there are no tentacles, but the Fistulidans present a floriform coronet of tentacles. Tentacles as exploratory, prehensory, and locomotive organs, exist in several other classes of animals. In none, however, are they more remarkable than in the Cephalopoda: in these animals they are used as arms for prehension, as legs for locomotion, as sails for wafting their possessors over the boundless deep, as oars for passing through its waves, as a rudder for directing their course, and as an anchor for fixing themselves.—*Kirby.*

TE′RBIUM. A metal having an earthy oxide. Symbols Tb.

TEREBE′LLA. A genus of Annelidans, or annulose animals, placed by Cuvier in the order Tubicola. They are inhabitants of the sea, and are met with generally in shallow water, on the coasts, and on shells, &c. The body is oblong, creeping, naked, often enclosed in a tube, furnished with lateral fascicles, or tufts, and small branchiæ; mouth placed before, furnished with lips without teeth, and protruding a clavate proboscis; feelers numerous, ciliate, capillary, seated round the mouth. Terebellæ not being provided by nature with any external shell, endeavour to furnish themselves with an armature. For this purpose they collect grains of sand, or fragments of decayed shells, or other substances, which they agglutinate together by means of a viscid exudation, so as to form a firm defensive covering, like a coat of mail.

TE′REBRA. A genus of turreted subulated marine univalves: the opening short, and notched in the lower part. The basis of the columella twisted: found fossil in the environs of Paris.

TE′REBRATING. A term applied to shells which form holes in rocks, wood, &c., and reside therein.

TEREBRA′TULA. A genus of the class Brachiopoda. Terebratulæ are marine bivalves found moored to rocks, shells, &c., at depths varying from ten to ninety fathoms. The valves are unequal and united with a hinge, but having no ligament: the summit of one, more salient than the other, is perforated to permit the passage of a fleshy pedicle, by means of which the animal attaches itself to rocks, shells, &c. The recent species are few, but the fossil are very numerous. In the fossil shell, the operculum which serves for the attachment of the animal to the object to which it is moored, can rarely be traced. The casts of some species of fossil Terebratulæ are of a most extraordinary form. These casts are said to have been first noticed by Pliny, who describes certain stones, some of which were white and others brown. Agricola next noticed these bodies, as having been found whilst digging near the fortress of Ehrenbreitstein, in Treves.

After various opinions had been offered respecting the true origin and nature of these fossils, Wolfart advanced the opinion that they were the casts of marine shells. While spiriferæ and orthoceratites have disappeared, Terebratulæ appear to have existed through all the changes which have affected our planet. They form an abundant genus discovered in the grauwacke series, not fewer than thirty species having been already determined by Sowerby, Goldfuss, and others, as contained therein. Among the organic remains of the carboniferous limestone, twenty-one species of Terebratulæ are enumerated. From the red sandstone group nine species have been obtained. From the oolite, including the lias, the immense number of sixty species have been determined. From the cretaceous group fifty-four species have been extracted.

TEREBRA'TULITE. A fossil terebratula.

TEREDI'NA. (from *teredo*, Lat.) A genus of acephalous testacea, belonging to the family Inclusa, or, according to Lamarck, Tubicolaria. The valves are equal and inequilateral, with a little hollow on the inside of each valve, and a small, free, shield-shaped piece on the hinge.

TERE'DO. (*teredo*, Lat. a little worm that eateth wood; τερηεών, from τερεώ, to bore, Gr.) A genus of marine bivalves, belonging to the family Inclusa, in Cuvier's arrangement, and to Tubicolaria, in Lamarck's. The Teredo is contained in the lower end of a cylindrical tubular shell, generally open at both ends, two opercula being adapted to the upper end: it is capable of penetrating wood. One species, Teredo gigantea, has been found in mud at the bottom of the ocean; it attains to a great size, one specimen measuring five feet four inches in length, with a circumference, at the base, of nine inches.

TERE'DO NAVA'LIS. The ship-worm. The name given to a species of Teredo from the circumstance of its insinuating itself into the timbers of the bottoms of ships, even although the oak is perfectly sound: it very soon completely destroys the timbers it attacks. This destructive creature was brought originally, by our ships, from tropical climates, but it has now become an inhabitant of most of the harbours of this country. The teredo navalis, or ship-worm, will destroy every thing constructed of timber that is under the surface of the water. Their object is not to devour the wood, but to make for themselves a cell, in which they may be safe from their enemies. They bore in the direction of the grain of the timber, deviating only to avoid the track of others. Fortunately these animals cannot exist in fresh water.

TE'RMINAL. (from *terminalis*, Lat.) In botany, applied to flowers and umbels proceeding from the extremity of the stem or branches.

TE'RNA FO'LIA. In botany, leaves growing three together in a whorl. This term, it must be remarked, is very different in its signification from *ternate*.

TE'RNARY. (*ternarius*, Lat.) Proceeding by threes; consisting of three. Applied to things arranged in order by threes: thus, in botany, a flower is said to have a ternary division of its parts when it has three sepals, three petals, three stamens, or twice or thrice as many.

TE'RNATE. (from *ternus*, Lat.) A term in botany, applied to compound leaves that consist of three leaflets on a petiole: the leaf of the strawberry affords a familiar example of a ternate leaf.

TER'OXIDE.  }  When oxygen is combined with any substance in the proportion of three equivalents, it is called a teroxide, or tritoxide.
TRITO'XIDE. }

TE'RRA SIE'NNA. The name given to an ochreous earth from its being brought from Sienna: it is a sort of brown bole, and is used as a pigment.

TE'RRAIN TERTIA'IRE. The name given by the French geologists to the tertiary strata. The Germans call these, Tertiärgebile.

TE'RRE VERTE. Green earth. The grün erde of Werner; talc zographique of Haüy; chlorite baldogée of Brongniart. An earth of a green colour, sometimes passing to olive. It consists, according to Klaproth, of silex 53, magnesia 2, potash 10, oxide of iron 28, water 6. Vauquelin states that it contains 7·00 of alumine. It is found in Germany, France, Italy, and North America. It is ground with oil, and used as a pigment.

TERRE'STIAL REFRA'CTION. The power which air possesses, in common with all transparent media, of refracting the rays of light, or bending them out of their straight course, rendering a knowledge of the constitution of the atmosphere important to the astronomer. Whenever a ray of light passes obliquely from a higher level to a lower one, or vice versâ, its course is not rectilinear, but concave downwards; and of course any object seen by means of such a ray, must appear deviated from its true course, whether that object be, like the celestial bodies, entirely beyond the atmosphere, or, like the summits of mountains, seen from the plains, or other terrestial stations, at different levels, seen from each other, immersed in it. Every difference of level, accompanied, as it must be, with a difference of density in the aërial strata, must also have, corresponding to it, a certain amount of refraction. This refraction between terrestial stations is termed *terrestial refraction*. The refraction of a terrestial object is estimated differently from that of a celestial body. It is measured by the angle contained between the tangent to the curvilineal path of the ray where it meets the eye, and the straight line joining the eye and the object. The quantity of terrestial refraction is obtained, by measuring contemporaneously the elevation of the top of a mountain above a point in the plain at its base, and the depression of that point below the top of the mountain. The distance between these two stations is the chord of the horizontal angle; and it is easy to prove that double the refraction is equal to the horizontal angle, increased by the difference between the apparent elevation and the apparent depression. Whence it appears that in the mean state of the atmosphere, the refraction is about the fourteenth part of the horizontal angle.—*Mrs. Somerville. Sir John Herschel.*

TER'TIARY. Third, a term applied to those formations which have been deposited subsequently to the chalk formation.

TE'RTIARY STRA'TA. A division of sedimentary formations called tertiary, as being of newer origin than the secondary, and characterized by distinct species of fossil animals and plants. They present a most decided contrast with the secondary and older strata in most of their essential characters. The most striking feature of these formations consists in the repeated alternations of marine deposites with those of fresh water. We are indebted to Cuvier and Brongniart, for the first detailed account of the nature and relations of a very important portion of the tertiary strata, namely, those which occur in the neighbourhood of Paris. These were found to fill a depression in the chalk, and to be composed of different materials, sometimes including the remains of marine, sometimes of fresh-water animals. The first discovery of the tertiary strata in the Isle of Wight and south-east of England, we owe to Mr. Webster. The whole of the tertiary accumulations are stratiform deposites, exhibiting various kinds of lamination and bedding. Previously to the commencement of the present century, the true nature of the tertiary formations was unknown, the chalk was considered the highest known rock, and the tertiary deposites as mere superficial sands, gravels, or clays.

The organic remains of the tertiary deposites of England agree with those of the Paris basin, but the mineral character of these deposites is extremely different, those rocks, in particular, which are common to the Paris basin and central France being wanting, or extremely rare, in England.

The tertiary system may be said to constitute a series of formations which link together the present and the past; while the more ancient tertiary deposites contain organic remains related to the secondary formations, the most recent contain many existing species of animals and plants, associated with forms now extinct. **The term** tertiary has been altogether disapproved of by some geological authors.

The tertiary strata have been sub-divided by Sir C. Lyell and M. Deshayes into four principal groups, to which Sir C. Lyell has assigned the terms eocene, miocene, pliocene, and pleistocene; each group being characterized by the relative proportion of recent and extinct species of shells therein contained.

Commencing from below upwards, we find included in the tertiary series. 1. The eocene, comprising marine and fresh-water deposits, including in the former, the calcaire grossier and the London clay; in the latter, the calcaire silicieux. 2. The miocene, comprising the faluns of the Loire, marine deposits, and sands, clays, lignites, &c., &c.; fresh-water deposits. 3. The pliocene, consisting of sub-apennine marl, sub-apennine yellow sand, English crag, marine deposits; and sands, clays, lignites, &c., fresh-water formations. 4. Pleistocene, in this, the most recent of the tertiary deposits, we have limestone, sand, clays, conglomerates, marls, &c., containing marine fossils, being marine stratifications; and sands, clays, sandstones, lignites, &c., containing fresh-water fossils, being fresh-water deposits.

TE′SSELATED. (*tesselatus*, Lat. wrought in chequer-work.) Chequered, like a chess board. In conchology, applied to shells that are coloured in regular and defined patches.

TE′SSULAR. A term applied to a system of crystallization, not susceptible of variation. The cube, tetrahedron, and several other forms belong to the tessular system.

TE′STA. (*testa*, Lat. a shell.)
1. Commonly applied to the shelly covering of testaceous animals.
2. In botany, the outer coat of the seed. The seed, or ripened ovulum, consists of coverings called integuments or seed-coats, the outer of which is called the *testa*.

TESTA′CEA. (from *testa*, Lat. a shell.) In the Linnæan system of natural history, the third order of Vermes. This order comprises all shell-fish, arranged by Linnæus under thirty-six genera. The testacea differ from the crustacea in their composition; the calcareous part of the shells of testacea being carbonate of lime, whereas in the shells of

2 K

crustacea it is phosphate of lime. The testacea also retain their shells as long as they live; the crustacea cast them annually, or, at least, periodically.

TESTA'CEOUS. (*testaceus*, Lat. *testacée*, Fr. *testaceo*, It.) Belonging to the order Testacea; having a strong thick shell, the calcareous portion of which consists of carbonate of lime.

TESTU'DINATE. (*testudinatus*, Lat.) Arched; vaulted; resembling the back of a tortoise.

TESTU'DO. (*testudo*, Lat.)
1. A genus of the order Chelonia; the tortoise. This genus has been divided into five subgenera.
2. The land tortoise, a sub-genus of the genus above mentioned.

TETRABRANCHIA'TA. The second order of the class Cephalaphora, comprising the nautilus, ammonite, &c.

TETRACAU'LODON. An extinct animal of the order Mammalia; allied to the mastodon, and referrible to the miocene period.

TETRADA'CTYLOUS. (from τετραδάκτυλος, Gr.) Having four toes.

TETRADYNA'MIAN. ⎱ A term applied to
TETRADYNA'MOUS. ⎰ plants that have six stamens, four of which are longer than the other two: the wall-flower is a familiar example.

TE'TRAGYN. In botany, a plant that has four pistils.

TETRAHE'DRAL. Having four triangles, equal and equilateral.

TETRAHE'DRON. ⎱ (from τέτρα, four,
TETRAË'DRON. ⎰ and ἕδρα, side, or base, Gr. *tétraèdre*, Fr.) The solid angles of a tetrahedron are formed by three equilateral plane triangles, and the solid is bounded by four equal and equilateral plane triangles, therefore it is a pyramid; a four-sided solid contained by four equal-sided triangles; a solid contained by four triangular surfaces.

THAL'ICTRUM. ⎱ The name
THALICTRO'IDES WEBSTERI. ⎰ given to a fossil plant resembling the meadow-rue, and found in the London clay.

THA'LLITE. The name given by Lemetherie to the Epidote of Haüy, or Pistazit of Werner.

THA'LLUS. (from θαλλὸς, a green leaf, Gr.) In botany, a name given to the frond, or leaf-like part, of certain plants. Lichens, are stemless, leafless plants, consisting of a tough wrinkled substance, called a *thallus*.

THA'LOGEN. The thalogens constitute a class of plants comprising four orders; algæ, fungi, lichens, and characeæ.

THE'CA. (from θήκη, Gr. *theca*, Lat.) A case or sheath. The sporules of plants are contained within thecæ.

THECODONTOSAU'RUS. A genus of fossil saurians, found in the magnesian limestone: the vertebræ are deeply concave at each end.

THELO'DUS. (from θηλὴ, mamilla, and ὀδοὺς, dens, Gr. mammillated tooth.) A genus of ichthyolites discovered in the Upper Ludlow rock, and thus named by Agassiz. The teeth of the Thelodus resemble in some respects those of the Lepidotus. One species only is described, namely, Thelodus parvidens.

THE'RMAL. (*thermal*, Fr. θερμὸς, warm, Gr.) A term applied principally to warm springs and waters. The temperatures of different thermal springs vary greatly from each other, but the same spring is found to be of a uniform temperature at all seasons of the year. Thermal waters are found to be, on the average, neither more nor less pure than springs of common temperature. It appears, from Dr. Daubeny's researches, that nitrogen gas is very common in hot springs, and, perhaps, very rare in cold waters. Thermal waters prove the extensive effects of subterranean heat, deriving their temperature from a deep-seated internal source of heat, and not from any local cause, or from chemical changes. Some thermal springs have flowed without any known diminution of temperature for nearly two thousand years. The evidence that many springs rise from considerable depths, and possess a temperature independent of solar influence, rests on their great heat, which varies from the boiling point of water downwards, to ordinary temperatures. It is impossible to account for this, otherwise than by supposing such heat communicated to the water in parts of the earth far beneath the surface, and removed from atmospheric influence. Many thermal springs contain silica, though this substance is of exceedingly difficult solution. Dr. Turner found that some of the thermal springs of India which yielded twenty-four grains of solid matter in a gallon, contained 21·5 per cent. of silica, 19 of chloride of sodium, 19 of sulphate of soda, 19 of carbonate of soda, 5 of pure soda, and 15·5 of water. Dr. Back gives the following as an analysis of the thermal waters of the geysers of Iceland: soda 5·56, alumina 2·80, silica 31·50, muriate of soda 14·42, and sulphate of soda 8·57.

THER'MOSCOPE. (from $\theta\epsilon\rho\mu\grave{o}s$, heat, and $\sigma\kappa o\pi\acute{\epsilon}\omega$, to explore, Gr.) An instrument for measuring the degrees of heat.

THIN OUT. This is a term used by geological writers to express the appearance of a stratum which gradually becomes thinner, till it wholly disappears.

THO'MSONITE. A mineral, thus named after Dr. Thomson, a variety of zeolite, crystallized in rectangular prisms. It is found near Dumbarton, in Scotland, in trap.

THORA'CIC. (*thorachique*, Fr. from *thorax*, Lat.) Pertaining to the chest. The name given to the duct into which the absorbents empty themselves, namely, the *thoracic* duct, and which terminates in the left subclavian vein.

THO'RAX. (*thorax*, Lat. $\theta\acute{\omega}\rho\alpha\xi$, Gr. *thorax*, Fr.) The cavity of the chest, containing the heart, lungs, &c., &c.

THORI'NA. An earth discovered by Berzelius, in Gadolinite. It bears a strong resemblance to zirconia, but differs from it in many particulars. After being heated to redness it is soluble in acids. No precipitate is caused by the addition of sulphate of potash to a solution of it. Thorina is colourless, and infusible after ignition. Dr. Ure states that a strong solution of the sulphate becomes a thick mass by boiling, but it is soluble in cold water; a property which particularly characterizes it.

THORI'NUM. } The metallic base of the
THORI'UM. } earth thorina.

THO'RITE. A mineral discovered in Norway, by Esmark, and named thorite by Berzelius. It is compact; of a black colour; brittle. Specific gravity 4·8. It is rare, and not used.

THU'LITE. A mineral of a peach-blossom colour, occurring in Norway; it is very rare, and considered to be a variety of epidote.

THU'MERSTONE. } A mineral, thus
THU'MMERSTONE. } named, by Kirwan, from its being found, in masses, near Thum, in Saxony. It is the Axinite of Haüy and Brongniart; the Axinit of Werner; La pierre de Thum of Brochant. See *Axinite*.

THYRSE. } (*thyrsus*, Lat.) In botany,
THY'RSUS. } a kind of inflorescence, as when the middle branches of a panicle are longer than the others. The horse-chesnut, lilac, &c., afford examples.

TI'BIA. The name given to the shin-bone, or large bone of the leg. It is said to have received its name from a supposed resemblance to a pipe or flute.

TIDE. (*ebbe* and *fluth*, Germ. *flusso del mare*, It.) The flow of the water in the ocean; the alternate rise and fall of the sea twice in the course of a lunar day, or in 24h. 50m. 28s. of solar time. The tides are a subject on which many persons find a great difficulty of conception. As the tides depend upon the action of the sun and moon, their rise and fall are classed among astronomical problems, of which they are the most difficult, and their explanation the least satisfactory. That the moon, by her attraction, should heap up the waters of the ocean under her, seems to most persons very natural, —that the same cause should, at the same time, heap them up at the opposite, appears to many a palpable absurdity. Yet nothing is more true, nor indeed more evident, when we consider that it is not by her *whole* attraction, but by the *differences* of her attractions, at the two surfaces and at the centre, that the waters are raised. In the semi-diurnal tides there are two phenomena particularly to be distinguished, one occurring twice in a month, and the other twice in a year. The first phenomenon is that the tides are much increased in the syzigies, or at the time of new and full moon. In both cases the sun and moon are in the same meridian: for when the moon is new they are in conjunction; when she is full they are in opposition. In each of these positions, their action is combined to produce the highest or spring-tides under that meridian. The neap-tides take place when the moon is in quadrature. The higher the sea rises in full tide, the lower it is in the ebb. The second phenomenon is the augmentation in the tides occurring at the time of the equinoxes, when the sun's declination is zero, which happens twice every year. The greatest tides take place when a new or full moon happens near the equinoxes, while the moon is in perigee. The height to which the tides rise is much greater in narrow channels than in the open sea, on account of the obstructions they meet with. The tides in the British channel sometimes, in some parts, rise as high as fifty feet; whereas on the shores of some of the islands near the centre of the Pacific ocean, they do not exceed one or two feet. Theoretically, all bodies of water, even large fresh water lakes, have tides; but they are so insignificant that inland seas, such as the Mediterranean and Black Seas, are generally termed tideless. One of the most remarkable circumstances in the theory of the tides is the assurance, that in consequence of the density of the sea being only one-fifth of the mean density of the earth, and the earth itself increasing in density towards its centre, the stability of the equilibrium of the ocean can never be subverted by any physical cause. A general inundation arising from the mere instability of the ocean is therefore impossible.—*Herschell*.

TIL′ESTONE. The name given to that formation which constitutes the third or lowest division of Sir R. Murchison's Old Red system. The tilestone, though of much smaller dimensions than the overlying formations, has very marked characters both in structure and fossil contents, and is very clearly defined by occupying a position in which it passes upwards into the cornstone and marls, and downwards into the Silurian rocks. The beds are finely laminated, hard, reddish or green, micaceous, quartzose sandstones, which split into tiles. Although the greenish colours prevail, these beds are usually associated with reddish shale, and the decomposition of the mass uniformly produces a red soil, by which character alone the outline of the division is easily defined; being always clearly separable from the upper beds of the Silurian system, which decompose into a grey surface. Organic remains are abundant, and indicate clearly the lines of deposit. The fossils consist of unpublished species of the following genera: Arca, Avicula, Bellerophon, Cucullœa, Lingula, Orthoceras, Terebratula, Turbo, Turritella, Trochus. This assemblage furnishes convincing proof that certain genera of molluscs, which have hitherto been supposed to be confined to the tertiary and secondary deposits, have co-existed with the genera Orthoceras, Terebratula, and Bellerophon, which peculiarly characterize the older strata.

The fact may now be regarded as established, that the Tilestones of England belong to a deposit contemporaneous with the ichthyolite beds of Caithness and Cromarty.—*Hugh Miller.*

The tilestones of England compose the *least* of the three divisions of the Old Red Sandstone system; their representative in Scotland forms by much the *greatest* of the three, and there seems to be zoological as well as lithological evidence that its formation must have occupied no brief period. The same *genera* occur in its upper as in its lower bed, but the *species* appear to be different.—*Hugh Miller.*

TIL'GATE BEDS, OR STRATA OF TILGATE FOREST. These form the central members of the Hastings beds, one of the three divisions of the Wealden formation. The Tilgate strata or beds consist of sand and friable sandstone, of various colours; of a compact bluish or greenish grey grit, in lenticular masses, its surface often covered with mamillary concretions, the lower beds frequently conglomeritic, and containing large quartz pebbles; and clay or marl, of a bluish grey colour, alternating with sand, sandstone, and shale. Its organic remains consist of bones and shells, rarely occurring, ferns and stems of vegetables. The Tilgate beds are met with in Sussex and Kent.

TIN. (*zinn*, Germ.) A metal of a white brilliant colour, slightly tinged with grey, being one of the simple or elementary bodies. Its specific gravity is 7·0 to 7·9. It fuses at a temperature of 442° Fahrenheit. It is of greater hardness than lead, but not so hard as gold. It is very malleable, and may be beaten out into leaves one two-thousandth of an inch in thickness. It is more tenacious than lead, and a wire of tin one-tenth of an inch diameter will sustain a weight of forty-seven pounds. It is very flexible, and, while being bent, it causes a crackling noise. Tin unites with many metals, forming valuable alloys. Its symbol Sn, from its Latin name Stannum. The bronze of the ancients consisted of 88 or 90 parts of copper, with 10 or 12 parts of tin. Bell metal consists generally of one-fifth of tin to four-fifths of copper. The gongs of the Chinese are formed of one-fifth of tin and four-fifths of copper.

Tin is mentioned repeatedly in the Pentateuch. It is generally believed that the Phœnicians came to Britain for tin, and, from the importance of the trade, that they concealed the situation whence it was obtained; it is certain that before the time of Herodotus tin was obtained from Cornwall. The period at which Cornish tin was first worked and exported would appear to be lost in the obscurity of past ages. Mr. Hawkins considers that the Phœnician colony of Gades, on the western coast of Spain, was the medium or entrepôt of the commercial intercourse between Phœnicia and Cornwall. Diodorus says, "we will now give an account of the tin which is produced in Britain."

Tin occurs in rocks of granite, gneiss, &c., in veins or fissures, called lodes; also in horizontal beds termed floors; and it is also found loosely scattered among pebbles.

TINSTONE. Oxide of tin; an ore containing tin. Tinstone sometimes yields nearly 80 per cent. of its weight in tin.

TI'TANITE. The ore or oxide of titanium; it is nearly a pure oxide, is of a brown colour, and is met with in granite and quartz.

TITA'NIUM. (from τίτανος, Gr.) One of the substances commonly known as metals. Titanium was first discovered by the Rev. Mr. Gregor in 1789; it is of a dark, copper-red colour, with a strong metallic lustre, which tarnishes by exposure to the atmosphere. Werner gave the name of Menak to titanium, from the circumstance of its having **been first found** at Menachan, in **Cornwall.**

TOAD-STONE. A provincial term for a species of wack, or basaltick rock, found in Derbyshire. Toad-stone is a pyrogenous or volcanic production that has been erupted in **a fluid** state. Its ordinary colours are brownish grey, purplish brown, bluish, or greenish; and its vesicles are either empty, or filled with carbonate of lime. Toadstone is found abundantly in Derbyshire, lying between beds of limestone; in some instances, beds of toadstone and limestone are found alternating with each other.

TONGUE-SHAPED. In botany, applied to leaves of an oblong, blunt, thick form, being generally of a cartila-ginous substance at the edges.

TO'PAZ. (from τοπάζιον, Gr. *topaz*, Fr.) **A** precious stone or gem, generally **of a yellow** colour. It is the silice fluatée alumineuse of Haüy; the topaze of Brongniart and Brochant. It is harder than quartz, hardness = 8, **with a** specific gravity of from 3·4 **to** 3·6. Is massive, in rounded pieces, and crystallised; most frequently crystallised. According to Haüy the primitive form of topaz is a rectangular octahedron. Mr. Allan says "its structure is lamellar at right angles to the axis of the prism; it also cleaves, though with difficulty, parallel to the sides of a right rhombic prism of about 124° 22' and 55° 38'; and it appears to yield to mechanical division on all the angles of the prism. There are many varieties of topaz, differing greatly in form and colour. The highly crystallised and transparent varieties are named precious topaz. In some places, as in Scotland, the topaz is found in alluvial earths. The Scotch pebble, called cairn-gorum, **is a** topaz. Generally, the topaz occurs in primary rocks, the finest specimens being obtained from the mountains of Brazil, and from the Uralian mountains of Asiatic Russia. Other minerals are sometimes sold **for** topaz, but it may generally be distinguished by the rhombohedron base of its crystals, straight foliated cross fracture, and longitudinally streaked lateral planes. The precious topaz consists of alumina, silica, and fluoric acid, with, sometimes, a small quantity of iron.

TOPA'ZOLITE. A pale yellow, nearly transparent, sub-variety of garnet, found in Piedmont. Its constituents are, silex 37, alumine 2, lime 29, glucine 4, iron 25, manganese 2.

TOR. A small round hill; called also Carn, or Karn.

TORNATE'LLA. A genus of oval, spirally grooved, marine, univalves, belonging to the family Plicacea. Recent tornatellæ are found in shallow water, creeping on sands, and leaving furrows. Several species are found in the oolite and superjacent strata.

TORO'SE. } (*torosus*, Lat.) Swelling
TO'ROUS. } into knobs or protuberances. A term used both in botany and conchology.

TORPE'DO. (*torpedo*, Lat. from *torpeo*, to benumb.) A subgenus of fishes, belonging to the genus Raia. The torpedo is found fossil in the tertiary formations. The torpedo is furnished with an electrical apparatus, resembling the voltaic battery, which it has the power of charging and discharging at pleasure. The benumbing effect producible by the torpedo depends on certain singularly constructed organs composed of membranous columns, filled from end to end with laminæ, separated from each other by a fluid.

TO'RRELITE. A new mineral brought from the United States, and thus named after Dr. Torrey.

TO'RTOISE. (*tortue*. Fr.) An order of the class Reptilia, or reptiles; tortoises are also termed Chelonians. The chelonians, or tortoises, were all included by Linnæus in one genus, namely, testudo; they are now divided into five subgenera. 1. Testudo, or land-tortoise; 2. Emys, or fresh-water tortoise; 3 Chelonia, or sea-tortoise; 4. Chelys; 5. Trionyx, or the tortoise with a soft shell.

The Testudo Indicus, a species of land tortoise, sometimes attains to an immense size, so as to require six or eight men to lift it from the ground, and affording as much as two hundred pounds of meat.

This order of reptiles, geologists inform us, began to exist at about the same period with the order of Saurians, and has continued from that time to the present. No fossil remains of the tortoise have been discovered in any strata not more recent than the coal formations.

TO'RALOSE. In entomology, a surface with but few elevations, scattered about, but these of considerable size.

TOU'RMALINE. } A mineral divided by
TOU'RMALIN. } some mineralogists into two sub-species, schorl and tourmaline; by others, tourmaline is regarded as a sub-species, or variety of schorl. The tourmalin of Werner; the tourmaline verte of Hany; the schorl électrique of Brochant.—It is of various colours, the shades of some of which are so dark as to approach nearly to black: hardness about 7·5. Sp. gr. from 3·0 to 3·2. By friction it yields vitreous electricity; by heating, vitreous electricity at one extremity and resinous electricity at the other, the termination, according to Haüy, which presents the greatest number of planes, exhibiting the positive, or vitreous, electricity, the termination which consists of the smaller number of planes exhibiting the negative, or resinous, electricity. It occurs but seldom massive, most frequently crystallized; its primary form is an obtuse rhomboid of 133° 50′, and 46° 10′. It occurs imbedded in granite, gneiss, mica slate, &c., in Scotland, Sweden, America, Spain, and other parts.

TRA'CHEA. (*tráchea*, Lat. from τραχὺς, Gr. rough.)
1. The windpipe, or that canal which leads from the throat to the lungs.
2. In botany, the vessels of plants in which the internal fibres run in a spiral direction; they are also called air-tubes.

TRACHEA'RIA. The third order of the class Arachnida; the acarus is an example.

TRACHE'LIPODS. (from τράχηλος, the neck, and πούς, a foot, Gr.) In Lamarck's arrangement, the third order of Molluscans: they have the greatest part of the body spirally convoluted, always inhabiting a spirivalve shell; the foot free, attached to the neck, formed for creeping. Trachelipods may be divided into herbivorous and carnivorous, the latter possessing a respiratory siphon, which the herbivorous have not. This order contains fourteen families, and upwards of seventy genera.

TRA'CHYTE. (from τραχύς, Gr. rough.) A kind of volcanic porphyry, usually containing crystals of glassy felspar, and excessively rough to the touch. It is found in the neighbourhood of all volcanic craters. Trachyte sometimes possesses a columnar structure; it is generally of a coarse grain, and with a degree of porosity. From this latter circumstance it easily breaks down, and forms frequently a conglomerate with other substances. It sometimes is found to contain augite and hornblende. In some districts trachyte seems little else than granite which has been again fused, and, having been exposed to different conditions, no longer presents the appearance of granite. Mr. Poulett Scrope proposes to divide trachyte as follows:—1. Compound trachyte, with mica, hornblende, or augite, sometimes both, and grains of titaniferous iron. 2. Simple trachyte, without any visible ingredients but felspar. 3. Quartziferous trachyte, when containing numerous crystals of quartz. 4. Silicious trachyte, when apparently much silex enters into its composition.

TRACHYDO'LERITE. A sort of intermediate lava. The name trachydolerite is proposed by Abich to be given to certain lavas where the minerals pass into one another, or are so intimately blended that they cannot be distinguished.

TRANSI'TION ROCKS. } The name
TRANSI'TION SERIES. } transition
TRANSI'TION FORMA'TIONS. } given by Werner, has been applied to certain rocks, from an opinion that they had been formed at a period when the globe was undergoing a great change, fitting it for the reception of organized beings: or it was adopted to express the theory that, at this period, the causes which had given rise to crystalline formations were still in action; while those which produced stratified sedimentary rocks, including organic remains, were only beginning to operate. The term, though no longer applicable in its original signification, is still retained. The rocks usually included in the transition series are the Dudley limestone, the Caradoc sandstones, and the British and Llandilo rocks; the whole possessing a thickness of upwards of two thousand four hundred yards, and containing, throughout, organic remains. The transition rocks rest upon the rocks called primary, and are themselves covered by the old red sandstone formation. Buckland observes, "it is most convenient to include within the transition series, all kinds of stratified rocks, from the earliest slates, in which we find the first traces of animal or vegetable remains, to the termination of the great coal formation. Upwards of six hundred species of fossils have been discovered in the transition rocks. They are mostly peculiar to these rocks, though some occur also in the carboniferous system, and are nearly all marine. About 66 per cent. belong to genera supposed to be extinct. The mineral character of the transition formations present alternations of slate and shale, with slaty sandstone, limestone, and conglo-

merated rocks; the latter bearing evidence of the action of water in violent motion; the former showing, by their composition and structure, and by the organic remains which they frequently contain, that they were for the most part deposited in the form of mud and sand, at the bottom of the sea."

TRANSI'TION LIME'STONE. (*calcaire de transition*, **calcaire intermédiare**, Fr. *uebergangskalkstein*, Germ.) A member of the Grauwacke group, called also Grauwacke Limestone.

TRANSLU'CENCY. (from *trans*, through, and *luceo*, to shine, Lat.) A term **used in mineralogy**, to express the property which some minerals possess, of permitting the passage **of** rays of light, but without sufficient transparency to perceive objects through the mineral.

TRANSLU'CENT. A mineral is said **to** be translucent when light evidently passes, but objects cannot be distinguished through the mineral.

TRAP. } (*trappa*, a stair, Sw.
TRAP ROCKS. } *trapp*, a step, Germ. probably from τράπεζα, Gr. a table.) Bergman gave the name of trap to basalt, which he divided into two families, namely, Common Trap, **and** Figurate Trap. The word **trap is** usually employed to designate **certain** volcanic rocks, frequently occurring in large tabular masses at different heights, forming a succession **of terraces or** steps. The **term is applied to various** igneous rocks **without any regard** to their constituent parts, **but** merely in reference to their form. Great alterations are produced **on** stratified rocks by the introduction of trap rocks; these consist in some cases in the greater degree of induration, by which loose grits pass into compact quartz rock, and shale into flinty slate; coal is converted into coke, or becomes charred, forming an ash-grey porous mass, which breaks into small columnar concretions, exactly resembling the coak obtained by baking coal in close iron cylinders. Limestone **is** often rendered highly crystalline and unfit for lime, when in the vicinity of trap-rock; and slate clay is turned into a substance resembling flinty slate or porcelain jasper. Many of the trap-rocks, like some lavas, afford on decomposition one of the most fertile soils with which we are acquainted. A slight acquaintance, says Sir R. Murchison, in his splendid work, "The Silurian System," with volcanic phenomena teaches us that they are the results of some general and deep-seated cause, which occasions eruptions of gaseous and earthy matters, or of lava, both under the atmosphere and beneath the ocean. It is perhaps to submarine volcanoes that all our British trap-rocks are referable. The most common ingredient in trap **as** well as in granitic rocks, is felspar.

TRAPE'ZIUM. (from τραπέζιον, a little table, Gr.)
1. In anatomy, the name given **to** one of the bones of the carpus, **or** wrist.
2. In geometry, a quadrilateral figure, whose four sides are not equal, and none of its sides parallel.

TRAPEZOI'DAL. In mineralogy, when the surface is composed of twenty-four trapeziums, all equal and similar.

TRASS. The name given to a tufaceous alluvium of the Rhine volcanoes. Lyell says, "this *trass* is unstratified; and its base consists almost entirely of pumice, in which are included fragments of basalt and other lavas, pieces of burnt shale, slate, and sandstone, and numerous trunks and branches of trees."

TRAU'MATE. The name given by the French geologists to grauwacke, a term about as euphonious as that they have substituted it for.

TRA′VERTIN. (*Lapis Tiburtinus*, Lat. *travertino*, It.) An Italian name for a concretionary limestone or calcareous precipitate, deposited by water holding in solution a considerable quantity of carbonate of lime. The water parting with some of its carbonic acid gas, which rises in bubbles to the surface, the lime becomes deposited. Thermal springs, in volcanic districts, are found to issue from the earth, so highly charged with carbonate of lime, as to overspread large tracts of country with beds of calcareous tufa, or travertin. Deposits of travertin are by no means uncommon from *cold* springs in the Apennines, particularly near the volcanic region of Southern Italy. In Italy, immense masses of travertin are being constantly formed from waters copiously charged with carbonate of lime. At the baths of San Fillippo a manufactory of medallions in *basso-relievo* is carried on. The water is first conducted into pits, where it frees itself from its grosser parts; it is then conducted by a tube to the top of a small chamber, and made to fall through a space of ten or twelve feet in height. The stream is broken in its descent by means of numerous sticks placed across, by which the spray is dispersed upon moulds, which are slightly soaped, and the result is a cast of the figures formed in the mould, of a solid, marble-like substance. The waters of the lake of the Solfatara, between Rome and Tivoli, are so supersaturated with carbonic acid gas, that they appear in a constant state of ebullition, from the extrication of the gas. The Coliseum, and the majority of the public buildings of Rome, are composed of travertin. The usual explanation of the deposition of travertin seems very probable. It supposes the carbonic acid to be derived from the volcanic regions beneath, which, passing with the water through the calcareous strata, dissolves as much lime as it can take up, giving off the excess of carbonic acid under diminished pressure in the atmosphere, and causing the carbonate of lime to be deposited. The carbonic acid found so abundantly in acidulous springs is ascribed by Von Buch, Brongniart, Boué, Von Hoff, and other geologists, to volcanic or igneous action at various depths beneath the surface. Masses of travertin are very abundant in districts which have been, or still are, subjected to volcanic action; they are, however, almost equally abundant in countries where there are no distinct appearances of volcanic action near the surface. All that seems necessary for their production is, that a spring should hold in solution a sufficient quantity of carbonate of lime, which, upon the escape of the excess of carbonic acid gas into the atmosphere, deposits its earthy residuum on any object over which it flows.

TRE′MOLITE. (from *Tremola*, a valley of St. Gothard, where it was first found.) A nearly white, grey, or bluish, variety of hornblende or hemiprismatic augite. It occurs in crystals, but most commonly in fibrous or radiated masses, composed of minute, imperfect prisms or fibres. There are several sub-varieties of tremolite, namely, common tremolite, glassy tremolite, fibrous tremolite, and Baikalite. Tremolite is found almost exclusively in primary rocks, but it sometimes occurs in secondary. Specific gravity from $2·9$ to $3·2$. Hardness from $5·0$ to $6·0$. Before the blow-pipe, tremolite fuses into a white glass, full of pores. Its consituents are silex, lime, magnesia, water and carbonic acid, and oxide of iron. Very differing analyses are, however, given by different chemists and mineralogists; some describe it as containing fluoric acid and alumina.

TRIA'NDRIA. (from τρεῖς, three, and ἀνήρ, a male, Gr.) The name of the third class in Linnæus's sexual system, consisting of plants with hermaphrodite flowers, having three stamens or male organs. This class is divided into three orders:—1. Monogynia. 2. Digynia. 3. Trigynia.

TRIAS, OR NEW RED SANDSTONE. (Sometimes called '*Poikilitic*,' from the prevalence of a variegated character in its sandstones, and marls.) On the continent where its several members are better developed than in England it has received the name of 'Trias,' as divisible into three great sub-divisions. The lowest sub-division is the Bunter sandstone; the second, or Muschelkalk, is deficient in England; the upper is the Keuper marls, which contain our great deposits of salt. Upon the northern and southern flanks of the Austrian Alps, at Hallstatt and St. Cassian, are deposits of great thickness, which are superimposed upon the upper members of the Trias, and are regarded as forming a passage between the Trias and the Lias; the bone beds upon the banks of the Severn at Westbury and Aust are believed to be the reduced equivalents of these great deposits.—*Lycett.*

TRIA'SSIC. "This," says Mr. Jukes, "is a very badly chosen name. The triassic formation consists of the Bunter sandstein, the Muschelkalk, and the Keuper. In the British Islands it is not easy to draw any boundary lines in the Trias, or New Red Sandstone."

TRICA'PSULAR. A plant that has three capsules to each flower.

TRICHI'TES. The name assigned by Mr. Lycett to a genus of fossil inequivalve bivalves, whose structure is fibrous, the substance of the test being of great thickness. The hinge is without teeth or hinge-plate, and the anterior extremity of the shell, which is pointed, forms an aperture; the general figure is sub-quadrate, the valves being irregular and undulated, but closing all round. This singular genus occurs in the oolitic system of rocks: the absence of teeth, and the character of the terminal extremity distinguish it from Catillus.

TRICU'SPIDATE. Three-pointed.

TRIDA'CNA. A genus of subtransverse inequilateral, equivalve, marine, bivalves, belonging to Lamarck's family of Tridacnacea, and found both recent and fossil. Recent tridacnæ are found at depths varying to seven fathoms, moored by a byssus to rocks, and on coral reefs. The shells of this genus are exceedingly beautiful, being radiately ribbed, the ribs adorned with vaulted foliations, and waved at the margins. The hinge is formed of two compressed and entering teeth. One species, the tridacna gigas, is met with of immense size. Fossil shells of this genus are very rare; it is said some large specimens have been obtained from the neighbourhood of Verona.

TRIDA'CTYLOUS. Having three toes.

TRIGONELLI'TES. A genus of shells described, and thus named, by Mr. Parkinson. A slightly rounded, trigonal, thick shell, gaping on each side. The anterior margin nearly on a straight line; the posterior in a gently waving, and the upper side in nearly a circular direction. The outer surface of each valve thickly pierced by foramina, which, passing nearly through its substance, give it the cancellous appearance of bone: the inner surface smooth, but marked with striæ, concentric with the upper margin. The hinge completely linear, without teeth; there being only an appropriate surface on the anterior margin of each valve for the attachment of the cartilage externally.

TRIGO'NIA. TRI'GON. TRY'GON. } (from τρίγωνον, triangular, Gr.) A genus of marine bivalves, found both fossil and living, belonging to the family Ostracea. The trigonia is a triangular or suborbicular, inequilateral, equivalve, transverse bivalve. One valve has two oblong, flat, diverging, hinge teeth, transversely grooved on each side; the other has four flat, oblong, diverging, hinge teeth, transversely grooved on one side only, disposed in pairs, receiving between their grooved sides the two hinge teeth of the opposite valve. Recent trigoniæ have hitherto been discovered near Australia only, in sandy mud. They are nearly related to the cardium, or cockle, the foot of the animal being bent, like that of the cockle, at an acute angle, so as, upon pressure, to form a very elastic organ.

TRILO'BATE. (from *tres*, three, and *lobus*, a lobe, Lat.) Divided into three lobes; having three lobes.

TRI'LOBITE. (So named from its being divided into three lobes or principal parts.) A family of fossil marine crustaceans, which appear to have become extinct at the close of the period during which the carboniferous series was formed, no traces of their remains having been discovered in any strata of a more recent period. For a long time, fossil trilobites were confounded with insects, under the name of Entomolithus paradoxus: several names have also been given to them, derived chiefly from the three lobular divisions, by which they are so characteristically marked, as well as from their being found sometimes in a coiled, sometimes in an extended state. By Bromel, the trilobite was named Lapis insectiferus; by Wolsterdorf, Conchitus trilobus; by Hermann, Pectunculites trilobus imbricatus; by Da Costa, Pediculus marinus; by Linnæus, Entomolithes paradoxus; by Baumur, Trigonella striata; and by Wilke, Entomolithus cancriformis marini. The trilobite is often called the Dudley fossil, from its having been first noticed in the transition limestone near that town. "The great extent to which trilobites are distributed over the surface of the globe, and their numerical abundance in the places where they have been discovered, are remarkable features in their history. They have been found throughout all northern Europe, and in numerous localities in North America, in the Andes, and at the Cape of Good Hope. The anterior segment of the trilobites is composed of a large semi-circular, or crescent-shaped shield, succeeded by a body composed of numerous segments folding over each other, like those in the tail of a lobster, and generally divided by two longitudinal furrows into three ranges of lobes. The nearest approach among living animals to the external form of trilobites is that afforded by the genus Serolis, in the class Crustacea."

TRILO'CULAR. (from *tres*, three, and *loculus*, a partition, Lat.) A term applied in botany to seed-vessels divided into three portions or cells.

TRINU'CLEUS. The name given by Lhwyd to a genus of trilobites found in the Caradoc sandstone and Llandeilo Flags.

TRI'ONYX. (*trionices*, pl.) A sub-genus of fresh water testudo, belonging to the order Chelonia. The soft-shelled tortoise. Several species are mentioned. The soft-shelled tortoises have no scales, the shell and sternum being merely enveloped in a soft skin; many of the pieces that are bony in the tortoise being replaced by a simple cartilage or membrane. Fossil remains of trionices are found in the fresh water Wealden formations of the secondary series. Remains also are abundantly found in the lacustrine deposits of the tertiary formations.

TRI′POLI. An admixture of silex and clay. It has obtained its name from having been originally brought from Tripoli, in Barbary. It is a mineral of a dull argillaceous appearance, occurring usually in friable or earthy masses. Its powder is fine, but dry and rough to the touch, and sufficiently hard to scratch metals, glass, &c. It is employed for the polishing of metals and stones. Rotten-stone and polishing slate, the Polierchiefer of Werner, appear to be varieties of Tripoli. By recent discoveries made by Prof. Ehrenberg it is ascertained that tripoli is entirely composed of millions of the skeletons or cases of microspic animalcules. At Bilin, in Bohemia a stratum of tripoli extends over a considerable space, being in some parts fourteen feet thick. This stone, when examined through a powerful microscope, is found to consist of the silicious cases of infusoria united together without any visible cement. From a calculation made by Ehrenberg, it would appear that in the Bilin tripoli there are 41,000 millions of individuals of the Gaillonella distans in every cubic inch.

TRISE′PALOUS. (from *tres*, three, and *sepal*.) A term used in botany for a calyx that has three sepals.

TRI′TON. A genus of the molluscous order of Nudibranchiata, inhabiting an oblong, thick, ribbed, or tuberculated spiral shell. The body is oblong; mouth with an involute spiral proboscis; tentacula twelve, six on either side, divided nearly to the base, the hind one cheliferous.

TROCHI′TA. A detached vertebra of a radiated animal. When several trochitæ are united, so as to form part of a column, the series is termed an entrochite.

TROCHO′TOMA. A genus of fossil trochiform univalves, distinguished from Trochus and Pleurotomaria by having a fissure upon the last whorl, which approaches the outer lip but does not reach it. Six species have been procured in the Great Oolite near Minchinhampton, and four in the Inferior Oolite of the same locality, one species is likewise recorded from the Lias of Normandy.—*Lycett.*

TRO′CHUS. (*trochus*, Lat. a top.) A genus of conical, spiral, thick, striated, marine, univalves, found both fossil and recent. Aperture transversely depressed, and somewhat quadrangular; columella oblique; operculum horny, with numerous whorls. Recent trochi are found in the ocean at depths varying to forty-five fathoms; they most commonly, however, are met with near the shore, creeping on rocks, sands, and gravel. One hundred and thirty-three species are described in Turton's Linné. Lamarck has separated from the genus trochus of Linnæus certain shells possessing peculiar characters, which he has arranged under two new genera, namely, Solarium and Monodonta. These genera, as well as trochus, are comprised in the family Turbinacea.

Very large casts of trochi are found in Oxfordshire, Gloucestershire, and Somethsire.

TRONGOTHE′RIUM. The name assigned to an extinct species of beaver, larger than any now known, whose fossil remains have been found in Russia.

TROGONTHE′RIUM. An extinct fossil genus of Rodents, nearly allied to, but much exceeding in size, the beaver. A relic of Trogontherium, discovered by the Rev. Mr. Green, of Barton, in the lacustrine formation at Ostend near Bacton in Norfolk, is now in the British Museum.

TRO'PHI. The name given by Kirby and Spence to the elementary parts which enter into the composition of the mouth of an insect.

TRUNCA'TION. A term used in mineralogy, implying that a segment is cut off or separated from the predominant form. The term may be applied either to an edge, or a solid angle of a crystal, and will leave a face more or less large in place of the edge or angle. A truncation is said to be oblique, when the face does not make equal angles with all the contiguous faces.

TU'BER. In botany, a fleshy irregular stem produced under ground, and distinguished from a root by its having eyes or buds, which the true root never possesses: the potatoe is a familiar example.

TUBER'CULAR. In mineralogy, a term applied to those minerals whose unevenness of surface arises from small and somewhat round elevations.

TUBICINE'LLA. (from *tubicen*, Lat. a trumpeter.) A genus of multivalve tubular shells, not spiral. Tubicinella is placed by Lamarck in the order Sessile cirripedes. The tube is cylindrical, and composed of six elongated valves, laterally united. The aperture circular, with a four-valved operculum.

TUBI'COLA. (from *tubus*, a tube, and *cola*, an inhabitant, Lat.) An order of Articulata, comprising Serpula, Sabella, Terebella, Amphitrite, Syphostoma, and Dentalium.

TUBI'PORA. }
TU'BIPORE. } Organ-pipe coral, consisting of tubes of a stony substance, each containing a polype. This genus is thus described by Parkinson: "a stony polypifer formed by cylindrical or oval tubes, communicating laterally with each other." A genus of corals or zoophytes belonging to the class Polypifera; subregnum Cyclo-neura or Radiata. There are several species. In one species, Tubipora musica, the tubes are placed parallel to each other, like the pipes of an organ, with transverse partitions at regular intervals: in another species, Sertularia, the tubes are joined together endwise, like the branches of a tree, leaving lateral apertures for the protrusion of the tentacula of each separate polype. Lamarck proposed to separate one species, namely, Tubipora catenulata, or chain coral, from the genus Tubipora, and to place it in a distinct genus, Catenipora. Tubipores are found among the earliest traces of organic bodies in the ancient strata. "The principal generic character of Tubipora as derived from their ascertained structure, is, that the animal substance contained in each tube so communicates with the whole mass by an intercurrent organization, as to render it one connected system."

TUBI'PORITE. A fossil tubipore. Many marbles and pebbles are beautifully marked by sections of the inclosed tubiporites.

TU'BULAR. In the shape of a hollow tube. In botany, applied to the florets of a compound flower, when they form a cylindrical tube, and are five-cleft.

TUBULA'RIA. (from *tubulus*, a hollow pipe, Lat.) A genus of corals belonging to the class Polypifera. Simple or branched tubes of a horny substance, each tube containing a polype. Tubulariæ are both fresh-water and marine.

TUBULA'RII. (*tubularius*, Lat.) A family of the order Coralliferi, class Polypi. The tubularii inhabit tubes of which the common gelatinous body traverses the axis, like the medulla of a tree, the tubes being open, either at their sides or summits, to allow the passages of the polypi.

TU'FA. (*tuf*, Fr. *túfo*, It.) An earthy precipitate deposited from water.

TUFF. } A name applied to
TU'FA, VOLCA'NIC. } several different substances the production of volcanic eruptions. Generally, an aggregate of sand, volcanic ashes, and fragments of scoria and lava, united by an argillaceous or muddy cement. Sometimes it is composed of volcanic ashes and sand, transported and deposited **by rain** water.

Tuff, or volcanic tufa, as distinguished from calcareous tufa, presents various shades of grey, brown, red, yellow, &c., or it is sometimes spotted. Hardness moderate; fracture dull and earthy.

TUFA'CEOUS. Having the appearance or texture of tufa.

TU'NGSTEN. (from *tung*, heavy, and *sten*, stone, Dan.)
1. A greyish white metal, brittle, and very hard. Specific gravity 17·4. Fusible in the most intense heat only, its infusibility equalling that of platinum. It has been obtained only in the form of grains of extreme hardness. It was discovered by Scheele, and by Werner has been named Scheel; by Haüy, Scheelin.
2. A mineral, of a grey or yellowish-grey colour, occurring in Bohemia, Sweden, and Cornwall, massive and disseminated. Its external lustre is shining and splendent; internal lustre shining and resinous. Specific gravity from 5·57 to 6·06. It can be scratched with a knife, and is easily broken. It is infusible, but before the blow-pipe it becomes opaque, and decrepitates. By digestion in nitric acid it is converted into a yellow powder, which is the oxide of tungsten. It consists of oxide of tungsten 77·75, lime 17.60, silex 3·00, according to Klaproth: Berzelius gives its analysis as consisting of oxide of tungsten 80·24, lime 19·40. It is the Scheelin calcaire of Haüy and Brongniart; the schwerstein of Werner; **the tungsten of Kirwan.**

TUNICA'TA. (*tunicatus*, coated, Lat.) The tunicated animals have no external shell nor internal solid parts, but are covered with a tough, elastic, homogeneous tunic, in the form of an enveloping sac with a respiratory and an anal orifice. This exterior sac is the analogue of the valves of conchifera, and has the muscular fibres of the lining mantle inserted into its inner surface.

TU'NICATED. Covered with one or more tunics or membranes.

TURBELLA'RIA. According to Busk, the sixth order of the class Scolecida.

TU'RBINATED. (*turbinatus*, Lat.) Of a spiral oblong form; in conchology, applied to shells, broad at the base, and becoming gradually narrower till they are pointed at the apex.

TURBINO'LIA. A genus of lamellated stony polypifers; free, (and in this differing from the genus caryophyllia, which is fixed,) simple, turbinated or cuneiformed; longitudinally striated on the exterior; the base pointed, the terminating cell stelliformly lamellated and **sometimes** oblong.

TU'RBO. (*turbo*, a top, Lat.) Plural, *turbines*. A genus of marine univalves, found on rocks and seaweeds, at depths varying to ten fathoms.

The turbo is a conoidal or slightly turreted shell, the aperture complete, rounded, and not toothed; the margins disjoined in the upper part; the columella smoothed at the base. In Turton's Linné one hundred and fifty-one species of turbines are described, sixty-nine of which are indigenous to Britain. Many species of turbines have been described as found fossil.

TURKO'IS. } (*turquoise*, Fr.) A gem
TURQUO'ISE. } of a blue or greenish colour, and opaque, found in roundish masses, from the size of a pea to that of an egg. The finest specimens are brought from Persia,

where they occur in small veins in slate-clay.

TU'RRILITE. (from *turris*, a tower, Lat. and λίθος, a stone Gr.) A spiral, turriculated, multilocular shell; the turns contiguous, and all visible. The chambers divided by sinuous septa, pierced by a siphuncle in their disks. The mouth round. The shells of this genus abound in the chalk marl, galt, and Shanklin sand. They are extremely thin, and their exterior is adorned and strengthened with ribs and tubercles. The outer chamber, which contained the animal, is large. Buckland states that turrilites do not appear until the commencement of the cretaceous formations, and that having thus suddenly appeared, they become as suddenly extinct, at the same period with the ammonites. De La Beche says "a turrilite has been mentioned, though with doubt, as occurring in the coral rag of the north of France. Several species are enumerated as having been discovered in the cretaceous group."

TURRITE'LLA. A genus of turreted, elongated, marine univalves, of the family Turbinacea, found both recent and fossil. Turritellæ are commonly known by the name of screw-shells. Recent turritellæ are found in sandy mud, at depths varying from five to twenty fathoms. Fossil turritellæ are found in the tertiary and secondary deposits.

TY'MPAN. } (*tympanum*, Lat.) A
TY'MPANUM. } cavity or chamber of the ear. It is sometimes also applied to a membrane that stretches across the cavity of the ear, called the drum of the ear.

# U

U'LNA. (*ulna*, Lat. from ὠλένη, Gr.) The cubit or large bone of the fore-arm.

ULTRAMARI'NE. Azure-stone; lapis lazuli. A pigment remarkable for the durability of its colour. See *Lapis Lazuli*.

U'MBEL. (*umbella*, Lat.) In botany, a peculiar form of inflorescence: an umbel consists of several flower-stalks or rays, nearly equal in length, which spread from one common centre, and the summits of which form a regular surface, either level, convex or globular; sometimes, but rarely, concave. An umbel is either simple or compound: a simple umbel has the stalks springing from the same part of the principal one, and each bears but one flower. A compound umbel has each ray or stalk terminating in another set of rays: the carrot, parsnip, parsley, &c., furnish familiar examples of compound umbels.

UMBELLA'TÆ. One of Linnæus' natural classes or orders of plants. The umbellatæ are plants whose flowers grow in umbels, with five petals, and two naked seeds, joined at top and separated below. The parsley, fennel, &c., are examples.

UMBELLI'FERÆ. A large order of plants, characterized by their flowers being in umbels. Calyx entire or five-toothed. Petals five, usually inflexed at the point. Stamens five, alternate with the petals. Ovary two-celled. Fruit consisting of two carpels, separable from a common axis. Seed pendulous. Herbaceous plants with fistular stems.

Umbili´cated. (*umbilicatus*, Lat.) In conchology, a term applied to such shells as have a depression in the centre, like a navel. Univalves that have the umbilicus covered in a greater or less degree by a thin process, are termed sub-umbilicated.

Umbili´cus. (*umbilicus*, Lat.) The navel. In conchology, a circular perforation in the base of the lower whorl, or body, of many spiral univalves, and common to most of the Trochi, in some of which it runs from the base to the apex.

U´mbo. (*umbo*, Lat.) A boss or protuberance. In conchology, that point of a bivalve shell situated immediately above the hinge. This word makes *umbones* in the plural.

U´mbonated. Bossed; knobbed in the centre.

Unconfo´rmable. In geology, a term applied to a stratum or strata lying in a different plane from the subjacent strata upon which they rest. Strata not lying parallel with those beneath them. Supposing certain strata to have been upheaved, so that their inclination is at an angle with the horizon, or even vertical, such strata may all be, notwithstanding, conformable one with another; if, however, upon these tilted strata, fresh strata be deposited, the more recent strata lying horizontally upon the subjacent vertical or inclined strata, then the superjacent strata are termed *unconformable*.

Unconfo´rmably. Not being in the same plane with those upon which they are deposited. Strata lie unconformably when placed upon others having a different line of direction or inclination.

U´nctuous. (*unctus*, Lat.) Greasy; soapy to the touch. In mineralogy certain minerals are said to be unctuous; soap-stone is a good example.

Unctuo´sity. Greasiness. A character belonging to certain minerals, which is very useful in assisting to distinguish them. Some minerals, when the finger is passed over their surface, or their powder is rubbed between the finger and thumb, feel as if they were coated with some greasy matter. The sensation is different from that produced by mere smoothness of surface.

Un´derlie. A term used in mining, or by miners. In speaking of the inclination of a fault it is better, says Jukes, not to use the term "dip" as if it were a bed, but to adopt that of "hade," or "underlie."

U´ngual. (from *unguis*, a nail or claw, Lat.) A name applied to such bones of the feet as have attached to them a nail, claw, or hoof.

U´nguical. The name given to the claw-bone of certain animals.

Ungui´culated. Clawed; possessing claws.

U´ngulate. (from *ungula*, a hoof, Lat.) **Shaped like the hoof of a horse.**

U´ngulite grit. A member of the Russian Lower Silurian rocks, and so named from a peculiar fossil it contains: it is a white or ferruginous sandstone, not much exceeding 100 feet in thickness.

Unica´psular. (from *unus*, one, and *capsula*, a capsule, Lat.) Having one capsule only to each flower.

Unilo´cular. (from *unus*, one, and *loculus*, a cell or partition.) Having one chamber or cell only. In conchology, applied to shells which are not divided by septa into chambers. In botany, applied to seed vessels not separated into cells.

U´nio. (*unio*, Lat. a pearl.) A genus of freshwater bivalve shells, placed by Lamarck in the family Nayades, and by Cuvier in the family Mytilacea. Uniones are equivalve, inequilateral, transverse, internally pearly, externally covered with an epidermis, bivalves; they are commonly called freshwater muscles.

U'NIVALVE. (from *unus*, one, and *valva*, a shell, Lat.) A shell which is complete in one piece. The Linnæan arrangement of shells consists of three orders, namely, multivalves, bivalves, and univalves. Univalves are far more numerous than either multivalves or bivalves, both in genera and species.

UNMA'LLEABLE. That cannot be extended by hammering; that cannot be hammered out into plates or laminæ.

UNO'RGANIZED. Having no parts instrumental to the motion or nourishment of the rest.

UNSTRA'TIFIED. A term applied to rocks that are not stratified, or not arranged in strata. When no traces of beds can be detected, and the rock merely forms a great mass of mineral matter, without other lines than those of cleavage, it is said to be unstratified. De la Beche observes, "the terms stratified and unstratified have been commonly considered as respectively synonymous with aqueous and igneous. Practically, this division is highly valuable; but theoretically, it is not so satisfactory, at least, if we are to infer that all rocks divided into tabular masses, one resting on another, must have been deposited, either chemically or mechanically, from water. We should be careful not to couple too far stratification with aqueous deposition, as sheets of igneous rocks may cover pre-existing sheets of similar rocks, and the result be stratification."

UPPER GREENSAND. (*Glauconie Crayeuse*, Fr. *Chloritische Kreide*, Germ.) A member of the cretaceous group. An arenaceous rock, for the most part very calcareous, containing grains of silicate of iron. The upper greensand is a marine deposit, consisting of marly stone and sand, with numerous green particles, frequently including concretions of chert, and, in some places, beds of stone. From the application of portions of this deposit in the construction of ovens, furnaces, &c., it has obtained the name of Firestone. The situation of the upper greensand is between the chalk without flints above, and the gault below. Its thickness is about 100 feet.

UPPER LUDLOW ROCK. A sub-division of the Ludlow formation, thus named by Sir R. Murchison. This great explorer of the silurian rocks says, "this sub-division of the Ludlow formation consists essentially of thin bedded, lightly coloured, and very slightly micaceous sandstones, in some parts highly argillaceous, and in others so calcareous as to assume the character of impure limestones. When deeply cut into, these beds are of greenish-grey or bluish-grey tints, but they rapidly weather to an ashen or, more rarely, to a rusty-brown colour."

UPPER SILURIAN ROCKS. Sir R. Murchison divides the Silurian rocks into upper Silurian and lower Silurian. The upper Silurian comprise the Ludlow formation, sub-divided into the upper Ludlow, the Aymestry limestone, the lower Ludlow, the Wenlock limestone, and the Wenlock shale: the lower Silurian rocks comprise the Caradoc sandstones and the Llandeilo flags. "The two formations," he says, "of Ludlow and Wenlock, possess so much of a common lithological aspect, and offer such intimate passages from one to the other in the distribution of the organic remains, that they form a very distinct natural sub-group, which I have termed upper Silurian."

U'PPER TE'RTIARY STRA'TA. These strata are considered to be of more recent origin than those usually denominated tertiary, and, by some geologists, have been termed quaternary. They are supposed to be of more recent origin than any of the strata composing the formations of the London and Paris basins.

U'RANITE. An ore of Uranium of a black or green colour, called also Pechblende.

URA'NIUM. A metallic substance of a grey colour, obtained with great difficulty from a mineral called pechblende. It is infusible. Specific gravity 9·0 to 9·50. It is found in Saxony and in Cornwall. It imparts a deep orange colour to the enamel of porcelain.

URCE'OLATE. (from *urceolus*, Lat. a water-pitcher.)
1. In conchology, a term applied to a shell that swells in the middle, and is therefore supposed to bear a resemblance to a water-pitcher.
2. In botany, applied to a calyx or corolla swelling out like a pitcher.

U'RCHIN. The hedgehog. The name of sea-urchin has been given to the echinus.

U'TRICLE. (*utriculus*, Lat. a little bottle.) In botany, a term applied to a species of capsule resembling a small bladder, or a capsule of one cell, which never opens by valves, and drops with the seed. It is thin and transparent, and is more frequently considered to be the external coat of the seed than a real capsule.

UTRI'CULAR. Resembling a utricle in form or shape. The pelvis of apiocrinites is of a tumid *utricular* form.

# V

VA'GINATED. (from *vagina*, a sheath, Lat.) A term used in botany, sheathed.

VAGINOPE'NNOUS. (from *vagina*, a sheath, and *penna*, a wing, Lat.) Sheath-winged; having the wings covered with hard cases.

VAGI'NULA. A genus of minute pyramidal shells, known only as fossils, and found in the tertiary strata near Bordeaux.

VAL'LEY OF ELE'VATION. This name is given by geologists to a valley which seems to have originated in a fracture of the strata, and a movement of the fractured part upwards, so that the strata dip from the valley on either side. Probably a very large proportion of mountain valleys might be arranged under this head; but, at present, geologists seem to have confined the application of the term to those which are bounded by hills of moderate height.

VAL'LEY OF DENUD'ATION. A name given by geologists to any valley where the strata are not far removed from the horizontal position on either side, and of which the former continuity cannot be doubted.

VALVA'TA. A genus of small freshwater univalves belonging to the family Peristomata. Valvatæ are found both recent and fossil.

VALVE. (*valva*, Lat. *valve*, Fr. *valva*, It.)
1. In conchology, the shell. When the whole shell of the animal is in one piece, it is called a univalve; when there are two shells, or the shelly covering consists of two pieces, as in the oyster, muscle, &c., they are called bivalves; and when the covering consists of more than two pieces, multivalves.
2. In anatomy, a folding door or lid attached to a hollow vessel by means of a hinge, which allows the

valve to open in one direction only for the passage of fluids, and prevents their retrograde motion, or regurgitation. Thus there are valves in the heart; valves in the veins; valves in the lymphatics, &c.

VANA′DIUM. A metal discovered by Sefström, in Sweden, and thus named after Vanadis, a Scandinavian deity. Its properties are not yet known.

VA′RIEGATED SAND′STONE. The New Red Sandstone of the English; the Grès Bigarré of the French, and the Bunter Sandstein of the German geologists. For a description see *New Red Sandstone*.

VARI′ETY. (*varietas*, Lat.) A subdivision of species, arising from accidental, or unimportant and trifling, differences.

VARI′OLITE. (from *variolæ*, Lat.) The name given to an amygdaloidal or porphyritic rock, merely in consequence of its spotted appearance.

VA′SCULAR. (from *vasculum*, Lat. a little vessel.) Containing vessels or tubes; full of vessels within which the fluids are confined, and by which their course and their velocity are regulated.

VE′GETABLES. The first appearance of vegetable existence may be discovered in the transition slate, which contains impressions of algæ or sea-weeds. A few fronds of ferns have been found in some of the transition rocks. "It appears," says Buckland, "that nearly at the same points in the progress of stratification, where the most striking changes take place in the remains of animal life, there are found also concurrent changes in the character of fossil vegetables."

VEIN. (*vena*, Lat.)
1. In anatomy, elastic tubes pervading every part of the body, and conveying dark or venous blood from the arteries to the heart. The veins are larger and more numerous than the arteries, and may be compared to rivers, which, collecting all the water that is not imbibed by the soil, and reconveying it into its general receptacle, the ocean, perform an analogous office in the economy of nature.

2. In geology and mineralogy, fissures in rocks filled up by mineral or metallic substances differing from the rocks in which they are situated. Mineral repositories of a flat or a tabular shape, which in general traverse the strata of mountains, crossing the strata, and having a different direction from them, and filled with mineral matter, differing from the nature of the rocks in which they occur. Bakewell says, "perhaps the reader may obtain a clearer notion of a metallic vein, by first imagining a crack or fissure in the earth a foot or more in width, and extending east and west on the surface, many hundred yards. Suppose the crack or fissure to descend to an unknown depth, not in a perpendicular direction, but sloping a little to the north or south. Now, let us again suppose each side of the fissure to become coated with mineral matter of a different kind from the rocks of which the fissure is made, and then the whole fissure to be filled by successive layers of various metallic and mineral substances; we shall thus have a type of a metallic vein. Its course from east to west is called its *direction*, and the dip from the perpendicular line of descent its *hading*. Thus it is said to *hade* or dip to the north or south."

Veins occur principally in the primary and transition rocks, but they are also found in the lowest of the secondary series.

As regards the geographical distribution of mineral veins, it is an established fact, that while extensive plain countries are totally destitute of them, there are few mountainous districts in which they are not abundantly found. It has been very generally observed that the character of metalliferous veins changes with the structure of the rock through which they pass. If the direction of a vein approaches to a vertical plane, it is called a rake vein, if to the horizontal, a pipe or flat vein.

The depth to which the metallic veins descend is not known, all large veins continuing beyond the reach of the deepest mine. They frequently contain totally different ores at different depths.

Veins vary in width from less than an inch to thirty feet and upwards; sometimes the same vein at one part contracting, so as to be almost lost, and then expanding to an immense width.

Werner supposed that veins had become filled by matter descending into them from above, in a state of aqueous solution: Hutton, on the other hand, imagined that their contents were injected from below, in a state of igneous fusion. A third hypothesis refers the filling of veins to a process of sublimation from subjacent matters of intensely heated mineral matter, into apertures and fissures of the superincumbent rocks. A fourth hypothesis attributes these metallic collections to segregation, or infiltration. Buckland says, "Whatever may have been the means whereby veins were charged with their precious contents; whether segregation or sublimation were the exclusive method by which the metals were accumulated; or, whether each of the supposed causes may have operated simultaneously or consecutively in their production; the existence of these veins remains a fact of the highest importance to the human race: and although the disturbances, and other processes in which they originated, may have taken place at periods long antecedent to the creation of our species, we may reasonably infer, that a provision for the comfort and convenience of the last, and most perfect creatures He was about to place upon its surface, was in the providential contemplation of the Creator, in His primary disposal of the physical forces, which have caused some of the earliest and most violent perturbations of the globe."

VEIN-STONE. 1. The earthy, stony, saline, or combustible substance, which contains the ore, or is mingled with it, without being chemically combined, is called the gangue, or *vein-stone*.

2. Vein-stones are the different stony substances with which the ore is intermixed, and which as a whole constitute the vein. Werner was of opinion that in the same vein the parts of the vein-stone nearest to the Saalbände are the oldest, those in the middle the most modern, and the intermediate parts of a middle age.

VENERICA'RDIA. A genus of equivalved, inequilateral, marine, oblong bivalves; ribbed longitudinally on the outside; two thick hinge-teeth disposed obliquely, and in the same direction: muscular impressions two. Venericardiæ are found recent and fossil. The recent are met with at depths of the ocean varying to fifty fathoms, in mud and sands. Fossil venericardiæ are found in the secondary and tertiary formations.

VE'NTRICLE. (from *ventriculus*, Lat.) A chamber or cavity; the stomach. A term applied to different cavities of the body. The heart contains two chambers distinguished as the right and left ventricles, as well as two others, termed auricles. Certain cavities found in the brain are also called ventricles.

VE'NTRICOSE. ⎱ 1. In botany, applied
VE'NTRICOUS. ⎰ to parts of plants which are distended, bellied, or swelling out in the middle.

2. In conchology, applied to shells, inflated, or swelling in the middle.

VENTRI'CULITE. A zoophyte found fossil in flints and in the chalk. The ventriculite, when living, must have been of a cyathiform figure, and composed of a tough, jelly-like substance, capable of expansion and contraction. The general form of the animal appears to have been that of a hollow inverted cone, having numerous ramose fibres proceeding from the base, by which it was attached to other bodies. Externally, it was composed of a reticulated integument, which seems to have been capable of expanding and contracting according to the impressions it received; and, internally, it possessed a surface covered with the apertures of numerous tubuli; in all probability the openings of vessels, by which nutrition was effected. The smaller extremity was attached to the rock by rootlike processes; the outer tissue consisted of a net work of cylindrical, perhaps tubular, fibres; the inner surface of the funnel-like cavity was studded with polypiferous cells or openings. The specimens enveloped in flint are usually of a cyathiform or turbinated shape, while those imbedded in chalk are more frequently expanded into a broad circular disk. When contracted into a cylindrical form, the ventriculite is from one to six inches in length: when expanded, its diameter generally exceeds nine inches: the thickness of its substance is rarely more than 0·2 inch.

VE'NUS. (from *Venus*, the goddess of beauty.) A genus of exceedingly beautiful marine bivalves. Equivalve, inequilateral, subglobose, transverse shells, the frontal margin flattened, with incumbent lips; hinge with three teeth, all of them approximate, the lateral ones divergent at the tip. The middle tooth, which is sometimes bifid, is placed straight, and the one on each side obliquely.

Cuvier places Venus in the family **Cardiacea**. The common clam is a true Venus.

VERDE-ANTI'QUE. A very beautiful marble, highly prized, and used for ornamental purposes. It is an aggregate of serpentine and white crystallized marble, irregularly mingled.

VE'RRUCOSE. ⎱ (*verrucosus*, full of warts,
VE'RRUCOUS. ⎰ from *verruca*, Lat.)

1. In entomology, applied to the surface of insects, when studded with large smooth elevations, resembling warts.

2. In conchology, applied to shells beset with excrescences resembling warts.

3. In botany, applied to the surface of stems, beset with hard tubercles or warts.

VERSICO'LOURED. Displaying different colours, indeterminately restricted.

VE'RTEBER. ⎱ (*vertebra*, Lat.) A bone
VE'RTEBRA. ⎰ of the spinal column, or back-bone. The different vertebræ composing the back-bone are distinguished into cervical, dorsal, and lumbar. The vertebral column or spine, from the constancy with which it has been found in all animals of this type, and from the uniformity of plan with which, amidst endless variations, it is modelled, has been chosen as the distinctive character of all that great assemblage of individuals denominated Vertebrata. In man, the number of vertebræ is twenty-four, namely, seven cervical, twelve dorsal, and five lumbar. In different animals the number of vertebræ varies exceedingly; the vertebral

column of the Ichthyosaurus consisted of more than one hundred joints. The cervical vertebræ alone of the Plesiosaurus were about thirty-three in number. In birds the number varies from nine to twenty-three, and in living reptiles from three to eight. In the Mososaurus, the whole number of vertebræ amounted to one hundred and thirty-three.

ERTEBRA'TA. The whole animal kingdom has been distributed into four great divisions, namely, Vertebrata, Mollusca, Articulata, and Radiata. The vertebrate section of the animal kingdom contains five great classes, all agreeing in these seven particular points: first, in the possession of a brain and a spinal chord, and these enclosed in a skull and vertebral column; secondly, in the possession of an internal bony skeleton; thirdly, in the number of their limbs not exceeding four; fourthly, in the possession of organs serving the purposes of hearing, seeing, smelling, and tasting; fifthly, in having a mouth with two jaws, placed one above the other, and not on opposite sides; sixthly, in having a muscular heart, and circulating red blood; seventhly, in the individual distinctiveness of the sexes. In the ascending series, the first of the true vertebrated animals is the class of Fishes; then follows Amphibia, comprehending the various forms of frogs, toads, and tritons; the third class is the Reptiles. These three classes are all cold-blooded. The fourth class comprises the birds; these form the first class in the ascending series of hot-blooded animals; the fifth, or highest class, comprehends Mammalia, and includes man.

E'RTEBRATED. } (*vertebratus*, Lat.)
E'RTEBRATE. } Possessing an osseous spinal column.

E'RTEX. (*vertex*, Lat.) The crown, or top of the head; the summit. In conchology, in some shells the most prominent part, in others the apex.

VE'RTICIL. (*vertillum*, Lat. a whirl.) In botany, a little whorl; thus when, instead of two opposite leaves, three or more are produced from points forming a ring on the stem, such ring is termed a verticil.

VERTICI'LLATE. } (*verticillatus*, Lat.)
VERTICI'LLATED. } Whorled; growing in rings or whorls. Leaves produced from three or more points, forming a ring on the stem, are called verticillate leaves. Verticillate leaves are considered as being produced by the non-developement of several adjacent internodia.

VE'SICLE. (*vesicula*, a little bladder, Lat.)
1. A small bladder filled with serum.
2. A small membranous cavity, either in animals or vegetables. The simplest and apparently the most elementary texture met with in vegetables is formed of exceedingly minute *vesicles*, the coats of which consist of transparent membranes of extreme tenuity. These vesicles vary from the one-thousandth to the thirtieth of an inch in diameter.

VESI'CULAR. Consisting of vesicles; containing vesicles; full of small cavities, hollows, or interstices.

VE'STIBULE. (*vestibulum*, Lat.) The name given to a cavity of the internal ear.

VESU'VIAN. (from *Vesuvius*.) The Idocrase of Haüy; the Vesuvienne of Brochant. Pyramidal garnet. A mineral of different shades of green, brown, red, and sometimes, though rarely, of a blue colour, occurring in granular distinct concretions, crystallized, and massive. The primitive form is a four-sided prism with square bases, and one side of the base is to the height nearly as 13 to 14; hence it differs but little from a cube, and is divisible into triangular prisms, for the integrant particles. It yields to cleavage readily, parallel to all its planes, with sufficient brilliancy to obtain incidences of 90′ by the reflective goniometer in every direction. Haüy has described eight modifications of its primitive form. It is somewhat harder than quartz. Specific gravity from 3·2 to 3·4. It is often translucent, sometimes transparent, and sometimes nearly or quite opaque. It possesses double refraction. Fracture uneven, inclining to small conchoidal. Before the blow-pipe, it fuses easily into a yellowish translucent glass, which afterwards becomes black. Its constituents vary, in some degree, according to the locality whence it has been obtained. Klaproth obtained from a specimen from the neighbourhood of Vesuvius, silex 35·50, lime 33·00, alumine 22·2, oxide of iron 7·50, oxide of manganese 0·25. It is found abundantly in the vicinity of Mount Vesuvius, in the ejected masses, where its crystals line the cavities of volcanic rocks, accompanied by garnet, hornblende, melanite, mica, and icespar, from which circumstance it has obtained its name, and in primary rocks in Ireland, Scotland, and other parts of Europe.

VEXI′LLUM. } (*vexillum*, a standard, or
VE′XIL. } banner, Lat.) In botany, the upper large petal of a papilionaceous flower.

VI′LLI. (from *villus*, a hair, Lat.)
1. In anatomy, fine small fibres, resembling a covering of down.
2. In botany, fine downy hairs, covering fruits, flowers, and plants.

VI′LLOUS. (*villosus*, Lat.) Downy; pubescent; shaggy.

VIO′LET QUARTZ. See *Amethyst*.

VI′SCID. } (*viscidus*, Lat.) Glutinous;
VI′SCOUS. } tenacious; clammy.

VI′SCUS. (*viscus*, Lat.) In the plural, viscera; this term is generally applied to the organs of digestion; sometimes only. Any organ of be denominated a vis

VI′TREOUS. (*vitreus*, L resembling glass.
1. In mineralogy, a denote a particular minerals, resembling
2. In anatomy, a te one of the humour. The *vitreous* humour than three-fourths of the eye; it is contai ceedingly delicate tex substance, and is si the crystalline lens.

VI′TREOUS SAND TUBES.

VI′VIANITE. The Vivia the Fer Phosphate of phate of iron. A min or blue colour. It are, protoxide of iro phoric acid 32, water

VIVI′PAROUS. (*vivipar vivus*, alive, and *pa* forth; *vivipare*, Fr.) termed viviparous, in properly so called, i bring forth their yo perfect.

VOLCA′NIC PRODU′CTIONS. numerous and divers scoria, enamel, and g by far the most in interesting volcanic The different states of vitreous, compact, o depend on the diffe stances under which Aqueous vapour is in abundant. Volcanic naturally divisible in classes; sub-aerial, a ous. The first, bei respects open to our is to a considerable stood. The second, part hidden from ex necessarily but little k recent observation has light upon it. " Va cations of mineral volcanic products have been proposed, among which the division into Trachytic and Basaltic seems to be that most commonly adopted; trachyte being considered as essentially composed of felspar, and containing crystals of glassy felspar; while basalt is supposed to be essentially composed of felspar, augite, and titaniferous iron. Lavas, however, present such **various mixtures** of different minerals, **that exact** classifications **of them would appear** exceedingly **difficult."** — *De la Beche.* The **principal gases hitherto** detected **consist, according to Dr. Daubeny, of muriatic acid** gas, sulphur **combined with oxygen or hydrogen,** carbonic acid gas, and nitrogen. The sublimations of Vesuvius are, according to Sir H. Davy, chloride of sodium, chloride of iron, sulphate of soda, muriate and sulphate of potash, and a small quantity of oxide of copper. The principal metallic substances in volcanic rocks are iron and titanium; but ores of antimony, copper, and manganese, have sometimes been found in the craters of volcanoes. Tellurium, gold, and mercury are also said to occur in some volcanic rocks. The island of Ischia, which is entirely **volcanic,** contains a mine of gold.

Volca'nic rocks. These are not deemed synonymous with Plutonic rocks. While **the** Plutonic rocks are supposed to have **been** formed by igneous action at great depths, the Volcanic have risen up from below, and have cooled from **a** melted state upon or near the surface. Volcanic rocks belong **to** every period.

Vo'lcanite. Another name for augite.

Volca'no. (from Vulcan, the god of fire.) An opening in the earth's surface made by internal fire. Volcanoes exist in all quarters of the globe, and, according to Jameson, about one hundred and ninety-three active volcanoes have been observed; of which 13 belong to Europe and its islands,—66 to Asia and its islands,—8 to the islands of Africa, —and 106 to America and its islands. Volcanoes have been long considered in the light of safety valves, and this was the opinion entertained of them by Strabo, and it can scarcely be doubted that the tranquillity of the incandescent fluid mass, composing the earth's centre, is owing to **these** numerous vents **for the passage of steam,** &c. Although **volcanoes** generally exist **in the** neighbourhood **of** the sea, **yet this is not** invariably **the** case, as was once supposed. In central Asia there is a volcanic region with an area of 2500 square geographical miles, at a distance of upwards of 300 leagues from the ocean.

Vo'lva. (*volva,* Lat.) In botany, **a** species of calyx. A term applied to a membranous wrapper or covering of the fungus tribe, which conceals their parts of fructification: in due time it bursts, and forms a ring upon the stalk.

Volva'ria. A genus of cylindrical convoluted shells known only in a fossil state. The spire is not extruded; aperture narrow, extending the length of the shell: the columella plicated at its base: outer lip dentated. Volvaria approaches very near to Bulla cylindrica.

Volu'ta. A genus of simple marine univalves, found in sands and mud at depths varying from seven to fourteen fathoms. Two species of voluta, V. luctator and V. bicorona are described as occurring in the London clay; one species V. luctator in the Bognor sandstone; and one species, V. ambigua, in the chalk marl.

Volu'tion. A spiral wreath or turn. The wreaths or turnings of the shells of univalves are termed volutions.

Vo′lvox. (*volvox*, Lat. from *volvo*, to roll.) A genus of globular animalcules. To the presence of one species of volvox, the volvox globator, a loricated animalcule, and to its great abundance in such situations, pools of stagnant water owe their green colour.

Vu′lcanist. One who supports the Vulcanian theory, namely, that all rocks are of igneous production. The Vulcanists were opposed to the Neptunians, who, on the other hand, maintained that all rocks were of aqueous origin.

Vul′canian the′ory. That theory which explained the formation of all rocks by the agency of fire.

Vu′lpinite. A mineral of a greyish-white colour; thus named from its being found at Vulpino, in Italy.

Vulva. (Lat.) In conchology, a spatulated mark in shells of the Venus tribe.

# W

Wacke. ⎱ A name given by the Germans to a soft earthy
Wacke′. ⎰ basalt, to which it is nearly allied, and of which it may be deemed a variety. Its colours are greenish grey, sometimes passing into blackish green, brown, and greyish black, with sometimes a shade of yellow or red. It is invariably opaque. It occurs in amorphous masses, compact or vesicular. Fracture uneven, or slightly conchoidal. Hardness moderate. It is easily broken, and may be cut by a knife. Specific gravity from 2·5 to 2·8. Before the blow-pipe it fuses into an opaque, porous mass. It appears to be intermediate between clay and basalt, often passing into basalt. It does not adhere to the tongue, which circumstance distinguishes it from clay, nor will it form a paste with water. It does not, as does marl, effervesce with acids. Wacke is included among the trap rocks. When wacke, being vesicular, contains within its cavities calcedony, agates, &c. it forms a variety of amygdaloid. It is found more abundantly in Germany than any other country, but it is not confined to Germany.

Wad. ⎱ Another name for black-
Wadd. ⎰ lead.

Wadd black. A name given to the earthy manganese of Devonshire: it is a hydrate of manganese, and has the peculiar property of taking fire when dry, moderately heated, and mixed with linseed oil.

Warm-blooded animals. In the ascending series of the four great divisions of the animal kingdom, the highest, or vertebrata, alone contains what are called warm-blooded animals. Of this division, consisting of five classes, two classes only, namely, aves, or birds, and mammalia, are warm-blooded; the remaining three are cold-blooded. In warm-blooded animals the circulation is two-fold, there being, in fact, two hearts, perfectly distinct, and separated by thick partitions, which do not permit any direct transmission of fluid from one to the other. These two hearts are joined together, and enclosed within one capsule or envelope. The following is the course of circulation in warm-blooded animals. From the left ventricle the blood is propelled into the aorta, the great artery of the body, to be by it forwarded into all the arterial ramifications of the whole system; from these arteries it passes on through the veins into the venæ cavæ, and by them is carried into the right auricle; from the right auricle it passes into the right venticle, and by the right ventricle is propelled into the pulmonary arteries, to be conveyed through the lungs, in its passage through which it becomes aerated, loses its dark and assumes a florid colour, and is once more arterial blood; it then passes into the pulmonary veins, and is conveyed into the left auricle, whence it is forced into the left ventricle, and once more into the aorta.

WA'TER. (*wasser*, Germ.) When pure, water is transparent, and destitute of colour, taste and smell. The specific gravity of water is always supposed $= 1.000$, and it is made the measure of the specific gravity of all other bodies. When water is cooled down to 32° Fah. it assumes the form of ice. When heated to the temperature of 212° Fah. it boils, and is converted into steam. Pure water consists of two parts of hydrogen and one of oxygen.

WA'VED. Variegated; undated.
1. In botany, applied to the margins of leaves, when bordered alternately with numerous minute segments of circles and angles.
2. In entomology, applied to insects when the margin of the body is marked with a succession of arched incisions.

WA'VELLITE. A rare mineral, first discovered in Devonshire by Dr. Wavell, and named after him. Its colours are either pure white or white tinged with grey, green, or yellow; lustre silky. Specific gravity from 2·25 to 2·70. It consists essentially of alumine, being composed of alumine 71·5, water 28·0, oxide of iron 0·5. Sometimes a trace of silex and lime is present, and Sir H. Davy discovered in Wavellite the presence of fluoric acid.

WEALD. (from *wald*, Germ. a wood.) The name given to a valley, or tract of country, lying between the North and South Downs of Kent and Sussex. In some of the older publications the Weald is called the Wild. At the close of his account of the organic remains of Tilgate Forest, Dr. Mantell says, "it may be remarked that the vast preponderance of the land and freshwater exuviæ over those of marine origin, observable in these strata, warrants the conclusion that the Hastings or Wealden beds were formed by a very different agent from that which effected the deposition of the Portland limestone below, and the sands and chalk above them. Whether the land of that time were an island or a continent, may not be determined; but that it was diversified by hill and valley, and enjoyed a climate of a higher temperature than any part of modern Europe, is more than probable. Several kinds of ferns appear to have constituted the immediate vegetable clothing of the soil. But the loftier vegetables were so entirely distinct from any that are now known to exist in European countries, that we seek in vain for anything at all analogous, without the tropics. The forests of Clathraria and Endogenitæ (the stems of which, like some of the recent arborescent ferns, probably attained a height of thirty or forty feet,) must have borne a much greater resemblance to those of tropical regions, than to any that now occur in temperate climates. Turtles of various kinds must have been seen on the banks of its rivers and lakes, and groups of enormous crocodiles, basking in the fens and shallows. The gigantic Megalosaurus, and yet more gigantic Iguanodon, must have been of such prodigious magnitude, that the existing animal creation presents us with no fit objects of comparison. Imagine an animal of the lizard tribe, three or four times as large as the largest crocodile, having jaws, with teeth equal in size to the incisors of the rhinoceros, and crested with horns; such a creature must have been the Iguanodon!"

WEALD-CLAY. (Argile veldienne of Brongniart; *Wealdthon*, Germ.) A tenacious blue clay, containing subordinate beds of sandstone and shelly limestone, with layers of septaria of argillaceous ironstone, forming the subsoil of the wealds of Sussex and Kent, and separating the Shanklin sand from the central mass of the Hastings beds.—*Dr. Mantell's Geology of the South-east of England.*

The Weald-clay contains argillaceous iron-stone, occurring in regular beds. This ore of iron was so valuable, when it was the practice to use wood-charcoal for smelting, that furnaces were formerly numerous along the verge of the Weald. The thickness of the Weald-clay is estimated at 150 or 200 feet in Western Sussex.

WEA'LDEN FORM'ATION. } The Wealden
WEA'LDEN STR'ATA. } formation, group, or strata, have been separated into three principal divisions.
1. The Wealden clay, above described.
2. The Hastings sands: grey, white, yellow, and reddish-brown sands, and friable sandstone, passing into limestone.
3. The Purbeck beds, called also Ashburnham beds, consisting of grey limestone, alternating with blue clay and sandstone shale. The whole of these are freshwater or fluviatile deposits. The wealden is covered by the marine cretaceous system, and reposes upon the uppermost member of the oolite, which is also a purely marine deposit.

This intercalation, says Sir C. Lyell, of a great freshwater formation between two others of marine origin is a remarkable fact, and attests, in a striking manner, the great extent of former revolutions in the position of sea and land. From these and other data, he says, it seems a legitimate deduction that the marine formations of an antecedent period (that of the oolite) had become land throughout a portion of the space now occupied by the South of England and the opposite coast of France; and that this land then sunk down, with its forests, and became submerged beneath the waters of a great river. The country may then have continued to subside, until a thickness of two thousand feet of fluviatile sediment had been gradually accumulated; and this deposit, or delta, by a continuation of the same depressing operations, may, in its turn, have become buried deep beneath the ocean of the chalk. Dr. Mantell may be considered the great geological champion and hero of the Wealden, for to his indefatigable exertions in that field, are owing some of the most splendid discoveries in palæontology. Until the appearance of Dr. Mantell's works on the Geology of Sussex, the peculiar relations of the sandstones and clays of the interior of Kent, Sussex, and Hampshire, were entirely misunderstood. No one supposed that these immense strata were altogether of a peculiar type, and interpolated amid the rest of the marine formations, as a local freshwater deposit.—*Prof. Phillips.*

Dr. Mantell observes that the Wealden may be considered as covering an area 200 miles in length, from west to east, and 220 miles from north-west to south-east, the total thickness averaging about 2000 feet. Of this series of deposits, clays, or argillaceous sediments, with limestone almost wholly composed of freshwater snail-shells, occupy the uppermost place; sand and sandstones, with shales and lignite, prevail in the middle; while in the lowermost, argillaceous beds, with shelly marbles or limestones, again appear; and, buried beneath the whole, is a petrified forest, in which the trees are still standing, and the vegetable mould undisturbed. The organic remains of the Wealden consist of leaves, stems, and branches of plants of a tropical character; bones of enormous reptiles of extinct genera; of crocodiles, turtles, flying reptiles, and birds; fishes of several genera and species, and shells of a fluviatile character. The vegetable remains belong, some of them, to plants which appear to have held an intermediate place between the Equiseta and Palms, as the Clathraria; while others approach to arborescent ferns, the species being very peculiar, and not known in any other deposit, whether of higher or inferior antiquity. For a knowledge of that enormous reptile, the Iguanodon, we are entirely indebted to the indefatigable and scientific researches of Dr. Mantell.

WEA'THERING. A term used to express the action of the atmosphere, rain, &c., on the surface of rocks. There is no rock, even the hardest, that does not bear some marks of what has been termed *weathering*. The amount of surface-change, so produced, is exceedingly variable, depending much on local causes. The tors of Dartmoor, Devon, may be referred to as excellent examples of the weathering of a hard rock. The *weathering* of these tors is so exceedingly slow, that the life of man will scarcely enable him to perceive a change; therefore the period requisite to produce their present appearances must have been very considerable. Variations in temperature much assist the chemical decomposing power of the atmosphere.

WEDGE-SHAPED. In botany, applied to leaves that are broad at the summit and gradually taper toward the base.

WEIGHT OF THE ATMOSPHERE. The air is an elastic fluid resisting pressure in every direction, and is subject to the law of gravitation. The pressure of the atmosphere is calculated to be about fifteen pounds to every square inch, so that the surface of the globe sustains a weight of 11,449,000,000 hundreds of millions of pounds.

WE'NLOCK FORMATION. The lower formation of the upper Silurian rocks, comprising the Wenlock limestone and Wenlock shale.

WE'NLOCK LIMESTONE. The upper subdivision of the Wenlock formation. The Wenlock limestone, says Sir R. Murchison is in every respect identical with the well known rock of Dudley, and contains the same organic remains. The colour of the rock is usually grey, but the crystalline varieties are sometimes dark blue, and more rarely pink, the mass being freckled with veins and strings of white crystallized carbonate of lime. The simple minerals hitherto observed in the Wenlock limestone of Shropshire, consist of crystallized carbonate of lime in various forms, sulphate of barytes, sulphurets of iron, peroxide of manganese, crystals of copper pyrites and bitumen. As regards the organic remains hitherto discovered, the most striking zoological feature consists in the vast number of contained corals. The most prevalent of these are Heliopora pyriformis; Catenipora escharoides; Stromatopora concentrica; Favosites gothlandica; Cyathophyllum turbinatum; Limaria clathrata, &c. Among the mollusks and conchifers may be enumerated

Euomphalus discors, Euomphalus funatus, and Euomphalus rugosus; Productus euglyphus and Productus depressus; Atrypa aspera and Atrypa tenuistriata; Terebratula imbricata and Terebratula cuneata; Nerita, Haliotis, &c. Orthocerata are also found. Trilobites are most abundant, the prevailing species being Asaphus caudatus and Calymene Blumenbachii. Some other species of trilobites appear to be peculiar to the Wenlock formation, as the Calymene variolaris, Calymene macrophthalma; Asaphus Stokesii, and the genus Acidaspis, so named by Sir R. Murchison.

WENLOCK SLATE. Called also Wenlock shale. An argillaceous, dark-grey or liver-coloured shale, constituting the lower member of the Wenlock formation, and containing nodules of sandstone. The Wenlock shale, or lower member of the Wenlock formation, is characterized by certain species of shells, the most abundant of which are Productus transversalis; Spirifer cardiospermiformis; and Spirifer trapezoidalis; Terebratula breviostra, Terebratula imbricata, and Terebratula interplicata; Asaphus longicaudatus; and Orthoceras attenuatum."—*Sir R. Murchison.*

WERNERITE. A rare mineral of a greenish grey, olive green, or, sometimes, white colour, occurring in eight-sided prisms, terminated by four-sided summits, whose faces form with the alternate lateral plates, on which they stand, an angle of about 121°. It is found at Arendal in Norway, and in Sweden and in Switzerland, and named after Werner. It consists of silex 45·5, alumine 33·5, lime 13·22, oxide of iron 5·75, oxide of manganese 1·47.

WHEEL-SHAPED. In botany, a term applied to a corolla of a salver-shaped form, having scarcely any tube.

WETHERE'LLIA. The name given by Mr. Bowerbank to a genus of fossil fruits found in the London clay. He says "it is perhaps the most abundant of all the fruit found in the Isle of Sheppey, and is well known throughout the island by the name of Coffee, to which some of the sections of the fruit, when separated from each other, bear a very strong resemblance."

WHET SLATE. } The Novaculite
WHET-STONE SLATE. } of Kirwan.
For a description, see *Hone.*

WHI'NSTONE. A provincial term applied to some of the trap rocks. In the western parts of Sussex, says Dr. Mantell, layers of chert or hornstone, provincially termed *whinstone* prevail in the sand near Petworth, &c. This stone is a compact mass of quartz, but not homogeneous, for it contains iron, and perhaps some other substance.

WHITE-STONE. Felspathic granite, called by the French Eurite, and by Werner Weiss-stein.

WHORL. In conchology, a wreath, volution, or turn of the spire of a univalve; the axis of revolution is termed the columella, and the turns of the spiral are denominated whorls.

2. In botany, a species of inflorescence, in which the flowers surround the stem in the form of a ring; also applied to leaves, when they arise in a circle round the stem.

WING.
1. The limb of a bird or insect by the aid of which it is able to fly.
2. In botany, a membranous appendage to some seeds, serving to waft them along in the air; applied also to the two side petals of a papilionaceous corolla.

WINGED. Having wings. In botany, applied to stems, when the angles are extended into leafy borders;

also to petioles having a leafy border on each side.

WI'THAMITE. A mineral, so named by Sir D. Brewster, in honour of its discoverer, H. Witham, Esq. It occurs at Glenco, in Argyleshire, in minute translucent, brilliant carmine red crystals, in form resembling epidote. Specific gravity = 3·1-3·3. Hardness = 6·0-6·5. —*Allan.*

WI'THERITE. So named after Dr. Withering, its discoverer. Carbonate of Barytes. The Baryte carbonatée of Haüy; the Witherit of Werner; the Barolite of Kirwan. Witherite, or native carbonate of barytes, is one of the rarer productions of the mineral kingdom. At Anglesark, in the county of Lancaster, it is found in veins traversing the independent coal formation, and accompanied by blende, galena, calamine, and heavy spar. It occurs also in Shropshire, in the lead mines, where it is met with in irregular masses, weighing from forty pounds to two or three hundred weight, imbedded in heavy spar. The name given to this substance by the miners is yellow spar, not that this is its real colour by day-light, but its transparency is so considerable that if a lighted candle be placed behind a mass of it, the whole will glow with a yellowish light, a circumstance by which the miners distinguish it from heavy spar. The colour of *witherite* is white with the slightest possible, if any, tinge of yellow; its fracture is broad striated, approaching to straight foliated; it is for the most part massive. The Anglesark witherite, according to Klaproth, contains, besides carbonate of Barytes, above two per cent. of carbonate of strontites, and a scarcely appreciable quantity of oxide of copper. A specimen analyzed by Mr. Aiken, gave, carbonate of barytes 96·3, carbonate of strontites 1·1, sulphate of barytes 0·9, silex 0·5, alumine and oxide of iron 0·25.

WOLF. (*wolf*, Germ.) The wolf affords an excellent illustration of the complete extinction of species. Wolves were formerly exceedingly numerous in Great Britain, and were met with in Ireland even so late as the beginning of the 18th century. At the present day, unless seen in a menagerie, or read of as still existing in other countries, and formerly in this, the natives of these islands might be perfectly unaware that the wolf ever had any existence.

WO'LFRAM. The name given by Werner to the ferruginous oxide of tungsten.

WO'LLASTONITE. A mineral, thus named in honour of Dr. Wollaston. Prismatic augite.

WOOD-COAL. Another name for brown coal.

WOOD-OPAL. Opalized wood. The Holz-opal of Werner; the Quartz résinite xyloïde of Haüy: Ligniform opal of Kirwan. A variety of opal, occurring in various vegetable forms, and being in reality opalized vegetable matter. Wood changed by silicious infiltration, in which the original structure is still preserved, often in its minutest parts, and the woody fibres appear rather masked by its silicious investment than destroyed. Wood opal is of various colours, white, grey, brown, and black. In fracture, lustre, and translucency, it scarcely differs from semi-opal; it may be regarded as intermediate between common opal and semi-opal. Prof. Jameson relates that many years ago, the trunk of a tree penetrated with opal was found in Hungary, which was so heavy that eight oxen were required to draw it. It is found in Hungary, the Faroe Islands, and Van Diemen's Land.

WOOD-STONE. The Holstein of Werner; Quartz agathe xyloïde of Haüy; Le bois petrifié of Brochant. Wood petrified with hornstone. Prof. Jameson places wood-stone as a subspecies of hornstone; Prof. Cleaveland terms it agatized wood, and says this substance appears to have been produced by the process, commonly called the petrifaction of wood. It is essentially composed of silicious earth, which, it is highly probable, has been gradually deposited, as the vegetable matter was decomposed and removed. Both its form and texture indicate its origin. Thus it presents, more or less distinctly, the form of the trunk, branches, roots, or knots, which once belonged to the vegetable. The colour of wood-stone is generally grey, shaded with blue, yellow, &c. The colours are often with spots, sometimes striped. Hardness nearly that of common quartz. Specific gravity 2·67. It occurs in sandstone or sandy loam, and is capable of being highly polished.

WOOD-TIN. The Etain oxidé concretioné of Haüy. The Kornishches Zinnerz of Werner. A variety of oxide of tin; fibrous oxide of tin. This has been hitherto found only in Cornwall and Mexico. It occurs in fragments which are generally rounded. Its colours are light or chesnut brown, reddish brown, and yellowish grey. It is opaque; of a fibrous texture; easily broken. Specific gravity from 6·4 to 6·7. It is infusible before the blowpipe, and irreducible. It consists of oxide of tin 91·0, oxide of iron 9.

WORTH SANDSTONE. So named from its being fully developed at Worth, near Crawley, in Sussex. A series of white and yellow sands, constituting the lowermost member of the Hastings beds. Its organic remains are principally ferns and arundinaceous plants.

WOR'THITE. The name assigned to an earthy mineral, occurring in boulders in Sweden, and Finland. It is met with in foliated crystalline masses of a white colour; transparent; lustre vitreous; scratches quartz.—*Allan.*

# X

XA'NTHITE. An earthy mineral, consisting of a congeries of small rounded grains, easily separable and may be crushed by the nail. Colour greyish or yellow. It has been found only at Amity, in Orange County, United States.

XI'PHIAS. (ξιφίας, from ξίφος, a sword, Gr.) The sword-fish.

XI'PHODON. (Sword-tooth.) The name assigned by Cuvier to a sub-genus of Anoplotheria. In the xiphodon the anterior molars are slender and trenchant, and the posterior ones below have, opposite the concavity of each of their crescents, a point which, in the course of wear, also takes the form of a crescent, so that then the crescents are double, as in the ruminants. — *Pidgeon.* The xiphodon has been hitherto found fossil only, and in post-cretaceous strata. It is a small and delicate, long and slender limbed anoplotherian animal. A second species, Xiphodon Geylensis, has been added by M. Gervais to the type species, Xiphodon Gracilis. The existence of the Xiphodon is considered to have been limited to the Eocene period.

XI'PHOID. (from ξίφος, a sword, and

εἶδος, form.) Resembling a sword: a term applied to the cartilage placed at the lower extremity of the sternum or breast-bone.

XIPHOS'URA. The seventh order of the class Crustacea, comprising the king-crab, or limulus, and the extinct genera Halycina and Bellinurus.

XULINOSPRIONI'TES. The name given to a genus of fossil fruits found in the London clay. Mr. Bowerbank describe them thus: "legumes valveless, woody, two-seeded," and observes "the pericarp of the fruits of this genus unites, in a singular manner, the characters of the legume and the drupe."

XYLO'PHAGI. (from ξύλον, wood, and φαγεῖν, to eat, Gr.) A family of coleopterous insects, comprising several genera.

# Y

YA'NOLITE. The name given by Lametherie to the Axinite of Haüy and the Thummerstone of Kirwan.

YELLOW-QUARTZ. (The Quartz hyalin Jaune of Haüy.) A variety or sub-species of quartz, of various shades of yellow, and nearly transparent. It has been also called citrine and false or Bohemian topaz. In Scotland it has obtained the name of Cairn-gorm.

YE'NITE. So named from Jena. Called also Lievrite. It is found in the Isle of Elba and in Norway. A mineral of a black or blackish-green colour, occurring crystallized and massive. It is opaque; scratches glass; **gives** sparks with steel. Specific gravity from 3·8 to 4. Longitudinal fracture foliated; cross fracture conchoidal or uneven: lustre resinous. **Before the blow-pipe** it fuses into a dull, opaque, black globule, strongly attracted by the magnet. It consists of silex 30·0, lime 12·5, oxide of iron 57·5.

YU'CCITES. A genus of plants, thus designated by Dr. Martins, and which, he says, constitutes a series allied to the palms, differing in structure from most of the monocotyledons, in having the stem broadly expanded above by a more or less perfect dichotomy.

Y'TTRIA. A name given by Ekeberg to a new earth discovered by Gadolin, in 1797, in the quarry of Yetterby, in Sweden. It has also been named Gadolinite, after Gadolin. The equivalent of Yttria, according to Berzelius, is 40·2.

Y'TTRIUM. The name given by Ekeberg to a metal forming the basis of Yttria. In that mineral it is combined with the oxides of iron and manganese, and a small portion of lime, and silica. When separated from these substances, it has the appearance of a fine white powder, without either taste or smell. It is infusible, and is insoluble in water. Its equivalent, according to Berzelius, is 32·2.

YTTROTA'NTALITE. The name given by Brochant to Ittrious oxide of Columbium. Yttrotantalite, like like Yttria, is found in the quarry of Ytterby, in Sweden. It is of a dark grey colour; shining, metallic lustre; found in reniform masses.

# Z

ZA'FFRE. An impure oxide of cobalt, obtained by roasting the ore of cobalt, by which process the arsenic and sulphur contained in the ore are driven off. Zaffre melted with silex and potash, and reduced to powder, constitutes the article known under the name of powder blue. So intense is the blue afforded by Zaffre, that one grain will give a full blue to 240 grains of glass.

ZA'MIA. A genus of the class Appendix palmæ, Diœcia, order Polyandria, natural order of Palmæ, Filices (Juss). *Generic character:* Male, — calyx; ament strobile-shaped, ovate, obtuse; scales horizontal, pellate, obovate, very blunt, one-flowered, thickened at the top, permanent. Corolla none. Stamina: filaments none. Anthers subglobular, clustered, accumulated in the lower surface of the scales, sessile, two-valved, opening above by a longitudinal cleft. Pollen, farinaceous. Female, — calyx; ament strobile-shaped, larger, ovate, imbricate; scales pedicelled, pellate, angular, finally distant, permanent. Corolla none. Pistil: germs two, irregular, angular, inserted into the margin under the pelta of the scales, solitary on each side, nodding. Style none. Stigma obtuse, obscurely cloven at the side. Pericarp: berries to each scale two ovate, barked at the base, fleshy, one-celled. Seed one in each berry, ovate. *Essential character:* Male, ament strobile-shaped, scales covered with pollen underneath. Female, ament strobile-shaped, with scales at each margin; berry solitary.

Fossil zamiæ have been discovered in the coal formation, and in the Wealden formation at Yaverland, on the south coast of the Isle of Wight.

ZA'MITE. A fossil zamia. M. Ad. Brongniart has referred the zamite, or fossil zamia, to a new genus, to which he has assigned the name of Mantellia nidiformis.

ZECH-STEIN. } (The Magnesian Lime-
ZETCHSTEIN. } stone of English; Calcaire Alpen of French; the Alpenkalkstein of German Geologists. The name Zechstein was given to this formation by Humboldt.) The second member of the red sandstone series, in the ascending order. The zech-stein is a calcareous deposit, or magnesian limestone, of a somewhat variable aspect; it is fossiliferous, and in it, as far as observations have yet gone, are found, for the first time, those shells known by the name of Productæ; Spirifers also now are found, for the first time, in the descending series; these and Productæ both abounding in the Carboniferous Limestone. The organic character of the zech-stein, as far as the researches of geologists have hitherto gone, nearly approaches that of the carboniferous group, a circumstance which will greatly tend to add to the difficulty of determining between the zechstein and the carboniferous limestone, when their geological position cannot be ascertained with certainty. Some geologists are of opinion that the connection between the two formations of red sandstone and zech-stein is so intimate, that the latter may be regarded as a subordinate formation to the former. The zech-stein lies immediately under the new red sandstone and

above the marl slate, or kupfer schiefer, of the magnesian limestone formation. It is a deposit not widely spread over the European area, and is principally known in Germany and England. Some authors comprise a series of deposits, formerly known to the German miners under a variety of terms, under the name of zechstein, these were the Asche, Stinkstein, Rauchwacke, Zechstein, and Kupferschiefer. The Magnesian Limestone of England may be regarded as the equivalent of the Zechstein of Germany.

ZE′OLITE. (from ζέω, to swell or foam, and λίθος, a stone, Gr.) The Mesotype of Haüy: Kouphon Spath of Mohs.

1. Under this name, some mineralogists comprise eleven subgenera of the mineral genus zeolite.

2. A translucent and, sometimes, transparent mineral of a white, yellow, or brownish-yellow colour, exhibiting double refraction. Zeolite is rendered electric by heat, one summit of its prisms becoming positive and the other negative; the latter is usually that summit which was connected with the gangue. It is found in distinct crystals, whose surfaces have a strong lustre, slightly pearly; and in masses composed of several fascicular groups of minute crystals; and in each group the crystals or fibres diverge, or even radiate, from one point, and at the surface frequently appear distinct from each other, or exhibit pyramidal terminations. Some zeolites phosphoresce by friction. Before the blowpipe, zeolite fuses with much ebullition or intumescence into a whitish spongy enamel, and it is from this property that it has obtained its name. When reduced to powder and thrown into nitric acid, it is converted into a jelly in the course of a few hours. This property of becoming gelatinous, as well as that of becoming electric by heat, sufficiently distinguish zeolite from stilbite, analcime, chabasite, harmotome, and prehnite. Zeolite most frequently occurs in amygdaloid, basalt, greenstone, and clinkstone poryhyry. It is also found in granite and gneiss.

ZE′RO. (*zéro*, Fr. *zéro*, It.) This word is of Italian derivation, and means a cipher or 0. The expression is used to denote a certain point or mark on the thermometrical scale. In the thermometers of Celsius and Reaumur, zero is the point at which water congeals. In Wedgewood's pyrometer, zero corresponds with $1077°$ of Fahrenheit's scale. The question has been propounded, "at what degree would a thermometer stand (supposing the thermometer capable of measuring so low) were the body to which it is applied totally deprived of caloric? or what degree of the thermometer corresponds to the real zero?" This question does not appear to have ever been satisfactorily answered. Dr. Crawford placed the real zero $1268°$ below 0. Mr. Kirwan fixed the real zero at $1048°$ below 0. Lavoisier and La Place placed the real zero at $2736°$ and $5803°$ below 0.

ZE′THUS. The name given by Pander to a genus of trilobites, added by himself to those previously described.

ZE′US. (*zeüs*, Lat.) A genus of fishes of the thoracic order, having the head compressed and sloping, the upper lip arched, the tongue subulated, the body compressed, thin, and shining, and the rays of the first dorsal fin ending in filaments. In Dr. Mantell's Geology of the South-East of England, a species of Zeus, found in the chalk, the Zeus Lewesiensis, is beautifully figured. This ichthyolite is from six to eight inches long, and its width is nearly equal to the length of the body.—*Dr. Mantell.*

ZINC. ⎱ (*zink*, Germ.) A metal of a
ZINK. ⎰ bluish-white colour, with a fine granular fracture. Zinc was not obtained in its metallic form till the sixteenth century, though its ores were known to the ancients, and used by them in the formation of their brass or bronze. It does not occur native; its most abundant ore is the sulphuret, called *Blende,* common in most veins which contain sulphurets of iron, lead, copper, &c. in every country. The structure of zinc is foliated. As regards its hardness, it may be easily cut with the knife. Specific gravity from 6·9 to 7·2. It is malleable, but its malleability is greatly increased by heating it to a temperature of 300 Fahrenheit. By exposure to the air it tarnishes and loses its lustre, but it is but little oxidated. United with copper it forms brass. The ores of zinc are few. Its presence may be determined by roasting the ore, and then fusing it by the blow-pipe on charcoal with filings of pure copper. If zinc be present, the copper will be converted into brass. The ore called calamine is a carbonate of zinc.

ZI′PHIUS. The name assigned by Cuvier to a genus of cetacea; it contains three species, and approximates to the cachalots and hyperoodontes. The head differs from that of the hyperoodon, in the maxillary bones not forming vertical partitions on the sides of the muzzle, and in the partition behind the nostrils not only rising vertically, but also curving, so as to form a kind of half cupola over these cavities.—*Pidgeon.*

ZI′RCON. (By some the word zircon is deemed of Indian origin, others derive it from the French word *jargon, espèce de diamant jaune.*) A mineral occurring in rounded grains or fragments, or in regular crystals. The primitive crystal of zircon is, according to Haüy, an obtuse octohedron; the common base of the pyramids is square. The measurements, as afforded by the reflecting goniometer, are 84° 20′ by 95° 40′. It is harder than quartz, and possesses double refraction in a high degree. It is transparent, or sometimes only translucent. Specific gravity from 4·3 to 4·7 Before the blow-pipe it is infusible, but loses its colour. It may be distinguished from garnet, idocrase, staurotide, &c., by its infusibility, specific gravity, and strong double refraction. There are two varieties of zircon, called zircon jargon and zircon hyacinth. Zircon jargon consists of zirconia 69, silica 26·5, oxide of iron 0·5. Zircon hyacinth of zirconia 70, silica 25, oxide of iron 0·5 Some mineralogists divide zircon into three subspecies, namely, zirconite, hyacinth, and jargon. The finest specimens are brought from the island of Ceylon. It occurs in primary and transition rocks, but is usually obtained from the sand of rivers.

ZIRCO′NIA. An earth, when pure, white and tasteless, supposed to be a compound of zirconium, its metallic basis, and oxygen. An oxide of zirconium.

ZIRCO′NITE. A subspecies of zircon, consisting, according to the analysis of Klaproth, of zirconia 69, silica 26·5, oxide of iron 0·5· It occurs in reddish-brown and nearly opaque prismatic crystals. It is harder than quartz, but softer than diamond. Sp. gr. from 4·5 to 4·7.

ZIRCO′NIUM. The metallic base of zircon.

ZOANTHA′RIA The third order of the class actinozoa, comprising six-starred corals, sea-anemone, beroe, &c.

ZO′DIAC. (*zodiaque,* Fr. *zódiaco,* It. *zodiacus,* Lat. ζωδιακὸς, Gr.) A broad circle or region in the heavens, remarkable, not from anything peculiar in its own composition, but from its being the area within which the apparent motion of the sun, moon, and all the great planets are confined. The centre of the zodiac is the ecliptic, which is inclined to the equinoctial at an angle of about 23° 28′, intersecting it at two opposite points, called the equinoctial points. The zodiac extends 9° on either side of the ecliptic.

ZODI′ACAL. Pertaining to the zodiac, as the zodiacal constellations, &c.

ZOI′SITE. } A mineral, thus named from
ZOI′ZITE.  } Baron Von Zois, its discoverer. A variety of Epidote, of a grey, brown, or yellowish colour. Prof. Jameson constitutes zoisite a species, which he divides into two subspecies, namely, common zoisite and friable zoisite: he adds, "it would probably be an improvement to arrange zoisite as a subspecies of epidote."

ZOO′LOGY. (from ζῶον, an animal, and λόγος, discourse, Gr.) That branch of natural history which treats of animals, their habits, structure, classification, &c.

ZOO′LOGY FOSSIL. That division of zoology which treats of fossil animals. The examination of the fossil remains of a former state of creation has demonstrated the existence of animals far surpassing in magnitude those now living, and brought to light many forms of being which have nothing analagous to them at present, and many others which afford interesting connecting links between existing genera.

ZOO′PHAGA. (from ζῶον, an animal, and φαγεῖν, to eat, Gr.) A tribe of animals which attack and devour *living* animals. The animals of this tribe have three kinds of teeth, namely, cutting teeth, canine teeth, and grinders; their paws are armed with claws; their muzzle is often set with whiskers, usually called

smellers; their mammary organs are dispersed; their intestines are less voluminous than those of herbivorous animals.

ZOO′PHAGOUS. Attacking and devouring *living* animals.

ZO′OPHYTE. (ζωόφυτα, from ζῶον, and φύω, Gr. *quæ media sunt naturæ inter animalia et plantas*.) Animal plants, corals, sponges, and other aquatic animals which have obtained the name from an opinion formerly entertained that they were intermediate between animals and vegetables. In consequence of their aggregation, which produces trunks and expansions of various forms, together with the simple nature of their organization, and the radiating disposition of their organs, resembling the petals of flowers, these animals owe their name of zoophytes or animal plants. But possessing the power of voluntary motion, enjoying the sense of touch, feeding on matters which they have swallowed, and digesting these in an internal cavity, they must, in every point of view, be considered to be animals. Many of the lowest zoophytes, which have no digestive sac, and no polypi, absorb their whole nourishment by the surface of their body, or by the parieties of canals which traverse their interior. Zoophytes present to the physiologist the simplest independent structures compatible with the existence of animal life; the means of their propagation and increase are the first of a series of facts on which a theory of generation must rise. Zoophytes are either free in the sea, or are attached for life after a very early period of growth. "We may compare," says Lyell, "the operation of zoophytes in the ocean to the effects produced on a smaller scale upon the land by the plants which generate peat. In corals, the more durable materials of the generation that has passed away serve as the foundation on which living animals are continuing to rear a similar structure.

ZOOPHYTO′LOGY. (from ζωόφυτα, a zoophyte, and λόγος, discourse, Gr.) That branch of natural history which treats of the structure, habits, &c. of zoophytes.

ZOO′TOMY. (from ζῶον, an animal, and τέμνειν, to cut.) A term employed to express the knowledge acquired by dissecting the bodies of animals. This science makes us acquainted with their organization, or with the structure and form of all their internal parts and organs. It points out the connections which subsist between the different parts of the animal machine, by which they are all enabled to co-operate towards the same great objects—the preservation of the individual, and the continuance of the race. It examines the changes which the organs undergo at different periods of life. It traces the modification of form and structure presented by the different organs and parts of the machine, in all the inferior tribes of animals, by which the whole organization of the species is always admirably adapted to the circumstances in which they are placed.—*Prof. Grant.*

ZO′STERA. A genus of plants growing by the sea-side, belonging to the class gynandria, order polyandria.

ZOSTERI′TES. Fossil plants of the genus zostera. Four species have been determined by Ad. Brongniart, namely, Z. bellovisana, Z. caulinicæfolia, Z. elongata, and Z. lineata: these have all been found in the chalk of the Isle d'Aix.

ZU′RLITE. The name given to a Vesuvian mineral.

RYDE:
JAMES BRIDDON, PRINTER,
CROSS STREET.

# WORKS

PUBLISHED BY

## CHARLES GRIFFIN AND COMPANY.

**Professor Ramsay's Manual of Roman Antiquities.** With numerous Engravings. New edition. Crown 8vo. 8s. 6d., cloth.

**Professor Ramsay's Elementary Manual of Roman Antiquities,** for the use of Beginners. Many Woodcuts. Crown 8vo. 4s., cloth.

**Professor Ramsay's Manual of Latin Prosody.** Second edition, Revised and greatly enlarged. Crown 8vo. 5s., cloth.

**Professor Ramsay's Elementary Manual of Latin Prosody,** for the use of Beginners. Crown 8vo. 2s., cloth.

**Professor Ramsay's Speech of Cicero for Aulus Cluentius,** with Prolegomena and Voluminous Notes. Second edition. Crown 8vo. 6s., cloth.

**Professor Ramsay's Selections from Ovid and Tibullus,** with Notes. Third edition. 12mo. 5s., cloth.

**Professor Senior's Treatise on Political Economy; the Science** which treats of the Nature, the Production, and Distribution of Wealth. Fourth Edition. Crown 8vo. 4s., cloth.

**Professor Nichol's Cyclopædia of the Physical Sciences;** comprising Acoustics, Astronomy, Dynamics, Electricity, Heat, Magnetism, Meteorology &c., &c. Second edition, enlarged. Maps and Illustrations. Large 8vo. £1 1s., half-bound, Roxburghe.

**Professor Phillip's Manual of Geology; Practical and Theoretical.** A New edition, entirely re-written *(In the Press).*

**Professor Eadie's Classified Bible: an Analytical Concordance** to the Holy Scriptures, or the Bible presented under distinct or classified Heads or Topics, with Synopsis and Index, Fourth edition, Revised. Post 8vo. 8s. 6d., cloth.

**Professor Eadie's Commentary on the Greek Text of the** Epistle of Paul to the Ephesians. Second edition, enlarged. 8vo. 14s., cloth.

**Professor Eadie's Commentary on the Greek Text of the** Epistle of Paul to the Colossians. 8vo. 10s. 6d., cloth.

**Professor Eadie's Commentary on the Greek Text of the** Epistle of Paul to the Philippians. 8vo. 10s. 6d., cloth.

**Professor Eadie's Complete Concordance to the Holy Scriptures,** on the basis of Cruden, with Introductory Essay. by the REV. DAVID KING, LL.D. Twenty-fourth edition. Post 8vo. 5s., cloth.

**Professor Eadie's Early Oriental History; comprising the** History of Egypt, Assyria, Persia, Media, Phrygia, and Phoenicia. Numerous Illustrations. Crown 8vo. 8s., cloth.

**Professor Craik's Manual of English Literature,** for the use of Colleges and Schools, selected from the larger work. Crown 8vo. 7s. 6d., cloth.

**Professor Fleming's Vocabulary of Philosophy, with Quota-** tions and References. Second edition. Small 8vo. 7s. 6d., cloth.

**Professor Maurice's Manuals—The Systems of Philosophy** Anterior to Christ. Fourth edition. Crown 8vo. 5s., cloth.

**Professor Maurice's Manuals—The Philosophy of the First** Six Centuries. Second edition. Crown 8vo. 3s. 6d., cloth.

**Professor Maurice's Manuals—The Philosophies of the Middle** Ages, from the Sixth to the Fourteenth Centuries. Crown 8vo. 5s., cloth.

**Professor Maurice's Manuals—Modern Philosophy, from the** Fourteenth Century to the French Revolution, with a Glimpse into the Nineteenth Century, Crown 8vo. 10s. 6d., cloth.

**Many Thoughts of Many Minds. Compiled and arranged by** HENRY SOUTHGATE, Tenth Thousand. Large 8vo., toned paper, 12s. 6d., cloth elegant, Richly Gilt.

*₊* It has been the aim of Mr. Southgate to produce a *resumé* of the finest passages in English Literature. He has scrupulously excluded from his volume all merely pretty conceits or sentimental fancies, and brought together only the thoughts conceived in power and fertile in suggestions to the reader's mind. NOTE—In this edition references to the original sources are given.

**Golden Leaves from the Works of the Poets and Painters.** Edited by ROBERT BELL, and Illustrated with nearly One Hundred Superb Steel Vignettes. One Large Volume 4to. *Printed in the Best Style.* 25s., cloth elegant. Richly Gilt.

*₊* The *Saturday Review* says :—" *Golden Leaves* is by far the most important book of the season. It is edited with something of a scientific aim in literature, and the illustrations are really works of art. The crimson and gold of the cover almost detract from the more solid gold within; but the rich paper, sumptuous typography, and admirable printing, of this noble volume might almost recall Dr. Dibdin from that paradise of bibliography where we trust he revels in the creamy tomes of unique vellum copies of the Sixteenthers, among which he twaddled and prattled while on earth."

**Handbook of Biography; containing a Complete Series of** Original Memoirs of the most Remarkable Individuals of all Times and Nations. By ALISON, BREWSTER, BURTON, CREASY, EADIE, KNIGHT, NICHOL, SPALDING, WORNUM, and other Contributors. New edition, enlarged. With many Portraits, Birthplaces, &c., &c. Thick Post 8vo. 10s. 6d., Half-bound, Roxburghe style.

**Ecclesiastical History—History of the Christian Church from** the Birth of Christ to the Present Day. By Right REV. SAMUEL HINDS, D.D., Bishop of Norwich; REV. J. H. NEWMAN, B.D.; Dean J. A. JEREMIE, D.D., Regius Professor of Divinity in the University of Cambridge; REV. J. B. S. CARWITHEN, B.D.; Right REV. DR. HAMPDEN, Bishop of Hereford; REV. J. E. RIDDLE, M. A.; REV. HENRY J. ROSE, B.D., &c., &c. 3 Volumes crown 8vo. £1 1s., cloth.

**Archbishop Whately's Treatise on Logic, with Synopsis and** Index. Crown 8vo. 3s., cloth.

**Archbishop Whately's Treatise on Rhetoric, with Synopsis** and Index. Crown 8vo. 3s. 6d., cloth.

**Archdeacon Hale's History of the Jews, from the time of** Alexander the Great to the Destruction of Jerusalem by Titus. Crown 8vo. 2s. 6d., cloth.

**Rev. A. J. D. D'Orsey's Spelling by Dictation, for the Civil** Service Examination. New edition. 18mo. 1s., cloth.

**WORKS PUBLISHED BY CHARLES GRIFFIN AND COMPANY.**

**Professor Eadie's Ecclesiastical Cyclopædia; or, a Dictionary** of Christian and Jewish Sects, Denominations, and Heresies.—History of Dogmas Rites, Ceremonies, Sacraments, &c.—Liturgies, Creeds, Confessions, Monastic and Religious Orders, &c., &c. Post 8vo. 8s. 6d., cloth.

**Professor Eadie's Biblical Cyclopædia; or, Dictionary of** Eastern Antiquities, Geography, Natural History, Sacred Annals and Biography, Biblical Literature, &c. With Maps and Numerous Illustrations. Ninth edition, revised. Post 8vo. 7s. 6d., cloth.

**Professor Eadie's Dictionary of the Bible, for the use of** Young Persons. With 150 Illustrations. Sixth edition. Small 8vo. 2s. 6d., cloth.

**Professor Airy's Treatise on Trigonometry, Revised by Pro-**fessor Blackburn. Second edition. Crown 8vo. 2s. 6d., cloth.

**Professor Thomson's Popular Dictionary of Chemistry, The-**oretical and practical, with its application to the Arts and Medicine. Numerous Illustrations. Second edition. Post 8vo. 8s. 6d., cloth.

**Professor Rankine's Manual of Applied Mechanics. Illus-**trated with Diagrams. Second edition, Crown 8vo. 12s. 6d., cloth.

**Professor Rankine's Manual of the Steam Engine and other** Prime Movers. Illustrations. Third edition. Crown 8vo. 12s. 6d., cloth.

**Professor Rankine's Manual of Civil Engineering, Surveys,** Earthworks, Foundations, Masonry, Carpentry, Canals, Rivers, Waterworks, &c., &c. Numerous Diagrams. Second edition. Crown 8vo. 16s., cloth

**Professor Faraday's Lectures on the Physical Forces;** delivered before a Juvenile Audience at the Royal Institution. Woodcuts. Second edition. Post 8vo. 3s. 6d., cloth.

**Professor Faraday's Lectures on the Chemical History of a** Candle; deliverd at the Royal Institution. With Illustrations. Post 8vo. 3s. 6d., cloth.

**Professor Schoedler and Medlock's Treasury of Popular** Science, Astronomy, Botany, Chemistry, Geology, Mineralogy, Natural Philosophy, Physiology, and Zoology. Many Hundred Woodcuts. Post 8vo. 7s. 6d., cloth.

**Professor Aitken's Science and Practice of Medicine. The** Second edition almost entirely re-written. Illustrated by Numerous Diagrams and Woodcuts. 2 Volumes 8vo. 28s., cloth.

*⁎* *The Lancet says:*—"A book in which the work of collection, arrangement, and exposition of all the new things in our calling is done to our hand . . . It will be seen that Dr. Aitken has boldly recognized the advance of pathology. . . . Evidently no fact or dogma has been accepted without careful consideration . . . We can strongly recommend this new work on the Practice of Medicine—for such it really is."

*⁎* *The British Medical Journal says:*—"Dr. Aitken's book is the most comprehensive of any that have in late years been published on the Practice of Medicine. . . . There is not one topic of pathological or practical interest that has been the subject of modern investigation and discovery which the author has omitted to notice."

**Professor Ansted's Natural History of the Inanimate Crea-**tion; recorded in the Structure of the Earth, the Plants of the Field, and the Atmospheric Phenomena. With numerous Illustrations. Large Post 8vo. 8s. 6d., cloth.

**Professor Dallas' Natural History of the Animal Kingdom;** being a Systematic, and Popular Description of the habits, Structure, and Classification of Animals. New edition, 600 Illustrations. Large Post 8vo. 8s. 6d., cloth.

**Professor Jeremie's History of the Christian Church in the** Second and Third Centuries. Crown 8vo. 4s., cloth.

**Lord Brougham's Lives of Philosophers of the Time of**
George III.; Comprising BLACK, WATT, PRIESTLEY, CAVENDISH, DAVY, SIMSON, ADAM SMITH, LAVOISIER, BANKS, and D'ALEMBERT. Post 8vo. 5s., cloth.

**Lord Brougham's Lives of Men of Letters of the Time of**
George III; Comprising VOLTAIRE, ROUSSEAU, HUME, ROBERTSON, JOHNSON, and GIBBON. Post 8vo. 5s., cloth.

**Lord Brougham's Sketches of British Statesmen of the Reign**
of George III. New edition, with the Letters of George III. to Lord North, *now first published*. Post 8vo. 5s., cloth.

**Lord Brougham's Sketches of British Statesmen of the Time**
of George III. and IV. New edition, enlarged by Numerous Fresh Sketches, and other additional matter. Post 8vo. 5s., cloth.

**Lord Brougham's Sketches of Foreign Statesmen of the Time**
of George III. New edition, greatly enlarged. Post 8vo. 5s., cloth.

**Lord Brougham's Natural Philosophy; comprising a Discourse**
of Natural Theology, Dialogues on Instinct, and Dissertation on the Structure of the Cells of Bees. Post 8vo. 5s., cloth.

**Lord Brougham's Rhetorical and Literary Dissertations; com-**
prising Discourse of Ancient Literature; Lord Rector's Address; Rhetorical Contributions to the "Edinburgh Review;" and Discourses on the Objects, Pleasures, and Advantages of Social and Political Science. Post 8vo. 5s., cloth.

**Lord Brougham's Speeches on Social and Political Subjects,**
with Historical Introductions, and his Lordship's Latest Corrections. 2 Volumes post 8vo. 10s., cloth.

**Lord Brougham's Historical and Political Dissertations; com-**
prising Balance of Power, Foreign Policy and Relations, War Measures, Penal Legislation, Revolutions, Reform, Right of Search, &c., &c. Post 8vo. 5s., cloth.

**Lord Brougham's The British Constitution; its History,**
Structure, and Working. New edition, with Latest Corrections and Additions. Post 8vo. 5s., cloth.

**Lord Brougham's Contributions to the "Edinburgh Review;"**
Political, Historical, and Miscellaneous. 3 Volumes 8vo. £1 16s. cloth.

**Lord Brougham's England and France under the House of**
Lancaster. New edition, 8vo. 10s. 6d. cloth.

**Lord Brougham's Tracts, Mathematical and Physical.** 2nd edition. Post 8vo. 7s. 6d. cloth.

**Lord Brougham's and Sir Charles Bell's Paley's Natural**
Theology, with their Notes and Dissertations. Small 8vo. 3s. 6d., cloth.

**Lord Brougham's Dialogues on Instinct and Sir Charles**
Bell's Animal Mechanics. Small 8vo. 2s. 6d., cloth.

Post 8vo, 7s., cloth.

# A DICTIONARY
## OF
# DOMESTIC MEDICINE
## AND
# HOUSEHOLD SURGERY

BY

SPENCER THOMSON, M.D. Edin., F.R.C.S.

*Twenty-fifth Thousand*

WITH ONE HUNDRED AND FIFTY ILLUSTRATIONS

---

*From the Author's Prefatory Address.*

WITHOUT entering upon that difficult ground which correct professional knowledge, and educated judgment, can alone permit to be safely trodden, there is a wide and extensive field for exertion, and for usefulness, open to the unprofessional, in the kindly offices of a *true* Domestic Medicine; the timely help and solace of a simple Household Surgery, or better still, in the watchful care, more generally known as "Sanitary Precaution," which tends rather to preserve health than to cure disease. "The touch of a gentle hand" will not be less gentle, because guided by knowledge, nor will the *safe* domestic remedies be less anxiously or carefully administered. Life may be saved, suffering may always be alleviated. Even to the resident in the midst of civilization, the "knowledge is power" to do good; to the settler and the emigrant, it is invaluable.

I know well what is said by a few, about injuring the medical profession, by making the public their own doctors. Nothing will be so likely to make "long cases" as for the public to attempt any such folly; but people of moderate means—who, as far as medical attendance is concerned, are worse off than the pauper—will not call in and fee their medical adviser for every slight matter, and in the absence of a little knowledge *will* have recourse to the prescribing druggist, or to the patent quackery which flourishes upon ignorance, and upon the mystery with which some would invest their calling. And not patent quackery alone, but professional quackery also, is less likely to find footing under the roof of the intelligent man, who, to common sense and judgment, adds a little knowledge of the whys and wherefores of the treatment of himself and family. Against that knowledge which might aid a sufferer from accident, or in the emergency of sudden illness, no humane man could offer or receive an objection.

NOTICES OF THE PRESS:—

*The London Journal of Medicine* says,—"The best and safest book on Domestic Medicine and Household Surgery which has yet appeared."

*The Dublin Journal of Medical Science* says,—"Dr. Thomson has fully succeeded in conveying to the public a vast amount of useful professional knowledge."

*The Christian Witness* says,—"The best production of the kind we possess."

*The Medical Times and Gazette* says,—"The amount of useful knowledge conveyed in this work is surprising."

---

CHARLES GRIFFIN & COMPANY, LONDON.

# THE
# BOOK OF DATES

### A Treasury of Universal Reference

Comprising the Principal Events of all Ages, from the Earliest Periods to the End of 1861, arranged Chronologically and Alphabetically, with Index of Names, in one large volume, post 8vo, cloth gilt, antique bevelled, price 7s. 6d.

*From the Preface.*

THE great objects in view in this work are to present simultaneously to the eye of the reader the chief contemporaneous events in the history of the world, divested, as much as possible, of reflection and inference, yet clearly showing the tendency and progress of each nation in connection with all; to give a full and fair record of facts, so arranged that the widening circle of their effects might be readily traced by a thinking mind; to furnish the well-read historian with a remembrancer, and the unread student with a book of reference, wherein he could, at a moment's notice, ascertain the events amid which an individual life was passed—the circumstances which rendered a particular era remarkable—the date of any important or curious fact in War, Statesmanship, Science, Art, Religious development, &c.,—wherein, too, he could trace their antecedents and sequences, though his memory might afford but a slight clue to the discovery of those precise periods, acts, or dates.

The work consists of two parts: in the first of which are exhibited, in synchronistic order, the principal events that have occurred in the several countries of the world; in the second is given an alphabetical arrangement of all the curious or important facts of history, so classified that, with a very small expenditure of time and care, the work may be used as a comprehensive guide to universal history.

The publishers have great confidence that they are giving to the scholar a *vade mecum* of considerable value—to the man of business a "ready reckoner" in history—to the general reader a treasury of facts—and to the man of letters a complete book of reference.

*From the Spectator.*—" A storehouse of facts and dates."
*From the Literary Gazette.*—" Admirably adapted for a work of reference."

**CHARLES GRIFFIN & COMPANY, LONDON.**

# THE
# HAND-BOOK OF BIOGRAPHY

Biographies of the most Distinguished Persons of all Times

A Complete Series of Original Memoirs of the most Remarkable Individuals of all Times and Nations, by Alison, Brewster, Burton, Creasy, Eadie, Knight, Nichol, Spalding, Wornum, **and** other Contributors. Third edition, revised and enlarged, with 100 engraved portraits, and 150 illustrations of Birthplaces, &c., in one thick volume, 8vo, half-bound Roxburghe style, gilt top, price 10s. 6d.

---

*The Critic says,*—"Messrs. GRIFFIN & Co.'s 'Hand-book of Biography' is, we believe, the most carefully compiled and satisfactory of this class of books which have of late years issued from the press. It differs from all popular biographical dictionaries in this—that the principal lives of each class of remarkable men have been entrusted to practical writers, who have cultivated the corresponding departments of learning. The result is, not only an eminently trustworthy, but an unusually interesting volume. It extends to nearly 900 pages, and has some 150 or more illustrations of birthplaces, monuments, and other memorials of departed greatness, and is a wonderful instance of cheapness, completeness, and excellence."

*The Art Journal says,*—"The 'Hand-book of Biography' now before us is in many respects a remarkable book. The best authorities have been chosen, and the most eminent of our living writers in art, science, and literature, have been secured to write these biographies. The names of Alison, of Brewster, of Nichol, with numerous others of equal standing in the walks of history and science, are a sufficient guarantee for the excellency of their biographies."

*The Spectator says,*—"This work differs from all single volumes of biography in several remarkable features. The great number of subjects which it contains, the variety, and indeed, the celebrity of the writers engaged, and the original character imparted to the larger notices. The first, and for the purposes of reference the most useful feature, is the number of persons noticed; the men engaged on the biographies are numerous and respectable, as may be seen from the enumeration on the title, and each writer takes that class of lives with which his studies have made him most familiar: Alison and Creasy, for example, undertake naval and military men."

---

CHARLES GRIFFIN & COMPANY, LONDON.

Second Edition, in crown 8vo, price 7s. 6d., cloth.

# PROFESSOR CRAIK'S
# MANUAL OF ENGLISH LITERATURE

AND OF

THE HISTORY OF THE ENGLISH LANGUAGE
FROM THE NORMAN CONQUEST.

With Numerous Specimens.

---

CONTENTS:

Introduction—The Languages of Modern Europe—*Original English* (Saxon or Anglo-Saxon)—*Second English* (Semi-Saxon)—*Third English* (Mixed or Compound English)—Chaucer, Spenser, Shakespeare, &c., &c.—Middle and Latter Part of the Seventeenth Century; Milton, Butler, Dryden, &c.—The Century between the English Revolution and the French Revolution: Swift, Pope, Addison, Defoe, Goldsmith, Johnson, &c., &c.—The Latter Part of the Eighteenth Century; Cowper, Burns, &c.—The Nineteenth Century; Wordsworth, Scott, Shelley, Byron, Crabbe, &c., &c.—Literature of the Present Day; Browning, Tennyson, Hood, &c.

In this volume, and in the larger work of which it is an abridgment, an attempt has been made to interweave with the history of our national Literature such an exposition of the revolutions and more gradual changes which the Language has undergone as may suffice to present at least a correct general view of the subject. It is shown that the English Language has assumed successively only *three* distinct constitutions or forms; and while each of them is precisely defined both in the principle of its structure and in its limits as to time, the simplest explanation is given of the causes which appear to have produced the conversion of the first into the second and of the second into the third.

In the survey that is taken of our Literature, it is believed that no writer of any considerable note has been passed over, and as full accounts have been given as the space at command would permit of all those of the highest rank, and of their lives as well of their works, from Chaucer down to our own day. Many more names, indeed, have been introduced than can, on the most liberal construction, be held to belong to what is properly to be called the Literature of the language; but much necessarily enters into the history of every thing which makes no part of the thing itself. All care has been taken throughout to be accurate in the important point of dates. Nor in the critical appreciation of the authors that come under review will there be found, it is hoped, any want of a sufficiently catholic spirit. Finally, large extracts have been given from many of our great writers both in verse and in prose; and here every possible aid has been contributed by glossarial foot-notes to the easy understanding of all obsolete or dialectic words and forms.

*The Spectator says*,—"The present volume is particularly adapted for the use of Students going up to competitive examinations. Numerous illustrations of the progress and development of the English Language are introduced, beginning with King Canute and ending with Thomas Hood."

*The Saturday Review says*,—"Professor Craik is always clear and straightforward, and deals not in theories but in facts; all that he says is sound and practical and eminently distinguished by good sense. We have philological books treating of our earliest Literature, and we have critical books treating of our latest Literature, but we do not know of any book which, like the present, embraces both."

CHARLES GRIFFIN & COMPANY, LONDON.